Solutions Manual to Fundamentals and Practice in Statistical Thermodynamics

Solutions Manual to Fundamentals and Practice in Statistical Thermodynamics

Jianzhong Wu
University of California, Riverside

John M. Prausnitz
University of California, Berkeley

WILEY

Published by John Wiley & Sons, Inc., Hoboken, New Jersey.
Published simultaneously in Canada.

Library of Congress Cataloging-in-Publication Data applied for:

Paperback ISBN: 9781394264094

Cover Design: Wiley
Cover Image: © Jorg Greuel/Getty Images

Set in 9.5/12.5pt STIXTwoText by Straive, Chennai, India

Contents

1

Microscopic Origin of Thermodynamics

Problem 1.1 Using the phase rule, justify whether the following thermodynamic systems are adequately defined:

(i) a pure liquid of known mass at given temperature and density;
(ii) a pure liquid of known mass at given density and pressure;
(iii) a two-phase system at given temperature, pressure, and the mole fractions of all chemical species in one phase;
(iv) an ionic liquid containing two types of cations and one type of anions at given temperature, pressure, and the ionic composition.

Solution

The phase rule indicates that, for a bulk system with C chemical species and P phases without chemical reaction, $N_f = C - P + 2$ intensive variables are needed to define a thermodynamic state. In addition, one extensive variable is required to define the system size.

(i) Yes, here $C = 1$ and $P = 1$, thus $N_f = 2$. Temperature and density are independent intensive variables, mass is an extensive quantity.
(ii) The specification could be problematic because pressure and density are conjugated variables. For example, liquid water exhibits a maximum density at 1 atm and 4 °C, implying that there could be two temperatures for one density at the same pressure.
(iii) No, the system is over specified. $P = 2$ then $N_f = C$. But temperature, pressure, and the compositions of one phase constitute $C + 1$ independent intensive variables.
(iv) Here $C = 2$ (two ion pairs), $P = 1$ and $N_f = 3$. The intensive variables are well defined but the system size is unknown.

Problem 1.2 Based on the fundamental thermodynamic relations, show

$$\left(\frac{\partial C_P}{\partial P}\right)_{T,N} = -T\left(\frac{\partial^2 V}{\partial T^2}\right)_{P,N}.$$

Solution

First, write the constant-pressure heat capacity in terms of entropy

$$C_P = T\left(\frac{\partial S}{\partial T}\right)_{P,N}.$$

Taking a partial derivative with respect to P at constant T and N leads to

$$\left(\frac{\partial C_P}{\partial P}\right)_{T,N} = T\left(\frac{\partial^2 S}{\partial T \partial P}\right)_{N}.$$

Solutions Manual to Fundamentals and Practice in Statistical Thermodynamics, First Edition. Jianzhong Wu and John M. Prausnitz.

From the Maxwell relation

$$\left(\frac{\partial S}{\partial P}\right)_{T,N} = -\left(\frac{\partial V}{\partial T}\right)_{P,N},$$

we then have

$$\left(\frac{\partial^2 S}{\partial T \partial P}\right)_N = -\left(\frac{\partial^2 V}{\partial T^2}\right)_{P,N} \Rightarrow \left(\frac{\partial C_P}{\partial P}\right)_{T,N} = -T\left(\frac{\partial^2 V}{\partial T^2}\right)_{P,N}.$$

Problem 1.3 Acoustic velocity, or the speed of sound, in a fluid can be derived from the first law of thermodynamics for steady-state flow processes

$$d(\hat{H} + v_s^2/2) = 0$$

in combination with the mass-balance equation

$$d(\hat{\rho} v_s) = 0.$$

Here, $\hat{H}$ is the specific enthalpy of the fluid, $\hat{\rho}$ is the mass density, and v_s is the velocity of the elastic wave, i.e., the sound speed.

(i) Derive the following expression from the 1st law and mass balance

$$v_s = \sqrt{\left(\frac{\partial P}{\partial \hat{\rho}}\right)_S}.$$

(ii) Show that the partial derivative at constant entropy S can be expressed in terms of measurable quantities

$$v_s = \sqrt{\frac{\gamma}{\hat{\rho}\kappa_T}},$$

where $\gamma \equiv C_P/C_V$ is the ratio of the constant-pressure and constant-volume heat capacities of the fluid, κ_T is the isothermal compressibility

$$\kappa_T \equiv -\frac{1}{V}\left(\frac{\partial V}{\partial P}\right)_T.$$

(iii) Predict the speed of sound in the atmosphere at 25 °C. Assuming atmosphere is an ideal gas with molar heat capacity $C_P = 29.3\,\text{J/(mol K)}$ and molecular weight $M = 29\,\text{g/mol}$.

Solution

(i) Combing the 1st law $d\hat{H} = -v_s dv_s$ with the fundamental equation $d\hat{H} = Td\hat{S} + \hat{V}dP$ leads to, at constant entropy,

$$d\hat{H} = \frac{1}{\hat{\rho}} dP = -v_s dv_s,$$

where $\hat{\rho} = 1/\hat{V}$. The expression for the sound speed is obtained by dividing the above equation by that from the mass balance, $v_s d\hat{\rho} = -\hat{\rho} dv_s$,

$$v_s = \sqrt{\left(\frac{\partial P}{\partial \hat{\rho}}\right)_S}.$$

(ii) First, we rewrite the partial derivative at constant entropy S in terms the molar volume V:

$$\left(\frac{\partial P}{\partial \hat{\rho}}\right)_S = -\hat{V}^2\left(\frac{\partial P}{\partial \hat{V}}\right)_S = -\frac{V}{\hat{\rho}}\left(\frac{\partial P}{\partial V}\right)_S.$$

To make a connection with measurable quantities, we use the following identities:

$$\left(\frac{\partial P}{\partial V}\right)_S = -\left(\frac{\partial P}{\partial S}\right)_V \left(\frac{\partial S}{\partial V}\right)_P \quad \text{(chain rule)}$$

and

$$\left(\frac{\partial P}{\partial S}\right)_V \left(\frac{\partial S}{\partial T}\right)_V = \left(\frac{\partial P}{\partial T}\right)_V; \quad \left(\frac{\partial S}{\partial V}\right)_P \left(\frac{\partial T}{\partial S}\right)_P = \left(\frac{\partial T}{\partial V}\right)_P.$$

Combining these equations leads to

$$\left(\frac{\partial P}{\partial V}\right)_S = -\frac{C_P}{C_V}\left(\frac{\partial P}{\partial T}\right)_V \left(\frac{\partial T}{\partial V}\right)_P = \frac{C_P}{C_V}\left(\frac{\partial P}{\partial V}\right)_T,$$

where we have used the following equations

$$\frac{C_P}{T} = \left(\frac{\partial S}{\partial T}\right)_P, \quad \frac{C_V}{T} = \left(\frac{\partial S}{\partial T}\right)_V, \quad \left(\frac{\partial P}{\partial T}\right)_V \left(\frac{\partial T}{\partial V}\right)_P = -\left(\frac{\partial P}{\partial V}\right)_T.$$

Therefore, the partial derivative at constant entropy S becomes

$$\left(\frac{\partial P}{\partial \hat{\rho}}\right)_S = -\frac{V}{\hat{\rho}}\frac{C_P}{C_V}\left(\frac{\partial P}{\partial V}\right)_T = \frac{\gamma}{\hat{\rho}\kappa_T}$$

and

$$v_s = \sqrt{\frac{\gamma}{\hat{\rho}\kappa_T}},$$

where $\gamma \equiv C_P/C_V$ is the ratio of the constant-pressure and constant-volume heat capacities of the fluid, and κ_T is the isothermal compressibility

$$\kappa_T \equiv -\frac{1}{V}\left(\frac{\partial V}{\partial P}\right)_T.$$

(iii) For an ideal gas, the compressibility is

$$\kappa_T^{\mathrm{IG}} = \frac{1}{V}\frac{NRT}{P^2} = \frac{1}{P}$$

and

$$\gamma = \frac{C_P}{C_V} = \frac{C_P}{C_P - R},$$

where R is the gas constant. At 1 atm and 25 °C, the mass density of atmosphere is

$$\hat{\rho} = \frac{MP}{RT} = 1185.4\,\mathrm{g/m^3},$$

where we have used the average molecular weight of air $M = 29\,\mathrm{g/mol}$. Substituting $\hat{\rho} = 1185.4\,\mathrm{g/m^3}$ along with the molar heat capacity $C_P = 29.3\,\mathrm{J/(mol\,K)}$ into the equation for the sound speed yields $v_s \approx 330\,\mathrm{m/s}$.

Problem 1.4 Consider a one-dimensional model for a thermodynamic system containing N identical particles. Assume that the internal energy is completely determined by temperature T, system total length L, and particle number N. Show that the fundamental equation of thermodynamics can be written as

$$dU = TdS + fdL + \mu dN,$$

where temperature, line tension (i.e., negative of the one-dimensional pressure), and chemical pressure are defined as, respectively,

$$T = \left(\frac{\partial U}{\partial S}\right)_{L,N},$$

$$f = \left(\frac{\partial U}{\partial L}\right)_{S,N},$$
$$\mu = \left(\frac{\partial U}{\partial N}\right)_{L,S}.$$

What are the analogs of the Maxwell relations for this one-dimensional system?

Solution

We may derive the fundamental equation of thermodynamics from $U = U(S, L, N)$, which is a linear homogeneous function

$$\begin{aligned} dU &= \left(\frac{\partial U}{\partial S}\right)_{L,N} dS + \left(\frac{\partial U}{\partial L}\right)_{S,N} dL + \left(\frac{\partial U}{\partial N}\right)_{L,S} dN \\ &= TdS + fdL + \mu dN. \end{aligned}$$

In analogy with the conventional fundamental equation for three-dimensional systems, we can derive the Maxwell relations from the second-order differentiations of the internal energy:

$$\left(\frac{\partial T}{\partial L}\right)_{S,N} = \left(\frac{\partial f}{\partial S}\right)_{L,N},$$
$$\left(\frac{\partial T}{\partial N}\right)_{L,S} = \left(\frac{\partial \mu}{\partial S}\right)_{L,N},$$
$$\left(\frac{\partial f}{\partial N}\right)_{L,S} = \left(\frac{\partial \mu}{\partial L}\right)_{N,S}.$$

Problem 1.5 Two-dimensional models are often used to describe the thermodynamics of molecular adsorption from a gas or a liquid solution on planar surfaces. For a one-component two-dimensional system, the fundamental equation is given by

$$dU = TdS - \varsigma dA + \mu dN,$$

where ς stands for surface pressure. Derive the Gibbs adsorption isotherm:

$$\left(\frac{\partial \varsigma}{\partial \mu}\right)_T = \frac{N}{A}.$$

Can you extend the Gibbs adsorption isotherm to multicomponent systems? What is the surface pressure if there is no adsorption?

Solution

For this system, the integrated form of the fundamental equation is

$$U = TS - \varsigma A + \mu N.$$

In comparison with its differential form, we can derive the Gibbs–Duhem equation for the two-dimensional system

$$SdT - Ad\varsigma + Nd\mu = 0.$$

At constant temperature, $SdT = 0$, thus

$$\left(\frac{\partial \varsigma}{\partial \mu}\right)_T = \frac{N}{A}.$$

For multicomponent systems, the Gibbs–Duhem equation becomes

$$SdT - Ad\varsigma + \sum_i N_i d\mu_i = 0.$$

In this case, the Gibbs adsorption isotherm becomes

$$\left(\frac{\partial \varsigma}{\partial \mu_i}\right)_{T,\mu_{j\neq i}} = \frac{N_i}{A}.$$

If there is no adsorption, $N_i = 0$ and the surface pressure vanishes, i.e., $\varsigma = 0$.

Problem 1.6 Consider gas adsorption on a planar surface at temperature T. Assume that the gas phase is ideal and that the adsorption can be described by Henry's law

$$N/A = k_H P,$$

where N is the number of gas molecules at the surface, A is the surface area, and k_H is Henry's constant. Show that the surface pressure is given by

$$\varsigma A = N k_B T,$$

where k_B is Boltzmann's constant.

Solution

For an ideal gas in the bulk, the chemical potential depends on pressure P

$$\mu = \mu_0 + k_B T \ln(P/P_0),$$

where μ_0 is the chemical potential of the ideal gas at system temperature and a reference pressure $P_0 = 1$ atm, P and P_0 take the same units.

The combination of the Gibbs isotherm with Henry's law predicts

$$\left(\frac{\partial \varsigma}{\partial \mu}\right)_T = \frac{N}{A} \Rightarrow \frac{1}{k_B T}\left(\frac{\partial \varsigma}{\partial \ln P}\right)_T = k_H P \Rightarrow \frac{1}{k_B T}\left(\frac{\partial \varsigma}{\partial P}\right)_T = k_H \Rightarrow \left(\frac{\partial \varsigma}{\partial P}\right)_T = k_H k_B T.$$

With the boundary condition $\varsigma = 0$ at $P = 0$, integrating the above equation leads to

$$\int_0^P \left(\frac{\partial \varsigma}{\partial P}\right)_T dP = \int_0^P k_H k_B T dP \Rightarrow \varsigma = k_H P k_B T.$$

Using Henry's law $N/A = k_H P$ again, we obtain an analog of the ideal-gas law for two dimensional systems,

$$\varsigma A = N k_B T.$$

Problem 1.7 Consider molecular adsorption from a liquid solution on a planar surface of area A and temperature T. Assume that the liquid phase is an ideal solution and that the adsorption of the solute molecules can be described by Henry's law

$$N/A = k_H C,$$

where N is the *moles* of solute molecules at the surface, k_H is Henry's constant, and C stands for the *molar* concentration of the solute molecules in the bulk solution. Show that the surface pressure is given by

$$\varsigma A = NRT,$$

where R is the gas constant.

Solution

The chemical potential of solute molecules in an ideal solution can be written as

$$\mu = \mu_0 + RT \ln(C/C_0),$$

where μ_0 is the chemical potential of the solute at a reference state of system temperature and $C_0 = 1$ mol/L. With Henry's law $N/A = k_H C$, the Gibbs isotherm predicts

$$\left(\frac{\partial \varsigma}{\partial \mu}\right)_T = \frac{N}{A} \Rightarrow \frac{1}{RT}\left(\frac{\partial \varsigma}{\partial \ln C}\right)_T = k_H C \Rightarrow \frac{1}{RT}\left(\frac{\partial \varsigma}{\partial C}\right)_T = k_H \Rightarrow \left(\frac{\partial \varsigma}{\partial C}\right)_T = k_H RT.$$

With the boundary condition $\varsigma = 0$ at $C = 0$, integrating the above equation leads to

$$\int_0^C \left(\frac{\partial \varsigma}{\partial C}\right)_T dC = \int_0^P k_H RTdC \Rightarrow \varsigma = k_H CRT.$$

Using Henry's law again, we obtain a two-dimensional analog of the ideal-gas law for the surface pressure

$$\varsigma A = NRT.$$

Problem 1.8 Consider ideal-gas adsorption on a planar surface such that gas molecules at the surface follows a two-dimensional equation of state

$$\varsigma(a - a_0) = k_B T,$$

where a is the surface area per molecule, and a_0 is the surface area occupied by each molecule. Show that the adsorption isotherm follows the Volmer isotherm

$$Pk_H = \frac{\theta}{1-\theta}\exp\left(\frac{\theta}{1-\theta}\right),$$

where $\theta = a_0/a$ stands for the surface coverage, and k_H is Henry's constant defined by $k_H = \theta/P$ as $P \to 0$. How would you extend the Volmer equation to adsorption on a planar surface from an ideal solution?

Solution

For an ideal gas at constant temperature T, the chemical potential varies with the bulk pressure P

$$d\mu = k_B T d\ln P.$$

The two-dimensional equation of state predicts that the surface pressure varies with the surface area per molecule

$$d\varsigma = -k_B T\frac{da}{(a-a_0)^2}.$$

Substituting these two equations into the Gibbs isotherm (Problem 1.5) yields

$$\left(\frac{\partial \varsigma}{\partial \mu}\right)_T = \frac{N}{A} \Rightarrow -\frac{1}{(a-a_0)^2}\left(\frac{\partial a}{\partial \ln P}\right)_T = \frac{1}{a}.$$

Integrating the above equation leads to

$$-\int_\infty^a \frac{a}{(a-a_0)^2}da = \int_0^P d\ln P \Rightarrow -\ln(a-a_0) + \frac{a_0}{a-a_0} = \ln P + K,$$

where K is an integration constant. Replacing a_0/a with θ gives

$$-\ln\left(\frac{1-\theta}{\theta}\right) + \frac{\theta}{1-\theta} = \ln(Pa_0) + K. \quad \text{(A)}$$

The integration constant can be fixed with the boundary condition $\theta = k_H P$ as $P \to 0$,

$$K = \ln(k_H/a_0).$$

Substituting K into Eq. (A) leads to the Volmer equation

$$Pk_H = \frac{\theta}{1-\theta}\exp\left(\frac{\theta}{1-\theta}\right).$$

For adsorption from an ideal solution, we replace P with the molar concentration of the solute

$$d\mu = k_B T d\ln C,$$

which leads to

$$Ck_H = \frac{\theta}{1-\theta}\exp\left(\frac{\theta}{1-\theta}\right),$$

where $\theta = a_0/a = k_H C$ as $C \to 0$.

Problem 1.9 Consider ideal-gas adsorption on a planar surface such that gas molecules at the surface follows a two-dimensional analog of the van der Waals equation of state

$$\varsigma = k_B T/(a-a_0) - c/a^2,$$

where a is the surface area per molecule, a_0 is the surface area occupied by each molecule, and c is an energy parameter that accounts for the intermolecular attraction at the surface. Show that the adsorption isotherm follows the Hill–Deboer equation

$$\ln\left[\frac{P(1-\theta)}{\theta}\right] - \frac{\theta}{1-\theta} = -\ln K_1 - \frac{K_2\theta}{k_B T},$$

where $\theta = a_0/a$ stands for the surface coverage, K_1 is defined by $K_1 = \theta/P$ as $P \to 0$ (viz., Henry's constant), and $K_2 = 2c/a_0$.

Solution

For an ideal gas at constant temperature T, the chemical potential varies with the bulk pressure P

$$d\mu = k_B T d\ln P.$$

The van der Waals-like equation predicts that the surface pressure varies with the surface area per molecule

$$d\varsigma = -k_B T\frac{da}{(a-a_0)^2} + \frac{2c}{a^3}da.$$

Substituting these two equations into the Gibbs isotherm gives (Problem 1.5)

$$\left(\frac{\partial\varsigma}{\partial\mu}\right)_T = \frac{N}{A} \Rightarrow \left[-\frac{1}{(a-a_0)^2} + \frac{2c}{k_B T a^3}\right]\left(\frac{\partial a}{\partial \ln P}\right)_T = \frac{1}{a}.$$

Integrating the above equation leads to

$$-\int_\infty^a \frac{a}{(a-a_0)^2}da + \int_\infty^a \frac{2c}{k_B T a^2}da = \int_0^P \ln P dP \Rightarrow$$

$$-\ln(a-a_0) + \frac{a_0}{a-a_0} - \frac{2c}{k_B T a} = \ln P + K,$$

where K is an integration constant. Replacing a_0/a with θ gives

$$-\ln\left(\frac{1-\theta}{\theta}\right) + \frac{\theta}{1-\theta} - \frac{2c\theta}{k_B T a_0} = \ln(Pa_0) + K, \tag{B}$$

where the integration constant can be fixed with the boundary condition $\theta = K_1 P$ as $P \to 0$,

$$K = \ln(K_1/a_0).$$

The Hill–Deboer isotherm is obtained by substituting K into Eq. (B)

$$\ln\left[\frac{P(1-\theta)}{\theta}\right] - \frac{\theta}{1-\theta} = -\ln K_1 - \frac{K_2\theta}{k_B T},$$

where $K_2 = 2c/a_0$.

Problem 1.10 A container includes a large number of gas molecules that do not interact with each other. Assume that the container wall is elastic and that the gas molecules can be represented by non-interacting classical particles. Does the system satisfy the ergodic hypothesis? Why?

Solution

For non-interacting classical particles in an elastic container, the kinetic energy is fixed by the initial condition. Because microstates in the ensemble are not mutually accessible to each other, the system is non-ergodic.

Problem 1.11 The Gibbs equation for entropy can be derived from the hypothesis that entropy is an ensemble average of some microscopic property depending only on the microstate probability, i.e.,

$$S = \sum_\nu p_\nu f(p_\nu),$$

where $f(p_\nu)$ may be understood as the microscopic counterpart of entropy at microstate ν. Show that this hypothesis leads to $f(p_\nu) = -k_B \ln p_\nu$. [Hint: Consider entropy additivity for independent thermodynamic systems.]

Solution

Consider the total entropy S_T of two independent systems A and B. Because entropy is extensive (viz., additivity), S_T can be written as

$$S_T = S_A + S_B,$$

where S_A and S_B are the entropies of individual systems. The microstate of the entire system is the sum of those corresponding to systems A and B, $\nu_T = \nu_A + \nu_B$. Because systems A and B are independent, the joint probability p_{ν_T} is a product of p_{ν_A} and p_{ν_B}, the microstate probabilities of individual systems,

$$p_{\nu_T} = p_{\nu_A} p_{\nu_B}.$$

If the entropy corresponds to an ensemble average of some microscopic property depending only on the microstate probability, we have

$$S_T = \sum_{\nu_T} p_{\nu_T} f(p_{\nu_T}) = \sum_{\nu_A}\sum_{\nu_B} p_{\nu_A} p_{\nu_B} f(p_{\nu_A} p_{\nu_B})$$

and

$$S_A + S_B = \sum_{\nu_A} p_{\nu_A} f(p_{\nu_A}) + \sum_{\nu_B} p_{\nu_B} f(p_{\nu_B}) = \sum_{\nu_A}\sum_{\nu_B} p_{\nu_A} p_{\nu_B} [f(p_{\nu_A}) + f(p_{\nu_B})],$$

where the second equal sign follows the probability normalization conditions,

$$\sum_{\nu_A} p_{\nu_A} = \sum_{\nu_B} p_{\nu_B} = 1.$$

Because the microstate distributions p_{ν_A} and p_{ν_B} are arbitrary, the entropy additivity requires

$$f(p_{\nu_A} p_{\nu_B}) = f(p_{\nu_A}) + f(p_{\nu_B}). \quad \text{(C)}$$

Differentiating both sides of Eq. (C) with respect to p_{ν_A} at fixed p_{ν_B} leads to

$$\frac{\partial f(p_{\nu_A}p_{\nu_B})}{\partial p_{\nu_A}} = \frac{\partial f(p_{\nu_A}p_{\nu_B})}{\partial p_{\nu_A}p_{\nu_B}} p_{\nu_B} = \frac{\partial f(p_{\nu_A})}{\partial p_{\nu_A}}. \tag{D}$$

Similarly, differentiation of Eq. (C) with respect to p_{ν_B} at constant p_{ν_A} gives

$$\frac{\partial f(p_{\nu_A}p_{\nu_B})}{\partial p_{\nu_B}} = \frac{\partial f(p_{\nu_A}p_{\nu_B})}{\partial p_{\nu_A}p_{\nu_B}} p_{\nu_A} = \frac{\partial f(p_{\nu_B})}{\partial p_{\nu_B}}. \tag{E}$$

A comparison of Eqs. (D) and (E) yields

$$p_{\nu_A}\frac{\partial f(p_{\nu_A})}{\partial p_{\nu_A}} = p_{\nu_B}\frac{\partial f(p_{\nu_B})}{\partial p_{\nu_B}}. \tag{F}$$

Because systems A and B are arbitrary, Eq. (F) suggests

$$\frac{\partial f(p_\nu)}{\partial \ln p_\nu} = K, \tag{G}$$

where K is a system-independent constant. Integration of Eq. (G) gives

$$f(p_\nu) = K \ln p_\nu + K', \tag{H}$$

where K' is an integration constant. In Eq. (H), entropy additivity requires $K' = 0$, and the universal constant K is not essential in the definition of the entropy (e.g., it was set as one in information entropy). In comparison with classical thermodynamics, it can be identified as $K = k_B$ by applying the Gibbs entropy to an ideal-gas system.

Problem 1.12 Suppose that a thermodynamic system has an initial distribution of microstates specified by $\{p_\nu^0\}$, where ν denotes microstates, and reaches a new microstate distribution $\{p_\nu\}$ at equilibrium. Show that

$$\sum_\nu p_\nu \ln p_\nu \geq \sum_\nu p_\nu \ln p_\nu^0,$$

where the equality holds if and only if $p_\nu = p_\nu^0$ for all ν. [Hint: $\ln x \leq x - 1$ for all $x > 0$ with equality if and only if $x = 1$.]

Solution

Using $\ln x \leq x - 1$ for all $x > 0$, we can prove the inequality by rearranging the terms

$$\sum_\nu p_\nu \ln(p_\nu^0/p_\nu) \leq \sum_\nu p_\nu(p_\nu^0/p_\nu - 1) = \sum_\nu p_\nu^0 - \sum_\nu p_\nu = 1 - 1 = 0.$$

Because $\ln x = x - 1$ only at $x = 1$, the equal sign holds if and only if $p_\nu = p_\nu^0$.

Problem 1.13 Consider a lattice model for a one-component gas such that each site can accommodate no more than one gas molecule. Assume that each microstate of the system is defined by a particular occupation of the lattice sites by the gas molecules.

(i) What is the number of microstates for the lattice system with n-sites containing N gas molecules?

(ii) Show that the lattice model predicts an entropy

$$S = -nk_B\left[x \ln x + (1 - x)\ln(1 - x)\right],$$

where $x = N/n$. [Hint: $\ln n! \approx n \ln n - n$.]

(iii) Assume that the system volume is proportional to the number of lattice sites, i.e., $V = nv_0$, where v_0 is the volume for each site. Show that, when $x \to 0$ (i.e., low gas density), the entropy change in response to the volume change from V_1 to V_2 is

$$\Delta S = Nk_B \ln(V_2/V_1).$$

(iv) Show that, when $x \to 0$, the lattice model satisfies the ideal-gas law $PV = Nk_BT$.
(v) How does the entropy change with temperature according to this model?

Solution

(i) The number of microstates for the lattice system is given by the number of combinations:

$$W = \binom{n}{N} = \frac{n!}{N!(n-N)!}.$$

(ii) The entropy is given by the Boltzmann equation

$$S/k_B = \ln W = \ln n! - \ln N! - \ln(n-N)!.$$

Using the Stirling approximation $\ln n! \approx n \ln n - n$, we can rewrite the above equation as

$$S/k_B = -N\ln(N/n) - (n-N)\ln(n-N)/n = -n\left[x\ln x + (1-x)\ln(1-x)\right],$$

where $x = N/n$.

(iii) When $x \to 0$, $|\ln x| \gg 1$, the system entropy becomes

$$S = -nk_B\left[x\ln x + (1-x)\ln(1-x)\right] \approx -Nk_B\ln(N/n).$$

Because the system volume is proportional to the number of lattice sites, we have $n = V/v_0$, where v_0 is the volume per lattice site. The above equation becomes

$$S = -Nk_B\ln(Nv_0/V). \tag{I}$$

Eq. (I) predicts that a change in volume from V_1 to V_2 at constant N leads to the entropy change

$$\Delta S = Nk_B\ln(V_2/V_1).$$

(iv) The Maxwell relation indicates

$$\left(\frac{\partial P}{\partial T}\right)_{V,N} = \left(\frac{\partial S}{\partial V}\right)_{T,N}.$$

According to Eq. (I)

$$\left(\frac{\partial S}{\partial V}\right)_{T,N} = \frac{Nk_B}{V}.$$

Using the boundary condition $P = 0$ as $T \to 0$, we obtain the ideal-gas law $PV = Nk_BT$ by integrating

$$\left(\frac{\partial P}{\partial T}\right)_{V,N} = \frac{Nk_B}{V}$$

with respect to T at constant V and N.

(v) This lattice model predicts that entropy does not change with temperature (Eq. I). In comparison with a real gas, the erroneous prediction can be attributed to the neglect of molecular energy.

Problem 1.14 Consider a lattice model for a one-component gas such that each lattice site can accommodate no more than one gas molecule. Each gas molecule on the lattice is able to take m orientations with equal energy. Assume that the system volume is proportional to the number of lattice sites, $V = nv_0$, where n is the number of lattice sites, and v_0 is the volume per lattice site.

(i) What is the entropy of the system in terms of the number of gas molecules N and the total number of lattice sites n.

(ii) Show that, when $x = N/n \to 0$ (i.e., low gas density), the entropy change in response to the volume change from V_1 to V_2 is

$$\Delta S = Nk_B \ln(V_2/V_1).$$

(iii) Show that, when $x \to 0$, the lattice model satisfies the ideal-gas law $PV = Nk_BT$.

Solution

(i) According to this model, each microstate is defined by a particular occupation of the lattice sites and the orientations of the gas molecules

$$W = \binom{n}{N} m^N = \frac{n!}{N!(n-N)!} m^N.$$

The Boltzmann equation predicts that the entropy is

$$\begin{aligned} S/k_B &= \ln W = \ln n! - \ln N! - \ln(n-N)! + N\ln m \\ &= -n\,[x\ln x + (1-x)\ln(1-x)] + N\ln m, \end{aligned}$$

where $x = N/n$.

(ii) When $x \to 0$, $|\ln x| \gg 1$, the system entropy becomes

$$S = -nk_B\,[x\ln x + (1-x)\ln(1-x) + N\ln m] \approx -Nk_B\ln(N/n) + Nk_B\ln m.$$

Because the system volume is proportional to the number of lattice sites, we have $n = V/v_0$, where v_0 is the volume per lattice site. The above equation becomes

$$S = -Nk_B\ln(Nv_0/V) + Nk_B\ln m. \tag{J}$$

Eq. (J) predicts that a change in volume from V_1 to V_2 at constant N leads to the entropy change

$$\Delta S = Nk_B\ln(V_2/V_1).$$

This problem affirms that entropy is a relative quantity, i.e., grouping m independent orientations for each molecule into one "coarse-grained" state does not change the relative entropy.

(iii) Following the Maxwell relation

$$\left(\frac{\partial P}{\partial T}\right)_{V,N} = \left(\frac{\partial S}{\partial V}\right)_{T,N},$$

and Eq. (J) for

$$\left(\frac{\partial S}{\partial V}\right)_{T,N} = \frac{Nk_B}{V},$$

we can obtain the ideal-gas law $PV = Nk_BT$ by integrating

$$\left(\frac{\partial P}{\partial T}\right)_{V,N} = \frac{Nk_B}{V}$$

with respect to T at constant V and N and the boundary condition $P = 0$ as $T \to 0$.

Problem 1.15 Repeat Problem 1.14 by assuming that the gas molecules do not interact with each other and that the m-orientations for each gas molecule on the lattice have different energies $\epsilon_i, i = 1,2, \dots, m$. How does the molecular orientation influence the relative entropy of the system?

Solution

(i) Each microstate is still defined by a particular occupation of the lattice sites and the orientation of the gas molecules. If the gas molecules do not interact completely, we cannot keep track of whether each site is occupied or empty. Therefore,

$$W = \frac{n^N}{N!}\mathfrak{m}^N.$$

where $\mathfrak{m}$ is the number of *accessible* orientations. In general, $\mathfrak{m}$ depends on the system temperature and may not be the same as m.
The entropy is then given by

$$S/k_B = \ln W = N\ln(n\mathfrak{m}) - \ln N! = N[\ln(n\mathfrak{m}/N) + 1].$$

(ii) When $x = N/n \to 0$, $|\ln x| \gg 1$, the equation for entropy remains the same because the molecules do not interact

$$S/k_B = N[\ln(n\mathfrak{m}/N) + 1].$$

If the system volume is proportional to the number of lattice sites, we have $n = V/v_0$, where v_0 is the volume per lattice site. The above equation becomes

$$S = -Nk_B\ln(Nv_0/V) + Nk_B(\ln \mathfrak{m} + 1). \tag{K}$$

Eq. (K) predicts that a change in volume from V_1 to V_2 at constant N leads to the entropy change

$$\Delta S = Nk_B\ln(V_2/V_1).$$

Again, entropy is a relative quantity, i.e., grouping m independent orientations for each molecule into one "coarse-grained" state does not change the relative entropy.

(iii) Following the Maxwell relation

$$\left(\frac{\partial P}{\partial T}\right)_{V,N} = \left(\frac{\partial S}{\partial V}\right)_{T,N},$$

and Eq. (K) for

$$\left(\frac{\partial S}{\partial V}\right)_{T,N} = \frac{Nk_B}{V},$$

we can thus obtain the ideal-gas law $PV = Nk_BT$ by integrating

$$\left(\frac{\partial P}{\partial T}\right)_{V,N} = \frac{Nk_B}{V}$$

with respect to T at constant V and N and the boundary condition $P = 0$ as $T \to 0$.

Because the gas molecules do not interact with each other, the molecular orientations may be considered as the internal degrees of freedom, which make only a fixed contribution to the total entropy ($Nk_B\ln\mathfrak{m}$) but have no effect on the relative entropy.

Problem 1.16 Consider a lattice model such that each site can be in one of two energy states, 0 or $\epsilon > 0$. Let n be the total number of lattice sites, and N the number of sites with energy ϵ.

(i) How would you define the microstates of this system?
(ii) Show that the lattice sites with different energies follow the Boltzmann distribution

$$\frac{N}{n-N} = \exp\left(-\frac{\epsilon}{k_B T}\right).$$

(iii) Show that the constant-volume heat capacity of the lattice system is given by

$$C_V = \frac{N(n-N)}{n}\frac{\epsilon^2}{k_B T^2}.$$

(iv) Can the absolute temperature be negative? Why?

Solution

(i) For this system, each microstate is defined by the energies of individual lattice sites.
(ii) The system temperature can be found from the fundamental equation

$$\frac{1}{T} = \left(\frac{\partial S}{\partial U}\right)_V, \tag{L}$$

where $V = nv_0$. The entropy of the lattice system is

$$S/k_B = -N\ln(N/n) - (n-N)\ln(n-N)/n,$$

and the internal energy is

$$U = N\epsilon.$$

Because

$$\left(\frac{\partial S}{\partial U}\right)_V = \frac{1}{\epsilon}\left(\frac{\partial S}{\partial N}\right)_V = \frac{1}{\epsilon}\left(\frac{\partial S}{\partial N}\right)_n = -\frac{k_B}{\epsilon}\ln\left(\frac{N}{n-N}\right).$$

Substituting the above equation into Eq. (L) yields

$$\frac{N}{n-N} = \exp\left(-\frac{\epsilon}{k_B T}\right).$$

(iii) The constant-volume heat capacity is defined as

$$C_V = \left(\frac{\partial U}{\partial T}\right)_V.$$

Because $U = \epsilon N$ and $V = nv_0$, the heat capacity of this system becomes

$$C_V = \epsilon\left(\frac{\partial N}{\partial T}\right)_n.$$

Noting

$$N = \frac{n}{1+\exp\left(\frac{\epsilon}{k_B T}\right)},$$

we have

$$\left(\frac{\partial N}{\partial T}\right)_n = \frac{n\exp\left(\frac{\epsilon}{k_B T}\right)}{\left[1+\exp\left(\frac{\epsilon}{k_B T}\right)\right]^2}\frac{\epsilon}{k_B T^2} = \frac{N(n-N)}{n}\frac{\epsilon}{k_B T^2}.$$

Thus

$$C_V = \frac{N(n-N)}{n}\frac{\epsilon^2}{k_B T^2}.$$

(iv) According to this model, the absolute temperature is negative when $N > n - N$ (see part ii). Negative temperature is possible because, under such conditions, the entropy falls as the system energy increases. The possibility of negative absolute temperatures was first proposed by Lars Onsager.[1]

Problem 1.17 Imagine that Maxwell's demon could control the direction of molecules passing through a hole between identical chambers A and B containing a one-component ideal gas. The demon allows all molecules, fast or slow, to pass from B to A but prevent them from passing from A to B. Eventually all molecules will be concentrated in A and a vacuum will be created in B so that a pressure difference is generated without doing any work. Explain whether the process violates the first or the second law of thermodynamics and discuss how this might be implemented with modern technology.

Solution

The system temperature is invariant during this process because the demon manipulates only the locations of ideal gas molecules without altering the molecular velocities. Overall, there is no change in the internal energy of the system. The entropy change for the compression of the ideal gas is

$$\Delta S = k_B N \ln(V/2V) = -Nk_B \ln 2,$$

where N is the number of gas molecules, and V is the volume for each chamber. The reduction of entropy, ΔS, is attributed to the information gained by the demon in recognizing the direction of each gas molecule

$$S_{info} = k_B \ln 2^N = Nk_B \ln 2.$$

Therefore, neither the first nor the second law was violated. Useful energy can be acquired if a device could distinguish the direction of molecular motions at any instant.[2]

Problem 1.18 In a computer, the Memory Address Register (MAR) is responsible for storing either the memory address from which data will be fetched to the Central Processing Unit (CPU), or the address to which data will be sent and stored. Consider a computer operation that reformats a memory register of n bits. Before the operation, the register as a whole could have existed in any of 2^n states. However, after the operation, the register is left in only one state. To maintain the temperature of the computer at this point, how much heat must be released?

Solution

The operation compressed 2^n logical states into one, leading to the reduction of entropy $\Delta S = -nk_B \ln 2$. The heat dissipated satisfies the Clausius inequality $\Delta S \geq Q/T$, i.e., the minimum heat released is $Q_{min} = nk_B T \ln 2$.

1 Onsager, L., "Statistical Hydrodynamics", *Il Nuovo Cimento*, 6: 279–287 (1949).

2 Toyabe S., Sagawa T., Ueda M. et al., "Experimental demonstration of information-to-energy conversion and validation of the generalized Jarzynski equality", *Nature Phys.*, 6, 988–992 (2010).

2

Statistical Ensembles and MD Simulation

Problem 2.1 Consider an isolated system with N distinguishable but noninteracting particles. Each particle can be in one of two possible energy states: one at ϵ, and the other at $-\epsilon$. Let E be the total energy of the system

(i) Derive equations for the number of particles at each energy state;
(ii) Derive the system entropy;
(iii) Derive the system temperature;
(iv) Verify the prediction of the third law of thermodynamics, i.e., $S \to 0$ when $T \to 0$ K;
(v) Show that at some values of E and N, the absolute temperature can be negative.

Solution:

(i) The microstates of an isolated system constitute a microcanonical ensemble. Here, each microstate can be described by the energy states of individual particles

$$\nu = (s_1, s_2, \dots, s_N)$$

where $s_i = \pm 1$. For this isolated system, the internal energy is the same as the total energy $U = E$

$$E_\nu = N_+\epsilon - N_-\epsilon = E,$$

where N_+ is the number of particles with energy ϵ, and N_- is the number of particles with energy $-\epsilon$. Because both $N = N_+ + N_-$ and the total energy are fixed, we can find the number of particles at each energy state:

$$N_+ = (N + E/\epsilon)/2 \text{ and } N_- = (N - E/\epsilon)/2.$$

(ii) For this system, the total number of microstates is same as the number of ways to assign the N particles at the two energy states:

$$W = \frac{N!}{N_+!N_-!}.$$

According to the Boltzmann equation, the system entropy is then given by

$$S = k_B \ln W = k_B \ln \frac{N!}{N_+!N_-!}.$$

Based on Stirling's approximation for factorials, $\ln N! = N \ln N - N$, we may rewrite the entropy as

$$S = -k_B \left[N_+ \ln(N_+/N) + N_- \ln(N_-/N)\right].$$

Solutions Manual to Fundamentals and Practice in Statistical Thermodynamics,
First Edition. Jianzhong Wu and John M. Prausnitz.

(iii) From the entropy and the internal energy, we derive the system temperature from the thermodynamic identity

$$T = \left(\frac{\partial U}{\partial S}\right)_{N,V} = \left(\frac{\partial E}{\partial N_+}\right)_{N,V}\left(\frac{\partial N_+}{\partial S}\right)_{N,V}.$$

From $\partial N_+/\partial E = 1/(2\epsilon)$ and $\partial S/\partial N_+ = k_B \ln(N_-/N_+)$, we find an analytical expression for the temperature

$$T = \frac{2\epsilon}{k_B \ln(N_-/N_+)}.$$

(iv) At $T = 2\epsilon/[k_B \ln(N_-/N_+)] = 0$ K, we have $N_+ = 0$. Substituting this into the entropy equation yields $S = 0$.

(v) If $\epsilon > 0$, the last equation from part (iii) predicts that the absolute temperature would be negative when $N_+ > N_-$, i.e., when $E > 0$.

Problem 2.2 Suppose that the model system considered in Problem 2.1 is immersed in a thermal bath at temperature T.

(i) Derive equations for the number of particles at each energy state, $< N_+ >$ and $< N_- >$;

(ii) Show that the system entropy can be expressed as

$$S = -k_B \left[< N_+ > \ln(< N_+ > /N) + < N_- > \ln(< N_- > /N)\right].$$

(iii) Derive an expression for the internal energy;

(iv) Verify the prediction of the third law of thermodynamics, i.e., $S \to 0$ when $T \to 0$ K;

(v) In absolute scale, can the system temperature be negative?

Solution:

Now we consider the same system but use the canonical ensemble, i.e., an ensemble corresponding to systems at fixed T and total number of particles N. In this case, each microstate is still specified by the energy states of individual particles. For a given microstate, the total energy is

$$E_\nu = \sum_{i=1}^{N} s_i \epsilon.$$

For this simple system, the canonical partition function can be evaluated analytically

$$Q \equiv \sum_\nu e^{-\beta E_\nu} = \prod_{i=1}^{N} \sum_{s_i=\pm 1} e^{-\beta s_i \epsilon} = (e^{-\beta\epsilon} + e^{\beta\epsilon})^N.$$

(i) The number of particles at each energy state is most conveniently expressed in terms of the mean energy

$$< E > = -\frac{\partial \ln Q}{\partial \beta} = -N\frac{-\epsilon e^{-\beta\epsilon} + \epsilon e^{\beta\epsilon}}{e^{-\beta\epsilon} + e^{\beta\epsilon}} = N\epsilon\frac{e^{-\beta\epsilon} - e^{\beta\epsilon}}{e^{-\beta\epsilon} + e^{\beta\epsilon}} = -N\epsilon \tanh(\beta\epsilon).$$

From $N = < N_+ > + < N_- >$ and $< E > =< N_+ > \epsilon - < N_- > \epsilon = E$, we find

$$< N_+ > = (N + < E > /\epsilon)/2 = \frac{Ne^{-\beta\epsilon}}{e^{-\beta\epsilon} + e^{\beta\epsilon}},$$

$$< N_- > = (N - < E > /\epsilon)/2 = \frac{Ne^{\beta\epsilon}}{e^{-\beta\epsilon} + e^{\beta\epsilon}}.$$

(ii) The system entropy is given by

$$\begin{aligned} S &= (< E > -F)/T = -N\epsilon \tanh(\beta\epsilon)/T + Nk_B \ln(e^{-\beta\epsilon} + e^{\beta\epsilon}) \\ &= (< N_+ > \epsilon - < N_- > \epsilon)/T + (< N_+ > + < N_- >)k_B \ln(e^{-\beta\epsilon} + e^{\beta\epsilon}) \\ &= - < N_+ > k_B \ln e^{-\beta\epsilon} - < N_- > k_B \ln e^{\beta\epsilon} + (< N_+ > + < N_- >)k_B \ln(e^{-\beta\epsilon} + e^{\beta\epsilon}) \\ &= -k_B \left[< N_+ > \ln(< N_+ > /N) + < N_- > \ln(< N_- > /N)\right]. \end{aligned}$$

(iii) The internal energy is $U =< E > = -N\epsilon \tanh(\beta\epsilon)$.

(iv) When $T \to 0$ K, $\beta \to \infty$ so that $e^{\beta\epsilon} \gg e^{-\beta\epsilon}$ for $\epsilon > 0$. Therefore, $\tanh(\beta\epsilon) = 1$ and the system entropy becomes

$$S = -N\epsilon/T + Nk_B(\beta\epsilon) = 0.$$

(v) At equilibrium, the system temperature is always the same as that of the heat bath, which cannot be negative.

Problem 2.3 Proposed by Loup Verlet in 1967, the Verlet algorithm is one of the most popular integration schemes used in molecular dynamics (MD) simulation. The numerical method can be derived from the Taylor expansions of the particle position at two consecutive time steps, i.e., from t_{n-1} to t_n and from t_n to t_{n+1}, such that the time reversal symmetry of Newton's equation of motion is preserved:

$$\mathbf{r}_{n+1} = \mathbf{r}_n + \mathbf{v}_n\delta t + \frac{1}{2!}\mathbf{a}_n\delta t^2 + \frac{1}{3!}\dot{\mathbf{a}}_n\delta t^3 + O(\delta t^4), \tag{A}$$

$$\mathbf{r}_{n-1} = \mathbf{r}_n - \mathbf{v}_n\delta t + \frac{1}{2!}\mathbf{a}_n\delta t^2 - \frac{1}{3!}\dot{\mathbf{a}}_n\delta t^3 + O(\delta t^4), \tag{B}$$

where $\mathbf{v}_n = d\mathbf{r}_n/dt$, $\mathbf{a}_n = d\mathbf{v}_n/dt$, and $\dot{\mathbf{a}}_n = d\mathbf{a}_n/dt$. Adding Eqs. (A) and (B) leads to

$$\mathbf{r}_{n+1} = 2\mathbf{r}_n - \mathbf{r}_{n-1} + \frac{\mathbf{f}_n}{m}\delta t^2 + O(\delta t^4), \tag{C}$$

where $\mathbf{f}_n$ is the net force on the particle at t_n, and m is the particle mass. On the other hand, subtracting Eq. (A) with (B) gives the velocity equation:

$$\mathbf{v}_n = \frac{\mathbf{r}_{n+1} - \mathbf{r}_{n-1}}{2\delta t} + O(\delta t^3). \tag{D}$$

(i) What are the advantages of the Verlet algorithm in comparison with Euler's method

$$\mathbf{r}_{n+1} = \mathbf{r}_n + \mathbf{v}_n\delta t + \frac{\mathbf{a}_n}{2!}\delta t^2 + O(\delta t^3), \qquad \mathbf{v}_{n+1} = \mathbf{v}_n + \mathbf{a}_n\delta t + O(\delta t^2)?$$

(ii) The three terms on the left side of Eq. (C) involves two large numbers (viz., $2\mathbf{r}_n$ and $\mathbf{r}_{n-1}$) and one small number, $(\mathbf{f}_n/m)\delta t^2$. The numerical issue can be avoided in the so-called leapfrog algorithm, a slight modification of Verlet's algorithm using the position at time step n and the velocity at time step $n - 1/2$ to update both position and velocity at the next step:

$$\mathbf{r}_{n+1} = \mathbf{r}_n + \mathbf{v}_{n+1/2}\delta t, \tag{E}$$

$$\mathbf{v}_{n+1/2} = \mathbf{v}_{n-1/2} + \frac{\mathbf{f}_n}{m}\delta t. \tag{F}$$

Show that the leapfrog algorithm are accurate up to the third order (δt^3) for both position and velocity.

(iii) Show that Eqs. (A) and (B) are equivalent to those from the velocity Verlet algorithm

$$\mathbf{r}_{n+1} = \mathbf{r}_n + \mathbf{v}_n\delta t + \frac{\mathbf{f}_n}{2m}\delta t^2, \tag{G}$$

$$\mathbf{v}_{n+1} = \mathbf{v}_n + \frac{(\mathbf{f}_n + \mathbf{f}_{n+1})}{2m}\delta t. \tag{H}$$

Does the velocity Verlet algorithm satisfy the time reversal symmetry of Newton's equation?

Solution:

(i) The Verlet algorithm has at least two advantages in comparison with the Euler method: (1) it is more accurate because the position is calculated up to the third order in δt instead of the second order; (2) More importantly, the Euler method is not time reversible thereby suffering from energy drift while the Verlet algorithm is time reversible.

(ii) Both Eqs. (E) and (F) follow directly from Eq. (D), which can be written as

$$\mathbf{r}_{n+1} = \mathbf{r}_{n-1} + 2\mathbf{v}_n\delta t + O(\delta t^3).$$

Using the step length of $\delta t/2$ instead of δt, we obtain

$$\mathbf{r}_{n+1} = \mathbf{r}_n + \mathbf{v}_{n+1/2}\delta t + O((\delta t/2)^3).$$

The same procedure can be used to derive the leapfrog velocity equation:

$$\mathbf{v}_{n+1/2} = \mathbf{v}_{n-1/2} + \frac{\mathbf{f}_n}{m}\delta t + O((\delta t/2)^3).$$

(iii) Note that the velocity Verlet algorithm predicts

$$\mathbf{r}_n = \mathbf{r}_{n-1} + \mathbf{v}_{n-1}\delta t + \frac{\mathbf{f}_{n-1}}{2m}\delta t^2,$$

Subtracting the above equation from Eq. (G) yields

$$\mathbf{r}_{n+1} - \mathbf{r}_n = \mathbf{r}_n - \mathbf{r}_{n-1} + (\mathbf{v}_n - \mathbf{v}_{n-1})\delta t + \frac{\mathbf{f}_n - \mathbf{f}_{n-1}}{2m}\delta t^2.$$

Using Eq. (H) for $\mathbf{v}_n - \mathbf{v}_{n-1}$, we have

$$(\mathbf{v}_n - \mathbf{v}_{n-1})\delta t = \frac{(\mathbf{f}_{n-1} + \mathbf{f}_n)}{2m}\delta t^2.$$

The combination of the above two equations gives

$$\mathbf{r}_{n+1} + \mathbf{r}_{n-1} = 2\mathbf{r}_n + \frac{\mathbf{f}_n}{m}\delta t^2.$$

This equation is identical to that given by the Verlet algorithm. Like the Verlet algorithm, the velocity Verlet algorithm satisfies the time reversal symmetry.

Problem 2.4 Exactly solvable models are often used in statistical mechanics for testing theoretical methods and simulation algorithms. In this exercise, we use the one-dimensional (1D–) harmonic oscillator as a benchmark to test the velocity Verlet algorithm. Consider a particle of mass m subject to a harmonic potential $\varphi(x) = kx^2/2$ where k represents the force constant, and x is the particle position. The equation of motion for the particle, $m\frac{d^2x}{dt^2} = -kx$, has the following exact solutions for the position and velocity:

$$x(t) = A\cos(\omega t + \phi), \qquad v(t) = -A\omega\sin(\omega t + \phi),$$

where $\omega = \sqrt{k/m}$ is the angular frequency, A is the oscillation amplitude, and ϕ denotes the phase. While k and m are intrinsic properties of the system, A and ϕ are set by the initial conditions. For simplicity, we adopt $1/\omega$ as the unit of time, A as the unit of length, $A\omega$ as the unit of velocity, kA as the unit of force, and kA^2 as the unit of energy. The initial condition is set as $x_0 = 1$ and $v_0 = 0$ at $t = 0$. Prepare a compute program to implement the velocity Verlet algorithm

$$\mathbf{r}_{n+1} + \mathbf{r}_{n-1} = 2\mathbf{r}_n + \frac{\mathbf{f}_n}{m}\delta t^2 \qquad \mathbf{v}_{n+1} = \mathbf{v}_n + \frac{(\mathbf{f}_n + \mathbf{f}_{n+1})}{2m}\delta t,$$

for the $1D$–harmonic oscillator and study the following:

(i) Compare the simulation and exact results for the particle position and velocity using integration time $\delta t = 0.1$ for the first 100 steps;

(ii) Suppose that the accuracy of the simulation can be measured by the mean-square error (MSE) of the particle position

$$MSE = \frac{1}{n_{step}} \sum_{n=1}^{n_{step}} [x_n - x(t_n)]^2$$

where n_{step} is the number of integration steps, and $x(t_n)$ stands for the particle position at time t_n predicted by the exact result. Plot MSE versus δt for $n_{step} = 1000$ and discuss how the integration time should be selected in MD simulation.

(iii) Set $\delta t = 0.1$, how does the simulation error propagate with time in terms of the MSE?

(iv) Set $\delta t = 0.1$, how does the system energy propagate with time? Compare the result with that from Euler's algorithm:

$$\mathbf{r}_{n+1} = \mathbf{r}_n + \mathbf{v}_n \delta t, \qquad \mathbf{v}_{n+1} = \mathbf{v}_n + \frac{\mathbf{f}_n}{m}\delta t.$$

Solution:

(i) Figure 2.1 compares the simulation and exact results for the particle position and velocity using integration time $\delta t = 0.1$ for the first 100 steps.

(ii) Figure 2.2 shows the MSE of the particle position versus time. Interesting, the log–log plot shows an inflection point about $\delta t = 0.1$ beyond which the MSE grows rapidly. The inflection point provides a good guidance for the selection of the integration time in MD simulation.

(iii) Figure 2.3 shows the simulation error versus time according to the Verlet algorithm. The error does not grow with the number of simulation steps, indicating the excellent performance of the Verlet method.

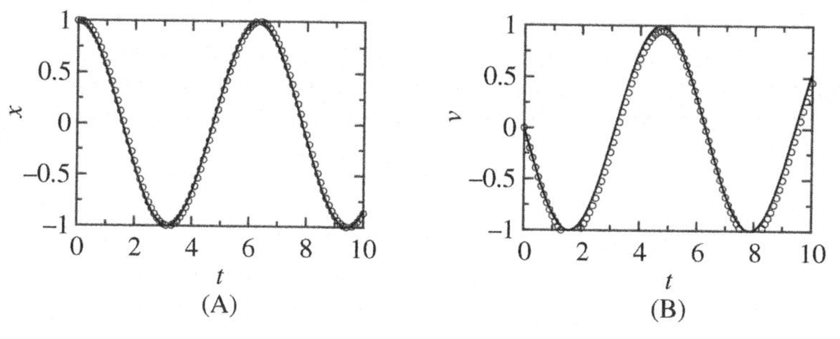

Figure 2.1 MD simulation of a harmonic oscillator implemented with the velocity Verlet algorithm. (A) Particle position; (B) Particle velocity. Points are simulation results with $\delta t = 0.1$, and the solid lines are from the analytical solution (Verlet.m).

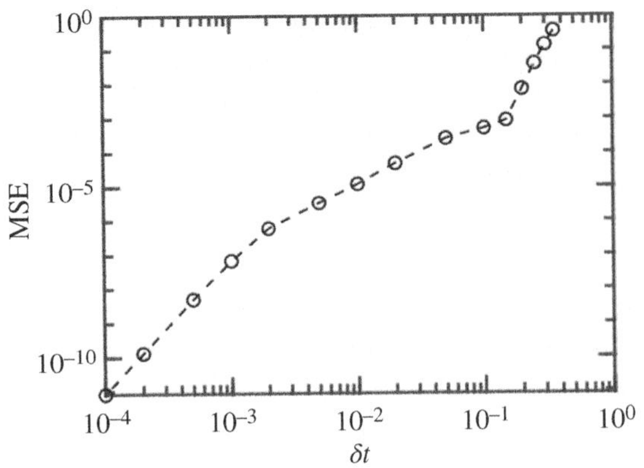

Figure 2.2 The mean-square error (MSE) of the particle position versus integration time (VerletMSE.m).

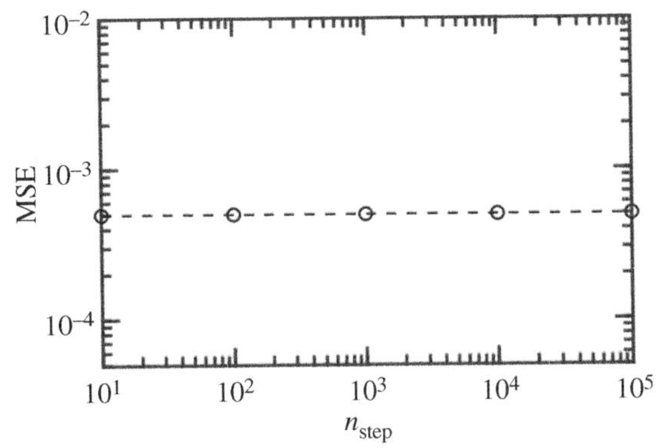

Figure 2.3 The mean-square error (MSE) of the particle position versus simulation steps (VerletMSE.m).

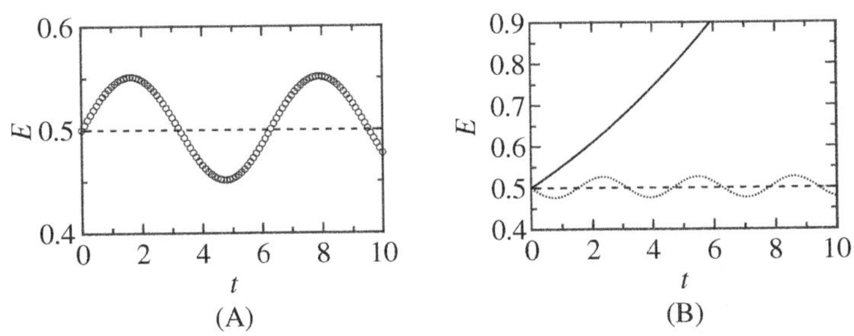

Figure 2.4 The energy of a 1D-harmonic oscillator calculated from Verlet's (A) and Euler's (B) algorithms. The horizontal line represents the exact results. In Euler's algorithm, the solid and dotted lines represent linear and second-order approximations in calculation of the particle position from Newton's equation (Euler.m).

(iv) Figure 2.4 compares the particle energy calculated from Verlet's and Euler's algorithms. With the linear approximation for the particle position, $\mathbf{r}_{n+1} = \mathbf{r}_n + \mathbf{v}_n\delta t$, the system energy grows with the number of the integration steps, suggesting that the algorithm is unstable. Interestingly, the second-order approximation, $\mathbf{r}_{n+1} = \mathbf{r}_n + \mathbf{v}_n\delta t + \mathbf{a}_n\delta t^2/2$, leads to perfect stability. In comparison with Verlet's algorithm (Figure 2.1), the second-order Euler algorithm gives similar particle position, but slightly less error in velocity and energy (Figure 2.5).

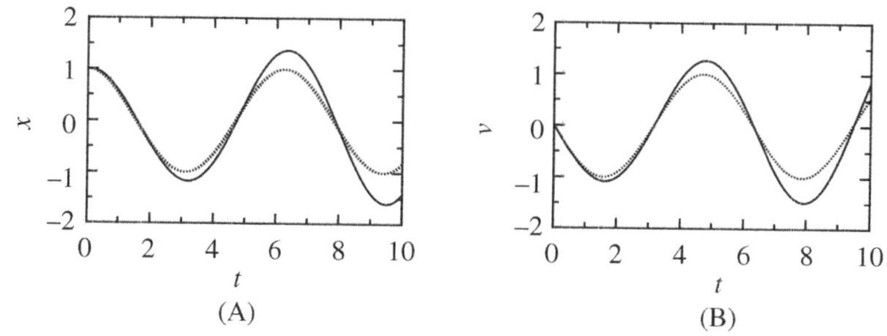

Figure 2.5 The particle position (A) and velocity (B) predicted by Euler's algorithm. The solid and dotted lines represent linear and second-order approximations in calculation of the particle position from Newton's equation (Euler.m).

Problem 2.5 The characteristic time of particle motions (τ) is often used to estimate the time step (δt) in MD simulation. A good rule of thumb is $\delta t \approx 0.1\tau$. Suppose that MD simulation is used to calculate the properties of liquid argon near its triple point ($T_t = 83.804$ K, $\rho_L = 35.475$ mol/L, $P_t = 0.06895$ MPa). Using the Lennard–Jones model ($\epsilon/k_B = 119.8$ K, $\sigma = 3.405$ Å), estimate the characteristic time of particle motions and proper time step in both absolute and appropriate dimensionless units. What is a proper duration of simulation?

Solution:

In a dense liquid, argon molecules collide with each other over a duration that may be approximated by the molecular diameter divided by the mean velocity. The latter can be estimated by the average kinetic energy

$$v_m = \sqrt{k_B T/m},$$

which yields $v_m \approx 132$ m/s at $T_t = 83.804$ K. Thus, we have $\tau = \sigma/v_m \approx 2.6 \times 10^{-12}$ s and $\delta t \approx 2.6 \times 10^{-13}$ s.

Because extremely large or small numbers are not convenient for computer operations, simulation programs often adopt dimensionless units for most physical quantities such that their numerical values are about the order of unity. For the Lennard–Jones model, the unit time may be deducted from the Verlet algorithm,

$$\tau_u \equiv \sigma\sqrt{m/\epsilon},$$

where m stands for the particle mass. For argon, $m = 39.948$ g/mol which gives a unit time of $\tau_u \approx 2.1564 \times 10^{-12}$. Therefore, in dimensionless units (time divided by τ_u), the characteristic time is 1.2 and the time step is about 0.1. In other words, $\delta t = 0.1$ would be considered as a reasonable choice for the time step.

The duration of simulation is determined by the relaxation time, $\tau_R \sim \sigma^2/D$, where D is diffusivity. At the triple point, the argon diffusivity in the liquid phase is about $D \sim 10^{-9}$ m^2/s, which predicts $\tau_R \sim 10^{-10}$ s. The approximate duration of simulation is about 10 times larger than the simulation time, $\sim 10^{-9}$ s or about 4000 MD steps per particle.

Problem 2.6 In Andersen's method for keeping a constant temperature during MD simulation, the frequency of velocity reassignment (viz., the probability of particle collision at each MD step) can be estimated by considering a simulation cell with volume V surrounded by a much larger heat bath at temperature T. When there is a temperature fluctuation in the simulation cell so that its average temperature is $T + \Delta T$, it will gain or lose energy at a rate proportional to the temperature difference ΔT.

(i) Show that the rate at which energy is absorbed/released by the system during time step Δt can be written as

$$\frac{\Delta E}{\Delta t} = -a\kappa \Delta T V^{1/3},$$

where κ is the thermal conductivity, and a is a dimensionless constant that depends on the shape of the simulation cell.

(ii) Show that the change in the total kinetic energy of the system due to the heat transfer can be compensated by rescaling the average velocity of particles

$$\frac{\Delta E}{\Delta t} = \frac{3}{2}\nu N k_B \Delta T,$$

where ν stands for the frequency of velocity reassignment, and N is the number of particles in the system.

(iii) Derive the frequency of velocity reassignment

$$\nu = \frac{2a\kappa}{3k_B \rho^{1/3} N^{2/3}},$$

where $\rho = N/V$.

(iv) Show that the frequency of velocity reassignment will be much smaller than the frequency of particle collisions when N is sufficiently large.

Solution:

(i) The rate of heat transfer is described by Fourier's law

$$\mathbf{j} = -\kappa \nabla T,$$

where $\mathbf{j}$ is the amount of energy flow per unit area per unit time. For a small temperature difference between the simulation cell and the heat bath, the total energy released/absorbed by the system over the duration of Δt is

$$\frac{\Delta E}{\Delta t} = \mathbf{j} \times \text{surface area} \approx -\kappa \times \frac{\Delta T}{\text{system size}} \times \text{surface area} = -a\kappa \Delta T V^{1/3},$$

where $V^{1/3}$ is obtained by dimensional analysis, and a is a dimensionless constant that is determined by the system geometry.

(ii) Before velocity reassignment, the average kinetic energy per particle is $3k_B(T + \Delta T)/2$, and after reassignment it is $3k_B T/2$. Therefore, each velocity reassignment leads to a change of the particle energy by $3k_B \Delta T/2$. With the frequency of velocity reassignment per integration step given by ν, the total energy change for N particles in the system is thus

$$\frac{\Delta E}{\Delta t} = \frac{3}{2}\nu N k_B \Delta T.$$

(iii) The frequency of velocity reassignment can be obtained by equaling the energy due to heat transfer and that due to the velocity reassignment

$$\nu = \frac{a\kappa \Delta T V^{1/3}}{\frac{3}{2} N k_B \Delta T} = \frac{2a\kappa}{3k_B \rho^{1/3} N^{2/3}}, \tag{I}$$

where $\rho = N/V$.

(iv) If the system being simulated contains only one particle, then all other particles in the thermodynamic system would comprise a heat bath. In that case, the frequency of velocity reassignment is the same as the particle collision frequency ν_c. Applying Eq. (I) to $N = 1$ gives

$$\nu_c = \frac{2a\kappa}{3k_B\rho^{1/3}}.$$

Therefore, the frequency of velocity reassignment scales with the number of particles being simulated according to

$$\nu = \nu_c/N^{2/3}.$$

When N is sufficiently large, the frequency of velocity reassignment will be much smaller than the frequency of particle collisions.

Problem 2.7 There is an increasing interest in the practical applications of so-called low-dimensional materials, for example, in studies of electrodes for energy storage, ion channels in biological membranes, and molecules in catalytic environments. Computer simulation is useful for understanding the thermodynamic properties and dynamical behavior of such systems which can be qualitatively different from those displayed in bulk systems. In this exercise, we use the one-dimensional Lennard–Jones (1D-LJ) model to simulate the structure and dynamic properties of spherical particles in a narrow pore.

(i) Prepare a MD program for simulating 1D-LJ fluids in the canonical ensemble using the velocity Verlet algorithm and Anderson thermostat; Use dimensionless units for all physical quantities, i.e., distance in σ, energy in ϵ, temperature in k_BT/ϵ, velocity in $\sqrt{k_BT/m}$, the particle number density in $\rho\sigma$, and time in $\sigma\sqrt{m/\epsilon}$. Show the simulation results at $k_BT/\epsilon = 1$ and $\rho\sigma = 0.7$.

(ii) Testing the simulation results with the Maxwell–Boltzmann velocity distribution for the particle velocity;

(iii) Calculate the radial distribution function of the 1D system, estimate the correlation length ξ, and discuss how ξ may vary with the system temperature and density.

(iv) Calculate the velocity autocorrelation function, estimate the correlation time τ, and discuss how τ may vary with the system temperature and density.

(v) Calculate the mean-square displacement and discuss how the diffusive behavior of the one-dimensional fluid differ from that corresponding to a bulk system.

Solution:

(i) The MD simulation has been implemented in Matlab program MDLJ1D.m. The simulation results are discussed in detail in the papers by, e.g.,

1) Bazhenov A. M. and Heyes D. M., *J. Chem. Phys.* 92, 1106 (1990);
2) Lepri S., Sandri P., and Politi A., *Eur. Phys. J. B.* 47, 549–555 (2005);
3) Bishop M., *J. Statistic. Phys.* 29, 623–629 (1982).

(ii) Figure 2.6 compares the particle velocity from MD simulation with the Maxwell–Boltzmann velocity distribution

(iii) Figure 2.7 shows the radial distribution function of the 1D system. The correlation length $\xi \approx 5\sigma$ can be estimated from the exponential decay of the peak values, $\sim \exp(-r/\xi)$. By running the simulation at different conditions, we see that ξ increases with the density but decreases with temperature.

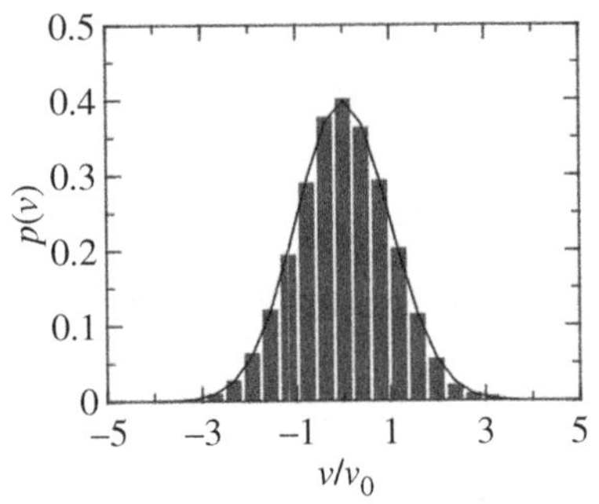

Figure 2.6 The particle velocity for a 1D-LJ fluid from MD simulation (bars) and the Maxwell–Boltzmann distribution (line). (MDLJ1D.m)

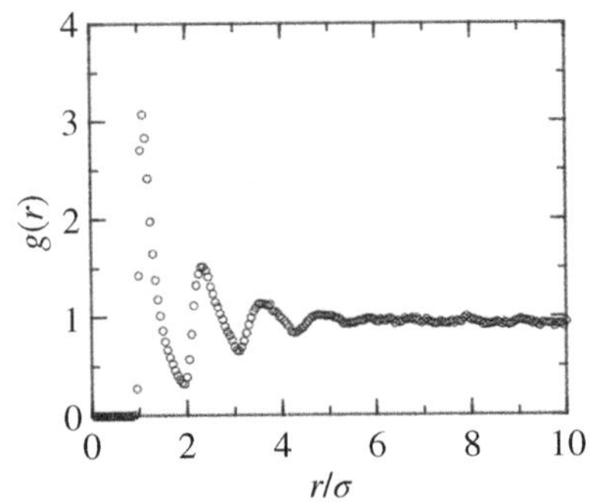

Figure 2.7 The radial distribution function of a 1D-LJ fluid calculated from MD simulation. (MDLJ1D.m)

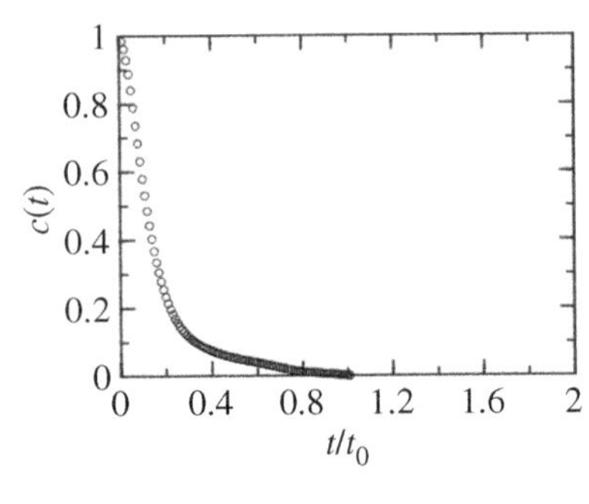

Figure 2.8 The velocity autocorrelation function of a 1D-LJ fluid calculated from MD simulation. (MDLJ1D.m)

(iv) Figure 2.8 shows the velocity autocorrelation function. The correlation time is not well defined because the one-dimensional Lennard–Jones system did not display an exponential decay.

(v) Figure 2.9 shows the mean-square displacement. It is related to the diffusion coefficient through the Green–Kubo formula

$$D = \int_0^\infty c(t)dt.$$

As mutual passage is prohibited for particles confined within a narrow channel, the motion of individual particles is highly correlated even at long time scales. The particle diffusion does not follow Fick's law because the displacement of a given particle over a long distance necessitates the motion of many other particles in the same direction.

Figure 2.9 The mean-square displacement of a 1D-LJ fluid calculated from MD simulation. (MDLJ1D.m)

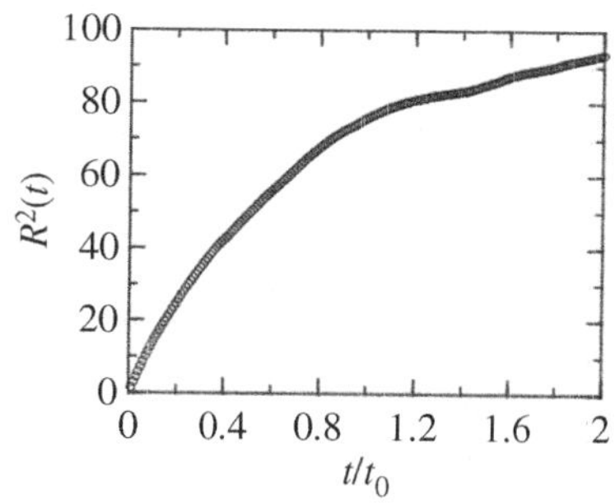

Problem 2.8 Show that Nosé's equations of motion

$$\frac{d\mathbf{r}_i}{dt} = \frac{\mathbf{p}_i}{ms^2}, \qquad \frac{d\mathbf{p}_i}{dt} = \mathbf{F}_i, \qquad \frac{ds}{dt} = \frac{p_s}{\varpi}, \qquad \frac{dp_s}{dt} = \sum_i \frac{p_i^2}{ms^3} - \frac{(3N+1)k_BT}{s},$$

can be transformed into the Nosé–Hoover equations

$$\frac{d\mathbf{r}_i}{dt'} = \frac{\mathbf{p}'_i}{m}, \qquad \frac{d\mathbf{p}'_i}{dt'} = \mathbf{F}_i - \xi\mathbf{p}'_i, \qquad \frac{ds}{dt'} = \xi s, \qquad \frac{d\xi}{dt'} = \frac{1}{\varpi}\left[\sum_i \frac{p'^2_i}{m} - (3N+1)k_BT\right],$$

with $\xi = p_s/\varpi$ by rescaling momentum and time: $\mathbf{p'}_i = \mathbf{p}_i/s$ and $t' = t/s$.

Solution:

Using $dt'/dt = 1/s$, we have

$$\frac{d\mathbf{r}_i}{dt} = \frac{d\mathbf{r}_i}{dt'}\frac{dt'}{dt} = \frac{d\mathbf{r}_i}{dt'}\frac{1}{s} \qquad \Longrightarrow \qquad \frac{d\mathbf{r}_i}{dt'} = s\frac{d\mathbf{r}_i}{dt} = s\frac{\mathbf{p}_i}{ms^2} = \frac{\mathbf{p}'_i}{m},$$

$$\frac{d\mathbf{p}_i}{dt} = \frac{d(\mathbf{p}'_i s)}{dt'}\frac{dt'}{dt} = \frac{d\mathbf{p}'_i}{dt'} + \frac{\mathbf{p}'_i}{s}\frac{ds}{dt'} \qquad \Longrightarrow \qquad \frac{d\mathbf{p}'_i}{dt'} = \mathbf{F}_i - \left(\frac{d\ln s}{dt'}\right)\mathbf{p}'_i = \mathbf{F}_i - \xi\mathbf{p}'_i,$$

$$\frac{ds}{dt} = \frac{ds}{dt'}\frac{dt'}{dt} = \frac{1}{s}\frac{ds}{dt'} \qquad \Longrightarrow \qquad \frac{ds}{dt'} = \frac{p_s}{\varpi}s = \xi s,$$

$$\frac{dp_s}{dt} = \frac{dp_s}{dt'}\frac{dt'}{dt} = \frac{1}{s}\frac{dp_s}{dt'} \qquad \Longrightarrow \qquad \frac{d\xi}{dt'} = \frac{1}{\varpi}\frac{dp_s}{dt'} = \frac{1}{\varpi}\left[\sum_i \frac{p'^2_i}{m} - (3N+1)k_BT\right].$$

Problem 2.9 A general form of the fluctuation-dissipation theorem may be derived by considering an arbitrary dynamic property of a thermodynamic system in response to the change in the total energy. Suppose that a system is prepared at a *nonequilibrium* condition by applying an external force f conjugated with dynamic quantity A such that the system energy is given by

$$\bar{E}_\nu = E_\nu + \Delta E_\nu,$$

where E_ν is the system energy without the perturbation, $\Delta E_\nu = -fA_\nu$, and ν denotes microstates. At time $t = 0$, the external force is turned off and the system is relaxing to equilibrium. Show that the dynamic quantity averaged over the initial nonequilibrium condition

$$\bar{A}(t) \equiv \frac{\sum_\nu A_\nu(t)e^{-\beta\bar{E}_\nu}}{\sum_\nu e^{-\beta\bar{E}_\nu}},$$

reaches to the equilibrium value linearly proportional to its fluctuation at the equilibrium condition, i.e.,

$$\Delta\bar{A}(t) = \beta f < \delta A(0)\delta A(t) >, \tag{J}$$

where $\Delta\bar{A}(t) \equiv \bar{A}(t) - < A >$, $\delta A(0) \equiv A_\nu(0) - < A >$, and $\delta A(t) \equiv A_\nu(t) - < A >$. Eq. (J) is a general form of the fluctuation-dissipation theorem.

Solution:
Under a small driving force f, the Boltzmann factor can be expanded in a Taylor series of $x \equiv \beta\Delta E_\nu$, i.e.,

$$e^{-\beta\bar{E}_\nu} = e^{-\beta E_\nu}e^{-x} = e^{-\beta E_\nu}(1 - x + x^2/2 + \cdots).$$

Accordingly, the dynamic quantity evolves corresponding to the nonequilibrium distribution at the initial condition:

$$\begin{aligned}
\bar{A}(t) &\equiv \frac{\sum_\nu A_\nu(t)e^{-\beta E_\nu}(1 - x + x^2/2 + \cdots)}{\sum_\nu e^{-\beta E_\nu}(1 - x + x^2/2 + \cdots)} \\
&= \frac{1}{\sum_\nu e^{-\beta E_\nu}} \frac{\sum_\nu A_\nu(t)e^{-\beta E_\nu}(1 - x + x^2/2 + \cdots)}{(1 - < x > + < x >^2/2 + \cdots)} \\
&= \frac{1}{\sum_\nu e^{-\beta E_\nu}} \left\{ \sum_\nu A_\nu(t)e^{-\beta E_\nu}(1 - x + \cdots) \right\} \{1 + < x > + \cdots\} \\
&= < A(t) > - < A(t)x > + < A(t) >< x > + \cdots \\
&= < A > - < A(t)x > + < A >< x > + \cdots,
\end{aligned}$$

where $< A(t) > = < A >$ because the average over the equilibrium distribution of microstates ($< \cdots >$) is time invariant. The desired equation is obtained by substituting with $x = -\beta f A_\nu(0)$ and taking the leading term in the above equation:

$$\Delta\bar{A}(t) \equiv \bar{A}(t) - < A > = \beta f(< A(t)A(0) > - < A >^2) = \beta f < \delta A(0)\delta A(t) > .$$

Problem 2.10 One of the simplest models to explain the fluctuation-dissipation theorem is provided by a one-dimensional harmonic system that contains only a single particle. Suppose that the potential energy is given by $\varphi(x) = \frac{1}{2}kx^2$ where k is a constant and x represents the particle coordinate. At temperature T, the probability density of the system taking the value of x is described by the Boltzmann equation

$$p(x) = \frac{1}{\mathcal{N}} e^{-\beta kx^2/2},$$

where $\beta = 1/(k_B T)$, and $\mathcal{N}$ is a normalization constant. Now consider how the system responds to a weak external field that is linearly dependent on the particle coordinate, $-fx$, where f is a constant.

(i) What is the probability density under the additional potential?
(ii) How does the additional potential change the mean position of the particle?
(iii) How does the additional potential change the fluctuation of the particle position?
(iv) Show that the deviation of the mean position is linearly proportional to the particle fluctuation in the original system.

Hint: Under the additional potential, the probability density is still described by the Boltzmann equation.

Solution:

(i) Under the additional potential, the probability density is still described by the Boltzmann equation but with a different energy:

$$p(x) = \frac{1}{N'} e^{-\beta(kx^2/2 - fx)}$$

where N' is a different normalization constant. This equation can be casted as a usual Gaussian form

$$p(x) = \frac{1}{\mathcal{N}''} \exp\left\{-\frac{\beta k}{2}(x - f/k)^2)\right\} = \frac{1}{\mathcal{N}''} \exp\left\{-\frac{(x - < x >_f)^2}{2 < x^2 >}\right\},$$

where $< x >_f = f/k$, $< x^2 > = 1/(\beta k)$, and $\mathcal{N}'' = \sqrt{2\pi < x^2 >}$.

(ii) The additional potential shifts the mean position of the particle by $< x >_f = f/k$.

(iii) The additional potential does not change the fluctuation of the particle position, $< x^2 > = 1/(\beta k)$.

(iv) The deviation of the mean position is linearly proportional to the particle fluctuation in the original system:

$$< x >_f = f/k = \beta f < x^2 > .$$

The above equation indicates that the particle position responds linearly to the driving force. Because βf is fixed, any linear equation describing the variation of the particle position is equally applicable to the fluctuation.

Problem 2.11 Lars Onsager won the 1968 Nobel Prize in Chemistry for his theoretical contributions to the thermodynamics of irreversible processes. A cornerstone of his pioneering work was the regression hypothesis, which states that the relaxation of a macroscopic quantity from non-equilibrium conditions is governed by the same equation describing microscopic fluctuations. Discuss Onsager's regression hypothesis in the context of the fluctuation-dissipation theorem and explain why it is applicable only to relaxation of dynamic quantities near equilibrium conditions.

Solution:

The fluctuation-dissipation theorem predicts that the relaxation of a macroscopic quantity is linearly proportional to equilibrium fluctuations

$$\Delta\bar{A}(t) = \beta f < \delta A(0)\delta A(t) > .$$

At given temperature and driving force f, this equation may be written as

$$\frac{\Delta\bar{A}(t)}{\Delta\bar{A}(0)} = \frac{< \delta A(0)\delta A(t) >}{< \delta A(0)^2 >},$$

which predicts that phenomenological equations describing the linear variation of dynamic quantities are equally applicable to the auto-correlation function $C(t) \equiv < \delta A(0)\delta A(t) >$. Near equilibrium assumption is used in the derivation of the fluctuation-dissipation theorem and in phenomenological linear equations.

Problem 2.12 The Green–Kubo equation provides a general formula for computing transport coefficients through MD simulation

$$\gamma = \int_0^{\infty} dt < \dot{\xi}(t)\dot{\xi}(0) >,$$

where γ represents a transport coefficient (within a multiplicative constant), ξ is the mechanical variable associated with the particular transport property under consideration, and $\dot{\xi}(t)$ signifies a time derivative. Show that the above equation can be equivalently written as

$$\gamma = \frac{1}{2}\lim_{t\to\infty}\frac{d}{dt} < (\xi(t) - \xi(0))^2 > .$$

Which equation is more convenient to use in MD simulation?

Solution:
First, change the integration boundary by taking the infinite time limit

$$\gamma = \int_0^{\infty} dt < \dot{\xi}(t)\dot{\xi}(0) > = \lim_{t\to\infty}\int_0^{t} dt' < \dot{\xi}(t')\dot{\xi}(0) > .$$

Let $\tau = t - t'$, so $t' = 0$ corresponds to $\tau = t$ and the above integration becomes

$$\gamma = -\lim_{t\to\infty}\int_t^{0} d\tau < \dot{\xi}(t-\tau)\dot{\xi}(t) > .$$

Now change τ back to t' but with t fixed, i.e., $t' = t - \tau$ and $dt' = -d\tau$ and the integration becomes

$$\gamma = \lim_{t\to\infty}\int_0^{t} dt' < \dot{\xi}(t')\dot{\xi}(t) > .$$

Integrating with respect to t' gives

$$\gamma = \lim_{t\to\infty} < (\xi(t) - \xi(0))\dot{\xi}(t) > = \frac{1}{2}\lim_{t\to\infty}\frac{d}{dt} < (\xi(t) - \xi(0))^2 > .$$

The derivative form is more suitable in MD simulation because numerical errors accumulate in the integration.

Problem 2.13 The Langevin equation is numerically more complicated to implement in MD simulation than Newton's equation because the change in velocity is related to a stochastic variable, which makes it not differentiable in conventional sense. Show that

(i) Integration of the Langevin equation over the time step from t_n to t_{n+1} leads to

$$\mathbf{v}_{n+1} - \mathbf{v}_n = \int_{t_n}^{t_{n+1}} dt'(\mathbf{f}/m) - \gamma_0(\mathbf{r}_{n+1} - \mathbf{r}_n) + \omega_{n+1}/m. \quad \text{(K)}$$

where $\mathbf{f} = -\nabla\varphi(\mathbf{r})$ is the conservative force, and

$$\omega_{n+1} \equiv \int_{t_n}^{t_{n+1}} dt'\xi(t'),$$

is a Gaussian random number with $< \omega_n > = 0$ and $< \omega_n \cdot \omega_l > = 6\gamma_0 m k_B T \delta t \delta_{nl}$, $\delta t = t_{n+1} - t_n$ is the step length, and δ_{nl} is the Kronecker-delta function.

(ii) Up to the second-order accuracy in the step length, the numerical integration of $d\mathbf{r}/dt = \mathbf{v}$ leads to

$$\mathbf{r}_{n+1} = \mathbf{r}_n + b\delta t\mathbf{v}_n + \frac{b\delta t^2}{2m}\mathbf{f}_n + \frac{b\delta t}{2m}\omega_{n+1}, \quad \text{(L)}$$

where

$$b \equiv \frac{1}{1 + \gamma_0\delta t/2},$$

(iii) Up to the second-order accuracy in the step length, Eq. (K) can be rewritten as

$$\mathbf{v}_{n+1} = \mathbf{v}_n + \frac{\delta t}{2m}(\mathbf{f}_n + \mathbf{f}_{n+1}) - \gamma_0(\mathbf{r}_{n+1} - \mathbf{r}_n) + \omega_{n+1}/m. \tag{M}$$

(iv) Eqs. (L) and (M) reduce to the velocity-Verlet algorithm when $\gamma_0 = 0$.

(v) How is ω_n, a contribution to the velocity change from the stochastic force, implemented in MD simulation?

Solution:

This problem is based on Grønbech-Jensen N. & Farago O. "A simple and effective Verlet-type algorithm for simulating Langevin dynamics", *Mol. Phys.* 111:8, 983–991 (2013).

(i) The main task is to show that ω_{n+1} satisfies the Gaussian distribution

$$< \omega_n > = \int_{t_{n-1}}^{t_n} dt' < \xi(t') > = 0,$$

and

$$\begin{aligned} < \omega_n \cdot \omega_l) > &= \int_{t_{n-1}}^{t_n} dt' \int_{t_{l-1}}^{t_l} dt'' < \xi(t') \cdot \xi(t'') > \\ &= \int_{t_{n-1}}^{t_n} dt' \int_{t_{l-1}}^{t_l} dt'' [6\gamma_0 m k_B T \delta t \delta(t' - t'')] \\ &= 6\gamma_0 m k_B T \delta t \delta_{nl}. \end{aligned}$$

(ii) Integration of $d\mathbf{r}/dt = \mathbf{v}$ leads to

$$\int_{t_n}^{t_{n+1}} dt'(d\mathbf{r}/dt') = \mathbf{r}_{n+1} - \mathbf{r}_n = \int_{t_n}^{t_{n+1}} dt'\mathbf{v}.$$

Up to the second-order accuracy in the step length δt, the integration over velocity is

$$\mathbf{r}_{n+1} - \mathbf{r}_n \approx \frac{\delta t}{2}(\mathbf{v}_{n+1} + \mathbf{v}_n).$$

Inserting $\mathbf{v}_{n+1}$ from Eq. (K) into the above equation yields

$$\mathbf{r}_{n+1} - \mathbf{r}_n = \delta t \mathbf{v}_n + \frac{\delta t}{2} \int_{t_n}^{t_{n+1}} dt'(\mathbf{f}/m) - \frac{\gamma_0 \delta t}{2}(\mathbf{r}_{n+1} - \mathbf{r}_n) + \frac{\delta t}{2}(\omega_{n+1}/m).$$

After the numerical integration for the force term and rearranging, we have

$$\mathbf{r}_{n+1} = \mathbf{r}_n + b\delta t \mathbf{v}_n + \frac{b\delta t^2}{2m}\mathbf{f}_n + \frac{b\delta t}{2m}\omega_{n+1}$$

where

$$b \equiv \frac{1}{1 + \gamma_0 \delta t/2}.$$

(iii) Up to the second-order accuracy in the step length dt, the integration of the force term in Eq. (K) gives

$$\int_{t_n}^{t_{n+1}} dt' \frac{\mathbf{f}}{m} \approx \frac{\delta t}{2m}(\mathbf{f}_n + \mathbf{f}_{n+1}).$$

Inserting the above equation into Eq. (K) leads to

$$\mathbf{v}_{n+1} = \mathbf{v}_n + \frac{\delta t}{2m}(\mathbf{f}_n + \mathbf{f}_{n+1}) - \gamma_0(\mathbf{r}_{n+1} - \mathbf{r}_n) + \omega_{n+1}/m.$$

(iv) As $\gamma_0 \to 0$, $b = 1$ so that Eqs. (L) and (M) reduce to the velocity-Verlet algorithm

$$\mathbf{r}_{n+1} = \mathbf{r}_n + \delta t \mathbf{v}_n + \frac{\delta t^2}{2m}\mathbf{f}_n,$$

$$\mathbf{v}_{n+1} = \mathbf{v}_n + \frac{\delta t}{2m}(\mathbf{f}_n + \mathbf{f}_{n+1}).$$

(v) The contribution of the stochastic force to velocity change, ω_n/m, is a vector with three components in x, y, z directions. In MD simulation, the random force in each direction is calculated from

$$\omega_n^{(x,y,z)} = \sqrt{2\gamma_0 m k_B T \delta t}\mathfrak{R},$$

where $\mathfrak{R}$ is a normally distributed random number with unit variance.

Problem 2.14 Prepare a computer program for the Brownian motion of a single particle in one dimension based on the results obtained from Problem 2.13. Validate your code by comparing the simulation results with the analytical solutions for

(i) the probability density for the particle velocity

$$p(v) = \sqrt{\frac{a}{\pi}}e^{-av^2},$$

where $a \equiv m/2k_BT(1 - e^{-2\gamma_0 t})$;

(ii) the mean-square displacement

$$R^2(t) = <|x(t) - x(0)|^2> = \frac{2k_BT}{m\gamma_0}\left(t - 1/\gamma_0 + e^{-\gamma_0 t}/\gamma_0\right);$$

(iii) the velocity autocorrelation function

$$c(t) = <v(t)v(0)> = \frac{k_BT}{m}e^{-\gamma_0 t}.$$

Hint: Identify characteristic time and length scales and use dimensionless units for particle position $x(t)$ and velocity $v(t)$.

Solution:

The Brownian motion has a characteristic time of $1/\gamma_0$. While there is no natural characteristic length scale, the velocity may be expressed in units of the mean velocity $v_0 = \sqrt{k_BT/m}$, which gives a unit length of $l_0 = v_0/\gamma_0$. Accordingly, we have

$$\frac{\omega}{m} = \frac{1}{m}\sqrt{2\gamma_0 m k_B T \delta t}\mathfrak{R} = \sqrt{\frac{2\gamma_0 k_B T \delta t}{m}}\mathfrak{R} = v_0\sqrt{2\gamma_0 \delta t}\mathfrak{R} = v_0\sqrt{2\delta\tau}\mathfrak{R},$$

where $\mathfrak{R}$ is a normally distributed random number with unit variance, $\tau = \gamma_0 \delta t$ is the dimensionless integration time. The equation of motion for the one-dimensional Brownian motion of a free particle ($\mathbf{f} = 0$)

$$\mathbf{r}_{n+1} = \mathbf{r}_n + b\delta t\mathbf{v}_n + \frac{b\delta t}{2m}\omega_{n+1},$$

can be written as:

$$x_{n+1} = x_n + b\delta\tau\frac{v_n}{\gamma_0 l_0} + \frac{b\delta\tau}{2}\frac{v_0}{\gamma_0 l_0}\sqrt{2\delta\tau}\mathfrak{R} = x_n + b\delta\tau u_n + \frac{b\delta\tau}{2}\sqrt{2\delta\tau}\mathfrak{R}, \quad \text{(N)}$$

where $x = r/l_0$ and $u = v/v_0$. Similarly, the velocity equation

$$\mathbf{v}_{n+1} = \mathbf{v}_n - \gamma_0(\mathbf{r}_{n+1} - \mathbf{r}_n) + \omega_{n+1}/m,$$

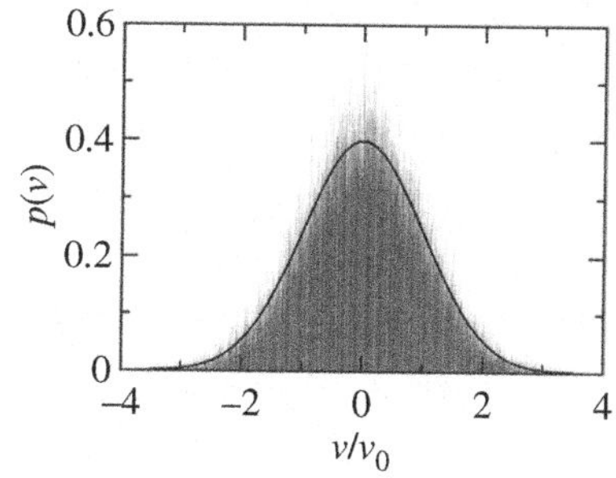

Figure 2.10 The probability distribution for the particle velocity calculated from MD simulation (bars) and from the exact equation (the solid line) (BrownianMD.m).

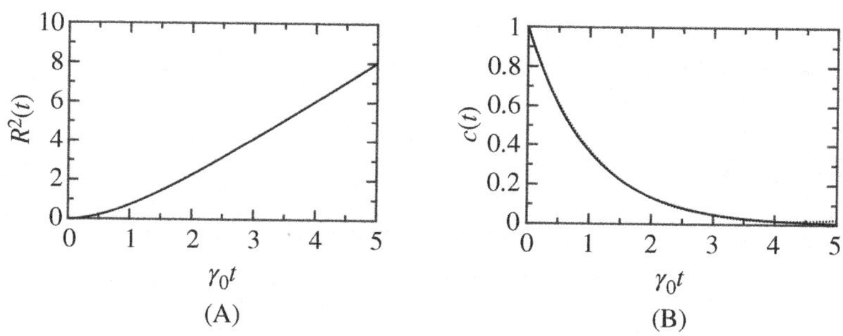

Figure 2.11 Comparison of the simulation (dotted lines) and analytical results (solid lines) for the mean-square displacement (A) and the velocity autocorrelation function (B) (BrownianMD.m).

becomes

$$
\begin{aligned}
u_{n+1} &= u_n - (x_{n+1} - x_n) + \sqrt{2\delta\tau}\Re = u_n - b\delta\tau u_n - \frac{b\delta\tau}{2}\sqrt{2\delta\tau}\Re + \sqrt{2\delta\tau}\Re \\
&= (1 - b\delta\tau)u_n + (1 - b\delta\tau/2)\sqrt{2\delta\tau}\Re = (1 - b\delta\tau)u_n + b\sqrt{2\delta\tau}\Re. \qquad \text{(O)}
\end{aligned}
$$

where the second equal sign follows Eq. (N), and as defined in Problem 2.13, $b = 1/(1 + \delta\tau/2)$. We can implement Eqs. (N) and (O) with a Matlab program.

Figures 2.10 and 2.11 compare the simulation and analytical results for the particle velocity, the mean-square displacement, and the velocity autocorrelation function. Here, the integration time step is $\delta\tau = 0.01$ in units of the characteristic time $1/\gamma_0$. The ensemble averages of dynamic variables are calculated from the averages over the simulation time. For example, we evaluate the velocity autocorrelation function from

$$
c(t) \equiv < v(t)v(0) > = \frac{1}{t_{MD}} \int_0^{t_{MD}} v(t' + t)v(t')dt',
$$

where t_{MD} stands for the duration of simulation.

Problem 2.15 In the canonical ensemble, the magnitude of energy fluctuations is related to the constant-volume heat capacity

$$
< \delta E^2 > = k_B T^2 C_V.
$$

Starting from this equation, verify that, for a classical system of N spherical particles at constant T and V, fluctuations in the total kinetic and potential energies are given by, respectively,

$$
< \delta K^2 > = 3N(k_B T)^2/2,
$$

$$
< \delta \Phi^2 > = k_B T^2 (C_V - 3Nk_B/2).
$$

Solution:
For a classical system, the kinetic and potential energies are uncorrelated. Thus, $< \delta E^2 > = < \delta K^2 > + < \delta\Phi^2 >$. Fluctuations in the kinetic energy can be calculated from the ideal-gas model

$$< \delta K^2 > = -\frac{\partial < K >}{\partial \beta} = -\frac{\partial 3Nk_BT/2}{\partial \beta} = 3N(k_BT)^2/2.$$

Accordingly, the ensemble average of fluctuations in the potential energy is given by

$$< \delta\Phi^2 > = k_BT^2C_V - 3N(k_BT)^2/2 = k_BT^2(C_V - 3Nk_B/2).$$

Problem 2.16 Cross fluctuations are important for describing the mutual influences of dynamic variables in thermodynamic systems.

(i) Show that, in the canonical ensemble, the cross fluctuations of any dynamic quantity A and the system energy E can be calculated from

$$< \delta A \delta E > = -\left(\frac{\partial < A >}{\partial \beta}\right)_V.$$

(ii) Considering a classical system of N spherical particles at constant T and V, show that the cross fluctuations of the total potential energy Φ and the system virial $\mathcal{V}$ are given by

$$< \delta\Phi\delta\mathcal{V} > = k_BT^2(V\gamma_V - Nk_B),$$

where $\gamma_V \equiv (\partial P/\partial T)_V$ is the thermal pressure coefficient.

Solution:

(i) In the canonical ensemble, the expectation of A is given by

$$< A > = \frac{\sum_\nu A_\nu \exp(-\beta E_\nu)}{\sum_\nu \exp(-\beta E_\nu)} = \frac{\sum_\nu A_\nu \exp(-\beta E_\nu)}{Q}.$$

Thus, the ensemble averages of A, $\delta A = A - < A >$, and $\delta E = E - < E >$ are related through

$$\left(\frac{\partial < A >}{\partial \beta}\right)_V = -\frac{\sum_\nu A_\nu E_\nu \exp(-\beta E_\nu)}{Q} + \frac{\sum_\nu A_\nu \exp(-\beta E_\nu) \sum_\nu' E_\nu' \exp(-\beta E_\nu')}{Q^2}$$
$$= - < AE > + < A >< E > = < \delta A \delta E >.$$

(ii) For a classical system of spherical particles, the virial equation predicts

$$PV = Nk_BT + < \mathcal{V} >.$$

Because the virial and kinetic energy are uncorrelated, we have

$$< \delta\Phi\delta\mathcal{V} > = < \delta E \delta\mathcal{V} > = -\left(\frac{\partial < \mathcal{V} >}{\partial \beta}\right)_V = -\left[\frac{\partial (PV - Nk_BT)}{\partial \beta}\right]_V$$
$$= -V\left(\frac{\partial P}{\partial \beta}\right)_V - N(k_BT)^2 = k_BT^2(V\gamma_V - Nk_B),$$

where $\gamma_V \equiv (\partial P/\partial T)_V$.

Problem 2.17 For a classical system of N spherical particles at constant volume V and temperature T, the instantaneous pressure may be defined as

$$\mathcal{P} = 2K/3V + \mathcal{V}/V$$

where K is the kinetic energy of translational motions, and $\mathcal{V}$ is the system virial given by

$$\mathcal{V} = -\frac{1}{3}\sum_{i}^{N} \mathbf{r}_i \cdot \nabla_{\mathbf{r}_i} \Phi.$$

Verify the following equations:

(i) The ensemble average of the instantaneous pressure can be written as

$$< \beta \mathcal{P} > = \frac{\partial \ln Q}{\partial V},$$

where $\beta = 1/(k_B T)$, and Q is the canonical partition function.

(ii) The fluctuations of instantaneous pressure is given by

$$< \delta \mathcal{P}^2 > = \frac{k_B T}{V}\left[\frac{2 < \mathcal{P} >}{3} - \frac{1}{\kappa_T} + \frac{N k_B T}{V} + \frac{1}{9V} < \mathbf{r}^N \cdot \nabla\nabla\Phi \cdot \mathbf{r}^N >\right],$$

where $\kappa_T \equiv -V^{-1}(\partial V/\partial P)_T$ is the isothermal compressibility, $\mathbf{r}^N \equiv \{\mathbf{r}_1, \mathbf{r}_2, \dots, \mathbf{r}_N\}$ and $\nabla\nabla\Phi$ denotes a matrix of second-order gradients.

(iii) The above equation for $< \delta \mathcal{P}^2 >$ allows for the computation of the isothermal compressibility through molecular simulation. Comment its computational efficiency in comparison to a "simpler" compressibility equation

$$\kappa_T = \frac{V}{N k_B T} + \frac{1}{k_B T}\int_0^\infty dr 4\pi r^2 [g(r) - 1],$$

where $g(r)$ stands for the radial distribution function.

(iv) The pressure-energy fluctuations satisfy

$$< \delta\Phi\delta\mathcal{P} > = k_B T^2(\gamma_V - \rho k_B)$$

where $\rho = N/V$ and $\gamma_V \equiv (\partial P/\partial T)_V$. In molecular simulation, this equation is often used to calculate thermal pressure coefficient γ_V.

Solution:

(i) For a system of N spherical particles, the canonical partition function may be written as

$$Q = Q^{IG} Q^{ex},$$

where Q^{IG} is the partition function of a monatomic ideal gas,

$$Q^{ex} \equiv \int d\mathbf{x}^N e^{-\beta\Phi(V^{N/3}\mathbf{x}^N)},$$

and $\mathbf{x} \equiv \mathbf{r}/V^{1/3}$ is a dimensionless vector. Thus, we have

$$\frac{\partial \ln Q}{\partial V} = \frac{\partial \ln Q^{IG}}{\partial V} + \frac{\partial \ln Q^{ex}}{\partial V} = \beta P^{IG} - \frac{\int d\mathbf{x}^N e^{-\beta\Phi(V^{N/3}\mathbf{x}^N)}(\partial\beta\Phi/\partial V)}{\int d\mathbf{x}^N e^{-\beta\Phi(V^{N/3}\mathbf{x}^N)}}$$
$$= \beta P^{IG} - < \partial\beta\Phi/\partial V > .$$

The partial derivative in the above equation can be evaluated as

$$\partial\Phi/\partial V = \sum_{i}^{N} \nabla_{\mathbf{r}_i}\Phi \cdot (\partial\mathbf{r}_i/\partial V) = \sum_{i}^{N} \nabla_{\mathbf{r}_i}\Phi \cdot [\partial(V^{1/3}\mathbf{x}_i)/\partial V] = \frac{1}{3V}\sum_{i}^{N} \nabla_{\mathbf{r}_i}\Phi \cdot \mathbf{r}_i = -\mathcal{V}/V.$$

Accordingly, we prove the identity

$$\frac{\partial \ln Q}{\partial V} = \beta P^{IG} + \beta < \mathcal{V} > /V =< \beta\mathcal{P} > .$$

(ii) From

$$\mathcal{P} = 2K/3V + \mathcal{V}/V \text{ and } <\mathcal{P}> = Nk_BT/V + <\mathcal{V}>/V,$$

we have

$$\begin{aligned}
<\delta\mathcal{P}^2> &= <\mathcal{P}^2> - <\mathcal{P}>^2 \\
&= \frac{4}{9V^2} <\delta K^2> + \frac{<\delta\mathcal{V}^2>}{V^2},
\end{aligned}$$

where

$$<\delta K^2> = -\frac{\partial <K>}{\partial\beta} = -\frac{\partial(3Nk_BT/2)}{\partial\beta} = 3N(k_BT)^2/2.$$

Let

$$Q^{ex} \equiv \int d\mathbf{x}^N e^{-\beta\Phi(V^{N/3}\mathbf{x}^N)} \text{ and } \beta F^{ex} \equiv \beta F - \beta F^{IG} = -\ln Q^{ex}.$$

From $\partial\Phi/\partial V = -\mathcal{V}/V$ or $\mathcal{V} = -k_BTV(\partial\beta\Phi/\partial V)$, we obtain

$$\frac{\partial \ln Q^{ex}}{\partial V} = -\frac{\int d\mathbf{x}^N e^{-\beta\Phi}(\partial\beta\Phi/\partial V)}{\int d\mathbf{x}^N e^{-\beta\Phi}} = -<\partial\beta\Phi/\partial V> = \beta <\mathcal{V}>/V,$$

and

$$\begin{aligned}
\frac{\partial^2 \ln Q^{ex}}{\partial V^2} &= \frac{\int d\mathbf{x}^N\left[-e^{-\beta\Phi}(\partial^2\beta\Phi/\partial V^2) + e^{-\beta\Phi}(\partial\beta\Phi/\partial V)^2\right]}{\int d\mathbf{x}^N e^{-\beta\Phi(V^{N/3}\mathbf{x}^N)}} - \left[\frac{\int d\mathbf{x}^N e^{-\beta\Phi}(\partial\beta\Phi/\partial V)}{\int d\mathbf{x}^N e^{-\beta\Phi(V^{N/3}\mathbf{x}^N)}}\right]^2 \\
&= -<\partial^2\beta\Phi/\partial V^2> + <(\partial\beta\Phi/\partial V)^2> - <\partial\beta\Phi/\partial V>^2 \\
&= -<\partial^2\beta\Phi/\partial V^2> + \beta^2(<\mathcal{V}^2> - <\mathcal{V}>^2)/V^2 \\
&= -<\partial^2\beta\Phi/\partial V^2> + \beta^2 <\delta\mathcal{V}^2>/V^2.
\end{aligned}$$

Thus

$$\begin{aligned}
<\delta\mathcal{V}^2> &= (k_BT)^2V^2\frac{\partial^2 \ln Q^{ex}}{\partial V^2} + k_BTV^2 <\partial^2\Phi/\partial V^2> \\
&= -(k_BT)^2V^2\frac{\partial^2 \beta F^{ex}}{\partial V^2} + k_BTV^2 <\partial^2\Phi/\partial V^2> \\
&= -(k_BT)^2V^2[\beta/(\kappa_TV) - N/V^2] + k_BTV^2 <\partial^2\Phi/\partial V^2> \\
&= -k_BTV/\kappa_T + N(k_BT)^2 + k_BTV^2 <\partial^2\Phi/\partial V^2>,
\end{aligned}$$

where $\kappa_T \equiv -V^{-1}(\partial V/\partial P)_T$ is the isothermal compressibility. Now the last term in the above equation can be evaluated from

$$\begin{aligned}
&\partial\Phi/\partial V = -\mathcal{V}/V, \\
&\partial^2\Phi/\partial V^2 = +\mathcal{V}/V^2 - (\partial\mathcal{V}/\partial V)/V.
\end{aligned}$$

Because

$$\begin{aligned}
\partial\mathcal{V}/\partial V &= -\frac{1}{3}\sum_i \frac{\partial\mathbf{r}_i}{\partial V}\cdot\nabla_{\mathbf{r}_i}\Phi - \frac{1}{9V}(\mathbf{r}^N\cdot\nabla\nabla\Phi\cdot\mathbf{r}^N) \\
&= -\frac{1}{9V}\sum_i \mathbf{r}_i\cdot\nabla_{\mathbf{r}_i}\Phi - \frac{1}{9V}(\mathbf{r}^N\cdot\nabla\nabla\Phi\cdot\mathbf{r}^N) \\
&= \frac{\mathcal{V}}{3V} - \frac{1}{9V}(\mathbf{r}^N\cdot\nabla\nabla\Phi\cdot\mathbf{r}^N),
\end{aligned}$$

where $\mathbf{r}^N \equiv \{\mathbf{r}_1, \mathbf{r}_2, \ldots, \mathbf{r}_N\}$ and $\nabla\nabla\Phi$ denotes a differential matrix, we have

$$\partial^2\Phi/\partial V^2 = \frac{2\mathcal{V}}{3V^2} + \frac{1}{9V^2}(\mathbf{r}^N \cdot \nabla\nabla\Phi \cdot \mathbf{r}^N).$$

Substituting the above equation into that for $< \delta\mathcal{V}^2 >$ gives

$$< \delta\mathcal{V}^2 > = -k_B TV/\kappa_T + N(k_B T)^2 + k_B TV^2\left(\frac{2 < \mathcal{V} >}{3V^2} + \frac{1}{9V^2} < \mathbf{r}^N \cdot \nabla\nabla\Phi \cdot \mathbf{r}^N >\right)$$

$$= k_B T\left[-V/\kappa_T + Nk_B T + \frac{2 < \mathcal{V} >}{3} + \frac{1}{9} < \mathbf{r}^N \cdot \nabla\nabla\Phi \cdot \mathbf{r}^N >\right].$$

Finally, the fluctuations of the instantaneous pressure is given by

$$< \delta\mathcal{P}^2 > = \frac{2N(k_B T)^2}{3V^2} + \frac{< \delta\mathcal{V}^2 >}{V^2}$$

$$= \frac{k_B T}{V}[-1/\kappa_T + \frac{5Nk_B T}{3V} + \frac{2 < \mathcal{V} >}{3V} + \frac{1}{9V} < \mathbf{r}^N \cdot \nabla\nabla\Phi \cdot \mathbf{r}^N >]$$

$$= \frac{k_B T}{V}\left[\frac{2}{3} < \mathcal{P} > -1/\kappa_T + \frac{Nk_B T}{V} + \frac{1}{9V} < \mathbf{r}^N \cdot \nabla\nabla\Phi \cdot \mathbf{r}^N >\right].$$

For an ideal gas, we have $1/\kappa_T = P$ and $< \mathcal{V} > = 0$ so the above equation becomes

$$< \delta\mathcal{P}^2 > = \frac{k_B T}{V}\left[\frac{2}{3} < \mathcal{P} > -P + \frac{Nk_B T}{V}\right] = \frac{2N(k_B T)^2}{3V^2}.$$

(iii) When applied in simulation, the "simpler" compressibility equation is less accurate because it is sensitive to the radial distribution function $g(r)$ at large values of r and to the system size. The fluctuation equation yields more reliable values of isothermal compressibility as the intermolecular potential is typically short ranged.

(iv) For a classical system, there is no correlation between the kinetic and potential energy. Following the results from Problem 2.16, we have

$$< \delta\Phi\delta\mathcal{P} > = \frac{2}{3} < \delta\Phi\delta K > /V + < \delta\Phi\delta\mathcal{V} > /V$$

$$= < \delta\Phi\delta\mathcal{V} > /V = k_B T^2(V\gamma_V - Nk_B)/V = k_B T^2(\gamma_V - \rho k_B),$$

where $\rho = N/V$.

Problem 2.18 The Jahn–Teller effect is related to the geometrical distortion of molecules or ions in a crystal that results from certain electron configurations. In a coarse-grained representation of such an effect, the distortion of each lattice site may be represented by a particle that may exist in one of three energy states

$$\epsilon_1 = \epsilon_2 = bx^2 - \frac{1}{2}cx,$$

$$\epsilon_3 = bx^2 + cx,$$

where $b > 0$ and $c > 0$ are constants, and $x \geq 0$ is a parameter that characterizes the uniform strain of the material.

(i) Assume that the lattice sites are independent to each other, find the canonical partition function for this system;

(ii) What is the reduced Helmholtz energy per lattice site, $a = \beta F/N$, as a function of temperature and x?

(iii) If parameter x is a small value but can change at constant T, what is its equilibrium value, $\overline{x}$? Assume that $a(x)$ can be expanded in a power series in x to the cubic oder, x^3.

(iv) Show that the system has a critical temperature T_c beyond which the three energy states become identical.

Solution:

(i) For a crystal with N independent lattice sites, the canonical partition function is

$$Q = q^N,$$

where q is the single-site partition function

$$q = \sum_{n=1}^{3} e^{-\beta\epsilon_n} = 2e^{-\beta(bx^2-cx/2)} + e^{-\beta(bx^2+cx)}.$$

(ii) The reduced Helmholtz energy per lattice site is $a = \beta F/N = -\ln q$.

(iii) At small x, the reduced Helmholtz energy per lattice site may be approximated by a power series in x up to the cubic order x^3:

$$a(x) = -\ln 3 + \left[\beta b - \frac{\beta^2 c^2}{4}\right] x^2 + \frac{\beta^3 c^3}{24} x^3 + \cdots.$$

The equilibrium value of x can be obtained by minimizing the free energy

$$\frac{\partial a(x)}{\partial x} \approx \left[2\beta b - \frac{\beta^2 c^2}{2}\right] x + \frac{\beta^3 c^3}{8} x^2 = 0,$$

which gives two roots, one is $\overline{x} = 0$, and the other is

$$\overline{x} = \frac{4}{\beta c}\left(1 - \frac{4b}{\beta c^2}\right) = \frac{4k_B T}{c}\left(1 - \frac{4k_B T b}{c^2}\right).$$

(iv) Because x is nonnegative, the nontrivial value of x exists only when $T < T_c = c^2/(4k_B b)$. Below T_c, the existence of two solutions signals a phase transition. When $T > T_c$, $\overline{x} = 0$ so that the three energy states are identical.

Problem 2.19 Show

$$\left(\frac{\partial \beta G}{\partial \beta}\right)_{N,\beta P} = \left(\frac{\partial \beta G}{\partial \beta}\right)_{N,P} - P\left(\frac{\partial G}{\partial P}\right)_{N,\beta}.$$

Solution:

For short notation, let $f \equiv \beta G$. At constant N, the differentiation of $f = f(\beta, \beta P)$ gives

$$df = \left(\frac{\partial f}{\partial \beta}\right)_{N,\beta P} d\beta + \left(\frac{\partial f}{\partial \beta P}\right)_{N,\beta} d(\beta P)$$

$$= \left[\left(\frac{\partial f}{\partial \beta}\right)_{N,\beta P} + P\left(\frac{\partial f}{\partial \beta P}\right)_{N,\beta}\right] d\beta + \beta\left(\frac{\partial f}{\partial \beta P}\right)_{N,\beta} dP.$$

Comparing the above equation with

$$df = \left(\frac{\partial f}{\partial \beta}\right)_{N,P} d\beta + \left(\frac{\partial f}{\partial P}\right)_{N,\beta} dP,$$

we obtain

$$\left(\frac{\partial f}{\partial \beta}\right)_{N,P} = \left(\frac{\partial f}{\partial \beta}\right)_{N,\beta P} + P\left(\frac{\partial f}{\partial \beta P}\right)_{N,\beta},$$

$$\left(\frac{\partial f}{\partial P}\right)_{N,\beta} = \beta\left(\frac{\partial f}{\partial \beta P}\right)_{N,\beta}.$$

Replacing f with βG leads to

$$\left(\frac{\partial \beta G}{\partial \beta}\right)_{N,\beta P} = \left(\frac{\partial \beta G}{\partial \beta}\right)_{N,P} - P\left(\frac{\partial G}{\partial P}\right)_{N,\beta}.$$

Problem 2.20 Suppose that a perfect crystal have N lattice sites and M interstitial locations. The Frenkel defects correspond to the migration of atoms from lattice sites to interstitial locations. Because a large positive energy ϵ is associated with each Frenkel defect, the number of displaced atoms n is typically much smaller than N or M.

(i) How many ways are there to remove n atoms from N sites and displace them at M interstitial locations?

(ii) Assume that the total volume is fixed and that the atoms have no degrees of freedom other than their locations, derive the canonical partition function for the crystal.

(iii) What is the average number of displaced atoms at temperature T?

(iv) Assume $N = M$ and $\epsilon = 1$ eV, what is the concentration of the displaced atoms at T=1000 K?

Solution:

(i) The degeneracy for generating n Frenkel defects is related to the number of ways to choose n atoms from N sites times the number of ways to place them at M interstitial locations:

$$W(n) = \frac{N!}{n!(N-n)!}\frac{M!}{n!(M-n)!}.$$

(ii) The canonical partition function is

$$Q = \sum_n W(n)e^{-n\beta\epsilon} = \sum_n e^{\ln W(n) - n\beta\epsilon}.$$

(iii) The average number of displaced atoms can be found from the maximum term method

$$\frac{\partial \ln W(n)}{\partial n} - \beta\epsilon = 0.$$

Using the Stirling approximation $\ln N! = N\ln N - N$ and the expression for $W(n)$, we obtain

$$\ln\frac{(N-n)(M-n)}{n^2} - \beta\epsilon = 0.$$

Because $n \ll M$ or N, the above equation predicts

$$n = \sqrt{MN}e^{-\beta\epsilon/2}.$$

(iv) When $N = M$, $\epsilon = 1$ eV, and $T = 1000$ K, $n/N = e^{-\beta\epsilon/2} \approx e^{-6} = 2.5 \times 10^{-3}$.

Problem 2.21 Consider the grand canonical ensemble for a two-component system, show that the numbers of molecules for different species are correlated through

$$\langle N_1 N_2\rangle - \langle N_1\rangle\langle N_2\rangle = k_B T\left(\frac{\partial\langle N_1\rangle}{\partial \mu_2}\right)_{V,T,\mu_1} = k_B T\left(\frac{\partial\langle N_2\rangle}{\partial \mu_1}\right)_{V,T,\mu_2}.$$

Solution:

From the grand partition function, we have

$$\langle N_1\rangle = \frac{1}{\Xi}\sum_\nu N_1 \exp\{-\beta E_\nu + \beta N_1\mu_1 + \beta N_2\mu_2\}$$

$$\langle N_2\rangle = \frac{1}{\Xi}\sum_\nu N_2 \exp\{-\beta E_\nu + \beta N_1\mu_1 + \beta N_2\mu_2\}.$$

Thus

$$k_B T \left(\frac{\partial \langle N_1 \rangle}{\partial \mu_2} \right)_{V,T,\mu_1}$$

$$= \left(\frac{\partial \langle N_1 \rangle}{\partial \beta \mu_2} \right)_{V,T,\mu_1} = \frac{1}{\Xi} \sum_\nu N_1 N_2 \exp\{-\beta E_\nu + \beta N_1 \mu_1 + \beta N_2 \mu_2\}$$

$$- \frac{1}{\Xi^2} \sum_\nu N_1 \exp\{-\beta E_\nu + \beta N_1 \mu_1 + \beta N_2 \mu_2\} \sum_\nu N_2 \exp\{-\beta E_\nu + \beta N_1 \mu_1 + \beta N_2 \mu_2\}$$

$$= \langle N_1 N_2 \rangle - \langle N_1 \rangle \langle N_2 \rangle .$$

Similarly, we have

$$k_B T \left(\frac{\partial \langle N_2 \rangle}{\partial \mu_1} \right)_{V,T,\mu_2}$$

$$= \left(\frac{\partial \langle N_2 \rangle}{\partial \beta \mu_1} \right)_{V,T,\mu_2} = \frac{1}{\Xi} \sum_\nu N_1 N_2 \exp\{-\beta E_\nu + \beta N_1 \mu_1 + \beta N_2 \mu_2\}$$

$$- \frac{1}{\Xi^2} \sum_\nu N_1 \exp\{-\beta E_\nu + \beta N_1 \mu_1 + \beta N_2 \mu_2\} \sum_\nu N_2 \exp\{-\beta E_\nu + \beta N_1 \mu_1 + \beta N_2 \mu_2\}$$

$$= \langle N_1 N_2 \rangle - \langle N_1 \rangle \langle N_2 \rangle .$$

Problem 2.22 Based on the fundamental equation for the grand potential, show the following useful relations:

(i)

$$\left(\frac{\partial P}{\partial \mu_i} \right)_{T,\mu_{j \neq i}} = \left(\frac{\partial \langle N_i \rangle}{\partial V} \right)_{T,\mu_i} .$$

(ii)

$$\rho_i = \left(\frac{\partial P}{\partial \mu_i} \right)_{T,\mu_{j \neq i}} ,$$

where $\rho_i = \langle N_i \rangle / V$.

(iii) In an ideal-gas mixture, the chemical potential for each species is given by

$$\mu_i = k_B T \ln(\rho_i \Lambda_i^3),$$

where Λ_i is a constant depending on temperature and the identify of chemical species i.

(iv) The fluctuation in the number of molecules in an open system is

$$\langle \delta N_i^2 \rangle = k_B T V \left(\frac{\partial^2 P}{\partial \mu_i^2} \right)_{V,T,\mu_{j \neq i}} .$$

(v) The cross fluctuation in the number of molecules is

$$\langle \delta N_i \delta N_j \rangle = k_B T \left(\frac{\partial \langle N_i \rangle}{\partial \mu_j} \right)_{V,T,\mu_{k \neq j}} = k_B T \left(\frac{\partial \langle N_j \rangle}{\partial \mu_i} \right)_{V,T,\mu_{k \neq i}} .$$

Solution:

(i) For a uniform multicomponent system, the fundamental equation for the grand potential is given by

$$d\Omega = -SdT - PdV - \sum_i \langle N_i \rangle d\mu_i,$$

which predicts Maxwell's relation

$$\left(\frac{\partial P}{\partial \mu_i}\right)_{T,\mu_{j\neq i}} = \left(\frac{\partial \langle N_i \rangle}{\partial V}\right)_{T,\mu_i}.$$

(ii) The fundamental equation also predicts

$$\langle N_i \rangle = -\left(\frac{\partial \Omega}{\partial \mu_i}\right)_{V,T,\mu_{j\neq i}}.$$

For a uniform system, $\Omega = -PV$, which can be substituted into the above equation to give

$$\langle N_i \rangle = V\left(\frac{\partial P}{\partial \mu_i}\right)_{V,T,\mu_{j\neq i}}.$$

For a uniform system, P is uniquely determined by T and $\{\mu_i\}$. Thus, we may drop subscript V in the above equation and find

$$\rho_i = N_i/V = \left(\frac{\partial P}{\partial \mu_i}\right)_{T,\mu_{j\neq i}}.$$

Alternatively, the above equation can be obtained from part (i). Because $\langle N_i \rangle = \rho_i V = \rho_i(T, \{\mu_i\})V$, we have

$$\left(\frac{\partial P}{\partial \mu_i}\right)_{T,\mu_{j\neq i}} = \left(\frac{\partial \langle N_i \rangle}{\partial V}\right)_{T,\mu_i} = \left(\frac{\partial \rho_i V}{\partial V}\right)_{T,\mu_i} = \rho_i.$$

(iii) Based on the ideal-gas law $P = \sum_i \rho_i k_B T$ and $\rho_i = \left(\frac{\partial P}{\partial \mu_i}\right)_{T,\mu_{j\neq i}}$ from part (ii), we have

$$\frac{1}{\rho_i} = \left(\frac{\partial \mu_i}{\partial P}\right)_{T,\mu_{j\neq i}} = \left(\frac{\partial \mu_i}{\partial \rho_i}\right)_{T,\mu_{j\neq i}} \left(\frac{\partial \rho_i}{\partial P}\right)_{T,\mu_{j\neq i}} = \frac{1}{k_B T}\left(\frac{\partial \mu_i}{\partial \rho_i}\right)_{T,\mu_{j\neq i}}.$$

Integrating the above equation with respect to ρ_i yields

$$\mu_i = \mu_i^0 + k_B T \ln \rho_i = k_B T \ln(\rho_i \Lambda_i^3).$$

Because T is held as a constant, the integration constant $\mu_i^0 \equiv k_B T \ln \Lambda_i^3$ depends on temperature and the identity of chemical species i.

(iv) The fluctuation in the number of molecules in an open system is

$$\langle \delta N_i^2 \rangle = \left(\frac{\partial < N_i >}{\partial \beta \mu_i}\right)_{V,\beta,\mu_{j\neq i}}.$$

Plugging

$$\langle N_i \rangle = V\left(\frac{\partial P}{\partial \mu_i}\right)_{V,T,\mu_{j\neq i}},$$

from part (ii) into the above equation gives

$$\langle \delta N_i^2 \rangle = k_B T V \left(\frac{\partial^2 P}{\partial \mu_i^2} \right)_{V,T,\mu_{j \neq i}} .$$

(v) The cross fluctuation in the number of molecules is

$$\langle \delta N_i \delta N_j \rangle = \langle N_i N_j \rangle - \langle N_i \rangle \langle N_j \rangle .$$

From the grand partition function, we have

$$\langle N_i \rangle = \frac{1}{\Xi} \sum_{\nu} N_i \exp\{-\beta E_\nu + \beta \sum_k N_k \mu_k\}.$$

Thus

$$k_B T \left(\frac{\partial \langle N_i \rangle}{\partial \mu_j} \right)_{V,T,\mu_{k \neq j}} = \left(\frac{\partial \langle N_i \rangle}{\partial \beta \mu_j} \right)_{V,T,\mu_{k \neq j}} = \frac{1}{\Xi} \sum_{\nu} N_i N_j \exp\{-\beta E_\nu + \beta \sum_k N_k \mu_k\}$$
$$- \frac{1}{\Xi^2} \sum_{\nu} N_i \exp\{-\beta E_\nu + \beta \sum_k N_k \mu_k\} \sum_{\nu} N_j \exp\{-\beta E_\nu + \beta \sum_k N_k \mu_k\}$$
$$= \langle N_i N_j \rangle - \langle N_i \rangle \langle N_j \rangle .$$

The final result is derived from the Maxwell relation (or by switching indices i and j)

$$\langle \delta N_i \delta N_j \rangle = k_B T \left(\frac{\partial \langle N_i \rangle}{\partial \mu_j} \right)_{V,T,\mu_{k \neq j}} = k_B T \left(\frac{\partial \langle N_j \rangle}{\partial \mu_i} \right)_{V,T,\mu_{k \neq i}} .$$

Problem 2.23 The following exercise is helpful to understand how different ensembles affect the averages and fluctuations of thermodynamic variables. Suppose that we know the expected value of a dynamic quantity A in an ensemble specified by extensive variable X as well as other parameters that will not be altered. The ensemble average of A is given by

$$< A >_X = \sum_{\nu} A_\nu W_\nu^X / W_X,$$

where W_ν^X is the appropriate statistical weight for microstate ν, and $W_X \equiv \sum_\nu W_\nu^X$ is the corresponding partition function. Let ξ be the conjugate intensive variable of X. From the viewpoint of classical thermodynamics, the variable change from X to ξ constitutes a Legendre transformation. In the new ensemble with X in the original ensemble replaced by ξ that is fixed as a constant, the partition function can be calculated from the Laplace transform

$$W_\xi = \int dX e^{-X\xi} W_X.$$

Show the following identifies in the ensemble with ξ being fixed:

(i) The expected value of A in the new ensemble

$$< A >_\xi = \int dX < A >_X W_X e^{-X\xi} / W_\xi.$$

(ii) Let $\Psi(\xi) \equiv -\ln W_\xi$, then we have

$$< X >_\xi = \partial \Psi(\xi) / \partial \xi,$$

and the mean-square fluctuation of X is

$$< \delta X^2 >_\xi = -\frac{\partial < X >_\xi}{\partial \xi} = -\partial^2 \Psi(\xi)/\partial \xi^2.$$

The relation between the fluctuations in X and the linear response of its mean $< X >$ to conjugate intensive variable ξ is known as the linear response theory.

(iii) To the leading order of the system size, the expected values of A from the two ensembles are related by

$$< A >_{\overline{X}} \approx < A >_\xi + \frac{1}{2} \frac{\partial^2 < A >_\xi}{\partial \xi \partial \overline{X}},$$

where $\overline{X} =< X >_\xi$.

(iv) In the thermodynamic limit, $N \to \infty$, the two ensembles predict the same expected values of A per particle, i.e.,

$$< a >_\xi =< a >_X,$$

where $a = A/N$.

(v) The mean-square fluctuations of A from the two ensembles are related by

$$< \delta A^2 >_X \approx < \delta A^2 >_\xi + \frac{\partial \xi}{\partial X} \left(\frac{\partial < A >_\xi}{\partial \xi} \right)^2.$$

(vi) For a thermodynamic system with N classical particles represented by a microcanonical ensemble, the mean-square fluctuation of the translational kinetic energy K is given by

$$\frac{< \delta K^2 >}{< K >} = k_B T \left(1 - \frac{3Nk_B}{2C_V} \right),$$

where T is the absolute temperature, and C_V is the constant-volume heat capacity.

Solution:

This problem and the solution are adapted from Lebowitz J. L. , Percus J. K. and Verlet L., *Phys. Rev.* 153, 250 (1967).

(i) With ξ being fixed as a constant, the expected value of A in the new ensemble is given by

$$< A >_\xi = \sum_\nu A_\nu W_\nu^\xi / W_\xi,$$

where W_ν^ξ / W_ξ represents the microstate probability. The partition function for this ensemble is

$$W_\xi = \sum_\nu W_\nu^\xi = \int dX W_X e^{-X\xi} = \int dX \sum_\nu W_\nu^X e^{-X\xi}$$
$$= \sum_\nu \int dX W_\nu^X e^{-X\xi}.$$

Therefore, the weight function for each microstate is

$$W_\nu^\xi = \int dX W_\nu^X e^{-X\xi}.$$

Substituting the above equation into the ensemble average leads to

$$< A >_\xi = \int dX \sum_\nu A_\nu (W_\nu^X e^{-X\xi} / W_X)(W_X / W_\xi)$$
$$= \int dX < A >_X W_X e^{-X\xi} / W_\xi.$$

(ii) From the relation between the two partition functions

$$W_\xi = \int dX e^{-X\xi} W_X,$$

we obtain

$$\partial W_\xi / \partial \xi = -\int dX X e^{-X\xi} W_X.$$

Let $\Psi(\xi) \equiv -\ln W_\xi$, then

$$\partial \Psi(\xi)/\partial \xi = -\frac{1}{W_\xi}(\partial W_\xi/\partial \xi) = +\frac{1}{W_\xi}\int dX X e^{-X\xi} W_X = < X >_\xi.$$

The mean-square fluctuation of X is

$$\begin{aligned} < \delta X^2 >_\xi &= < X^2 >_\xi - < X >_\xi^2 \\ &= \frac{1}{W_\xi}\int dX X^2 e^{-X\xi} W_X - \left[\frac{1}{W_\xi}\int dX X e^{-X\xi} W_X\right]^2 \\ &= -\frac{\partial < X >_\xi}{\partial \xi} = -\partial^2 \Psi(\xi)/\partial \xi^2. \end{aligned}$$

(iii) The expectation of A in the ensemble with X fixed can be written as a Taylor expansion with respect to X near its equilibrium value

$$< A >_X = < A >_{\overline{X}} + \delta X \frac{\partial < A >_{\overline{X}}}{\partial \overline{X}} + \frac{1}{2!}\delta X^2 \frac{\partial^2 < A >_{\overline{X}}}{\partial \overline{X}^2} + \frac{1}{3!}\delta X^3 \frac{\partial^3 < A >_{\overline{X}}}{\partial \overline{X}^3} + \cdots$$

where $\overline{X} = < X >_\xi$ is the ensemble average of X when ξ is fixed, and $\delta X \equiv X - \overline{X}$. Now apply the ensemble average to the above equation over all possible values of X with ξ being fixed:

$$\begin{aligned} < A >_\xi &= < A >_{\overline{X}} + < \delta X >_\xi \frac{\partial < A >_{\overline{X}}}{\partial \overline{X}} \\ &+ \frac{1}{2!} < \delta X^2 >_\xi \frac{\partial^2 < A >_{\overline{X}}}{\partial \overline{X}^2} + \frac{1}{3!} < \delta X^3 >_\xi \frac{\partial^3 < A >_{\overline{X}}}{\partial \overline{X}^3} + \cdots. \end{aligned}$$

Using $< \delta X >_\xi = 0$, $< \delta X^2 >_\xi = -\partial \overline{X}/\partial \xi$, and $< \delta X^3 >_\xi = \partial^2 \overline{X}^2/\partial \xi^2$, we obtain

$$< A >_\xi = < A >_{\overline{X}} - \frac{1}{2}\left(\frac{\partial \overline{X}}{\partial \xi}\right)\frac{\partial^2 < A >_{\overline{X}}}{\partial \overline{X}^2} + \frac{1}{3!}\left(\frac{\partial^2 \overline{X}}{\partial \xi^2}\right)\frac{\partial^3 < A >_{\overline{X}}}{\partial \overline{X}^3} + \cdots.$$

Note that, because X is an extensive variable, the derivatives on the right side of the above equation decay on the order of $1/N$. By keeping the leading order $(1/N)$, we have

$$< A >_\xi \approx < A >_{\overline{X}} - \frac{1}{2}\left(\frac{\partial \overline{X}}{\partial \xi}\right)\frac{\partial^2 < A >_{\overline{X}}}{\partial \overline{X}^2}.$$

Rearranging the above equation gives

$$\begin{aligned} < A >_{\overline{X}} &\approx < A >_\xi + \frac{1}{2}\frac{\partial \overline{X}}{\partial \xi}\frac{\partial^2 < A >_{\overline{X}}}{\partial \overline{X}^2} \\ &\approx < A >_\xi + \frac{1}{2}\frac{\partial^2}{\partial \xi \partial \overline{X}}\left[< A >_\xi + \frac{1}{2}\frac{\partial \overline{X}}{\partial \xi}\frac{\partial^2 < A >_{\overline{X}}}{\partial \overline{X}^2}\right] \\ &\approx < A >_\xi + \frac{1}{2}\frac{\partial^2 < A >_\xi}{\partial \xi \partial \overline{X}}, \end{aligned}$$

where the higher-order terms have been neglected, and $\overline{X} = < X >_\xi$ means that $\xi = \xi(\overline{X})$. Using this inverse function, we can rewrite the above equation as

$$< A >_{\overline{X}} \approx < A >_\xi + \frac{1}{2}\frac{\partial}{\partial \xi}\left[\frac{\partial \xi}{\partial X}\frac{\partial < A >_\xi}{\partial \xi}\right].$$

(iv) Following part (iii), we have the expected values of A per unit particle

$$< a >_{\overline{X}} \approx < a >_\xi + \frac{1}{2N}\frac{\partial^2 < A >_\xi}{\partial \xi \partial \overline{X}}.$$

In the thermodynamic limit, $N \to \infty$, while all other terms remain finite, the above equation becomes

$$< a >_X = < a >_\xi.$$

Here, we take an equal sign because all higher-oder terms vanish as $1/N \to 0$.

(v) Now consider the mean-square fluctuation of A following the same procedure above

$$< \delta A^2 >_{\overline{X}} \equiv < A^2 >_{\overline{X}} - < A >_{\overline{X}}^2 .$$

Following part (iii), we have

$$< A^2 >_{\overline{X}} \approx < A^2 >_\xi + \frac{1}{2}\frac{\partial}{\partial \xi}\left[\frac{\partial \xi}{\partial X}\frac{\partial < A^2 >_\xi}{\partial \xi}\right],$$

and

$$< A >_{\overline{X}}^2 \approx < A >_\xi^2 + < A >_\xi \frac{\partial}{\partial \xi}\left[\frac{\partial \xi}{\partial X}\frac{\partial < A >_\xi}{\partial \xi}\right] + \frac{1}{4}\left\{\frac{\partial}{\partial \xi}\left[\frac{\partial \xi}{\partial X}\frac{\partial < A >_\xi}{\partial \xi}\right]\right\}^2.$$

Note that the middle term in the above equation is

$$\begin{aligned} < A >_\xi \frac{\partial}{\partial \xi}\left[\frac{\partial \xi}{\partial X}\frac{\partial < A >_\xi}{\partial \xi}\right] &= \frac{\partial}{\partial \xi}\left\{< A >_\xi \left[\frac{\partial \xi}{\partial X}\frac{\partial < A >_\xi}{\partial \xi}\right]\right\} - \frac{\partial \xi}{\partial X}\left[\frac{\partial < A >_\xi}{\partial \xi}\right]^2 \\ &= \frac{1}{2}\frac{\partial}{\partial \xi}\left[\frac{\partial \xi}{\partial X}\frac{\partial < A >_\xi^2}{\partial \xi}\right] - \frac{\partial \xi}{\partial X}\left[\frac{\partial < A >_\xi}{\partial \xi}\right]^2. \end{aligned}$$

Therefore, the mean-square fluctuation of A becomes

$$\begin{aligned} < \delta A^2 >_{\overline{X}} \approx & < \delta A^2 >_\xi + \frac{\partial \xi}{\partial X}\left[\frac{\partial < A >_\xi}{\partial \xi}\right]^2 \\ & + \frac{1}{2}\frac{\partial}{\partial \xi}\left[\frac{\partial \xi}{\partial X}\frac{\partial < \delta A^2 >_\xi}{\partial \xi}\right] - \frac{1}{4}\left\{\frac{\partial}{\partial \xi}\left[\frac{\partial \xi}{\partial X}\frac{\partial < A >_\xi}{\partial \xi}\right]\right\}^2, \end{aligned}$$

where $< \delta A^2 >_\xi \equiv < A^2 >_\xi - < A >_\xi^2$. Because the three terms from the first line of the above equation scale with the system size ($\sim N$) while the two terms from the second line are size independent ($\sim N^0$), we can drop the bottom line and rewrite the above equation as

$$< \delta A^2 >_{\overline{X}} \approx < \delta A^2 >_\xi + \frac{\partial \xi}{\partial X}\left[\frac{\partial < A >_\xi}{\partial \xi}\right]^2.$$

(vi) For a classical system represented by the canonical ensemble, the translational kinetic energy K is given by

$$< K >_\beta = 3Nk_BT/2 = 3N/(2\beta),$$

where N is the number of particles in translational motions. The mean-square fluctuation of the translational kinetic energy is

$$< \delta K^2 >_\beta = -\partial < K >_\beta / \partial\beta = -\partial(3N/2\beta)/\partial\beta = 3N/(2\beta^2).$$

Following the identity from part (v), we have

$$\begin{aligned} < \delta K^2 >_{\bar{E}} &=< \delta K^2 >_\beta + \left(\frac{\partial\beta}{\partial\bar{E}}\right)\left(\frac{\partial < K >_\beta}{\partial\beta}\right)^2 \\ &= 3N/(2\beta^2) - \frac{1}{k_B T^2 C_V}[3N/(2\beta^2)]^2 \\ &= 3N/(2\beta^2)\left(1 - \frac{3Nk_B}{2C_V}\right) \end{aligned}$$

which leads to

$$\frac{< \delta K^2 >_{\bar{E}}}{< K >_\beta} = k_B T\left(1 - \frac{3Nk_B}{2C_V}\right),$$

where T is the absolute temperature, $\bar{E} =< E >_\beta$, and $C_V = (\partial\bar{E}/\partial T)$ is the constant-volume heat capacity. Note that $C_V \geq 3Nk_B/2$, the mean-square fluctuation of the translational kinetic energy is always nonnegative and disappears for an ideal gas of spherical particles.

Problem 2.24 The partition functions of conjugate ensembles are related by the Laplace transform

$$W_\xi = \int dX e^{-X\xi} W_X,$$

where X is an extensive variable, and ξ is the conjugate intensive variable of X. Accordingly, at a fixed value of ξ, the probability density of the system as a function of X is

$$p(X) = W_X e^{-X\xi}/W_\xi.$$

Both X and ξ are expressed in dimensionless forms. Show that, in the ensemble with ξ being fixed,

(i) The probability density $p(X)$ exhibits a maximum at $X =< X >_\xi$;

(ii) When the system is sufficiently large, $p(X)$ can be approximated by the Gaussian distribution

$$p(X) = p(\overline{X})\exp\left[-\frac{(X-\overline{X})^2}{2\sigma_X^2}\right],$$

where $\overline{X} =< X >_\xi$ and $\sigma_X^2 \equiv< (X-\overline{X})^2 >_\xi$.

(iii) $p(\overline{X}) = 1/\sqrt{2\pi\sigma_X^2}$;

(iv) In the thermodynamic limit, the probability density becomes a Dirac-delta function, i.e.,

$$\lim_{N\to\infty} p(X) = \delta(X-\overline{X}).$$

Solution:

(i) The probability density $p(X)$ exhibits a maximum when

$$\frac{dp(X)}{dX} = 0,$$

or equivalently,

$$\frac{d \ln p(X)}{dX} = 0.$$

Because

$$\ln p(X) = \ln W_X - X\xi - \ln W_\xi = -\Phi(X) - X\xi + \Psi(\xi),$$

where $\Phi(X) = -\ln W_X$ and $\Psi(\xi) = -\ln W_\xi$, we have

$$\frac{d \ln p(X)}{dX} = -\frac{d\Phi(X)}{dX} - \xi = 0,$$

which has a solution $X = < X >_\xi$ according to the Legendre transformation.

(ii) A Taylor expansion of $\ln p(X)$ with respect to X gives

$$\ln p(X) = \ln p(\overline{X}) + \left.\frac{d \ln p(X)}{dX}\right|_{X=\overline{X}} (X - \overline{X}) + \frac{1}{2!} \left.\frac{d^2 \ln p(X)}{dX^2}\right|_{X=\overline{X}} (X - \overline{X})^2 + \cdots .$$

At $X = \overline{X}$, the first-order derivative of $\ln p(X)$ with respect to X vanishes, and the second-order derivative of $\ln p(X)$ with respect to X is related to the mean-square fluctuation of X:

$$\frac{d^2 \ln p(X)}{dX^2} = -\frac{d\xi}{dX} = \frac{1}{\sigma_X^2},$$

where $\sigma_X^2 \equiv < (X - \overline{X})^2 >_\xi$. When the system is sufficiently large, higher-order terms become negligible so that $p(X)$ can be approximated by the Gaussian distribution

$$p(X) = p(\overline{X}) \exp\left[-\frac{(X - \overline{X})^2}{2\sigma_X^2}\right].$$

(iii) $p(\overline{X}) = 1/\sqrt{2\pi\sigma_X^2}$ can be derived from the normalization of the probability density

$$\int dXp(X) = 1.$$

(iv) In the thermodynamic limit, $\sqrt{\sigma_X^2}/N \to 0$. Therefore, the probability density becomes a Dirac-delta function, i.e.,

$$\lim_{N\to\infty} p(X) = \delta(X - \overline{X}).$$

3

Ideal Gases and Single-Molecule Thermodynamics

Problem 3.1 The "particle in a box" model is commonly used to represent the microstates affiliated with the translational degrees of freedom for particle motions. To identify these microstates, consider a particle moving freely in x direction over the range $0 \leq x \leq L$. Assume that the wave function for the single-particle motion is described by the Schrödinger equation

$$-\frac{\hbar^2}{2m}\frac{d^2\psi(x)}{dx^2} = \epsilon_T\psi(x), \tag{A}$$

where $\hbar = h/2\pi$ is the reduced Planck's constant, m is the particle mass, and ϵ_T stands for the kinetic energy due to the one-dimensional translational motion.

(i) Solve Eq. (A) with the boundary condition $\psi(0) = \psi(L) = 0$.

(ii) Show that the energy levels are given by

$$\epsilon_T = \frac{n^2h^2}{8mL^2},$$

with $n = 1, 2, 3, \ldots$.

(iii) Generalize the results for the translational motion of a single particle in three dimensions.

Solution:

(i) The second-order homogeneous differential equation has the general solution

$$\psi(x) = A\sin(ax) + B\cos(ax),$$

where $a = \sqrt{2m\epsilon_T}/\hbar$, A and B are constants. From $\psi(0) = \psi(L) = 0$, we find $B = 0$ and $\sin(aL) = 0$. The latter is satisfied only when $aL = n\pi$ with n being an integer. Further, constant A can be fixed from the normalization condition

$$\int_0^L \psi^2(x)dx = 1. \tag{B}$$

Substituting $\psi(x) = A\sin(ax)$ into Eq. (B) yields

$$1 = \int_0^L A^2\sin^2(ax)dx = \frac{A^2}{a}\int_0^{aL}\sin^2(\alpha)d\alpha = \frac{A^2}{a}\times\frac{aL}{2} = \frac{A^2L}{2}.$$

Therefore, the wave function for the translational motion is given by

$$\psi(x) = \sqrt{\frac{2}{L}}\sin\left(\frac{n\pi x}{L}\right).$$

Because both a and L are nonnegative and $\psi(x) = 0$ over the entire range of x does not satisfy Eq. (B), n takes only positive values, i.e., $n = 1, 2, 3, \ldots$.

Solutions Manual to Fundamentals and Practice in Statistical Thermodynamics,
First Edition. Jianzhong Wu and John M. Prausnitz.

(ii) The energy levels are obtained from $a = \sqrt{2m\epsilon_T}/\hbar$ and $aL = n\pi$:

$$\epsilon_T = \frac{(a\hbar)^2}{2m} = \frac{(nh/2L)^2}{2m} = \frac{n^2h^2}{8mL^2}.$$

(iii) For a particle in a three-dimension box, the translational motions in x, y, and z directions are independent. As a result, the wave function is given by the product of those corresponding to each direction

$$\psi(x, y, z) = \psi(x) \times \psi(y) \times \psi(z),$$

and the kinetic energy includes three components from x, y, and z directions

$$\epsilon_T = \frac{n_x^2 h^2}{8mL_x^2} + \frac{n_y^2 h^2}{8mL_y^2} + \frac{n_x^2 h^2}{8mL_y^2},$$

where $n_{x,y,z} = 1, 2, 3, \ldots$ are quantum numbers for the translational motion in (x, y, z) directions, respectively, and $L_{x,y,z}$ are the corresponding ranges for the particle displacement.

Problem 3.2 Predict the entropy, Helmholtz energy, chemical potential, and heat capacity of argon gas at 25 °C and 1 atm using the statistical-thermodynamic equations for a monatomic ideal gas. Compare your result with the entropy of argon obtained from calorimetric experiment, 155 J/(mol K).

Solution:

- The entropy of a monatomic ideal gas is described by the Sackur–Tetrode equation, which agrees well with the experimental result:

$$S/Nk_B = \ln(T^{5/2}W^{3/2}/P) - 1.16486 = \ln(298.15^{5/2}\, 39.948^{3/2}/1) - 1.16486 = 18.61.$$

$$\Rightarrow S = 154.74 \text{ J/(mol} \cdot \text{K)}.$$

- Helmholtz energy:

$$F/Nk_BT = \ln(\rho\Lambda^3) - 1$$

$$= \ln\left[\frac{P}{k_BT} \times \left(\frac{h}{2\pi mk_BT}\right)^{3/2}\right] - 1 \approx -17.11.$$

$\Rightarrow F \approx -42.41$ kJ/mol.

- Internal energy:

$$U/Nk_BT = 3/2. \Rightarrow U \approx 3.71 \text{ kJ/mol}.$$

- Heat capacity:

$$C_V/Nk_B = 3/2. \Rightarrow C_V \approx 12.47 \text{ J/(mol} \cdot \text{K)}.$$

Problem 3.3 In evaluating the translational partition function, the continuous approximation becomes problematic when $\rho\Lambda^3 \sim 1$.

(i) Estimate the temperature below which the Sackur–Tetrode equation breaks down for predicting the entropy of argon as an ideal gas at 1 atm.

(ii) What are the number density and the thermal wavelength of the argon gas at this temperature? and how are these numbers compared with those at 300 K?

Solution:

(i) At given pressure P, the number density can be predicted from the ideal-gas law $\rho = P/k_BT$. Accordingly, the quantum effects become significant when

$$\rho\Lambda^3 = \frac{P}{k_BT}\Lambda^3 = \frac{P}{k_BT}\left(\frac{h^2}{2\pi k_BTm}\right)^{3/2} \sim 1 \Rightarrow T \sim \frac{1}{k_B}\left(\frac{Ph^3}{(2\pi m)^{3/2}}\right)^{2/5}.$$

Plugging into the above equation with $P = 1$ atm=101325 Pa and $m = 39.948 \times 1.6735575 \times 10^{-27}$ kg yields $T \sim 0.47$ K. The quantum effect for the translational motion of argon molecules becomes relevant at extremely low temperature!

(ii) At $T \sim 0.47$ K, the number density of the argon gas is

$$\rho = \frac{P}{k_BT} = 1.5557 \times 10^{28}\ \text{m}^{-3},$$

and the thermal wavelength is

$$\Lambda = \left(\frac{h^2}{2\pi k_BTm}\right)^{1/2} = 0.4\ \text{nm}.$$

For comparison, at $T = 300$ K, the number density is $\rho = 2.4463 \times 10^{25}\ \text{m}^{-3}$, the thermal wavelength is $\Lambda = 0.015885$ nm, and $\rho\Lambda^3 = 9.8062 \times 10^{-8}$. We see that the classical (continuous) approximation is well justified for argon gas near room temperature.

Problem 3.4 The electronic energy of atoms and ions can be determined from atomic spectroscopy and the data have been well tabulated (e.g., the National Institute of Standards and Technology (NIST) Atomic Spectra Database). The table below shows the first few electronic states of Li in terms of electron configuration, term symbol for the electronic state, degeneracy and energy level.

(i) Based on the first few electronic energy levels, calculate the electronic partition function q_e for a Li atom at 300 and 10^6 K. How are the numbers compared with the single-particle translational partition function q_T with volume $V = 1\ \text{m}^3$ at these temperatures?

(ii) How does the electronic partition function contribute to the entropy for Li as an ideal gas at 300 K and 1 atm and at 10^6 K and 1 atm?

Configuration	Term	Degeneracy (g_i)	Energy level (ϵ_i, cm^{-1})
$1s^22s$	$^2S_{1/2}$	2	0.00
$1s^22p$	$^2P_{1/2}$	2	14903.66
	$^2P_{3/2}$	4	14904.00
$1s^23s$	$^2S_{1/2}$	2	27206.12
$1s^23p$	$^2P_{3/2}$	4	30925.38
	$^2P_{1/2}$	2	30925.38
$1s^23d$	$^2D_{3/2}$	4	31283.08
	$^2D_{3/2}$	6	31283.12

Solution:

(i) The electronic partition function can be calculated from the degeneracy (g_i) and energy of electronic state (ϵ_i):

$$q_e = \sum_i g_i e^{-\beta\epsilon_i} = g_0 + g_1 e^{-\beta\epsilon_1} + \cdots,$$

where g_0 stands for the degeneracy at the ground state. At $T = 300$ K, $q_e \approx 2.0$, which is essentially identical to g_0. At $T = 10^6$ K, $q_e \approx 25.09$; it is increased due to significant contributions from the higher-energy levels.

At $T = 300$ K and $V = 1$ m^3, the single–particle translational partition function is

$$q_T = \frac{V}{\Lambda^3} = V \times \left(\frac{2\pi m k_B T}{h^2}\right)^{3/2} = 2.495 \times 10^{32}.$$

And at $T = 10^6$ K, $q_T = 4.802 \times 10^{37}$. Clearly, $q_T \gg q_e$ at both conditions.

(ii) The electronic partition function contributes to the entropy through

$$S_e = -\frac{\partial}{\partial T}(k_B T \ln q_e).$$

Evaluating the above equation numerically leads to $S_e = 5.763$ J/(mol K) at 300 K and 27.09 J/(mol K) at 10^6 K. For comparison, the translational entropy is $S_T = 154.9$ and 323.5 J/(mol K), respectively. At 300 K, the electronic entropy is related to the ground-state degeneracy, i.e., $S_e \approx k_B T \ln g_0$.

Problem 3.5 Consider gas molecules passing through a small hole as shown schematically in Figure 3.1. Assume that the gas molecules do not interact with each other and that the planar wall has negligible thickness. Based on the Maxwell–Boltzmann distribution for the molecular velocity, show that the flux of gas molecules, i.e., the number of gas molecules passing through the hole per unit area and unit time, is given by the Knudsen equation

$$J = \rho\sqrt{\frac{k_B T}{2\pi m}}$$

where ρ is the number density, k_B is the Boltzmann constant, T is the absolute temperature, and m is the mass of a gas molecule.

Solution:

Only molecules moving toward the wall can pass through the hole. Let the velocity in that direction be positive, i.e., $v_z^+ > 0$. For a molecule with velocity v_z^+, it reaches the wall within a differential time dt only if it is located at a distance smaller than $v_z^+ dt$. Therefore, the number of gas molecules passing through the hole is $\rho A v_z^+ dt$ where A is cross-section area. As v_z follows the Maxwell–Boltzmann distribution, the number of molecules escaping from the hole is obtained by the average over v_z

$$dN = \int_0^\infty dv_z p(v_z)\rho A v_z dt = \rho A dt \int_0^\infty dv_z \sqrt{\frac{\beta m}{2\pi}} e^{-\beta m v_z^2/2} v_z$$

$$= \rho A dt \sqrt{\frac{\beta m}{2\pi}}\left(\frac{1}{\beta m}\right)\int_0^\infty dx e^{-x} = \rho A dt \sqrt{\frac{k_B T}{2\pi m}},$$

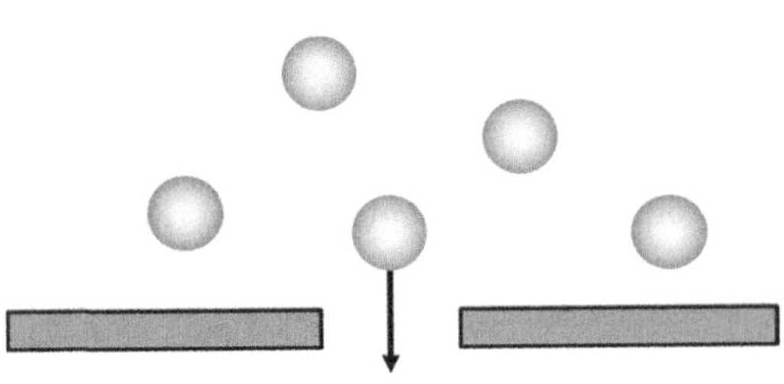

Figure 3.1 Effusion of gas molecules through a hole of negligible thickness.

where $x = \beta m v_z^2/2$. The flux of gas effusion is thus given by

$$J = \frac{dN}{Adt} = \rho\sqrt{\frac{k_B T}{2\pi m}}.$$

Problem 3.6 Liquid evaporation is a nonequilibrium process in which there is a net transport of liquid molecules across a liquid–vapor interface into the vapor phase. Approximately, the rate of condensation (number of molecules per unit area per unit time) can be estimated from the Hertz–Knudsen (HK) equation, which is related to the number of the gas molecules impinging on the liquid surface multiplied by an efficiency factor α

$$J = \alpha\rho\sqrt{\frac{k_B T}{2\pi m}},$$

where ρ is the number density of the gas molecules, and m is the molecular mass. At equilibrium, the rate of gas molecule condensation is the same as the rate of evaporation. Assume $\alpha = 0.01$, estimate how long it takes to dry up a pond of $A = 100$ m^2 in surface area and $h = 10$ m in depth open to 50 °C dry air. At 50 °C, the saturation pressure for liquid water is $P^{sat} = 1.227 \times 10^4$ Pa. The water density is about $d \approx 1000$ kg/m^3, and molecular weight is $W = 18$ g/mol.

Solution:

When liquid water is at equilibrium with its own vapor, the rate of evaporation is the same as the rate of condensation

$$J = \alpha\rho\sqrt{\frac{k_B T}{2\pi m}}.$$

The number density of water molecules in the vapor phase can be estimated from the ideal-gas law

$$\rho = \frac{P^{sat}}{k_B T} = \frac{1.227 \times 10^4}{1.3806 \times 10^{-23}(2731.15 + 50)} = 2.7502 \times 10^{24}\ \text{m}^{-3}.$$

Substituting ρ into the HK equation yields

$$J = \alpha\rho\sqrt{\frac{k_B T}{2\pi m}} = 0.01 \times 2.7502 \times 10^{24}\sqrt{\frac{1.3806 \times 10^{-23} \times 3231.15}{2\pi \times 18 \times 1.6736 \times 10^{-27}}} = 4.2224 \times 10^{24}\ \text{m}^2/\text{s}.$$

Therefore, the total time of evaporation to dry up the pond is given by

$$t = \frac{A \times h \times d}{J \times A \times W/N_A} = \frac{100 \times 10 \times 1000}{4.2224 \times 10^{24} \times 100} \times \frac{6.02214 \times 10^{23}}{18 \times 10^{-3}} \times \frac{1}{3600} = 22.01\ \text{hr}.$$

Comments: The HK equation ignores coupling between the vapor and liquid phases during evaporation (i.e., the evaporation and condensation fluxes depend on both the conditions in the liquid and in the vapor phases). More accurate expressions have been developed to predict the rate of evaporation.[1]

Problem 3.7 Historically, effusion was considered for the separation of 235U and 238U isotopes in the early development of atom bombs. To use the effusion method, the uranium isotopes are first converted to uranium hexafluoride (UF_6) gas, which is forced to diffuse through porous barriers repeatedly, each time becoming a little more enriched in slightly lighter 235U isotope. The 235U to 238U ratio in a typical uranium ore is 1:140. Based on the Knudsen equation, estimate how many separation stages are required to obtain 90% 235U.

1 Persad A. H. and Ward C. A.,"Expressions for the Evaporation and Condensation Coefficients in the Hertz-Knudsen Relation", *Chem. Rev.* 116 (14), 7727–7767 (2016).

Solution:
According to the Knudsen equation, the effusion flux is proportional to $\rho/\sqrt{m}$. After i stages of effusion, the isotope composition is thus given by

$$\frac{\rho_L^{(i+1)}}{\rho_H^{(i+1)}} = \frac{\rho_L^{(i)}}{\rho_H^{(i)}} \times \sqrt{\frac{m_H}{m_L}} = \frac{\rho_L^{(0)}}{\rho_H^{(0)}} \times \left(\frac{m_H}{m_L}\right)^{i/2},$$

where $m_H = 238 + 6 \times 19 = 352$, and $m_L = 235 + 6 \times 19 = 349$. For $\rho_L^{(i+1)}/\rho_H^{(i+1)} = 0.9/0.1$ and $\rho_L^{(0)}/\rho_H^{(0)} = 1/140$, the above equation predicts

$$i = 2 \times \frac{\ln(9 \times 140)}{\ln(352/349)} = 1668.$$

Problem 3.8 Consider two particles with mass m_1 and m_2 linked together by a bond potential $v_B(r)$, where r is the center-to-center distance.

(i) Prove that the relative motions of the two particles can be represented by that of one particle with an effective mass $\mu = m_1 m_2/(m_1 + m_2)$.

(ii) Assume that the inter-particle potential is described by

$$v_B(r) = -D_e + \frac{1}{2}k(r - r_e)^2.$$

where $-D_e$ stands for the bond formation energy. Show that the relative motion of the two particles can be represented by a harmonic oscillator with frequency

$$\nu = \frac{1}{2\pi}\sqrt{\frac{k}{\mu}}.$$

(iii) Show that, at fixed separation $r = r_e$, the rotation of the two particles is equivalent to that of one particle with effective mass μ.

Solution:

(i) The acceleration for the relative velocity between the two particles, $\mathbf{v} = \mathbf{v}_2 - \mathbf{v}_1$, is given by

$$\mathbf{a} = \frac{d\mathbf{v}}{dt} = \mathbf{a}_2 - \mathbf{a}_1 = \frac{\mathbf{F}_2}{m_2} - \frac{\mathbf{F}_1}{m_1},$$

where $\mathbf{F}_1$ and $\mathbf{F}_2$ are the forces on particle 1 and 2, respectively. Because $\mathbf{F}_2 = -\mathbf{F}_1 = -\nabla v_B(r)$, the above equation becomes

$$\mathbf{a} = \mathbf{F}_2\left(\frac{1}{m_2} + \frac{1}{m_1}\right) = \frac{\mathbf{F}_2}{\mu},$$

which is the same as Newton's equation for the motion of a single particle with an effective mass $\mu = m_1 m_2/(m_1 + m_2)$.

(ii) Relative to particle 1, the motion of particle 2 has two components, one corresponds to the vibration in the radial direction under bond potential $v_B(r)$, and the other is the free rotation at a fixed separation from particle 1. As discussed above, Newton's equation for the motion in the radial direction is

$$\mu\frac{d^2r}{dt^2} = \mathbf{F}_2 = -k(r - r_e),$$

which corresponds to that for a harmonic oscillator of effective mass μ. The solution to the above differential equation is

$$r(t) - r_e = A\cos(\omega t + \phi),$$

where $\omega = \sqrt{k/\mu}$ is the angular frequency, and ϕ is the phase angle. Hence, the frequency of the harmonic oscillator is

$$\nu = \frac{\omega}{2\pi} = \frac{1}{2\pi}\sqrt{\frac{k}{\mu}}.$$

(iii) When $r = r_e$, there is no net force on particle 2. Therefore, its rotation relative to particle 1 is equivalent to that of a rigid rotor. The moment of inertia, $I = \mu r_e^2$, is the same as that defined by two particles with a fixed separation r_e rotating around the center of mass position $\mathbf{r}_c = (m_1\mathbf{r}_1 + m_2\mathbf{r}_2)/(m_1 + m_2)$

$$\begin{aligned} I &= m_1 d_1^2 + m_2 d_2^2 = m_1|\mathbf{r}_1 - \mathbf{r}_c|^2 + m_2|\mathbf{r}_2 - \mathbf{r}_c|^2 \\ &= m_1\left|\mathbf{r}_1 - \frac{m_1\mathbf{r}_1 + m_2\mathbf{r}_2}{m_1 + m_2}\right|^2 + m_2\left|\mathbf{r}_2 - \frac{m_1\mathbf{r}_1 + m_2\mathbf{r}_2}{m_1 + m_2} - \mathbf{r}_c\right|^2 \\ &= \frac{m_1 m_2}{m_1 + m_2}|\mathbf{r}_1 - \mathbf{r}_2|^2 = \mu r_e^2, \end{aligned}$$

where d_1 and d_2 are the distances of particles 1 and 2 from the center of mass.

Problem 3.9 Consider an ideal gas mixture with two types of molecules 1 and 2. Show that the relative velocity between a type 1 molecule and a type 2 molecule satisfies the Maxwell–Boltzmann distribution

$$p(\mathbf{v}) = \left(\frac{\mu}{2\pi k_B T}\right)^{3/2} \exp\left(-\frac{\mu v^2}{2k_B T}\right),$$

where $\mu = m_1 m_2/(m_1 + m_2)$. What is the average speed of the relative velocity? Hint: Express the velocities of the two molecules in terms of the relative velocity $\mathbf{v_r} = \mathbf{v}_2 - \mathbf{v}_1$ and the center of mass velocity $\mathbf{v}_c = (m_1\mathbf{v}_1 + m_2\mathbf{v}_2)/(m_1 + m_2)$.

Solution:

In terms of the relative velocity $\mathbf{v_r} = \mathbf{v}_2 - \mathbf{v}_1$ and the center of mass velocity $\mathbf{v}_c = (m_1\mathbf{v}_1 + m_2\mathbf{v}_2)/(m_1 + m_2)$, $\mathbf{v}_1$ and $\mathbf{v}_2$ are given by

$$\mathbf{v}_1 = \mathbf{v}_c - \frac{m_2}{m_1 + m_2}\mathbf{v}_r; \qquad \mathbf{v}_2 = \mathbf{v}_c + \frac{m_1}{m_1 + m_2}\mathbf{v}_r.$$

Accordingly, the total kinetic energy of the two molecules can be written as

$$K = \frac{1}{2}m_1\mathbf{v}_1^2 + \frac{1}{2}m_2\mathbf{v}_2^2 = \frac{1}{2}(m_1 + m_2)\mathbf{v}_c^2 + \frac{1}{2}\mu\mathbf{v}_r^2.$$

The probability for the relative velocity is

$$p(\mathbf{v}) = \int d\mathbf{v}_1 \int d\mathbf{v}_2 \delta(\mathbf{v} - \mathbf{v}_2 + \mathbf{v}_1) p(\mathbf{v}_1) p(\mathbf{v}_2), \tag{C}$$

where $\delta(\mathbf{v})$ is the Dirac delta function, $p(\mathbf{v}_1)$ and $p(\mathbf{v}_2)$ are the Maxwell–Boltzmann probabilities for molecules 1 and 2, respectively. To evaluate the integration in Eq. (C), we change the variables from $\mathbf{v}_1$ and $\mathbf{v}_2$ to $\mathbf{v}_r$ and $\mathbf{v}_c$. The determinant of the Jacobian matrix for the coordinate change is

$$\begin{vmatrix} 1 & -\frac{m_2}{m_1+m_2} \\ 1 & \frac{m_1}{m_1+m_2} \end{vmatrix} = \frac{m_1}{m_1 + m_2} + \frac{m_2}{m_1 + m_2} = 1,$$

Accordingly, Eq. (C) becomes

$$p(\mathbf{v}) = \int d\mathbf{v}_r \int d\mathbf{v}_c \delta(\mathbf{v} - \mathbf{v}_r)\left(\frac{m_1}{2\pi k_B T}\right)^{3/2}\left(\frac{m_2}{2\pi k_B T}\right)^{3/2} \exp\left(-\frac{(m_1 + m_2)\mathbf{v}_c^2 - \mu\mathbf{v}_r^2}{2k_B T}\right)$$

$$= \left(\frac{m_1}{2\pi k_B T}\right)^{3/2} \left(\frac{m_2}{2\pi k_B T}\right)^{3/2} \left(\frac{2\pi k_B T}{m_1 + m_2}\right)^{3/2} \exp\left(-\frac{\mu v^2}{2k_B T}\right)$$
$$= \left(\frac{\mu}{2\pi k_B T}\right)^{3/2} \exp\left(-\frac{\mu v^2}{2k_B T}\right).$$

Alternatively, we can obtain the Maxwell–Boltzmann distribution simply considering the motion of an effective particle as discussed in Problem 3.8.

The average relative speed is given by

$$\langle|\mathbf{v}|\rangle = \int d\mathbf{v}p(\mathbf{v})|\mathbf{v}| = \left(\frac{\mu}{2\pi k_B T}\right)^{3/2} \int d\mathbf{v}|\mathbf{v}| \exp\left(-\frac{\mu v^2}{2k_B T}\right)$$
$$= \left(\frac{\mu}{2\pi k_B T}\right)^{3/2} \int_0^\infty dv 4\pi v^3 \exp\left(-\frac{\mu v^2}{2k_B T}\right)$$
$$= 2\pi \left(\frac{\mu}{2\pi k_B T}\right)^{3/2} \left(\frac{2k_B T}{\mu}\right)^2 \int_0^\infty dx x e^{-x}$$
$$= \frac{2}{\sqrt{\pi}} \left(\frac{2k_B T}{\mu}\right)^{1/2} = \sqrt{\frac{8k_B T}{\pi\mu}},$$

where $x = \mu v^2/2k_B T$ and $\int_0^\infty dx x e^{-x} = 1$.

Problem 3.10 The rotational partition function for diatomic and linear polyatomic molecules can be derived by applying the Schrödinger equation to a rigid rotor, i.e., a particle of mass μ freely rotating around the origin with fixed distance r.

(i) Starting from the Schrödinger equation

$$-\frac{\hbar^2}{2\mu}\nabla^2\psi = \epsilon_R \psi,$$

where $\hbar = h/2\pi$ is the reduced Planck's constant, and ϵ_R is the kinetic energy of rotation, show that the wave function of a rigid rotor is given by the spherical harmonics

$$\psi(\theta,\phi) = Y_n^m(\theta,\phi) = (-1)^m \sqrt{\frac{2n+1}{4\pi}\frac{(n-m)!}{(n+m)!}} P_n^m(\cos\theta) e^{im\phi},$$

where $n = 0, 1, 2, \ldots$, $m = -n, -n+1, \ldots, n-1, n$, θ is the zenith angle, ϕ is the azimuthal angle, and $P_n^m(x)$ are the associated Legendre functions.

(ii) Show that the energy levels of a rigid rotor are

$$\epsilon_R = \frac{n(n+1)h^2}{8\pi^2 I}$$

where $n = 0, 1, 2, \ldots$ is a rotational quantum number. What is the degeneracy of each rotational energy level?

(iii) Show that, at low temperature ($T \ll \theta_R$), the rotational partition function has the leading terms

$$q_{rot} = 1 + 3e^{-2\theta_R/T} + 5e^{-6\theta_R/T} + 7e^{-12\theta_R/T} + \cdots,$$

where $\theta_R \equiv (h^2/8\pi^2 I k_B)$ is the rotational temperature.

(iv) Using the Euler–Maclaurin formula

$$\sum_{n=L}^{M} f(n) = \int_L^M f(x)dx + \frac{1}{2}[f(L) + f(M)] + \sum_{k=1}^{\infty} \frac{B_{2k}}{(2k)!}[f^{(2k-1)}(M) - f^{(2k-1)}(L)],$$

where n is integer between $[L, M]$, $f(x)$ is an analytical function with $f^{(k)}$ being the kth derivative of f, and B_k is the kth Bernoulli number, show that the rotational partition function can be expressed as a power series of the inverse temperature

$$q_{rot} = \frac{T}{\theta_R}\left[1 + \frac{1}{3}\frac{\theta_R}{T} + \frac{1}{15}\left(\frac{\theta_R}{T}\right)^2 + \frac{4}{315}\left(\frac{\theta_R}{T}\right)^3 + \cdots\right].$$

The first few Bernoulli numbers are $B_2 = 1/6, B_4 = -1/30, B_6 = 1/42, \ldots$.

Solution:

(i) A rigid rotor has one end fixed at the origin and the other end moving at the surface of a sphere of radius r. In spherical coordinates, the Laplacian is given by

$$\nabla^2 = \frac{\partial^2}{\partial r^2} + \frac{2}{r}\frac{\partial}{\partial r} + + \frac{1}{r^2}\left[\frac{1}{\sin\theta}\frac{\partial}{\partial\theta}\left(\sin\theta\frac{\partial}{\partial\theta}\right)\frac{1}{\sin^2\theta}\frac{\partial^2}{\partial\phi^2}\right].$$

where the zenith angle θ is referred to the angle with respect to polar axis, and azimuthal angle ϕ is the angle of rotation from the initial meridian plane. As r is fixed, the Schrödinger equation becomes

$$-\frac{\hbar^2}{2I}\left[\frac{1}{\sin\theta}\frac{\partial}{\partial\theta}\left(\sin\theta\frac{\partial}{\partial\theta}\right)\frac{1}{\sin^2\theta}\frac{\partial^2}{\partial\phi^2}\right]\psi = \epsilon_R\psi, \tag{D}$$

where $I = \mu r^2$ is called the moment of inertia. Eq. (D) corresponds to a special form of the spherical harmonic differential equation

$$\left[\frac{1}{\sin\theta}\frac{\partial}{\partial\theta}\left(\sin\theta\frac{\partial}{\partial\theta}\right)\frac{1}{\sin^2\theta}\frac{\partial^2}{\partial\phi^2} + n(n+1)\right]y = 0, \tag{E}$$

where $n = 0, 1, 2, \ldots$. The solution to Eq. (E) is called spherical harmonics, $y = Y_n^m(\theta, \phi)$, with

$$Y_n^m(\theta, \phi) = (-1)^m\sqrt{\frac{2n+1}{4\pi}\frac{(n-m)!}{(n+m)!}}P_n^m(\cos\theta)e^{im\phi},$$

where $n = 0, 1, 2, \ldots$, $m = -n, -n+1, \ldots, n-1, n$, θ is the zenith angle, ϕ is the azimuthal angle, and $P_n^m(\cos\theta)$ are the associated Legendre functions.

Note that the rigid rotor model does not account for the quantum nature of nuclei (specifically, the nuclear spins). For a diatomic molecule consisting of two identical atoms with an even mass number (e.g., the deuteron), the rotational wave function must be symmetric in the nuclear coordinates. The symmetry is satisfied by the spherical harmonics with even values of j. If the nuclei have an odd mass number (e.g., the proton) the wave functions must be antisymmetric, satisfied by the spherical harmonics with odd values of j. The nuclear spins effects are important for understanding the thermodynamic properties of parahydrogen and orthohydrogen. See McDowell for more details.[2]

(ii) A comparison of Eqs. (D) and (E) indicates

$$\epsilon_R = \frac{n(n+1)\hbar^2}{2I} = \frac{n(n+1)h^2}{8\pi^2 I}.$$

Because there are $2n+1$ wave functions, $Y_n^m(\theta, \phi)$ with $m = -n, -n+1, \ldots, n-1, n$, yields the same energy, the degeneracy of the energy level is $g_n = 2n+1$.

2 McDowell R. S., "Rotational partition functions for linear molecules", *J. Chem. Phys.*, 88 (1): 356–361 (1988).

(iii) The rotational partition function is given by

$$q_{rot} = \sum_{n=0}^{\infty} g_n \exp\left[-\frac{n(n+1)h^2}{8\pi^2 I k_B T}\right] = \sum_{n=0}^{\infty}(2n+1)\exp[-n(n+1)\theta_R/T],$$

where $\theta_R \equiv h^2/(8\pi^2 I k_B)$. When $T \ll \theta_R$, q_{rot} is dominated by the small values of n. Writing out the terms related to $n = 0, 1, 2, 3, \ldots$ explicitly gives

$$q_{rot} = 1 + 3e^{-2\theta_R/T} + 5e^{-6\theta_R/T} + 7e^{-12\theta_R/T} + \cdots$$

(iv) Let $x = n(n+1)$ and $F(x) = \exp[-x\theta_R/T]$, then

$$\int_0^{\infty} F(x)dx = \int_0^{\infty}(2n+1)\exp[-n(n+1)\theta_R/T]dn = T/\theta_R.$$

Now apply the Euler–Maclaurin formula for $f(x) = (2x+1)\exp[-x(x+1)\theta_R/T]$ with $L = 0$ and $M = \infty$:

$$\sum_{n=0}^{\infty} f(n) = T/\theta_R + \frac{1}{2}[f(\infty)+f(0)] + \frac{B_2}{2}[f^{(1)}(\infty)-f^{(1)}(0)]$$
$$+\frac{B_4}{4!}[f^{(3)}(\infty)-f^{(3)}(0)] + \frac{B_6}{6!}[f^{(5)}(\infty)-f^{(5)}(0)] + \cdots. \quad \text{(F)}$$

Note that $f^{(k)}(\infty) = 0$ for all values of k. We only need to evaluate the derivatives of $f(x)$ at $x = 0$:

$$f(0) = 1$$
$$f^{(1)}(0) = 2 - (\theta_R/T)$$
$$f^{(3)}(0) = -(\theta_R/T)[12 - 12(\theta_R/T) + (\theta_R/T)^2]$$
$$f^{(5)}(0) = -(\theta_R/T)^2[-120 + 180(\theta_R/T) - 30(\theta_R/T)^2 + (\theta_R/T)^3].$$

Plugging the above expressions into Eq. (F) yields

$$q_{rot} = \sum_{n=0}^{\infty}(2n+1)\exp[-n(n+1)\theta_R/T]$$
$$= \frac{T}{\theta_R} + \frac{1}{2} + \frac{1/6}{2}[0-(2-\theta_R/T)]$$
$$- \frac{1/30}{4!}(\theta_R/T)[12 - 12(\theta_R/T) + (\theta_R/T)^2]$$
$$+ \frac{1/42}{6!}(\theta_R/T)^2[-120 + 180\theta_R/T - 30(\theta_R/T)^2 + (\theta_R/T)^3] + \cdots.$$

Combing the terms in the above equation leads to

$$q_{rot} = \frac{T}{\theta_R}\left[1 + \frac{1}{3}\frac{\theta_R}{T} + \frac{1}{15}\left(\frac{\theta_R}{T}\right)^2 + \frac{4}{315}\left(\frac{\theta_R}{T}\right)^3 \cdots\right].$$

Problem 3.11 The rotational motions of a nonlinear polyatomic molecule are considerably more complicated than that corresponding to a diatomic or linear polyatomic molecule. In general, the moment of inertia is described by a 3×3 tensor that depends on the mass and position of multiple atoms:

$$\mathbf{I} = \begin{bmatrix} I_{xx} & I_{xy} & I_{xz} \\ I_{yx} & I_{yy} & I_{yz} \\ I_{zx} & I_{zy} & I_{zz} \end{bmatrix} = \begin{bmatrix} \sum_k m_k(y_k^2+z_k^2) & -\sum_k m_k x_k y_k & -\sum_k m_k x_k z_k \\ -\sum_k m_k y_k x_k & \sum_k m_k(x_k^2+z_k^2) & -\sum_k m_k y_k z_k \\ -\sum_k m_k z_k x_k & -\sum_k m_k z_k y_k & -\sum_k m_k(x_k^2+y_k^2) \end{bmatrix},$$

where m_k stands for atomic mass, (x_k, y_k, z_z) is the atomic position relative the molecular center of mass, and the summation over index k applies to all atoms from the polyatomic molecule. The simplest scenario is called a spherical top, which applies to molecules such as CH_4 and C_{60} where the three principal moments of inertia (i.e., the eigenvalues of the moment tensor) are identical $I_A = I_B = I_C = I$. In this case, the Schrödinger equation predicts that the rotation has quantified energy levels ϵ_R and degeneracy g_n

$$\epsilon_R = \frac{n(n+1)\hbar^2}{2I} \quad \text{and} \quad g_n = (2n+1)^2.$$

Verify that, at high temperature, the rotational partition function can be approximated by

$$q_{rot} \approx \frac{\sqrt{\pi}}{\sigma}\left(\frac{T}{\theta_R}\right)^{3/2}, \tag{G}$$

where $\theta_R \equiv \hbar^2/(2Ik_B)$, and σ stands for the molecular symmetry number, i.e., the number of ways in which the molecule can achieve through rotation to reach the same orientation in space (e.g., $\sigma = 12$ for methane, corresponding to the threefold symmetry about each of the 4 C–H bonds). Eq. (G) can be generalized for nonlinear molecules either with a symmetric top ($I_A = I_B \neq I_C$) or with an asymmetric top ($I_A \neq I_B \neq I_C$)

$$q_{rot} \approx \frac{\sqrt{\pi}}{\sigma}\left(\frac{T^3}{\theta_{R,A}\theta_{R,B}\theta_{R,C}}\right)^{1/2},$$

with $\theta_{R,i} \equiv \hbar^2/(2I_i k_B)$ and $i = A, B$, or C.

Solution:

The rotational partition function is defined as

$$q_{rot} = \frac{1}{\sigma}\sum_{n=0}^{\infty} g_n \exp\left[-\frac{n(n+1)\hbar^2}{2Ik_BT}\right] = \frac{1}{\sigma}\sum_{n=0}^{\infty}(2n+1)^2 \exp\left[-\frac{n(n+1)\theta_R}{T}\right], \tag{H}$$

where the symmetry number σ is introduced *ad hoc* to correct the repeated counting of indistinguishable configurations. At high temperature (viz., $T \gg \theta_R$), the summation in Eq. (H) is dominated by the large values of n. As a result, we may approximate $(2n+1) \approx 2n$ and $n(n+1) \approx n^2$. In that case, Eq. (H) becomes

$$q_{rot} \approx \frac{4}{\sigma}\sum_{n=0}^{\infty} n^2 \exp(-n^2\theta_R/T), \tag{I}$$

Now replace the summation over n by integration and let $n^2\theta_R/T = x^2 + y^2 + z^2$, Eq. (I) becomes

$$q_{rot} \approx \frac{4}{\sigma}\int_0^{\infty} n^2 \exp(-n^2\theta_R/T)dn = \frac{1}{\pi\sigma}\left(\frac{T}{\theta_R}\right)^{3/2}\left(\int_{-\infty}^{\infty} e^{-x^2}dx\right)^3 = \frac{\sqrt{\pi}}{\sigma}\left(\frac{T}{\theta_R}\right)^{3/2},$$

where the second equal sign follows from $4\pi n^2 dn = (T/\theta_R)^{3/2}dxdydz$.

Problem 3.12 According to the classical physics, the kinetic energy of a rigid rotor is determined by the three principal moments of inertia $\mathbf{I} = (I_A, I_B, I_C)$ and corresponding angular velocity $\boldsymbol{\omega} = (\omega_A + \omega_B + \omega_C)$:

$$K = \frac{1}{2}(I_A\omega_A^2 + I_B\omega_B^2 + I_C\omega_C^2).$$

Show that the angular velocity ω follows the Maxwell–Boltzmann distribution

$$p(\boldsymbol{\omega}) = \sqrt{\frac{I_A I_B I_C}{(2\pi k_B T)^3}} \exp\left[-\frac{1}{2k_BT}(I_A\omega_A^2 + I_B\omega_B^2 + I_C\omega_C^2)\right],$$

and that each component of the angular velocity has an average kinetic energy of $k_BT/2$.

Solution:
At a given temperature, the microstates of an isolated rigid rotor follow the Boltzmann distribution

$$p(\boldsymbol{\omega}) = \frac{e^{-\beta K}}{q} = \frac{1}{q}\exp\left[-\frac{1}{2k_BT}(I_A\omega_A^2 + I_B\omega_B^2 + I_C\omega_C^2)\right].$$

The proportionality constant q can be fixed by the normalization condition

$$1 = \int d\boldsymbol{\omega}p(\boldsymbol{\omega}) = \int_{-\infty}^{\infty} d\omega_A \int_{-\infty}^{\infty} d\omega_B \int_{-\infty}^{\infty} d\omega_C p(\boldsymbol{\omega}).$$

Using the integration formula $\int_{-\infty}^{\infty} dxe^{-x^2} = \sqrt{\pi}$, we obtain

$$q = \sqrt{\frac{(2\pi k_BT)^3}{I_AI_BI_C}}.$$

Therefore, the angular velocity $\boldsymbol{\omega}$ follows the Maxwell–Boltzmann distribution

$$p(\boldsymbol{\omega}) = \sqrt{\frac{I_AI_BI_C}{(2\pi k_BT)^3}}\exp\left[-\frac{1}{2k_BT}(I_A\omega_A^2 + I_B\omega_B^2 + I_C\omega_C^2)\right].$$

The average kinetic energy is given by

$$\langle K\rangle = \int d\boldsymbol{\omega}p(\boldsymbol{\omega})(I_A\omega_A^2 + I_B\omega_B^2 + I_C\omega_C^2)/2 = -\frac{d\ln q}{d\beta} = 3k_BT/2.$$

As the three components of the angular velocity are independent of each other, each component makes a contribution of $k_BT/2$, as predicted by the equipartition theorem.
Note that q is not the rotational partition function. The latter is a dimensionless quantity given by $q_{rot} = 8\pi^2(I_AI_BI_C)q/h^3$.[3]

Problem 3.13 The microstates associated with bond vibration in a diatomic molecule is conventionally represented by the harmonic oscillator model, i.e., a particle of effective mass μ subject to a harmonic potential

$$v(x) = \frac{1}{2}k_sx^2,$$

where k_s is the spring constant, x is the deviation from the equilibrium bond length, and the effective mass is related to those of the two bonded atoms $\mu = m_1m_2/(m_1 + m_2)$. The microstate energy ϵ_V is defined by the solutions of the Schrödinger equation

$$-\frac{\hbar^2}{2m}\frac{d^2\psi(x)}{dx^2} + \frac{1}{2}k_sx^2\psi(x) = \epsilon_V\psi(x),$$

where $\hbar = h/2\pi$ is the reduced Planck's constant.

(i) Show that the wave function of the harmonic oscillator is given by the Hermite functions

$$\psi(x) = \frac{1}{\sqrt{2^nn!}}\left(\frac{\mu\omega}{\pi\hbar}\right)^{1/4}e^{-\frac{m\omega x^2}{2\hbar}}H_n\left(\sqrt{\frac{\mu\omega}{\hbar}}x\right),$$

where $n = 0, 1, 2, \ldots, \omega = \sqrt{k_s/\mu}$ is the angular frequency of the oscillator, and $H_n(x)$ are called the Hermite polynomials.

3 Desloge E. A., "Classical partition function of a rigid rotator", *Am. J. Phys.*, 52, 261–262 (1984).

(ii) Show that the vibrational energy levels are given by

$$\epsilon_V = (n + 1/2)\hbar\omega.$$

What is the degeneracy of each vibrational energy level? Why does not the vibrational energy vanish at the ground state?

(iii) Plot the wave function for the four lowest energy levels $n = 0, 1, 2,$ and 3 and discuss their significance.

(iv) Show that the vibrational partition function is

$$q_{vib} = \frac{1}{2\sinh(\theta_\upsilon/2T)},$$

where $\theta_\upsilon = \hbar\omega/k_B$ is the vibrational temperature.

(v) Discuss experimental/computational ways to determine the angular frequency.

Solution:

(i) In mathematics, the Hermite functions, also called the Hermite–Gaussian functions, are defined by the solutions of the following differential equation

$$\frac{d^2y(x)}{dx^2} + (2n + 1 - x^2)y(x) = 0, \tag{J}$$

where $n = 0, 1, 2, \ldots$ are integers. These functions have the general form of

$$y(x) = (2^n n!\sqrt{\pi})^{-1/2} e^{-x^2/2} H_n(x).$$

The first few Hermite polynomials $H_n(x)$ are given by

$$\begin{aligned}
H_0(x) &= 1\\
H_1(x) &= 2x\\
H_2(x) &= 4x^2 - 2\\
H_3(x) &= 8x^3 - 12x\\
H_4(x) &= 16x^4 - 48x^2 + 12\\
&\ldots
\end{aligned}$$

In terms of the angular frequency ω (see Problem 3.8), the Schrödinger equation for the harmonic oscillator can be written as

$$\frac{\hbar^2}{2m}\frac{d^2\psi(x)}{dx^2} + \left(\epsilon_V - \frac{1}{2}\mu\omega^2x^2\right)\psi(x) = 0. \tag{K}$$

Let $z = x/\sqrt{\hbar/\mu\omega}$, Eq. (K) becomes

$$\frac{\hbar^2}{2\mu}\frac{m\omega}{\hbar}\frac{d^2\psi(z)}{dz^2} + (\epsilon_V - \frac{1}{2}\mu\omega^2\frac{\hbar}{\mu\omega}z^2)\psi(z) = 0. \tag{L}$$

Dividing Eq. (L) by $\hbar\omega/2$ leads to

$$\frac{d^2\psi(z)}{dz^2} + \left(\frac{2\epsilon_V}{\hbar\omega} - z^2\right)\psi(z) = 0. \tag{M}$$

A comparison of Eq. (M) with Eq. (J) indicates

$$\psi(x) = \left(\frac{\mu\omega}{\hbar}\right)^{1/4} y\left(\sqrt{\frac{\mu\omega}{\hbar}}x\right) = \left(\frac{\mu\omega}{\hbar}\right)^{1/4}(2^n n!\sqrt{\pi})^{-1/2}e^{-\frac{\mu\omega x^2}{2\hbar}}H_n\left(\sqrt{\frac{\mu\omega}{\hbar}}x\right),$$

where a factor of $(\mu\omega/\hbar)^{1/4}$ has been added to make the wave function satisfy the normalization condition

$$\int \psi^2(x)dx = 1.$$

The energy term satisfies

$$\frac{2\epsilon_V}{\hbar\omega} = (2n+1).$$

(ii) As discussed above, the vibrational energy levels are related to the eigenvalues of the Hermite differential equation

$$\epsilon_V = (n+1/2)\hbar\omega.$$

where $n = 0, 1, 2, \ldots$ are integers. The degeneracy of the vibrational energy is $g = 1$ because each integer corresponds to one energy level.
A harmonic oscillator has a zero-point energy $\epsilon_{V,0} = \hbar\omega/2$ at $n = 0$. Its physical meaning is that, according to the uncertainty principle, the particle position is uncertain even at the ground state. Therefore, its momentum, and hence the kinetic energy, cannot be exactly zero.

(iii) Figure 3.2 shows the wave function for the four lowest energy levels of the harmonic oscillator, i.e., $n = 0, 1, 2$, and 3. Here, $a = \sqrt{\hbar/\mu\omega}$ has the units of length, and $b = 1/\sqrt{a}$ makes the wave function dimensionless. As well documented in the textbooks of quantum mechanics, the wave function square represents the probability, i.e., the probability for locating the particle in the harmonic model.
Figure 3.2 indicates that the particle displacement from its equilibrium position is comparable to parameter a. For example, from the vibrational wavenumber of the oxygen molecule 1580 cm^{-1} and the speed of light 2.997×10^{10} cm/s, we can calculate the angular frequency ω

$$\omega = 2\pi \times 2.997 \times 10^{10} = 2.9763 \times 10^{10}\ \text{s}^{-1}.$$

Thus, parameter a has the value of

$$a = \sqrt{\frac{\hbar}{\mu\omega}} = \sqrt{\frac{6.62607004 \times 10^{-34}}{2\pi \times 8 \times 1.6735575 \times 10^{-27} \times 2.9763 \times 10^{10}}} = 1.2895 \times 10^{-11}\ \text{m}.$$

It is smaller than the O–O covalent bond length 1.208×10^{-10} m by almost one order of magnitude.

(iv) The vibrational partition function is

$$q_{vib} = \sum_{n=0}^{\infty} \exp\left[-\frac{(n+1/2)\hbar\omega}{k_B T}\right]$$
$$= e^{-\theta_v/2T} \sum_{n=0}^{\infty} \exp(-n\theta_v/T) = \frac{e^{-\theta_v/2T}}{1 - e^{-\theta_v/T}} = \frac{1}{2\sinh(\theta_v/2T)}$$

where $\theta_v = \hbar\omega/k_B$.

(v) Experimentally, the angular frequency of bond vibration can be measured with infrared or Raman spectroscopy. It can also be estimated from the computation of the molecular energy as a function of the bond length and fit the results with the harmonic model.

Problem 3.14 Estimate the fraction of iodine molecules in an ideal gas at the excited vibrational states ($n > 0$) at room temperature ($T = 298.15$ K). The vibrational temperature for iodine is $\theta_v = 310$ K.

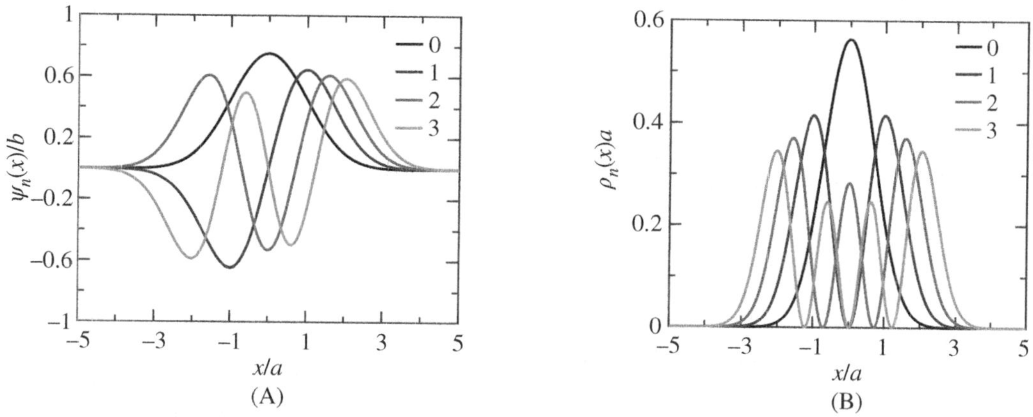

Figure 3.2 A. The wave functions corresponding to the four lowest energy levels of the harmonic oscillator, i.e., $n = 0, 1, 2,$ and 3. B. The probability for locating the particle at different energy levels (HarmonicOsilator.m).

Solution:

The fraction of iodine molecules at the ground state is related to the vibrational partition function q_{vib}

$$p_0 = \frac{e^{-\theta_v/2T}}{q_{vib}} = 1 - e^{-\theta_v/T}$$

Therefore, the fraction in the excited states is

$$p^* = 1 - p_0 = e^{-\theta_v/T} = e^{-310/298.15} \approx 0.35.$$

Problem 3.15 Predict the molecular partition function, molar entropy, and constant-pressure molar heat capacity for HF at 25 °C and 1 atm based on the following parameters:

	W, g/mol	Electronic state	θ_v, K	θ_R, K	D_0, kJ/mol	r_e, Å
HF	20.006	1Σ+	5953.1	30.14	569.7	0.91685

Compare the theoretical results with experimental values of $S = 173.78$ J/(mol K) and $C_P = 29.14$ J/(mol K).

Solution:

At 25 °C, the thermal wavelength for HF is

$$\Lambda = \sqrt{\frac{h^2}{2\pi m k_B T}} = \sqrt{\frac{(6.62607 \times 10^{-34})^2}{2\pi \times 20.006 \times 1.6736 \times 10^{-27} \times 1.3806 \times 10^{-23} \times 298.15}}$$
$$= 2.2517 \times 10^{-11} \text{ m}.$$

and the molecular density at 1 atm can be calculated from the ideal-gas law

$$\rho = \frac{P}{k_B T} = \frac{101325}{1.3806 \times 10^{-23} \times 298.15} = 2.4615 \times 10^{25}.$$

The single-molecule partition function is given by

$$q = q_e q_T q_{rot} q^*_{vib} e^{D_0/k_B T},$$

where

$$e^{D_0/k_B T} = \exp\left(\frac{569.7 \times 1000}{8.314 \times 298.15}\right) = 6.4139 \times 10^{99},$$

the electronic partition function is

$$q_e = g_0 + g_1 e^{-\beta\epsilon_{e,1}} + \cdots \approx g_0 = 1,$$

the translational partition function is

$$q_T = \frac{k_B T}{P\Lambda^3} = \frac{1.3806 \times 10^{-23} \times 298.15}{101325 \times (2.2517 \times 10^{-11})^3} = 3.5586 \times 10^6,$$

the rotational partition function is

$$q_{rot} \approx \frac{T}{\sigma\theta_R} = \frac{298.15}{1 \times 30.14} = 10.2324,$$

and the vibrational partition function is

$$q^*_{vib} = \frac{1}{1 - \exp(-\theta_v/T)} = \frac{1}{1 - \exp(-5953.1/298.15)} = 1.0000.$$

Plugging into the equation for q leads to

$$q = 1 \times 3.5586 \times 10^6 \times 10.2324 \times 1 \times 6.4139 \times 10^{99} = 2.3355 \times 10^{107}.$$

Accordingly, the molar entropy is given by

$$S = R\left[-\ln(\rho\Lambda^3) + \frac{5}{2} + \ln\left(\frac{T}{\sigma\theta_R}\right) + \frac{\theta_v/T}{e^{\theta_v/T} - 1} - \ln\left(1 - e^{-\theta_v/T}\right) + \ln g_0\right]$$

$$= 8.314 \times \Bigg\{ -\ln[2.4615 \times 10^{25} \times (2.2517 \times 10^{-11})^3] + 2.5$$

$$+ \ln\left(\frac{298.15}{1 \times 30.14}\right) + \frac{5953.1/298.15}{e^{5953.1/298.15} - 1} - \ln\left(1 - e^{-5953.1/298.15}\right) + \ln(1)\Bigg\}$$

$$= 173.58\ \mathrm{J/(mol\,K)},$$

which is in good agreement with the experimental value of 173.78 J/(mol K).

At height temperature, the heat capacity is

$$C_P = R\left[\frac{7}{2} + \left(\frac{\theta_v}{T}\right)^2 \frac{e^{\theta_v/T}}{(e^{\theta_v/T} - 1)^2}\right]$$

$$= 8.314 \times \left[3.5 + \left(\frac{5953.1}{298.15}\right)^2 \frac{e^{5953.1/298.15}}{(e^{5953.1/298.15} - 1)^2}\right] = 29.1006\ \mathrm{J/(mol\,K)},$$

which is also in good agreement with the experimental value of 29.14 J/(mol K).

Problem 3.16 Plot the constant-pressure molar heat capacity for HF as an ideal gas from 0 to 10^5 K. Assume that the electrons remain unexcited at all temperatures.

Solution:

Figure 3.3 presents the constant-pressure heat capacity versus temperature for HF. The heat capacity of a diatomic ideal gas is

$$C_P = R\left[\frac{5}{2} + C_{v,rot}/R + \frac{(\theta_v/T)^2}{[2\sinh(\theta_v/2T)]^2}\right]$$

where the rotational heat capacity has analytical expressions in terms of the reduced temperature $t \equiv T/\theta_R$:

$$C_{v,rot}/R = \begin{cases} \dfrac{12e^{-2/t} - 36e^{-4/t}}{t^2} & t < 0.5, \\[2ex] 1 + \dfrac{1}{45t^2} + \dfrac{16}{945t^3} - \dfrac{121}{4725t^4} & t > 0.5, \end{cases}$$

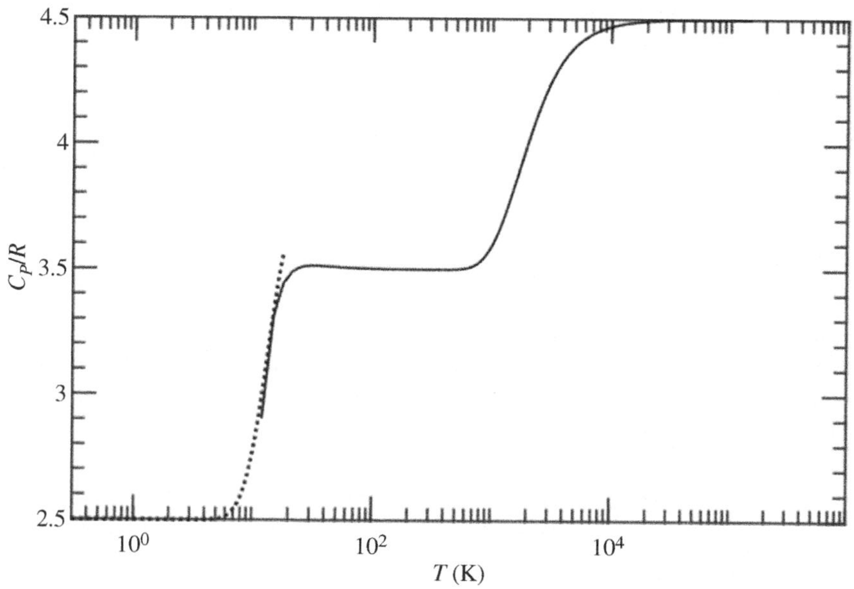

Figure 3.3 Constant-pressure molar heat capacity for HF at low pressure. The dotted line shows the heat capacity at the low-temperature limit (HF.m).

Problem 3.17 Predict the entropy of methyl radical, CH_3*, as an ideal gas at 298.15 K and 1 atm based on the following data for the molecular weight, structure, electronic and vibrational energy levels:

(1) Molecular weight: $W = 15.0345$,
(2) CH bond length: $b = 1.0767$ Å,
(3) Electronic energy levels (cm^{-1}): 46239, 59972, 66536, 69853.44 72508,
(4) Vibrational energy levels (cm^{-1}): 3004.43, 606.453, 3160.821,1396, 3160.821,1396.

Data from the NIST Chemistry WebBook (https://webbook.nist.gov/cgi/inchi?ID=C2229074&Mask=800).

Solution:

The ideal-gas entropy includes four contributions:

$$S = S_e + S_T + S_R + S_v,$$

corresponding to the electronic, translational, rotational, and vibrational entropies, respectively.

- The electronic partition function is given by

$$q_e = g_0 + g_1 e^{-\theta_{e,1}/T} + \cdots .$$

Methyl radical has one unpaired electron, which results in a degeneracy of $g_0 = 2$ at the group state. Because the electronic temperatures are much larger than 298.15 K (e.g., $\theta_{e,1} = 46239 \times 1.986 \times 10^{-23}/1.3806 \times 10^{-23} = 66513\text{K} \gg T$), the contributions to q_e from the exited states are negligible. Therefore, $q_e \approx g_0 = 2$, and the electronic entropy is

$$S_e = R \ln 2 = 5.7631 \text{ J/(mol K)}$$

where $R = 8.314$ J/(mol K) is the gas constant.

- The translational entropy can be calculated from the Sackur–Tetrode equation

$$S_T = R[\ln(T^{5/2}W^{3/2}/P) - 1.16486]$$
$$= 8.314 \times [\ln(298.15^{5/2}15.0345^{3/2}/1) - 1.16486] = 142.539 \text{ J/(mol K)}.$$

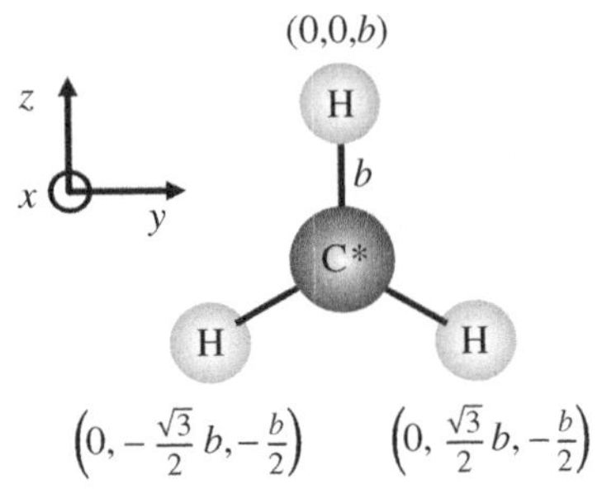

Figure 3.4 Schematic structure of methyl radical. The parentheses give the coordinates of different atoms.

- The rotational entropy depends on the symmetry number σ and the principal moments of inertia I_A, I_B, and I_C. As shown in Figure 3.4, methyl radical has a planar structure with all four atoms positioned on the same plane. The center of mass position coincides with that of the carbon atom, and the coordinates of the three hydrogen atoms can be expressed in terms of the C–H bond length b.

The CH_3 structure belongs to the D_{3h} point group and has a symmetry number $\sigma = 6$. Meanwhile, the inertial tensor can be calculated from the positions of hydrogen atoms and mass m_H:

$$I_{xx} = \sum_{k=1}^{3} m_H(y_k^2 + z_k^2) = 3m_H b^2; \quad I_{yy} = \sum_{k=1}^{3} m_H(x_k^2 + z_k^2) = 3m_H b^2/2;$$

$$I_{zz} = \sum_{k=1}^{3} m_H(x_k^2 + y_k^2) = 3m_H b^2/2; \quad I_{xy} = I_{yx} = -\sum_{k=1}^{3} m_H x_k y_k = 0;$$

$$I_{yz} = I_{zy} = -\sum_{k=1}^{3} m_H y_k z_k = 0; \quad I_{xz} = I_{zx} = -\sum_{k=1}^{3} m_H z_k x_k = 0.$$

Plugging in the numbers leads to $I_A(= I_{xx}) = 5.8204 \times 10^{-47}$ kg/m^2, and $I_B = I_C(= I_{yy} = I_{zz}) = 2.9102 \times 10^{-47}$ kg/m^2. The rotational temperature can be calculated from the three principal moments of inertia

$$\begin{aligned}\theta_R &= \frac{h^2}{8\pi^2 k_B}(I_A I_B I_C)^{-1/3} \\ &= \frac{(6.62607 \times 10^{-34})^2}{8\pi^2 \times 1.3806 \times 10^{-23}}(5.8204 \times 10^{-47} \times 2.9102 \times 10^{-47} \times 2.9102 \times 10^{-47})^{-1/3} \\ &= 10.9844 \text{ K}.\end{aligned}$$

With the high-temperature approximation, the rotational entropy is then given by

$$\begin{aligned}S_R &= R\left\{\ln\left[\frac{\sqrt{\pi}}{\sigma}(T/\theta_R)^{3/2}\right] + \frac{3}{2}\right\} \\ &= 8.314 \times \left\{\ln\left[\frac{\sqrt{\pi}}{6}(298.15/10.9844)^{3/2}\right] + 1.5\right\} = 43.5037 \text{ J/(mol K)}.\end{aligned}$$

- The vibrational entropy is

$$S_v/R = \sum_{i=1}^{6}\left[\frac{\theta_{v,i}}{T}\frac{1}{e^{\theta_{v,i}/T} - 1} - \ln(1 - e^{\theta_{v,i}/T})\right].$$

With $\theta_v = (4.3217, 0.8724, 4.5467, 2.0081, 4.5467, 2.0081) \times 10^3$ K calculated from the vibrational energy levels, the above equation predicts

$$S_v = 1.9896 \text{ J/(mol K)}.$$

Adding up the four contributions above gives the molar entropy of the ideal gas

$$S = 193.7955 \text{ J/(mol K)}.$$

These numbers indicate that the entropy of methyl radical as an ideal gas is dominated by the translational (74%) and rotational (22%) contributions. The vibrational contribution is only about 1%, and the degeneracy of the ground-state energy contributes about 3%.

Problem 3.18 Predict the single-molecule partition function for water at 298.15 K and 1 atm based on the following parameters

	W, g/mol	g_0	θ_R, K	σ	θ_v, K	D_0
H_2O	18.0153	1	40.1, 20.9, 13.4	2	5360, 5160, 2290	917.6, kJ/mol

Solution:

At 298.15 K, the thermal wavelength for H_2O is

$$\Lambda = \sqrt{\frac{h^2}{2\pi m k_B T}} = \sqrt{\frac{(6.62607 \times 10^{-34})^2}{2\pi \times 18.0153 \times 1.6736 \times 10^{-27} \times 1.3806 \times 10^{-23} \times 298.15}}$$
$$= 2.3728 \times 10^{-11} \text{ m}.$$

The single-molecule partition function is given by

$$q = q_e q_T q_{rot} q_{vib} e^{D_e/k_B T} = q_e q_T q_{rot} q^*_{vib} e^{D_0/k_B T}, \tag{N}$$

where q^*_{vib} is the vibrational partition function excluding the zero-point energy.

- The bond dissociation contribution

$$e^{D_0/k_B T} = \exp\left(\frac{917.6 \times 1000}{8.314 \times 298.15}\right) = 5.8403 \times 10^{160},$$

- The electronic partition function

$$q_e = g_0 + g_1 e^{-\beta \epsilon_{e,1}} + \cdots \approx g_0 = 1.$$

- The translational partition function

$$q_T = \frac{k_B T}{P\Lambda^3} = \frac{1.3806 \times 10^{-23} \times 298.15}{101325 \times (2.3728 \times 10^{-11})^3} = 3.0409 \times 10^6.$$

- The rotational partition function

$$q_{rot} \approx \frac{T}{\sigma(\theta_{rot,1}\theta_{rot,2}\theta_{rot,3})^{1/3}} = 43.0526.$$

- The vibrational partition function

$$q^*_{vib} = \prod_{i=1}^{3} \frac{1}{1 - \exp(-\theta_{vib,i}/T)} = 1.0005.$$

Plugging into Eq. (N) gives

$$q = q_e q_T q_{rot} q^*_{vib} e^{D_0/k_B T} = 7.6496 \times 10^{168}.$$

Problem 3.19 Predict the equilibrium constant for the hydrogenation reactions of CO_2 to CH_4 in the gas phase at 300 and 500 K.

$$CO_2 + 4H_2 \rightleftharpoons CH_4 + 2H_2O.$$

Compare the theoretical results with experimental value at 298 K, $K = 7.79 \times 10^{19}$.

	W, g/mol	g_0	θ_R, K	σ	θ_v, K	D_0, kJ/mol
H_2	2	1	87.54	2	6331.1	430.554
CO_2	44	1	0.561	2	3360, 954(2), 1890	1596.2
H_2O	18	1	40.1, 20.9, 13.4	2	5360, 5160, 2290	917.6
CH_4	16	1	7.54, 7.54, 7.54	12	4170, 2180(2), 4320(3), 1870(3)	1640.5

Solution:
The equilibrium constant of the gas-phase reaction can be determined from the single-molecule partition functions of underlying chemical species as pure ideal gases at system temperature and 1 atm (viz., reference states):

$$K = \frac{q_{CH_4} q_{H_2O}^2}{q_{CO_2} q_{H_2}^4}.$$

With the molar parameters calculated from quantum-mechanical calculations, it is straightforward to calculate the partition functions at both temperatures

	q (300 K)	q (800 K)
H_2	2.0147×10^{80}	8.2332×10^{34}
CO_2	2.8468×10^{287}	3.7365×10^{115}
H_2O	7.8197×10^{167}	5.8932×10^{69}
CH_4	4.2815×10^{293}	1.0292×10^{117}

Accordingly, the equilibrium constant is $K = 5.5822 \times 10^{20}$ at 300 K and $K = 20.8187$ at 800 K. At $T = 300$ K, the theoretical prediction is close to the experimental value of $K = 7.79 \times 10^{19}$ at 298 K. The equilibrium constant decreases as temperature rises because the reaction is strongly endothermic.

Problem 3.20 Predicting the thermodynamic properties of chemical systems such as equilibrium constants from first principles is extremely sensitive to the input from quantum-mechanical calculations. Fortunately, the results from quantum-mechanical predictions have been improving over the years, which makes *ab initio* thermodynamics become more and more reliable. In this exercise, predict the various contributions to the molecular partition function of CO at 298.15 K and 1 atm based on the following four sets of "well-known" parameters:

	W, g/mol	g_0	θ_R, K	σ	θ_v, K	D_0
CRC103*	28.01	1	2.78	1	3121.3	1076.6, kJ/mol
Dill and Bromberg	28.01	1	2.77	1	3122	1070.1, kJ/mol
McQuarrie	28.01	1	2.77	1	3103	255.8, kcal/mol
Hill	28.01	1	2.77	1	3070	9.14, eV

*CRC Handbook of Chemistry and Physics 103rd Edition, 2022; Dill K. A. and Bromberg S., Molecular Driving Forces, 2011, Garland Science; McQuarrie D.A., Statistical Mechanics. 2000, University Science Books; Hill T.L., An Introduction to Statistical Thermodynamics. 1986, New York: Dover Publications.

Solution:

The following table shows various terms for the single-molecule partition function of CO. The largest discrepancy from different sets of model parameters arises from the dissociation energy. The translational partition function depends only on the molecular weight. For small molecules, the vibrational partition function is typically very close to 1.

	CRC103	Dill and Bromberg	McQuarrie	Hill
q_e	1	1	1	1
q_T	5895363.765	5895363.765	5895363.765	5895363.765
q_{rot}	107.5822	107.698	107.9693	107.9693
q^*_{vib}	1.0000284	1.0000283	1.0000302	1.0000337
e^{D_0/k_BT}	4.09×10^{188}	2.97×10^{187}	3.34×10^{187}	3.20×10^{154}
q	2.60×10^{197}	1.89×10^{196}	2.13×10^{196}	2.04×10^{163}

Problem 3.21 Predict the molar entropy and chemical potential of hydrogen gas at 298.15 K and 1 atm based on the following parameters

	W, g/mol	g_0	θ_R, K	σ	θ_v, K	D_0, kJ/mol
H_2	2	1	87.54	2	6331.1	430.554

Solution:

At 298.15 K, the thermal wavelength for H_2 is

$$\Lambda = \sqrt{\frac{h^2}{2\pi mk_BT}} = \sqrt{\frac{(6.62607\times10^{-34})^2}{2\pi\times2.016\times1.6736\times10^{-27}\times1.3806\times10^{-23}\times298.15}}$$
$$= 7.0932\times10^{-11}\text{ m},$$

and the molecular density at 1 atm can be predicted from the ideal-gas law

$$\rho = \frac{P}{k_BT} = \frac{101325}{1.3806\times10^{-23}\times298.15} = 2.4615\times10^{25}.$$

- The molar entropy of hydrogen gas is

$$S = R\left[-\ln(\rho\Lambda^3) + \frac{5}{2} + \ln\left(\frac{T}{\sigma\theta_R}\right) + \frac{\theta_v/T}{e^{\theta_v/T}-1} - \ln\left(1-e^{-\theta_v/T}\right) + \ln g_0\right]$$
$$= 117.5873 + 1.1086\times10^{-7} + 12.7407 = 130.3280\text{ J/(mol K)},$$

which is in good agreement with the experimental value of 130.68 J/(mol K) (CODATA Review value).

- The molar enthalpy of hydrogen gas is

$$H = RT\left[\frac{7}{2} + \frac{e^{\theta_v/T}}{e^{\theta_v/T}-1}\right] - D_0 = -421.88\text{ kJ/mol}.$$

- The chemical potential is

$$\mu = H - TS = -460.73\text{ kJ/mol}.$$

Problem 3.22 Consider the ionization of hydrogen atoms in the gas phase

$$H \rightleftharpoons H^+ + e^-.$$

(i) What is the equilibrium constant at $T = 10000$ K?
(ii) When the above system is in equilibrium at 10000 K and 0.5 atm, what is the fraction hydrogen atoms ionized?
(iii) What is the Gibbs energy of reaction ΔG^0 for the ionization of hydrogen atoms at 25 °C?

Assume that the translational partition function of electrons can be described by the classical approximation, i.e., it has a form the same as that for classical particles. The ionization energy of the hydrogen atom is $D_0 = 13.59844$ eV.

Solution:

(i) The equilibrium constant for the ionization of hydrogen atoms is related to the single-molecule partition functions

$$K = \frac{q_{H^+} q_{e^-}}{q_H},$$

where

$$q_{H^+} = \frac{k_B T}{P_0 \Lambda_{H^+}^3} \qquad q_{e^-} = g_0 \times \frac{k_B T}{P_0 \Lambda_e^3} \qquad q_H = g_0 \times \frac{k_B T}{P_0 \Lambda_H^3} e^{D_0/k_B T} \qquad \Lambda = \sqrt{\frac{h^2}{2\pi m k_B T}}.$$

The proton spin degeneracy can be ignored because it appears in both H and H^+ and cancels in the equilibrium constant.

Plugging into the above equations with the numbers $T = 10000$ K, $P_0 = 101325$ Pa, $g_0 = 2$, $m_e = 9.1093837 \times 10^{-31}$ kg, $m_{H^+} = 1.67262192 \times 10^{-27}$ kg, and $D_0 = 13.53$ eV gives

$$K = \frac{q_{H^+} q_{e^-}}{q_H} = \frac{2.5887 \times 10^8 \times 6.5805 \times 10^3}{3.6965 \times 10^{15}} = 4.6085 \times 10^{-4}.$$

(ii) At $P = 0.5$ atm, the fraction ionized hydrogen atoms can be predicted from the charge neutrality condition $\rho_{H^+} = \rho_{e^-}$ and the mass-action law

$$K = \frac{P_{H^+} P_{e^-}}{P_H P_0} = \frac{k_B T \rho_{H^+} \rho_{e^-}}{P_0 \rho_H} = \frac{k_B T}{P_0} \frac{\rho_{H^+}^2}{\rho_H},$$

where $P_0 = 1$ atm. In terms of the fraction hydrogen atoms ionized α,

$$\rho_{H^+} = \alpha \rho_H^0, \qquad \rho_H = (1 - \alpha)\rho_H^0,$$

where ρ_H^0 is the initial density of hydrogen atoms in the system. Thus, we can rewrite the mass-action law as

$$K = \frac{\rho_H^0 k_B T}{P_0} \frac{\alpha^2}{1 - \alpha}.$$

On the other hand, the ideal-gas law predicts

$$P = (\rho_{H^+} + \rho_{e^-} + \rho_H) k_B T = \rho_H^0 k_B T (1 + \alpha).$$

Combing the above two equations yields

$$\frac{P_0 K}{P} = \frac{\alpha^2}{1 - \alpha^2} \qquad \Rightarrow \qquad \alpha = \sqrt{\frac{P_0 K/P}{1 + P_0 K/P}} = 0.0303.$$

(iii) At 25 °C, the equilibrium constant is $K = 6.9449 \times 10^{-231}$. Thus, the Gibbs energy of reaction is

$$\Delta G^0 \equiv \mu^0_{H^+} + \mu^0_{e^-} - \mu^0_H = -RT \log K = 1313.7 \text{ kJ/mol}.$$

Problem 3.23 Predict the equilibrium constant and the fraction of hydrogen molecules dissociated at 3000 K and 5 atm for hydrogen dissociation reaction:

$$H_2 \rightleftharpoons 2H.$$

What is the Gibbs energy of reaction ΔG^0 for hydrogen dissociation at 25 °C? The dissociation energy of a hydrogen molecule is $D_0 = 430.554$ kJ/mol.

Solution:

The MATLAB file is available for numerical details (H2dissoc.m). Briefly, the equilibrium constant for hydrogen dissociation can be predicted from the single-molecule partition functions of hydrogen molecule and hydrogen atom

$$K = \frac{q_H^2}{q_{H_2}} = 0.0296,$$

where

$$q_{H_2} = q_e q_T q_{rot} q^*_{vib} e^{D_0/k_B T} = 2.2578 \times 10^{16} \quad \text{and} \quad q_H = q_e q_T = 2.5851 \times 10^7.$$

The fraction of hydrogen dissociation at 3000 K and 10 atm can be calculated from the mass-action law and the ideal-gas equation of state

$$K = \frac{\rho^0_{H_2} k_B T}{P_0} \frac{4\alpha^2}{1-\alpha} \quad \text{and} \quad P = \rho^0_{H_2} k_B T(1+\alpha),$$

which lead to

$$\alpha = \sqrt{\frac{P_0 K/4P}{1 + P_0 K/4P}} = 0.0384.$$

In the above equations, $P_0 = 1$ atm.

At 25 °C, the equilibrium constant is $K = 1.1254 \times 10^{-71}$. Thus, the Gibbs energy of reaction is

$$\Delta G^0 \equiv 2\mu^0_H - \mu^0_{H_2} = -RT \log K = 404.98 \text{ kJ/mol}.$$

Problem 3.24 The standard hydrogen electrode (SHE) is defined as a platinum (Pt) metal submerged in an aqueous solution with 1 M H^+ concentration in equilibrium with a hydrogen gas at T = 25 °C and P = 1 atm. Based on the Born–Haber cycle as shown in Figure 3.5, predict the "absolute" electric potential

$$E \equiv (\mu_e - \mu_e^0)/e,$$

where e stands for the unit charge, μ_e is the electron chemical potential in the Pt metal, and μ_e^0 is the chemical potential of noninteracting electrons (viz., ideal gas) at $T = 25$ °C and $P = 1$ atm. Compare the theoretical result with the common value used in the literature $E_{SHE} = -4.44 \pm 0.02$ V. (Hint: The dissociation free energy of a hydrogen molecule and the ionization energy of a hydrogen atom can be derived from the single-molecule partition functions discussed in Problems 3.23

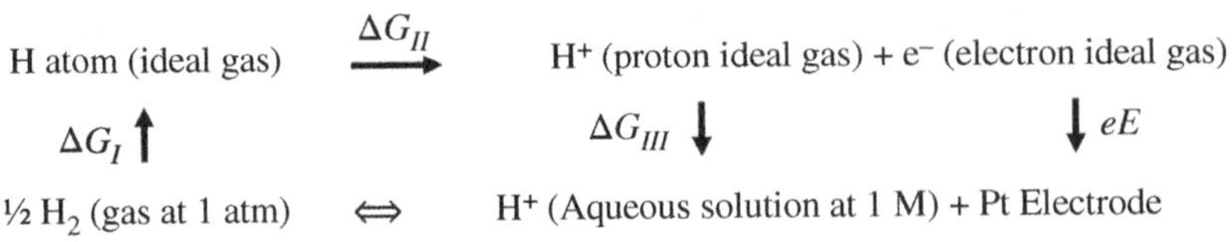

Figure 3.5 The Born–Haber cycle for determining the electric potential of the standard hydrogen electrode (SHE).

and 3.22. According to first-principles solvation-included electronic structure calculations, the hydration free energy of proton is $G_{sol} = -1097.9$ kJ/mol.[4]

Solution:
For the hydrogen gas in contact with the Pt electrode in an aqueous solution with 1 M H^+ concentration, the chemical equilibrium is established when

$$\frac{1}{2}\mu_{H_2} = \mu_{H^+}(1 \text{ M aqueous solution}) + \mu_e$$

The Born–Haber cycle suggests that

$$\begin{aligned}\mu_e &= \frac{1}{2}\mu_{H_2} - \mu_{H^+}(1 \text{ M aqueous solution})\\ &= \left[\frac{1}{2}\mu_{H_2} - \mu_H(\text{ideal gas})\right] + \left[\mu_H(\text{ideal gas}) - \mu_{H^+}(\text{ideal proton gas}) - \mu_e^0\right]\\ &\quad + \left[\mu_{H^+}(\text{ideal proton gas}) - \mu_{H^+}(1 \text{ M aqueous solution})\right] + \mu_e^0\\ &= \mu_e^0 - \Delta G_I - \Delta G_{II} - \Delta G_{III}.\end{aligned}$$

From Problems 3.23 and 3.22, we have

$$2 \times \Delta G_I = 404.98 \text{ kJ/mol} \qquad \text{and} \qquad \Delta G_{II} = 1313.7 \text{ kJ/mol}.$$

Meanwhile, ΔG_{III} is related to the proton solvation free energy and the difference in the translational entropy between the proton in the ideal-gas state and that in $C = 1$ M aqueous solution

$$\Delta G_{III} = G_{\text{sol}} + RT\ln\left(\frac{CRT}{P}\right) = -1090.0 \text{ kJ/mol}.$$

Therefore, the "absolute" electric potential of the SHE is

$$E \equiv (\mu_e - \mu_e^0)/e = \left[-\Delta G_I - \Delta G_{II} - \Delta G_{III}\right]/e = -4.4174 \text{ V}.$$

The number agrees well with the literature value of $E_{SHE} = -4.44 \pm 0.02$ V. For numerical details, a MATLAB file is available (H_electrode.m).

Problem 3.25 Consider the canonical ensemble for a one-component ideal gas in a uniform electric field $\mathbf{E_0}$ introduced by two parallel electrodes with opposite charges as shown schematically in Figure 3.6. Each gas molecule has polarizability α and permanent electric dipole moment d. Assume that the number density of the ideal-gas molecules is extremely low such that they do not interact with the charged surfaces and that the gas molecules are independently polarized.

(i) Show that the external potential for each gas molecule is given by

$$u^{ext} = -\frac{1}{2}\alpha E_0^2 - d\cos\theta E_0,$$

where $E_0 = |\mathbf{E}|$ and $d = |\mathbf{d}|$.

---

4 Zhan C. G. and Dixon D. A., "Absolute hydration free energy of the proton from first-principles electronic structure calculations", *J. Phys. Chem. A.* 105 (51), 11534–11540 (2001).

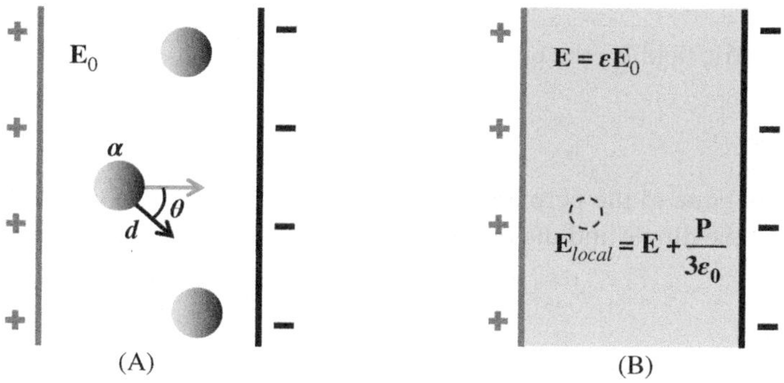

Figure 3.6 Molecular (A) and continuous (B) representations of an ideal gas in an external electric field $\mathbf{E_0}$. Each gas molecule experiences an external potential u^{ext} that depends on the local electric field $\mathbf{E}_{local}$, molecular polarizability α, permanent dipole moment $\mathbf{d}$, and the dipole orientation θ. If the ideal gas is described as continuum with dielectric constant ϵ, the local electric field $\mathbf{E}_{local}$ is related to the electric field in the medium $\mathbf{E}$ and the polarization density $\mathbf{P}$ through the Lorentz equation $\mathbf{E}_{local} = \mathbf{E} + \mathbf{P}/3\epsilon_0$.

(ii) Show that the single-molecule partition function for the ideal gas in external electric field $\mathbf{E_0}$ can be written as

$$q = q_0 e^{\alpha\beta E_0^2/2}\left(\frac{\sinh y}{y}\right),$$

where q_0 is the single-molecule partition function without the external potential, $\beta = 1/(k_B T)$, and $y \equiv \beta d E_0$.

(iii) Verify that the ideal-gas law remains valid in the presence of the electric field, i.e., $P = \rho k_B T$ where $\rho = N/V$ is the number density of the ideal-gas molecules in the system.

(iv) Show that the mean dipole moment of each ideal-gas molecule is given by

$$\mathbf{m} = \alpha\mathbf{E_0} + \mathbf{d}\mathcal{L}(y),$$

where $\mathcal{L}(y) = \coth y - 1/y$ is the Langevin function.

(v) In the continuous model, the dielectric constant is defined as $\epsilon \equiv E/E_0$. Show that the assumption of independent polarizations of the ideal-gas molecules leads to

$$\epsilon - 1 = \frac{\rho}{\epsilon_0}\left(\alpha + \frac{d^2}{3k_B T}\right).$$

(vi) The polarization of each ideal-gas molecule introduces its own electric field which subsequently affects the polarization of other molecules in the system. In classical electrostatics, the polarization effect is described by the Lorentz relation

$$\mathbf{E}_{local} = \mathbf{E} + \frac{\mathbf{P}}{3\epsilon_0},$$

where $\mathbf{E}_{local}$ stands for the electric field at the position of a particular molecule. Based on the Lorentz relation and the ideal-gas model for the average polarization density, $\mathbf{P} = \rho\mathbf{m}$, show that the dielectric constant of the system can be described by the Debye–Langevin equation:

$$\frac{\epsilon - 1}{\epsilon + 2} = \frac{\rho}{3\epsilon_0}\left(\alpha + \frac{d^2}{3k_B T}\right).$$

Solution:

(i) For each gas molecule in an electric field E_0, the energy of polarization is

$$-\int_0^{E_0} \alpha E_0 dE_0 = -\frac{1}{2}\alpha E_0^2.$$

Meanwhile, the electric potential due to the permanent dipole is $-dE_0 \cos\theta$, where θ is the angle between the directions of the dipole and the electric field. Thus, the external potential for each molecule is

$$u^{ext} = -\frac{1}{2}\alpha E_0^2 - d\cos\theta E_0.$$

(ii) As the electric potential is independent of the molecular degrees of freedom, the single-molecule partition function is augmented with an additional term due to the degree of freedom related to the dipole orientation

$$\begin{aligned} q_\theta &= \frac{1}{2}\int_0^\pi \sin\theta d\theta \exp(-\beta u^{ext}) \\ &= \frac{1}{2}\int_0^\pi \sin\theta d\theta \exp(\alpha\beta E_0^2/2 + \beta d E_0 \cos\theta) \\ &= \frac{\exp(\alpha\beta E_0^2/2)}{2}\int_{-1}^{1} d\xi e^{y\xi} = e^{\alpha\beta E_0^2/2}\left(\frac{\sinh y}{y}\right), \end{aligned}$$

where the prefactor 1/2 is introduced such that $q_\theta = 1$ in the absence of the external field, and $y \equiv \beta d E_0$. Therefore, the partition function for the ideal-gas molecules in external electric field $\mathbf{E_0}$ can be written as

$$q = q_0 q_\theta = q_0 e^{\alpha\beta E_0^2/2}\left(\frac{\sinh y}{y}\right),$$

where q_0 is the single-molecule partition function without the external potential.

(iii) The canonical partition function for the entire system is

$$Q = \frac{q^N}{N!} = \frac{(q_T q_{intra} q_\theta)^N}{N!},$$

where $q_T = V/\Lambda^3$ is the translational partition function, and q_{intra} is the partition function affiliated with the intramolecular degrees of freedom. Because both q_{intra} and q_θ are independent of the system volume, the pressure is given by

$$P = k_B T\left(\frac{\partial \ln Q}{\partial V}\right)_{N,T,E_0} = N k_B T\left(\frac{\partial \ln V}{\partial V}\right) = \rho k_B T,$$

where $\rho = N/V$ is the number density of the ideal-gas molecules.

(iv) The mean dipole moment of each ideal-gas molecule is given by

$$\begin{aligned} \mathfrak{m} &= \langle \alpha E_0 + d\cos\theta \rangle \\ &= \frac{1}{2q_\theta}\int_0^\pi \sin\theta d\theta \exp(-\beta u^{ext})(\alpha E_0 + d\cos\theta) \\ &= k_B T\left(\frac{\partial \ln q_\theta}{\partial E_0}\right)_T = k_B T\left(\frac{\partial[\frac{1}{2}\alpha\beta E_0^2 + \ln(\sinh y/y)]}{\partial E_0}\right)_T = \alpha E_0 + d\mathcal{L}(y), \end{aligned}$$

where $y \equiv \beta d E_0$, and $\mathcal{L}(y) = \coth y - 1/y$ is the Langevin function.

(v) The electric field inside the ideal gas **E** includes two contributions, one is the external electric field $\mathbf{E}_0$, and the other is the electric field due to the polarization of individual molecules. With the assumption that the molecular polarization is independent of each other, the latter is $\mathbf{P}/\epsilon_0 = \rho\mathbf{m}/\epsilon_0$, where ρ is the molecular number density, $\mathbf{m}$ is the average dipole of individual molecules, and ϵ_0 is the permittivity of the free space. Thus, the dielectric constant is

$$\epsilon = \frac{E_0 + P/\epsilon_0}{E_0} = \frac{1}{E_0}\left\{E_0 + \frac{\rho}{\epsilon_0}[\alpha E_0 + d\mathcal{L}(y)]\right\} = 1 + \frac{\rho}{\epsilon_0}\left[\alpha + \frac{\beta d^2 \mathcal{L}(y)}{y}\right].$$

In the linear response regime, the electric energy of the average dipole is small compared with $k_B T$, i.e., $y \sim 0$. Thus, $\mathcal{L}(y) \sim y/3$ or $\mathcal{L}(y)/y \approx 1/3$, and the above equation for the dielectric constant can be written as

$$\epsilon - 1 = \frac{\rho}{\epsilon_0}\left(\alpha + \frac{d^2}{3k_B T}\right).$$

The above equation indicates that the deviation of the dielectric constant from 1 is due to two effects: the induced polarization and the orientation of the permanent dipole moments. It predicts that the dielectric constant increases as the molecular polarizability α or the molecular electric dipole moment d increases.

(vi) According to classical electrostatics, the polarization density is related to the dielectric constant ϵ

$$\mathbf{P} = (\epsilon - 1)\epsilon_0 \mathbf{E},$$

where **E** is the electric field in the medium (viz., the macroscopic field), $\chi = \epsilon - 1$ is called electric susceptibility. On the other hand, the ideal-gas model predicts that the polarization density is related to the average dipole moment for each molecule induced by the local electric field $\mathbf{E}_{local}$ (small y)

$$\mathbf{P} = \rho\mathbf{m} = \rho\left(\alpha + \frac{d^2}{3k_B T}\right)\mathbf{E}_{local}. \quad \text{(O)}$$

According to the Lorentz relation, the local electric field is

$$\mathbf{E}_{local} = \mathbf{E} + \frac{\mathbf{P}}{3\epsilon_0} = \left(1 + \frac{\epsilon - 1}{3}\right)\mathbf{E} = \left(\frac{\epsilon + 2}{3}\right)\mathbf{E}.$$

Combining the above three equations gives

$$\mathbf{P} = (\epsilon - 1)\epsilon_0 \mathbf{E} = \rho\left(\alpha + \frac{d^2}{3k_B T}\right)\left(\frac{\epsilon + 2}{3}\right)\mathbf{E},$$

which leads to the Debye–Langevin equation

$$\frac{\epsilon - 1}{\epsilon + 2} = \frac{\rho}{3\epsilon_0}\left(\alpha + \frac{d^2}{3k_B T}\right).$$

Without the permanent dipole, the Debye–Langevin equation reduces to the Clausius–Mossotti equation

$$\frac{\epsilon - 1}{\epsilon + 2} = \frac{\rho\alpha}{3\epsilon_0}.$$

Comments: Although the Debye–Langevin equation is derived from the ideal-gas model, Eq. (O) remains valid if the polarizations of individual molecules are independent of each other. Therefore, the Debye–Langevin equation is often applied in various modified forms to condense phases including liquid mixtures and solutions.

Problem 3.26 Estimate the dielectric constant of water vapor at 100 °C and 1 atm using the Debye–Langevin equation. The polarizability volume of a water molecule is $\alpha' = 1.48 \times 10^{-24}$ cm^3, and the dipole moment is $d = 1.84$ Debye. Note that $\alpha = 4\pi\epsilon_0\alpha'$ and 1 Debye = 3.335640952×10^{-30} Coulomb×meter. Compare your result with the experimental value $\epsilon = 1.00589$.

Solution:
Rearranging the Debye–Langevin equation gives

$$\epsilon = \frac{1 + 2\mathcal{A}}{1 - \mathcal{A}}$$

where

$$\mathcal{A} = \frac{\rho}{3\epsilon_0}\left(\alpha + \frac{d^2}{3k_BT}\right).$$

At $T = 100$ °C and $P = 1$ atm, the number density of water molecules can be estimated from the ideal-gas law

$$\rho = \frac{P}{k_BT} = \frac{101325}{1.38064852 \times 10^{-23} \times 373.15} = 1.9668 \times 10^{25}\ \text{m}^{-3}.$$

Plugging in the numbers with all variables in SI units gives

$$\mathcal{A} = \frac{1.9668 \times 10^{25}}{3 \times 8.85418782 \times 10^{-12}}\left(4\pi \times 8.85418782 \times 10^{-12} \times 1.48 \times 10^{-30} + \frac{(1.84 \times 3.335640952 \times 10^{-30})^2}{3 \times 1.38064852 \times 10^{-23} \times 373.15}\right) = 0.0019.$$

Thus, the dielectric constant is

$$\epsilon = \frac{1 + 2 \times 0.0019}{1 - 0.0019} = 1.0058,$$

which is close to the experimental value $\epsilon = 1.00589$.

Problem 3.27 Consider chemisorption of an ideal gas on a surface with N_s independent binding sites. Assume that the adsorption does not change the molecular flexibility and that molecules binding to the surface can be described by the harmonic oscillator model. Show that parameter b in the Langmuir isotherm can be predicted from

$$b = \frac{q_e^s q_{vib}^s e^{-\beta\epsilon}}{q_e q_{rot}}\left(\frac{P_0\Lambda^3}{k_BT}\right).$$

In the above equation, q_e^s is the electronic partition function of an adsorbed molecule; q_{vib}^s is the vibrational partition function for the surface binding *without* the zero-point-energy contribution

$$q_{vib}^s = [(1 - e^{-\theta_x/T})(1 - e^{-\theta_y/T})(1 - e^{-\theta_z/T})]^{-1},$$

where $\theta_{x,y,z}$ are the surface vibrational temperatures; ϵ is the surface binding energy; Λ is the thermal wavelength of the gas molecules; $P_0 = 1$ atm; q_e and q_{rot} are the electronic and rotational partition functions in the gas phase. How would you predict the surface vibrational temperatures?

Solution:
A gas molecule attached to the surface by chemisorption has electronic and vibrational degrees of freedom. Thus, the single-molecule partition function is given by

$$q_s = q_e^s q_{vib}^s q_{vib} e^{-\beta\epsilon}.$$

Because the surface binding energy accounts for the zero-point energy of a molecule bonded with the surface, the surface vibrational partition function q^s_{vib} has the form of

$$q_{x,y,z} = \frac{1}{1 - e^{-\theta_{x,y,z}/T}},$$

in each dimension. The Langmuir parameter is defined as $b = q_s e^{\beta\mu_0}$. With the expression $q_0 = e^{-\beta\mu_0}$ from Section 3.5, we have

$$b = q_s e^{\beta\mu_0} = \frac{q^s_e q^s_{vib} q_{vib} e^{-\beta\epsilon}}{q^0} = \frac{q^s_e q^s_{vib} q_{vib} e^{-\beta\epsilon}}{q_e q_{rot} q_{vib}} \left(\frac{P_0\Lambda^3}{k_B T}\right) = \frac{q^s_e q^s_{vib} e^{-\beta\epsilon}}{q_e q_{rot}} \left(\frac{P_0\Lambda^3}{k_B T}\right).$$

The surface vibrational temperatures can be calculated from $\theta_{x,y,z} = h\nu_{x,y,z}/k_B$ with

$$\nu_{x,y,z} = \frac{1}{2\pi}\sqrt{\frac{k_{x,y,z}}{m}},$$

where $k_{x,y,z}$ are the spring constants, and m is the molecular mass.

Problem 3.28 The Langmuir model of gas adsorption assumes that gas molecules are localized once they are attached to a planar surface. This assumption is reasonable if the adsorption energy ϵ is significantly larger than $k_B T$. Otherwise, the adsorbed molecules will have two-dimensional translational degrees of freedom (Figure 3.7).

(i) Derive the single-molecule canonical partition function of a localized molecule on the surface using the harmonic oscillator model for surface tethering;
(ii) Derive the single-molecule canonical partition function of a mobile molecule on the surface using the harmonic oscillator model for surface tethering and the particle-in-a-box model for the two-dimensional translational motions on the surface;
(iii) Derive the grand partition function for the adsorption of mobile gas molecules on the surface;
(iv) Compare the adsorption isotherms derived from the localized and mobile models. Which model predicts more adsorption?

Solution:

(i) For a gas molecule that forms covalent bonds with the surface atoms, the harmonic oscillator model predicts that the single-molecule canonical partition function is

$$q_s = q^s_e q^s_{vib} q_{vib} e^{-\beta\epsilon}$$

where q^s_e is the electronic partition function of the adsorbed molecule, q_{vib} is the vibrational partition function for the gas molecule at the isolated state, $q^s_{vib} = q_x q_y q_z$ is the vibrational partition function for the adsorbed molecule as a whole oscillating relative to the surface, and ϵ is the surface binding energy.

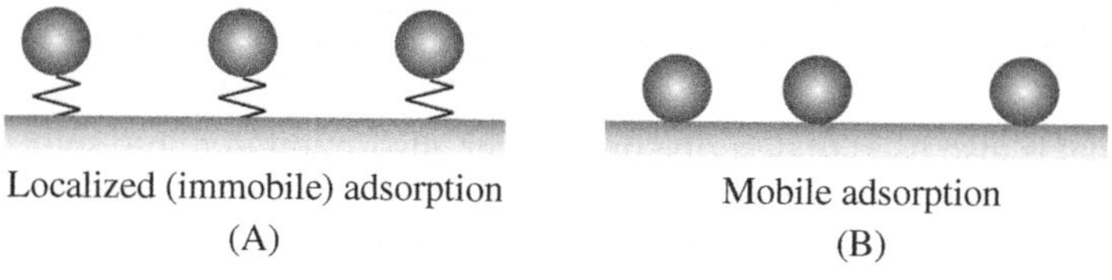

Figure 3.7 Two extreme cases of gas adsorption. A. In the localized adsorption model, the adsorbed molecules are tethered to surface sites. B. In the mobile adsorption model, the adsorbed molecules can move freely on the surface.

With the zero-point energy included in the surface binding energy ϵ, the surface vibrational partition functions are given by

$$q_{x,y,z} = \frac{1}{1 - e^{-\theta_{x,y,z}/T}},$$

where $\theta_{x,y,z}$ are the vibrational temperatures of the adsorbed gas molecule.

(ii) For a gas molecule that moves freely on the surface, the single-molecule canonical partition function includes contributions due to translational motions in horizontal (x, y) directions as well as vibrations in the perpendicular z direction:

$$q_s = q_e^s q_{vib} q_z e^{-\beta\epsilon} A/\Lambda^2,$$

where A stands for the surface area, and $\Lambda = h/\sqrt{2\pi m k_B T}$ is the thermal wave length with m being the mass of the translating adsorbate. A/Λ^2 may be understood as the number of ways to place a gas molecule on the surface.

(iii) For adsorption with gas molecules freely moving on the surface, the grand partition function is

$$\Xi = \sum_N \frac{1}{N!} q_s^N e^{\beta N \mu} = e^{q_s \lambda}$$

where $N!$ accounts for the indistinguishability of gas molecules, and $\lambda = e^{\beta\mu}$.

(iv) The number of mobile gas molecules on the surface is given by

$$\langle N \rangle_{mobile} = \frac{\partial \ln \Xi}{\partial \beta\mu} = q_s \lambda = q_e^s q_{vib} q_z \lambda e^{-\beta\epsilon} A/\Lambda^2.$$

Plugging into the above equation with the ideal-gas chemical potential $\mu = \mu_0 + k_B T \ln P$ gives

$$\langle N \rangle_{mobile} = q_e^s q_{vib} q_z \lambda e^{-\beta(\epsilon - \mu_0)} AP/\Lambda^2.$$

For comparison, the Langmuir model predicts the number of localized molecules

$$\langle N \rangle_{localized} = \frac{N_s q_s \lambda}{1 + q_s \lambda} = \frac{N_s bP}{1 + bP},$$

where $q_s = q_e^s q_{vib} q_x q_y q_z e^{-\beta\epsilon}$ and $b = q_e^s q_{vib} q_x q_y q_z e^{-\beta(\epsilon - \mu_0)}$.

The relative adsorption is

$$\frac{\langle N \rangle_{mobile}}{\langle N \rangle_{localized}} = \frac{q_e^s q_{vib} q_z \lambda e^{-\beta\epsilon} A/\Lambda^2}{N_s q_s \lambda}(1 + q_s \lambda) = \frac{q_e^s q_{vib} q_z \lambda e^{-\beta\epsilon} A/\Lambda^2}{N_s q_e^s q_{vib} q_x q_y q_z e^{-\beta\epsilon} \lambda}(1 + q_s \lambda)$$

$$= \frac{a/\Lambda^2}{q_x q_y}(1 + q_s \lambda) = \frac{a}{\Lambda^2}(1 - e^{-\theta_x/T})(1 - e^{-\theta_y/T})(1 + q_s \lambda),$$

where $a \equiv A/N_s$ is the surface area per adsorption site.

Typically, the vibrational temperatures are much larger than system temperature, $\theta_{x,y} \gg T$, and the average separation between the adsorption sites is larger than the thermal wavelength, $a > \Lambda^2$. As a result, the number of adsorbed molecules predicted by the mobile model is larger than that predicted by the localized model.

Problem 3.29 The Langmuir model is also used to describe monolayer adsorption in dilute solutions. Image a solid surface in contact with a dilute solution at temperature T. The surface can bind with the solute molecules in the solution with some binding sites independent of each other. To account for the molecular excluded volume effect, each surface site can bind at most one solute molecule with energy ϵ. For simplicity, we assume that the solute molecules have no internal degree of freedom either in the bulk solution or at the surface.

(i) Derive internal energy, entropy, and the number of adsorbed molecules using the grand partition function;
(ii) Suppose that the canonical ensemble is used to describe the same system, how is the equation for entropy compared with that derived from the grand canonical ensemble?
(iii) In a dilution solution, the solute chemical potential can be written as $\mu = \mu^0 + k_BT \ln C$, where μ^0 is the chemical potential at a reference state at unit concentration, and C is the solute concentration, verify the Langmuir adsorption isotherm

$$\theta \equiv N/n = \frac{KC}{1+KC},$$

where $K \equiv e^{-\beta(\epsilon-\mu^0)}$, N is the average number of adsorbed molecules, and n is the total number of binding sites.
(iv) If each solute molecule at the surface can vibrate relative to its binding site, how would you modify the adsorption isotherm?

Solution:

(i) We use the two-dimensional lattice model to represent solute molecules on the surface. Without worrying about the internal degree of freedom, we can describe the microstates according to the occupancy of the surface sites:

$$\nu = \{s_1, s_2, \ldots, s_n\},$$

where $s_i = 0$ or 1 denotes a site being empty or occupied by a solute molecule, respectively. At each microstate, the total energy and the number of solute molecules adsorbed on the surface are, respectively,

$$E_\nu = \sum_{i=1}^{n} s_i\epsilon, \qquad N_\nu = \sum_{i=1}^{n} s_i.$$

The grand partition function for the two-dimensional system is thus given by

$$\begin{aligned}\Xi &= \sum_\nu \exp\left[-\beta E_\nu + \beta\mu N_\nu\right] \\ &= \sum_{i=1}^{n}\sum_{s_i=0,1} \exp\left[-\beta\sum_{i=1}^{n} s_i\epsilon + \beta\mu\sum_{i=1}^{n} s_i\right] = \left[1+e^{-\beta(\epsilon-\mu)}\right]^n.\end{aligned}$$

From Ξ, we can derive the internal energy and the average number of molecules at the surface:

$$U =< E_\nu >= -\left(\frac{\partial \ln \Xi}{\partial \beta}\right)_{V,\beta\mu} = \frac{n\epsilon e^{-\beta(\epsilon-\mu)}}{1+e^{-\beta(\epsilon-\mu)}} = n\theta\epsilon,$$

$$N =< N_\nu >= \left(\frac{\partial \ln \Xi}{\partial \beta\mu}\right)_{V,\beta} = \frac{ne^{-\beta(\epsilon-\mu)}}{1+e^{-\beta(\epsilon-\mu)}} = n\theta,$$

where the surface coverage is defined as

$$\theta \equiv N/n = \frac{e^{-\beta(\epsilon-\mu)}}{1+e^{-\beta(\epsilon-\mu)}}.$$

The entropy of the adsorbed molecules is given by

$$\begin{aligned}S &= U/T - \mu N/T + k_B \ln \Xi \\ &= nk_B\left\{\frac{\beta(\epsilon-\mu)e^{-\beta(\epsilon-\mu)}}{1+e^{-\beta(\epsilon-\mu)}} + \ln\left[1+e^{-\beta(\epsilon-\mu)}\right]\right\} \\ &= -nk_B\left\{\theta\ln\theta + (1-\theta)\ln(1-\theta)\right\}.\end{aligned}$$

(ii) In canonical ensemble, we consider N noninteracting molecules on the surface. According the lattice model, the system energy is a constant $E = N\epsilon$. Therefore, the canonical partition function is

$$Q = \sum_{\nu} e^{-\beta E_\nu} = W(E)e^{-\beta N\epsilon},$$

where $W(E)$ is the number of microstates when the system has energy E. For this system, $W(E)$ is the same as the number of ways to arrange N molecules on n lattice sites:

$$W(E) = \frac{n!}{N!(n-N)!}.$$

Using standard thermodynamic relations and the Stirling equation ($\ln n! = n \ln n - n$), we can derive the system entropy

$$\begin{aligned} S &= (U-F)/T = N\epsilon/T + k_B \ln Q \\ &= k_B \ln W(E) = k_B[\ln n! - \ln N! - ln(n-N)!] \\ &= -k_B[N \ln(N/n) + (n-N)\ln(n-N)/n] \\ &= -nk_B \{\theta \ln\theta + (1-\theta)\ln(1-\theta)\}. \end{aligned}$$

Not surprisingly, the entropy derived from the canonical ensemble is the same as that from the grand canonical ensemble.

(iii) From $\mu = \mu^0 + k_B T \ln C$, we have

$$\theta = \frac{e^{-\beta(\epsilon-\mu)}}{1+e^{-\beta(\epsilon-\mu)}} = \frac{Ce^{-\beta(\epsilon-\mu^0)}}{1+Ce^{-\beta(\epsilon-\mu^0)}} = \frac{KC}{1+KC},$$

where $K \equiv e^{-\beta(\epsilon-\mu^0)}$.

(iv) If the solute molecules at the surface have an internal degree of freedom (e.g., vibration), we would be able to define an internal partition function for each adsorbed molecule

$$q_{int} = \sum_k e^{-\beta\epsilon_k},$$

where k represents an internal state, and ϵ_k is the corresponding energy. The adsorption isotherm remains the same but the binding constant is modified as $K = q_{int}e^{-\beta(\epsilon-\mu^0)}$.

Problem 3.30 Suppose that a surface with n binding sites is in contact with a bulk system containing two types of particles with chemical potentials μ_A and μ_B. Assume that each surface site can bind at most one particle and that the surface sites are independent to each other. Let ϵ_A is the energy when a surface site is occupied by particle A, and ϵ_B is the energy when a surface site is occupied by particle B.

(i) Derive an expression for the grand partition function of the surface as an open system;
(ii) How are the surface coverages of A and B dependent on temperature?

Solution:

(i) Each microstate is defined by occupation numbers of the surface sites,

$$\nu = \{s_1, s_2, \dots, s_n\}$$

where $s_i = -1, 0$ or $1, i = 1, 2, \ldots, n$ denotes a site being occupied by particle A, empty, or particle B, respectively. Accordingly, the energy for each site is

$$E_\nu^{(i)} = s_i \left[\frac{s_i - 1}{2}\epsilon_A + \frac{s_i + 1}{2}\epsilon_B\right],$$

and the occupation number for A and B are

$$N_{\nu,A}^{(i)} = s_i \frac{s_i - 1}{2},$$
$$N_{\nu,B}^{(i)} = s_i \frac{s_i + 1}{2}.$$

The grand partition function is then given by

$$\Xi = \sum_{i=1}^{n} \sum_{s_i=-1,0,1} \exp\left[-\beta E^{(i)} - \beta\mu_A \sum_i N_{\nu,A}^{(i)} - \beta\mu_B \sum_i N_{\nu,B}^{(i)}\right]$$
$$= \left[1 + e^{-\beta(\epsilon_A - \mu_A)} + e^{-\beta(\epsilon_B - \mu_B)}\right]^n.$$

(ii) The surface coverages for particle A is

$$\theta_A = < N_{\nu,A}^{(i)} > = \frac{e^{-\beta(\epsilon_A - \mu_A)}}{1 + e^{-\beta(\epsilon_A - \mu_A)} + e^{-\beta(\epsilon_B - \mu_B)}},$$

and that for B is

$$\theta_B = < N_{\nu,B}^{(i)} > = \frac{e^{-\beta(\epsilon_B - \mu_B)}}{1 + e^{-\beta(\epsilon_A - \mu_A)} + e^{-\beta(\epsilon_B - \mu_B)}}.$$

Problem 3.31 The Langmuir-type model is applicable to ligand binding in certain biological systems (e.g., binding tryptophan to Trp aporepressor). Suppose that a protein has n independent binding sites for a ligand with molar concentration C_L in the solution. Assume that all protein binding sites have the same binding affinity.

(i) Derive an expression for the grand partition function of the protein binding with the ligand;
(ii) Predict the average number of ligand molecules bound with the protein, N, based on the grand partition function;
(iii) The Scatchard plot refers to a linear correlation between N/C_L and N. Discuss the physical significance of the slope and interception.

Solution:

(i) In terms of ligand binding with a protein, each microstate is defined by the occupancy of individual sites similar to that in the Langmuir model

$$\nu = \{s_1, s_2, \ldots, s_n\},$$

where $s_i = 0$ or $1, i = 1, 2, \ldots, n$, denotes a site being empty or occupied by a ligand molecule, respectively. For each site, the energy due to the ligand binding may be described by $E_\nu^{(i)} = s_i\epsilon$ and the occupation number is $N_\nu^{(i)} = s_i$. Thus, the total energy and the number ligand molecules associated with the protein are, respectively,

$$E_\nu = \sum_{i=1}^{n} s_i\epsilon,$$
$$N_\nu = \sum_{i=1}^{n} s_i.$$

Accordingly, the grand partition function is given by

$$\Xi = \sum_{\nu} \exp\left[-\beta E_{\nu} + \beta\mu N_{\nu}\right]$$

$$= \sum_{i=1}^{n} \sum_{s_i=0,1} \exp\left[-\beta \sum_{i=1}^{n} s_i \epsilon + \beta\mu \sum_{i=1}^{n} s_i\right] = \left[1 + e^{-\beta(\epsilon-\mu)}\right]^n.$$

(ii) From the grand partition function Ξ, we can derive the average number of ligand molecules associated with the protein:

$$N = \langle N_{\nu} \rangle = \left(\frac{\partial \ln \Xi}{\partial \beta\mu}\right)_{V,\beta} = \frac{ne^{-\beta(\epsilon-\mu)}}{1 + e^{-\beta(\epsilon-\mu)}}.$$

(iii) Typically, the ligand concentration in the solution is small so its chemical potential can be expressed as that corresponding to an ideal solution

$$\mu = \mu^0 + k_B T \ln C_L,$$

where μ^0 is the chemical potential of the ligand at a reference state of unit concentration. As a result, the fraction of occupancy is

$$\theta = N/n = \frac{C_L K}{1 + C_L K}.$$

where $K \equiv e^{-\beta(\epsilon-\mu_0)}$ is known as the binding affinity. Rearranging this equation shows a linear correlation between N/C_L and N

$$\frac{N}{C_L} = \frac{nK}{1 + C_L K} = \frac{nK(1 + C_L K) - nK \times C_L K}{1 + C_L K} = nK - NK.$$

Therefore, N/C_L versus N (i.e., the Scatchard plot) yields a straight line with a slope of $-K$ and interception nK. As defined above, K depends on the (free) energy of binding as well as the chemical potential of the ligand molecule in the reference state, and n is the maximum number of ligand molecules that may bind with the protein.

Problem 3.32 The Scatchard plot (discussed in Problem 3.31) is applied to ligand binding to a protein with identical and fully independent sites. It is an idealized model because in reality binding with the protein sites are often correlated with each other. Another idealized model is that ligand binding is fully cooperative, i.e., the number of ligand molecule associated with a protein is either zero or n, the total number of binding sites available. Let C_L be the ligand concentration in the solution, and ΔE be the total binding energy when all sites are occupied.

(i) Derive an expression for the grand partition function of the fully cooperative ligand binding;
(ii) Predict the average number of ligand molecules associated with the protein per binding, θ, based on the grand partition function;
(iii) The Hill plot is referred to as a linear correlation between $\ln[\theta/(1-\theta)]$ and $\ln C_L$. Discuss the physical significance of the slope and interception.

Solution:

(i) For fully cooperative ligand binding, the protein exists only in two microstates, either all sites empty or all sites fully occupied by the ligand molecules. Therefore, the grand partition function is given by

$$\Xi = \sum_{\nu} \exp\left[-\beta E_{\nu} + \beta\mu N_{\nu}\right] = 1 + e^{-\beta\Delta E + n\beta\mu}.$$

(ii) From the grand partition function Ξ, we can derive the average number of ligand molecules associated with the protein:

$$N = < N_v > = \left(\frac{\partial \ln \Xi}{\partial \beta\mu}\right)_{V,\beta} = \frac{ne^{-\beta\Delta E+n\beta\mu}}{1+e^{-\beta\Delta E+n\beta\mu}}.$$

(iii) The ligand chemical potential can be expressed in terms of its concentration in the bulk solution

$$\mu = \mu^0 + k_B T \ln C_L$$

where μ^0 is the chemical potential of the ligand at a reference state of unit concentration. As a result, the average number of ligand molecules per site is

$$\theta = N/n = \frac{C_L^n K}{1 + C_L^n K},$$

where $K \equiv e^{-\beta(\Delta E - n\mu_0)}$. Rearranging this equation gives

$$\frac{\theta}{1-\theta} = C_L^n K.$$

Therefore, the Hill plot, i.e., $\ln[\theta/(1-\theta)]$ versus $\ln C_L$ yields a straight line with a slope of n and interception $\ln K$. As defined above, K depends on the (free) energy of simultaneous binding as well as the chemical potential of the ligand molecule in the reference state, and n is the maximum number of ligand molecules that may bind with the protein.

Problem 3.33 For a diblock copolymer of Gaussian chains A and B, the intra-chain correlation functions are defined as

$$\omega_{i,j}(r) = \frac{1}{m_A m_B}\sum_{n=1}^{m_A}\sum_{m=1}^{m_B}\langle \delta[\mathbf{r} - (\mathbf{r}_n - \mathbf{r}_m)]\rangle,$$

where i,j denotes block A or B, m_A and m_B are the number of Kuhn segments in blocks A and B, respectively. Show that the three-dimensional Fourier transform of $\omega_{i,j}(r)$ is given by

$$\hat{\omega}_{i,j}(k) = \begin{cases} 2(e^{-x_i} - 1 + x_i)/x_i^2 & i = j \\ (e^{-x_i} - 1)(e^{-x_i} - 1)/(x_i x_j) & i \neq j \end{cases},$$

where $x_i \equiv k^2 R_{g,i}^2$, and $R_{g,i}$ is the radius of gyration for block i.

Solution:

The Fourier transform of $\omega_{i,j}(r)$ yields

$$\hat{\omega}_{i,j}(k) \equiv \int d\mathbf{r}\omega_{i,j}(r)e^{i\mathbf{k}\cdot\mathbf{r}} = \frac{1}{m_A m_B}\sum_{n=1}^{m_A}\sum_{m=1}^{m_B}\langle e^{i\mathbf{k}\cdot(\mathbf{r}_n-\mathbf{r}_m)}\rangle.$$

Using the formula for Gaussian average

$$\langle e^{i\mathbf{k}\cdot(\mathbf{r}_n-\mathbf{r}_m)}\rangle = e^{-\langle|\mathbf{r}_n-\mathbf{r}_m|^2\rangle k^2/6},$$

we can rewrite $\hat{\omega}_{i,j}(k)$ as

$$\hat{\omega}_{i,j}(k) = \frac{1}{m_A m_B}\sum_{n=1}^{m_A}\sum_{m=1}^{m_B} e^{-\langle|\mathbf{r}_n-\mathbf{r}_m|^2\rangle k^2/6}.$$

Note that the end-to-end distance of a Gaussian chain is given by

$$\langle |\mathbf{r}_n - \mathbf{r}_m|^2 \rangle = \begin{cases} |n-m|b_i^2 & i=j \\ nb_A^2 + mb_B^2 & i \neq j \end{cases},$$

where b stands for the Kuhn length, and the indices of the polymer segments n and m start from the joint point of the two blocks. For $i = j$, we have

$$\hat{\omega}_{i,j}(k) = \frac{1}{m_i^2}\sum_{n=1}^{m_i}\sum_{m=1}^{m_i} e^{-|n-m|b_i^2k^2/6} \approx \frac{1}{m_i^2}\int_0^{m_i} dx \int_0^{m_i} dy e^{-|x-y|b_i^2k^2/6}$$

$$= \frac{1}{m_i^2}\left\{\int_0^{m_i} dx \int_0^{x} dy e^{-(x-y)b_i^2k^2/6} + \int_0^{m_i} dx \int_x^{m_i} dy e^{-(y-x)b_i^2k^2/6}\right\}$$

$$= \frac{1}{m_i^2}\int_0^{m_i} dx \left\{\frac{e^{-ax}}{a}(e^{ax}-1) - \frac{e^{ax}}{a}(e^{-am_i} - e^{-ax})\right\}$$

$$= \frac{1}{am_i^2}\int_0^{m_i} dx \left\{2 - e^{-ax} - e^{-a(m-x)}\right\} = \frac{1}{am_i^2}\left\{2m_i + \frac{e^{-am_i}-1}{a} - e^{-am}\frac{e^{am}-1}{a}\right\}$$

$$= \frac{1}{am_i^2}(2m_i - 2/a + 2e^{-am_i}/a) = \frac{2}{a^2m_i^2}(am_i - 1 + e^{-am_i}) = 2(e^{-x_i} - 1 + x_i)/x_i^2,$$

where $a \equiv b_i^2k^2/6$ and $x_i \equiv am_i = k^2m_ib_i^2/6 = k^2R_{g,i}^2$. Similarly, for $i \neq j$, we have

$$\hat{\omega}_{i,j}(k) = \frac{1}{m_i^2}\sum_{n=1}^{m_i}\sum_{m=1}^{m_j} e^{-(nb_i^2+ml_j^2)k^2/6} \approx \frac{1}{m_im_j}\int_0^{m_i} dx \int_0^{m_j} dy e^{-(xb_i^2+yl_j^2)k^2/6}$$

$$= \frac{1}{m_im_j}\int_0^{m_i} dx e^{-a_ix}\int_0^{m_j} dy e^{-a_jy} = (e^{-x_i}-1)(e^{-x_i}-1)/(x_ix_j),$$

where $a_i \equiv b_i^2k^2/6$ and $x_i \equiv a_im_i = k^2m_ib_i^2/6 = k^2R_{g,i}^2$.

Problem 3.34 Verify that the square of the radius of gyration for a worm-like chain (WLC) of persistence length ξ_p and contour length L is given by

$$R_g^2 = \frac{L\xi_p}{3} - \xi_p^2 + \frac{2\xi_p^3}{L} - \frac{2\xi_p^4}{L^2}\left(1 - e^{-L/\xi_p}\right).$$

Solution:

According to Eq. (3.215) in Section 3.11, the radius of gyration for a WLC with m Kuhn segments is given by

$$R_g^2 = \frac{1}{m^2}\int_0^m ds(m-s) < \mathbf{R}^2(s) > . \tag{P}$$

For a WLC with the contour length sb, where b is the Kuhn length, Eq. (3.211) predicts that the square of end-to-end vector satisfies

$$< \mathbf{R}^2(s) >= 2sb\xi_p + 2\xi_p^2\left(e^{-sb/\xi_p} - 1\right).$$

Accordingly, integrating the right of Eq. (P) gives

$$R_g^2 = \frac{1}{m^2}\int_0^m ds(m-s)\left[2sb\xi_p + 2\xi_p^2\left(e^{-sb/\xi_p}-1\right)\right]$$

$$= \frac{2b\xi_p}{m^2}\int_0^m ds(m-s)s + \frac{2\xi_p^2}{m^2}\int_0^m ds(m-s)\left(e^{-sb/\xi_p}-1\right)$$

$$= \frac{2b\xi_p m^3}{6m^2} + \frac{2\xi_p^2}{m^2} \frac{\left(e^{-L/\xi_p} - 1\right) + L/\xi_p - (L/\xi_p)^2/2}{(b/\xi_p)^2}$$

$$= \frac{L\xi_p}{3} + \frac{2\xi_p^4}{L^2}\left(e^{-L/\xi_p} - 1\right) - \xi_p^2 + \frac{2\xi_p^3}{L},$$

where $L = mb$.

Problem 3.35 The elasticity of a single polymer chain can be described by using a constant temperature and constant tension ensemble as shown schematically in Figure 3.8. Consider a freely-jointed chain (FJC) with one end fixed and the other end subject to a constant force f in the direction of the end-to-end vector, $\mathbf{R} = \mathbf{r}_m - \mathbf{r}_1$, m is the number of polymer segments. Because the bond lengths are fixed, the polymer configuration is uniquely defined by the orientations of individual bonds, i.e., the polar and azimuthal angles for each bond, $\{\theta_i, \phi_i\}$, $i = 1, 2, \ldots, m-1$.

(i) Show that the end-to-end distance is given by

$$R = b\cos\theta_1 + b\cos\theta_2 + \cdots + b\cos\theta_{m-1},$$

where b is the bond length.

(ii) Assume that the polymer energy is linearly proportional to the force $u(R) = -fR$, evaluate the canonical partition function

$$Z = \sum_{\nu} \exp\left[-\frac{u(R)}{k_B T}\right],$$

where ν stands for the microstates of the system.

(iii) Show that the average end-to-end distance due to the external force is given by

$$\langle R\rangle = \left(\frac{\partial \ln Z}{\partial \beta f}\right)_T = L\left[\coth\left(\frac{fb}{k_B T}\right) - \frac{k_B T}{fb}\right],$$

where $\beta = 1/(k_B T)$, and $L = (m-1)b$ is the contour length.

(iv) Plot the reduced force ($\beta f b$) versus the relative extension distance (R/L). Show that for small elongations, the polymer elasticity follows Hooke's law

$$f = \frac{3k_B T}{b}\frac{\langle R\rangle}{L},$$

and that, for strong stretching (i.e., $R \sim L$),

$$f = \frac{k_B T}{b}\frac{L}{L - \langle R\rangle}.$$

(v) What is the relative fluctuation of the chain length in the absence of the external force? Hint: $< R^2 > -\langle R\rangle^2 = (\partial\langle R\rangle/\partial\beta f)$.

Figure 3.8 A freely-jointed chain (FJC) with one end fixed and the other end subject to a constant force $\mathbf{f}_z$ along the direction of the end-to-end vector (viz., z-direction).

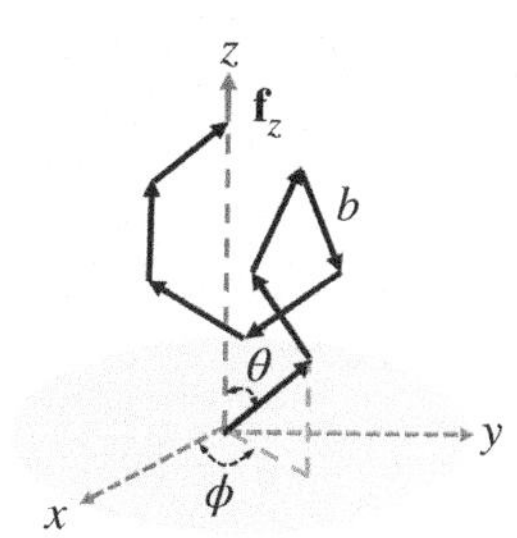

Solution:

(i) Because the direction of polymer extension aligns to that of the external force, the end-to-end distance is the same as the z-component of the end-to-end vector

$$\begin{aligned} R &= z_m - z_1 = z_m - z_{m-1} + z_{m-1} - z_{m-2} + \cdots + z_2 - z_1 \\ &= b\cos\theta_{m-1} + b\cos\theta_{m-2} + \cdots + b\cos\theta_1 \\ &= b\cos\theta_1 + b\cos\theta_2 + \cdots + b\cos\theta_{m-1}. \end{aligned}$$

Without the external force, the end segment has equal probability in the z and $-z$ directions so that the average of R disappears, i.e., $\langle R\rangle = 0$.

(ii) In this system, the microstates are described by the configurations of the polymer chain, as represented by $\{\theta_i, \phi_i\}$. Therefore, the partition function for the polymer chain can be written as

$$\begin{aligned} Z &= Z_0 \left\{ \prod_{i=1}^{m-1} \frac{1}{4\pi} \int_0^{\pi} \sin\theta_i d\theta_i \int_0^{2\pi} d\phi_i \right\} \exp[\beta f R] \\ &= \frac{Z_0}{2^{m-1}} \prod_{i=1}^{m-1} \int_0^{\pi} d\theta_i \sin\theta_i \exp(\beta f b \cos\theta_i), \end{aligned} \quad \text{(Q)}$$

where Z_0 is the partition function of the FJC with no external force, and 4π is the normalization constant for the angular distributions. Note that the system energy is reduced when the polymer extends in the direction of the force.

The integration in Eq. (Q) can be evaluated analytically

$$\int_0^{\pi} d\theta_i \sin\theta_i \exp(\beta f b \cos\theta_i) = \int_{-1}^{1} dx e^{\beta f b x} = \frac{2\sinh(\beta f b)}{\beta f b}.$$

Thus, the partition function becomes

$$Z = Z_0 \left[\frac{\sinh(\beta f b)}{\beta f b}\right]^{m-1}.$$

(iii) The average end-to-end distance is given by

$$\langle R\rangle = \left(\frac{\partial \ln Z}{\partial \beta f}\right)_T = b(m-1)\left[\coth(\beta f b) - \frac{1}{\beta f b}\right] = L\mathcal{L}(\beta f b),$$

where $\mathcal{L}(x) = \coth x - x^{-1}$ is known as the Langevin function.

(iv) Figure 3.9 displays the reduced force ($\beta f b$) versus the relative extension distance (R/L). From the Taylor expansion for small values of x, $\coth x = 1/x + x/3 + \cdots$, we have $\mathcal{L}(x) = \coth x - 1/x \approx x/3$. Thus

$$\frac{\langle R\rangle}{L} = \frac{fb}{3k_B T} \Rightarrow \frac{fb}{k_B T} = \frac{3\langle R\rangle}{L}.$$

At large x, $\coth x \approx 1$ and $\mathcal{L}(x) \approx 1 - 1/x$ so that

$$\langle R\rangle/L = 1 - \frac{k_B T}{fb} \Rightarrow \frac{fb}{k_B T} = \frac{1}{1 - \langle R\rangle/L}.$$

(v) In the presence of an external force, the mean-square fluctuation of the chain length is given by

$$\delta^2 \equiv \langle R^2\rangle - \langle R\rangle^2 = \frac{\partial^2 \ln Z}{\partial(\beta f)^2} = (\partial\langle R\rangle/\partial\beta f).$$

Figure 3.9 The reduced force βfb versus $\langle R\rangle/L$. The dashed lines show the asymptotic results at the weak and strong stretching limits. (DNAextension.m)

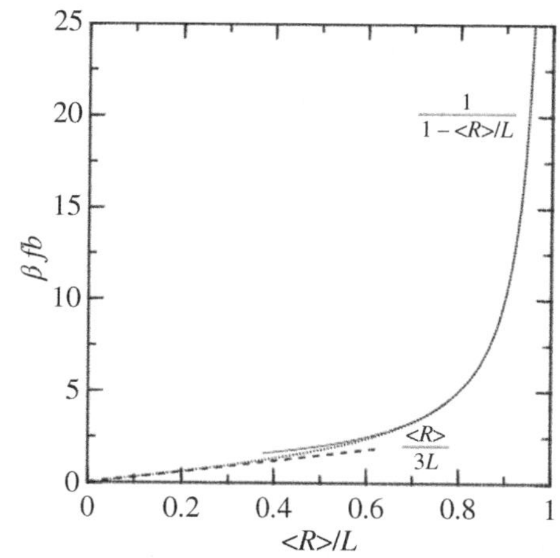

As $f \to 0$, the polymer extension is linearly proportional to the external force

$$\langle R\rangle = \frac{\beta fbL}{3}.$$

Therefore, the relative fluctuations of the chain length is given by

$$\delta/L = \sqrt{bL/3}/L \approx 1/\sqrt{3m}.$$

Problem 3.36 The elasticity of rubbery materials can be understood in terms of the conformation of individual polymer chains. To illustrate, consider the elongation of a rubber from its original length L_0 at relaxation to a stretched length L under an external force τ. Approximately, the elongation does not change the total volume of the rubbery material. To account for the additional degree of freedom due to the anisotropic extension, the fundamental equation of thermodynamics for an elastic material can be written as

$$dF = -TdS - pdV + \tau dL, \quad \text{(R)}$$

where τ is shown as a scalar with the understanding that dL occurs in the same direction of the force. Eq. (R) provides a starting point to describe the elastic behavior of rubbery materials with statistical thermodynamics.

(i) Assume that the rubber elongation from L_0 to L takes place in the z direction and that the cross section of the rubber is uniform. How are the dimensions in the x and y directions, $L_x^0 \times L_y^0$ at the relaxation state, changed by the elongation in the z direction if the rubber volume remains unchanged.

(ii) How does the rubber elongation affect the end-to-end distance of the polymer chain?

(iii) Assume that the rubber can be represented by an affine network (Figure 3.10) of freely-joined chains (FJC) each with m segments of equal bond length b. What is the change in the Helmholtz energy for each chain if they are randomly orientated in the polymer network?

(iv) How is the elastic force of the rubber related to the characteristics of individual chains?

Solution:

(i) The total volume of the rubbery material is

$$V = L_x^0 \times L_y^0 \times L_z^0 = L_x \times L_y \times L_z,$$

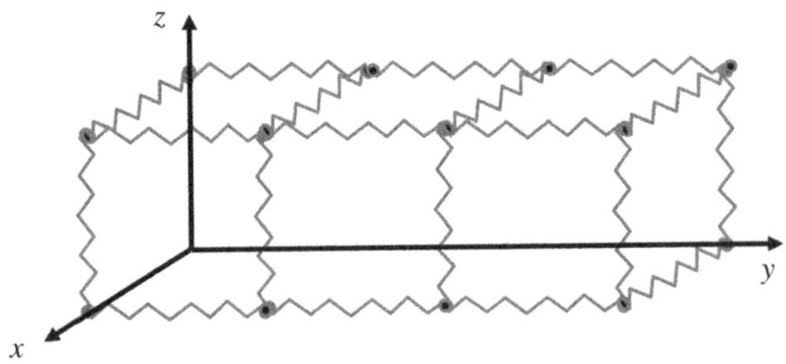

Figure 3.10 A rubbery material represented by an affine network where polymer chains are cross-linked with a cubic structure.

After elongation in the z direction from $L_z^0 = L_0$ to $L_z = L \equiv \lambda L_0$, the dimensions in the x and y directions must contract accordingly in order to keep the total volume unchanged

$$L_x = L_x^0/\sqrt{\lambda}; \qquad \text{and} \qquad L_y = L_y^0/\sqrt{\lambda}.$$

(ii) The affine network model assumes that the polymer chains are aligned along the x, y, and z directions as in a cubic structure. The elongation extends the end-to-end distance for all polymer chains parallel to the z direction while compresses those chains in the x and y directions:

$$\frac{R_z}{R_z^0} = \frac{L_z}{L_z^0} = \lambda; \qquad \frac{R_x}{R_x^0} = \frac{L_x}{L_x^0} = \frac{1}{\sqrt{\lambda}}; \qquad \text{and} \qquad \frac{R_y}{R_y^0} = \frac{L_y}{L_y^0} = \frac{1}{\sqrt{\lambda}},$$

where R_x^0, R_y^0, and R_z^0 are the mean-square root of the end-to-end vector $\mathbf{R}$ of a single polymer chain along the x, y, and z directions, respectively.

(iii) According to the FJC model, the Helmholtz energy of a polymer chain is related to the end-to-end distribution (Eq. (3.166))

$$F(\mathbf{R}) = \frac{3k_BT}{2}\left(\frac{R}{R_0}\right)^2,$$

where $R_0 = mb^2$. If the chains are randomly oriented in the polymer network, the change in the Helmholtz energy per chain due to the deformation in the z-direction is

$$\Delta F_z = \frac{3k_BT}{2}\left[\left(\frac{R_z}{R_0}\right)^2 - \left(\frac{R_z^0}{R_0}\right)^2\right] = \frac{3k_BT}{2}\left(\frac{R_z^0}{R_0}\right)^2(\lambda^2 - 1),$$

and those in the x and y-directions are

$$\Delta F_x = \frac{3k_BT}{2}\left(\frac{R_x^0}{R_0}\right)^2(1/\lambda - 1), \qquad \Delta F_y = \frac{3k_BT}{2}\left(\frac{R_y^0}{R_0}\right)^2(1/\lambda - 1).$$

Note that $(R_x^0)^2 = (R_y^0)^2 = (R_z^0)^2 = R_0^2/3$ for the polymer chains before extension. Therefore, on average the change in Helmholtz energy per polymer chain is

$$\Delta F_{chain} = \Delta F_x + \Delta F_y + \Delta F_z = \frac{k_BT}{2}(\lambda^2 + 2/\lambda - 3).$$

(iv) The elastic force of the rubber is related to the total Helmholtz energy

$$\tau = N\left(\frac{\partial F_{chain}}{\partial L}\right)_{T,V} = \frac{N}{L_0}\frac{\partial F_{chain}}{\partial \lambda} = \frac{Nk_BT}{L_0}(\lambda - 1/\lambda^2)$$

where N is the total number of chains in the polymer network.

Problem 3.37 Smith et al.[5] reported a modified freely-jointed chain (MFJC) model for the extension and force relation of single polymer chains

$$\langle R\rangle = L\left[\coth(\beta fb) - \frac{1}{\beta fb}\right](1 + f/k), \tag{S}$$

where k stands for the stretch modulus, accounting for the variation of the bond length in response to the external force. It was found that the MFJC model provides an excellent description of the elastic behavior of single-stranded (ss) DNA chains in comparison with experimental data at 25 °C with parameters $b = 15$ Å, $L = 27$ μm, and $k = 800$ pN.

(i) Show that the modified FJC equation is consistent with Hooke's law of elasticity in the limits of weak and strong stretching.
(ii) Estimate the reversible work to stretch a ssDNA chain from the fully relaxed state to 50% of its contour length.
(iii) Estimate the reversible work to stretch a ssDNA chain from the fully relaxed state to 100% of its contour length.
(iv) Assume that the stretch modulus is a constant, how does the reversible work of ssDNA stretching change as the temperature increases?

Solution:

(i) For a small force, $1 + f/k \approx 1$, the force–extension relation from the MFJC model is thus the same as that from the original FJC

$$f \approx \frac{3k_BT}{b}\frac{\langle R\rangle}{L}.$$

In the limit of strong stretching (viz., a large force), the MFJC model predicts

$$\langle R\rangle/L \approx \left[1 - \frac{1}{\beta fb}\right](f/k) \approx f/k \Rightarrow f \approx k\langle R\rangle/L.$$

Therefore, the modified FJC is consistent with Hooke's law at the limits of both small and large forces (but with different spring constants).

(ii) At $\langle R\rangle = 0.5L$, the stretching force can be calculated from the modified FJC equation, $f_{1/2} \approx 4.8879$ pN. As the ssDNA is stretched from $\langle R\rangle = 0$ to $\langle R\rangle = L/2$, the reversible work is

$$\Delta G = \int_0^{L/2} fd\langle R\rangle = f_{1/2}L/2 - \int_0^{f_{1/2}} \langle R\rangle df$$

$$= f_{1/2}L/2 - \int_0^{f_{1/2}} L\left[\coth(\beta fb) - \frac{1}{\beta fb}\right](1 + f/k)df \approx 3.0134 \times 10^{-17} \text{ joules.}$$

(iii) At $\langle R\rangle = L$, the stretching force can be calculated from the modified FJC equation, $f_L \approx 17.5811$ pN. As the ssDNA is stretched from $\langle R\rangle = 0$ to $\langle R\rangle = L$, the reversible work is

$$\Delta G = \int_0^{L} fd\langle R\rangle = f_L L - \int_0^{f_L} \langle R\rangle df$$

$$= f_L L - \int_0^{f_L} L\left[\coth(\beta fb) - \frac{1}{\beta fb}\right](1 + f/k)df \approx 126.6829 \times 10^{-17} \text{ joules.}$$

(iv) At the same percentage of extension, we have

$$f \approx \frac{3k_BT}{b}\frac{\langle R\rangle}{L},$$

suggesting that the force thus the reversible work increases with temperature.

5 Smith S. B., et al., "Overstretching B-DNA: The elastic response of individual double-stranded and single-stranded DNA molecules", *Science*, 271, 795–799 (1996).

Problem 3.38 Shokri et al. reported a quantitative model to describe the force-induced meting of a double-stranded (ds) DNA into single-stranded (ss) DNA.[6] The essential idea is that the force–extension curve for a partially melted DNA is given by a linear combination of those corresponding to ssDNA and dsDNA chains, i.e., the DNA length per base pair is given by

$$b = (1 - f_a) \times b_{dsDNA} + f_a \times b_{ssDNA},$$

where f_a denotes the fractional DNA melted, b_{dsDNA} and b_{ssDNA} are the per base pair length for dsDNA and ssDNA, respectively. Assume that the force–extension relation for dsDNA is described by the WLC model

$$\frac{f\xi_p}{k_BT} = \frac{b_{dsDNA}}{b_{dsNDA}^{max}} - \frac{1}{4} + \frac{1}{4(1 - b_{dsDNA}/b_{dsDNA}^{max})^2},$$

and that for the ssDNA is described by the modified freely-jointed chain (MFJC) model

$$b_{ssDNA} = b_{ssDNA}^{max}\left[\coth\left(\frac{fb_0}{k_BT}\right) - \frac{k_BT}{fb_0}\right](1 + f/k),$$

plot the force–extension curve (i.e., b versus f) for a partially melted DNA with $f_a = 0.5$ and compare the results with those for pure ssDNA and pure dsDNA. The following parameters are available from fitting to the experimental data at $T = 298.15$ K: for dsDNA, $\xi_p = 48$ nm, $b_{dsDNA}^{max} = 0.34$ nm; for ssDNA, $b_0 = 1.5$ nm, $k = 800$ pN, and $b_{ssDNA}^{max} = 0.56$ nm. What was the main conclusion of the single-molecule experiment?

Solution:

Figure 3.11 shows the force–extension relations predicted by the polymer models. For a partially melted DNA, the force on the dsDNA is twice that on ssDNA.

The polymer model supports force-induced dsDNA melting instead of a double-stranded form of overstretched DNA (vis., S-DNA).

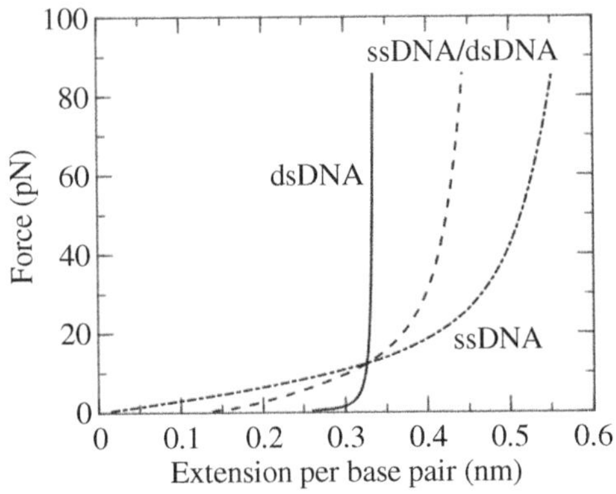

Figure 3.11 The force–extension relations predicted by the modified freely-jointed chain (MFJC) model for a single-stranded (ss) DNA and the worm-like chain (WLC) model for a double-stranded DNA. The partially melted DNA can be described by a linear combination of the MFJC and WLC models. (dsDNA.m)

6 Shokri et al., "DNA overstretching in the presence of glyoxal: Structural evidence of force-induced DNA melting", *Biophys. J.* 95, 1248–1255 (2008).

4

Thermodynamics of Photons, Electrons, and Phonons

Some Useful Mathematical Relations

(i) ***Gamma function***

The gamma function generalizes the factorial function to noninteger values. It is denoted by the Greek letter gamma, $\Gamma(z)$, and defined as

$$\Gamma(z) = \int_0^{\infty} x^{z-1} e^{-x} dx,$$

where z is a complex number with a positive real part. It can be shown that the gamma function satisfies the general relation

$$\Gamma(z+1) = z\Gamma(z).$$

For a positive integer n, the gamma function is equal to the factorial of $n-1$, i.e., $\Gamma(n) = (n-1)!$. Table 4.1 shows some specific values of the gamma function used in this chapter.

(ii) ***Polylogarithm function***

The polylogarithm function, denoted by $\mathrm{Li}_s(z)$, is a mathematical function that is defined as:

$$\mathrm{Li}_s(z) = \frac{1}{\Gamma(s)} \int_0^{\infty} \frac{t^{s-1}}{e^t/z - 1} dt = \sum_{k=1}^{\infty} \frac{z^k}{k^s},$$

where z is a complex number, and s is a real number. The polylogarithm function has the following properties:

(a) The derivative of a polylogarithm function satisfies

$$\frac{\partial \mathrm{Li}_s(z)}{\partial z} = \frac{\mathrm{Li}_{s-1}(z)}{z}.$$

(b) Asymptotic expansions at large and small values of z can be written as

$$-\mathrm{Li}_s(-z) = \begin{cases} z - \frac{z^2}{2^s} + \frac{z^3}{3^s} + \cdots, & z \ll 1, \\ \frac{(\ln z)^s}{\Gamma(s+1)} + \frac{\pi^2}{6}\frac{(\ln z)^{s-2}}{\Gamma(s-1)} + \frac{7\pi^4}{360}\frac{(\ln z)^{s-4}}{\Gamma(s-3)} + \cdots, & z \gg 1. \end{cases}$$

(c) Some special polylogarithm functions are

$$\mathrm{Li}_1(z) = -\ln(1-z);\ \mathrm{Li}_0(z) = \frac{z}{1-z};\ \mathrm{Li}_{-1}(z) = \frac{z}{(1-z)^2};\ \mathrm{Li}_{-2}(z) = \frac{z(1+z)}{(1-z)^3}.$$

$\mathrm{Li}_s(1) = \zeta(s)$ is the Riemann zeta function.

Solutions Manual to Fundamentals and Practice in Statistical Thermodynamics,
First Edition. Jianzhong Wu and John M. Prausnitz.

Table 4.1 A few specific values of the Γ-function.

Γ(−3/2)	Γ(−1/2)	Γ(1/2)	Γ(3/2)	Γ(5/2)	Γ(7/2)	Γ(1)	Γ(2)	Γ(3)	Γ(4)
$-4\sqrt{\pi}/3$	$-2\sqrt{\pi}$	$\sqrt{\pi}$	$\sqrt{\pi}/2$	$3\sqrt{\pi}/4$	$15\sqrt{\pi}/8$	1	1	2	6

(iii) ***Useful integration formula***

(a)

$$\int_0^\infty \frac{x^4 e^x}{(e^x-1)^2}dx = \frac{4\pi^4}{15}.$$

(b)

$$\int_0^\infty \frac{x^2 e^x}{(e^x-1)^2}dx = \frac{\pi^2}{3}.$$

(c)

$$\int_0^\infty \frac{x^3}{e^x-1}dx = \frac{\pi^4}{15}.$$

(d)

$$\int \frac{x^k}{e^x-1}dx = \int x^k \mathrm{Li}_0(e^{-x})dx = -\sum_{n=0}^{k} x^{k-n}\mathrm{Li}_{n+1}(e^{-x})\frac{\Gamma(k+1)}{\Gamma(k+1-n)}.$$

(e)

$$\int_x^\infty \frac{x^2}{e^x-1}dx = x^2\mathrm{Li}_1(e^{-x}) + 2x\mathrm{Li}_2(e^{-x}) + 2\mathrm{Li}_3(e^{-x}).$$

(f)

$$\int_x^\infty \frac{x^3}{e^x-1}dx = x^3\mathrm{Li}_1(e^{-x}) + 3x^2\mathrm{Li}_2(e^{-x}) + 6x\mathrm{Li}_3(e^{-x}) + 6\mathrm{Li}_4(e^{-x}).$$

(g)

$$\int_x^\infty x^2\ln(e^x-1)dx = x^3\mathrm{Li}_1(e^{-x}) + 4x^2\mathrm{Li}_2(e^{-x}) + 8x\mathrm{Li}_3(e^{-x}) + 8\mathrm{Li}_4(e^{-x}).$$

Problem 4.1 Consider a system with two identical particles that each may exist in one of two single-particle quantum states. How many microstates that the two particles may exist if they are (a) classical particles, (b) bosons, and (c) fermions?

Solution:

(a) If the two identical particles are classical, they can be in any of the two single-particle states. Therefore, there are three possible microstates, i.e., both in the first single-particle state, both in the second single-particle state, or one in each single-particle state.

(b) If the two identical particles are bosons, they are indistinguishable and can be in the following three possible states: both in the first single-particle state, both in the second single-particle state, or one in each single-particle state.

(c) If the two identical particles are fermions, they must obey the Pauli exclusion principle, which states that no two identical fermions can occupy the same quantum state simultaneously. Therefore, there is only one way to distribute the two fermions among the two single-particle states, i.e., one in the first state and one in the second state.

Problem 4.2 How many microstates may the two protons in a hydrogen molecule exist? What are the compositions of the spin isomers if these microstates are randomly distributed?

Solution:
Protons are fermions with a spin of $\hbar/2$. The two protons in a hydrogen molecule may exist in 4 microstates: the triplet state (with the degeneracy of 3) when the two spins are aligned in the same direction, $|\uparrow\rangle|\uparrow\rangle$, $|\downarrow\rangle|\downarrow\rangle$, $\frac{1}{\sqrt{2}}(|\uparrow\rangle|\downarrow\rangle + |\downarrow\rangle|\uparrow\rangle)$, and a singlet state when the spins are aligned in anti-parallel, $\frac{1}{\sqrt{2}}(|\uparrow\rangle|\downarrow\rangle - |\downarrow\rangle|\uparrow\rangle)$. Hydrogen in the triplet state is referred to as orthohydrogen, and that in the singlet state as parahydrogen. Under the random distribution, a hydrogen gas consists of approximately 75% orthohydrogen and 25% parahydrogen.

Problem 4.3 In a thermodynamic system of many particles, the quantum effects on the translational motion become significant when the thermal de Broglie wavelength Λ is comparable to the average inter-particle spacing. The condition $\rho\Lambda^3 = 1$ thus defines a so-called "degeneracy temperature"

$$T_0 = \frac{h^2}{2\pi m k_B}\rho^{2/3}$$

beyond which the quantum effects may be neglected. Assume that the molar volume of the system is 22.4 L/mol, estimate the degeneracy temperature for (a) hydrogen gas; (b) nitrogen gas; and (c) electron gas.

Solution:
The number density corresponding to the molar volume is

$$\rho = \frac{N_A \cdot 10^3}{22.4} = \frac{6.02214 \times 10^{23} \cdot 10^3}{22.4} = 2.6885 \times 10^{25}\text{m}^{-3}.$$

Substituting particle mass $m = 3.3734 \times 10^{-27}$, 4.6882×10^{-26}, and 9.1094×10^{-31} kg for hydrogen, nitrogen, and electron into

$$T_0 = \frac{h^2}{2\pi m k_B}\rho^{2/3}$$

gives the degeneracy temperatures of 0.0097, 0.1346, and 498.6 K, respectively. The degeneracy temperature is different for different systems, and depends on the density and mass of the particles, as well as the confinement of the system.

Problem 4.4 Based on the grand potential for noninteracting bosons derived in the text (Eq. (4.22))

$$\Omega = -\frac{k_B T V}{\Lambda^3}\text{Li}_{5/2}(z),$$

where $z = e^{\beta\mu}$ represents fugacity, and Λ is the thermal de Broglie wavelength, show the following expressions for

(i) pressure

$$P = \frac{k_B T}{\Lambda^3}\text{Li}_{5/2}(z),$$

(ii) internal energy

$$U = \frac{3k_BTV}{2\Lambda^3}\mathrm{Li}_{5/2}(z),$$

(iii) the average number of boson particles

$$N = \frac{V}{\Lambda^3}\mathrm{Li}_{3/2}(z),$$

(iv) entropy

$$S = \frac{k_BV}{\Lambda^3}\left[\frac{5}{2}\mathrm{Li}_{5/2}(z) - (\ln z)\mathrm{Li}_{3/2}(z)\right].$$

Solution:

We may follow standard statistical–mechanical equations to derive the above thermodynamic variables. Based on the analytical equation for the grand potential

$$\Omega = -\frac{k_BTV}{\Lambda^3}\mathrm{Li}_{5/2}(z),$$

where $z = e^{\beta\mu}$ and $\Lambda = h/\sqrt{2\pi mk_BT} = h\beta^{1/2}/\sqrt{2\pi m}$, it is straight forward to show

(i) pressure

$$P = -\left(\frac{\partial\Omega}{\partial V}\right)_{T,\mu} = \frac{k_BT}{\Lambda^3}\mathrm{Li}_{5/2}(z);$$

(ii) internal energy, which is also related to the grand potential

$$U = \left(\frac{\partial\beta\Omega}{\partial\beta}\right)_{V,\beta\mu}.$$

From

$$\beta\Omega = -\frac{V}{\Lambda^3}\mathrm{Li}_{5/2}(z) \quad \text{and} \quad \frac{\partial\Lambda}{\partial\beta} = \frac{k_BT}{2}\Lambda,$$

we obtain

$$U = \left(\frac{\partial\beta\Omega}{\partial\Lambda}\right)_{V,\beta\mu}\cdot\left(\frac{\partial\Lambda}{\partial\beta}\right)$$
$$= \frac{3V}{\Lambda^4}\mathrm{Li}_{5/2}(z)\frac{k_BT}{2}\Lambda = \frac{3Vk_BT}{2\Lambda^3}\mathrm{Li}_{5/2}(z) = \frac{3PV}{2};$$

(iii) the average number of particles

$$N = -\left(\frac{\partial\Omega}{\partial\mu}\right)_{T,V} = -\left(\frac{\partial\Omega}{\partial z}\right)_{T,V}\cdot\left(\frac{\partial z}{\partial\mu}\right)_{T,V} = \frac{k_BTV}{\Lambda^3}\cdot\frac{\partial\mathrm{Li}_{5/2}(z)}{\partial z}\cdot\left(\frac{\partial z}{\partial\mu}\right)_{T,V}$$
$$= \frac{k_BTV}{\Lambda^3 z}\mathrm{Li}_{3/2}(z)\cdot(\beta z) = \frac{V}{\Lambda^3}\mathrm{Li}_{3/2}(z),$$

where we have used $(\partial z/\partial\mu)_{T,V} = \beta z$ and the formula for the derivative of a polylogarithm function

$$\frac{\partial\mathrm{Li}_s(z)}{\partial z} = \frac{\mathrm{Li}_{s-1}(z)}{z};$$

(iv) entropy

$$S = (U + PV - N\mu)/T = (5U/3 - N\mu)/T$$
$$= \frac{5k_B}{2\Lambda^3}\mathrm{Li}_{5/2}(z) - \frac{Vk_B\ln z}{\Lambda^3}\mathrm{Li}_{3/2}(z) = \frac{k_BV}{\Lambda^3}\left[\frac{5}{2}\mathrm{Li}_{5/2}(z) - (\ln z)\mathrm{Li}_{3/2}(z)\right].$$

Problem 4.5 Show that, in the classical limit, the thermodynamic properties of noninteracting bosons are given by:

(i) chemical potential

$$\mu = k_B T \ln(\rho\Lambda^3),$$

(ii) grand potential

$$\Omega = -Nk_B T,$$

(iii) pressure

$$P = \frac{Nk_B T}{V},$$

(iv) internal energy

$$U = \frac{3}{2}Nk_B T,$$

(v) entropy

$$S = -Nk_B[\ln \rho\Lambda^3 - 5/2].$$

Solution:

In the classical limit, $e^{-\beta\mu} \gg 1$ thus $z \to 0$ and $\mathrm{Li}_s(z) \to z$.

(i) From the expression for average number of boson particles

$$N = \frac{V}{\Lambda^3}\mathrm{Li}_{3/2}(z) = \frac{V}{\Lambda^3}z = \frac{V}{\Lambda^3}e^{\beta\mu},$$

we obtain

$$z = e^{\beta\mu} = \frac{N\Lambda^3}{V}.$$

Let $\rho = N/V$, the above equation reduced to $\mu = k_B T \ln(\rho\Lambda^3)$, which is the familiar equation for the chemical potential of a monatomic ideal gas.

(ii) In the classical limit, the grand potential becomes

$$\Omega = -\frac{k_B TV}{\Lambda^3}\mathrm{Li}_{5/2}(z) = -\frac{k_B TVz}{\Lambda^3} = -\frac{k_B TV}{\Lambda^3}e^{\beta\mu} = -\frac{k_B TV}{\Lambda^3} \cdot \frac{N\Lambda^3}{V} = -Nk_B T.$$

(iii) In the classical limit, the pressure of noninteracting bosons is given by the ideal-gas law

$$P = -\frac{\Omega}{V} = \frac{Nk_B T}{V}.$$

(iv) The internal energy becomes identical to that of a monatomic ideal gas

$$U = \frac{3k_B T}{2\Lambda^3}\mathrm{Li}_{5/2}(z) = \frac{3k_B T}{2\Lambda^3}z = \frac{3k_B T}{2\Lambda^3} \cdot \frac{N\Lambda^3}{V} = \frac{3}{2}Nk_B T.$$

(v) The entropy is reduced to

$$S = \frac{k_B V}{\Lambda^3}\left[\frac{5}{2}\mathrm{Li}_{5/2}(z) - (\ln z)\mathrm{Li}_{3/2}(z)\right] = \frac{k_B V}{\Lambda^3}[z - z\ln z]$$

$$= \frac{k_B V}{\Lambda^3} \cdot \frac{N\Lambda^3}{V}\left[\frac{5}{2} - \ln\left(\frac{N\Lambda^3}{V}\right)\right] = -Nk_B[\ln \rho\Lambda^3 - 5/2],$$

which is the same as the Sackur–Tetrode equation.

Problem 4.6 Based on the grand potential for noninteracting fermions

$$\Omega = \frac{k_B TV}{\Lambda^3} \mathrm{Li}_{5/2}(-z),$$

where $z = e^{\beta\mu}$, show the following expressions for

(i) pressure

$$P = -\frac{k_B T}{\Lambda^3} \mathrm{Li}_{5/2}(-z),$$

(ii) internal energy

$$U = -\frac{3k_B TV}{2\Lambda^3} \mathrm{Li}_{5/2}(-z),$$

(iii) the average number of particles

$$N = -\frac{V}{\Lambda^3} \mathrm{Li}_{3/2}(-z),$$

(iv) entropy

$$S = -\frac{k_B V}{\Lambda^3} \left[\frac{5}{2}\mathrm{Li}_{5/2}(-z) - (\ln z)\mathrm{Li}_{3/2}(-z)\right].$$

Solution:

The above thermodynamic relations can be derived by following standard statistical–mechanical equations. Based on the analytical equation for the grand potential

$$\Omega = \frac{k_B TV}{\Lambda^3} \mathrm{Li}_{5/2}(-z),$$

where $z = e^{\beta\mu}$ and $\Lambda = h/\sqrt{2\pi m k_B T} = h\beta^{1/2}/\sqrt{2\pi m}$, we obtain

(i) pressure

$$P = -\left(\frac{\partial \Omega}{\partial V}\right)_{T,\mu} = -\frac{k_B T}{\Lambda^3} \mathrm{Li}_{5/2}(-z),$$

(ii) internal energy

$$U = \left(\frac{\partial \beta\Omega}{\partial \beta}\right)_{V,\beta\mu}.$$

Substituting into the above equation with

$$\beta\Omega = \frac{V}{\Lambda^3} \mathrm{Li}_{5/2}(-z) \quad \text{and} \quad \frac{\partial \Lambda}{\partial \beta} = \frac{k_B T}{2} \Lambda,$$

leads to

$$U = \left(\frac{\partial \beta\Omega}{\partial \Lambda}\right)_{V,\beta\mu} \cdot \left(\frac{\partial \Lambda}{\partial \beta}\right)$$
$$= -\frac{3V}{\Lambda^4} \mathrm{Li}_{5/2}(-z) \frac{k_B T}{2} \Lambda = -\frac{3Vk_B T}{2\Lambda^3} \mathrm{Li}_{5/2}(-z) = \frac{3PV}{2},$$

(iii) the average number of particles

$$N = -\left(\frac{\partial \Omega}{\partial \mu}\right)_{T,V} = -\left(\frac{\partial \Omega}{\partial z}\right)_{T,V} \cdot \left(\frac{\partial z}{\partial \mu}\right)_{T,V} = -\frac{k_B TV}{\Lambda^3} \cdot \frac{\partial \mathrm{Li}_{5/2}(-z)}{\partial z} \cdot \left(\frac{\partial z}{\partial \mu}\right)_{T,V}$$
$$= -\frac{k_B TV}{\Lambda^3 z} \mathrm{Li}_{3/2}(-z) \cdot (\beta z) = -\frac{V}{\Lambda^3} \mathrm{Li}_{3/2}(-z),$$

where we have used $(\partial z/\partial \mu)_{T,V} = \beta z$ and the formula for the derivative of a polylogarithm function

$$\frac{\partial \mathrm{Li}_s(z)}{\partial z} = \frac{1}{z}\mathrm{Li}_{s-1}(z),$$

(iv) entropy

$$S = (U + PV - N\mu)/T = (5U/3 - N\mu)/T$$
$$= -\frac{5k_B}{2\Lambda^3}\mathrm{Li}_{5/2}(-z) + \frac{Vk_B \ln z}{\Lambda^3}\mathrm{Li}_{3/2}(-z) = -\frac{k_BV}{\Lambda^3}\left[\frac{5}{2}\mathrm{Li}_{5/2}(-z) - (\ln z)\mathrm{Li}_{3/2}(-z)\right].$$

Problem 4.7 Show that, in the classical limit, the thermodynamic properties of noninteracting fermions are the same as those corresponding to bosons, i.e., both are reduced to the equations corresponding to an ideal gas of spherical particles:

(i) chemical potential

$$\mu = k_BT\ln(\rho\Lambda^3),$$

(ii) grand potential

$$\Omega = -Nk_BT,$$

(iii) pressure

$$P = \frac{Nk_BT}{V},$$

(iv) internal energy

$$U = \frac{3}{2}Nk_BT,$$

(v) entropy

$$S = -Nk_B[\ln \rho\Lambda^3 - 5/2].$$

Solution:

In the classical limit, $e^{-\beta\mu} \gg 1$ hence $z \to 0$ and $-\mathrm{Li}_s(-z) \to z$.

(i) As

$$N = -\frac{V}{\Lambda^3}\mathrm{Li}_{3/2}(-z) = \frac{V}{\Lambda^3}z = \frac{V}{\Lambda^3}e^{\beta\mu},$$

hence

$$z = e^{\beta\mu} = \frac{N\Lambda^3}{V}.$$

Let $\rho = N/V$ so that $\mu = k_BT\ln(\rho\Lambda^3)$.

(ii) The grand potential of noninteracting fermions is given by

$$\Omega = \frac{k_BTV}{\Lambda^3}\mathrm{Li}_{5/2}(-z) = -\frac{k_BTVz}{\Lambda^3} = -\frac{k_BTV}{\Lambda^3}e^{\beta\mu} = -\frac{k_BTV}{\Lambda^3}\cdot\frac{N\Lambda^3}{V} = -Nk_BT.$$

(iii) The ideal-gas law is recovered from

$$P = -\frac{\Omega}{V} = \frac{Nk_BT}{V}.$$

(iv) The internal energy becomes

$$U = -\frac{3k_BT}{2\Lambda^3}\mathrm{Li}_{5/2}(-z) = \frac{3k_BT}{2\Lambda^3}z = \frac{3k_BT}{2\Lambda^3}\cdot\frac{N\Lambda^3}{V} = \frac{3}{2}Nk_BT.$$

(v) The entropy is obtained from

$$S = -\frac{k_B V}{\Lambda^3}\left[\frac{5}{2}\mathrm{Li}_{5/2}(-z) + (\ln z)\mathrm{Li}_{3/2}(-z)\right] = \frac{k_B V}{\Lambda^3}[z - z\ln z]$$

$$= \frac{k_B V}{\Lambda^3}\cdot\frac{N\Lambda^3}{V}\left[\frac{5}{2} - \ln\left(\frac{N\Lambda^3}{V}\right)\right] = -Nk_B[\ln \rho\Lambda^3 - 5/2],$$

which is the same as the Sackur–Tetrode equation.

Problem 4.8 What is the change in entropy when one mole of noninteracting bosons expand isothermally to double the volume at $z_0 = 0.5$. How does the entropy change vary with the initial fugacity z_0?

Solution:

The average number of bosons depends on the system volume V and fugacity z

$$N = \frac{V}{\Lambda^3}\mathrm{Li}_{3/2}(z).$$

At $z_0 = 0.5$, the reduced density of bosons is

$$\rho\Lambda^3 = \mathrm{Li}_{3/2}(0.5) \approx 0.6248.$$

After the isothermal expansion to double the volume, the reduced density is halved while Λ is unchanged. Therefore, the fugacity can be solved from

$$\mathrm{Li}_{3/2}(z) = \frac{1}{2}\mathrm{Li}_{3/2}(z_0) \approx 0.3124,$$

which gives $z \approx 0.2796$. Thus, the entropy change due to the isothermal expansion is

$$\Delta S = S - S_0 = \frac{2k_B V}{\Lambda^3}\xi(z) - \frac{k_B V}{\Lambda^3}\xi(z_0)$$

$$= [2\xi(z) - \xi(z_0)]\frac{k_B V}{\Lambda^3} = [2\xi(z) - \xi(z_0)]\frac{Nk_B}{\mathrm{Li}_{3/2}(z_0)},$$

where $\xi(z) \equiv \frac{5}{2}\mathrm{Li}_{5/2}(z) - (\ln z)\mathrm{Li}_{3/2}(z)$. In dimensionless units, the above equation becomes

$$\Delta S/R = \frac{2\xi(z) - \xi(z_0)}{\mathrm{Li}_{3/2}(z_0)},$$

where is $R = k_B N_A$ is the ideal-gas constant. Plugging the numbers gives $\Delta S/R \approx 0.7218$.

As shown in Figure 4.1, the entropy change varies with the initial fugacity z_0 (e.g., $\Delta S/R = 0.7715$ when $z_0 = 0.9$). By contrast, the isothermal expansion of a classical ideal gas to double the volume is a constant. At $z_0 = 0$, the system reaches the classical limit so that the result predicted by the classical ideal gas law, $\Delta S/R = \ln 2 \approx 0.6931$, is reproduced. Note that the entropy change for the expansion of noninteracting bosons is always larger than that for a classical ideal gas. When $z_0 > 1$, the system undergoes the Bose–Einstein condensation.

Problem 4.9 What is the change in entropy when one mole of noninteracting fermions expand isothermally to double the volume at $z_0 = 0.5$. How does the entropy change with the initial fugacity z_0?

Solution:

At $z_0 = 0.5$, the reduced density of fermions is

$$\rho\Lambda^3 = -\mathrm{Li}_{3/2}(-0.5) = 0.4299.$$

Figure 4.1 The entropy change versus the initial fugacity (z_0) when one mole of noninteracting bosons expand isothermally to double the volume. At $z_0 = 0$, the result is the same as that predicted by the classical ideal gas law (BosExpansion.m).

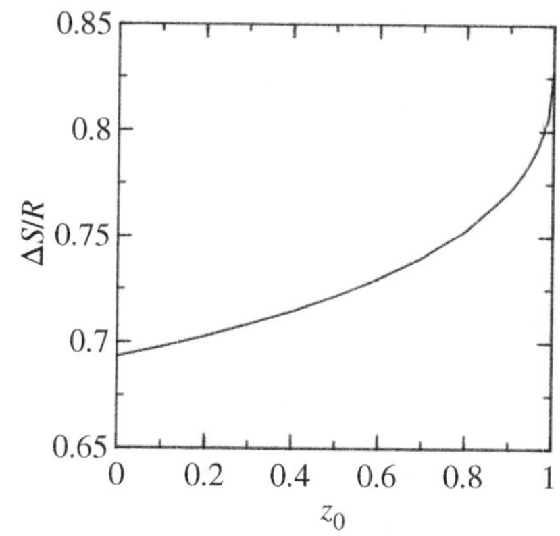

After isothermal expansion to double the volume, the reduced density is halved while Λ remains unchanged. Hence, the fugacity becomes

$$\mathrm{Li}_{3/2}(-z) = \mathrm{Li}_{3/2}(-z_0)/2 = 0.2149,$$

which gives $z = 0.2319$. Thus, the entropy change is

$$\Delta S = S - S_0 = -\frac{2k_BV}{\Lambda^3}\chi(z) + \frac{k_BV}{\Lambda^3}\chi(z_0)$$
$$= -[2\chi(z) - \chi(z_0)]\frac{k_BV}{\Lambda^3} = [2\chi(z) - \chi(z_0)]\frac{Nk_B}{\mathrm{Li}_{3/2}(-z_0)},$$

where $\chi(z) = \frac{5}{2}\mathrm{Li}_{5/2}(-z) - (\ln z)\mathrm{Li}_{3/2}(-z)$. In dimensionless units,

$$\Delta S/R = \frac{2\chi(z) - \chi(z_0)}{\mathrm{Li}_{3/2}(-z_0)},$$

where is $R = k_BN_A$ is the ideal-gas constant. Plugging the numbers gives $\Delta S/R = 0.6746$.

As for bosons, the entropy change depends on the initial fugacity z_0 (e.g., $\Delta S/R = 0.6632$ when $z_0 = 0.9$). At $z_0 = 0$, the system reaches the classical limit so that the result predicted by the classical ideal gas law, $\Delta S/R = \ln 2 \approx 0.6931$, is reproduced. As shown in Figure 4.2, the entropy change for the expansion of noninteracting fermions is always less than that for a classical ideal gas. Unlike bosons, fermions do not experience condensation so that the fugacity is not bounded. In the limit of $z_0 \to \infty$ ($T = 0$ K), there is no entropy increase due to the expansion, i.e., $\Delta S/R = 0$.

Figure 4.2 The entropy change versus fugacity (z_0) when one mole of noninteracting fermions expand isothermally to double the volume. At $z_0 = 0$, the result is the same as that predicted by the classical ideal gas law (FermiExpansion.m).

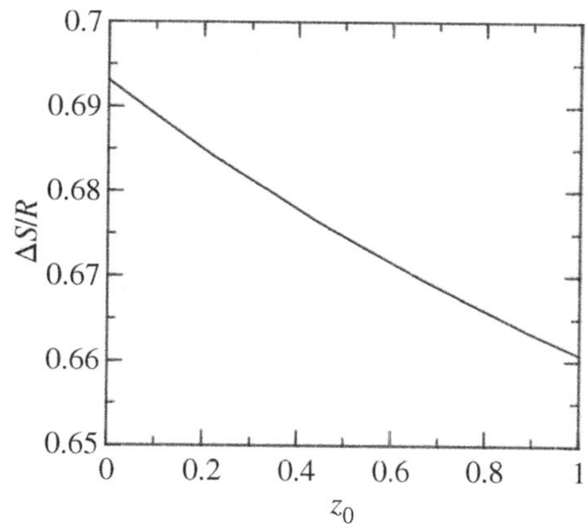

Problem 4.10 Calculate the entropy of mixing at constant temperature and pressure for equal moles of

(i) two boson gases;
(ii) two fermion gases;
(iii) two classical ideal gases.

In all cases, assume that the fugacity of both species before the mixing is $z_0 = 0.5$.

Solution:

(i) Consider isothermal and isobaric mixing of one mole of boson A with one mole of boson B. Before the mixing, bosons A and B have the same temperature T, pressure P, and fugacity z_0. Because

$$P = k_B T \mathrm{Li}_{5/2}(z)/\Lambda^3,$$

bosons A and B must have the same thermal wavelength (more precisely, the same temperature, pressure, and fugacity imply that the two bosons must have the same mass). From

$$N = V \mathrm{Li}_{3/2}(z)/\Lambda^3 \quad \text{and} \quad N_A = N_B,$$

we can also infer that systems A and B must also have the same volume V.
After mixing at constant temperature and pressure, the relation $N = V\mathrm{Li}_{3/2}(z)/\Lambda^3$ applies to both species A and B, and $N_A = N_B$ implies that the fugacities of the two bosons in the mixture must be equal, $z_A = z_B = z$ (but they are not the same as their initial value z_0 due to the volume change). Because the bosons are assumed noninteracting, the pressure of the mixture is additive

$$P = \frac{k_B T}{\Lambda^3}\mathrm{Li}_{5/2}(z_0) = \frac{k_B T}{\Lambda^3}\mathrm{Li}_{5/2}(z_A) + \frac{k_B T}{\Lambda^3}\mathrm{Li}_{5/2}(z_B).$$

Therefore, the new fugacity can be obtained from $\mathrm{Li}_{5/2}(z_0) = 2\mathrm{Li}_{5/2}(z)$, which gives $z = 0.2796$. The volume of the mixture is thus given by

$$V = N\Lambda^3/\mathrm{Li}_{3/2}(z) = 2N\Lambda^3/\mathrm{Li}_{3/2}(z_0),$$

which is twice of the initial volume of A or B.
The entropy of mixing can then be calculated from the summation of the entropy change for the expansion of A and B to double the volume as discussed in Problem 4.8:

$$\frac{\Delta S}{(N_A + N_B)k_B} = 0.7218.$$

(ii) We can repeat the above procedure but with the thermodynamic functions for fermions to give

$$\frac{\Delta S}{(N_A + N_B)k_B} = 0.6746.$$

The number is the same as that discussed in Problem 4.9.

(iii) The entropy of mixing for equal moles of classical particles is a constant

$$\frac{\Delta S}{(N_A + N_B)k_B} = \ln 2 \approx 0.6931.$$

Problem 4.11 Show that the Gibbs paradox can be resolved by using the analytical expressions of entropy of non-interacting quantum particles. Does the paradox hold in the classical limit?

Solution:
Consider the isothermal and isobaric "mixing" of two systems 1 and 2 of identical non-interacting bosons, one with initial volume V_1 and the other with volume V_2. Because pressure P is a unique function of fugacity z and temperature T, isothermal and isobaric mixing implies that the fugacity of the bosons in subsystem 1 is the same as that of subsystem 2 and they remain unchanged after mixing. Therefore, the entropy of mixing is

$$\Delta S = S_M - S_1 - S_2 = \frac{k_B(V_M - V_1 - V_2)}{\Lambda^3}\xi(z),$$

where $\xi(z) = \frac{5}{2}\mathrm{Li}_{5/2}(z) - (\ln z)\mathrm{Li}_{3/2}(z)$. From the conversation of mass $N_M = N_1 + N_2$ and $N = V\mathrm{Li}_{3/2}(z)/\Lambda^3$, we have $V_M = V_1 + V_2$. Thus, $\Delta S = 0$.

The above procedure can be repeated for the entropy of mixing for two systems of non-interacting fermions. For both cases, we obtain $\Delta S = 0$ in the classical limit, affirming that there is no entropy increase due to the isothermal–isobaric mixing of identical particles.

Problem 4.12 Consider the density of a noninteracting boson gas as a function of the fugacity z

$$\rho\Lambda^3 = \mathrm{Li}_{3/2}(z).$$

As shown in Figure 4.3, the polylogarithm function $\mathrm{Li}_{3/2}(z)$ has real values only when $z \leq 1$. Therefore, there exists a maximum number density beyond which the bosons will condense into a single-particle state with zero momentum (viz., the ground state with zero energy). This phenomenon is known as the Bose–Einstein condensation.

(i) Verify that $(\partial\rho/\partial\mu)_T$ diverges at $z = 1$;

(ii) At a given temperature, the Bose–Einstein condensation takes place at density

$$\rho_c = \zeta(3/2)/\Lambda^3,$$

where $\zeta(3/2) \approx 2.612$ is the Riemann zeta function.

(iii) At a given density, the Bose–Einstein condensation takes place at temperature

$$T_c = \frac{h^2}{2\pi m k_B}\left[\rho/\zeta(3/2)\right]^{2/3}.$$

(iv) Sketch the $\rho \sim T$ phase diagram for the Bose–Einstein condensation.

(v) At $T < T_c$, the fraction of the Bose–Einstein condensate (BEC) is given by

$$N_0/N = 1 - (T/T_c)^{3/2}.$$

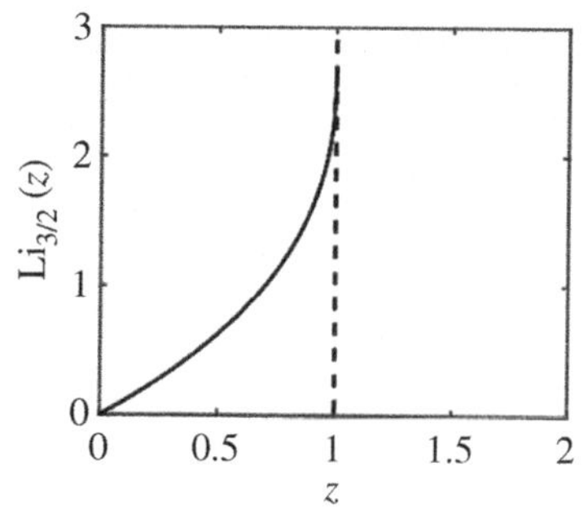

Figure 4.3 The reduced density of a boson gas, $\rho\Lambda^3 = \mathrm{Li}_{3/2}(z)$, versus the fugacity z.

Solution:

(i) First, we express the partial derivative using the chain rule

$$\left(\frac{\partial \rho}{\partial \mu}\right)_T = \left(\frac{\partial \rho}{\partial z}\right)_T \cdot \left(\frac{\partial z}{\partial \mu}\right)_T = \frac{1}{\Lambda^3}\frac{\partial \mathrm{Li}_{3/2}(z)}{\partial z} \cdot \left(\frac{\partial z}{\partial \mu}\right)_T.$$

Substituting $(\partial z/\partial \mu)_T = \beta z$ and

$$\frac{\partial \mathrm{Li}_{3/2}(z)}{\partial z} = \frac{1}{z}\mathrm{Li}_{1/2}(z)$$

into the above equation leads to

$$\left(\frac{\partial \rho}{\partial \mu}\right)_T = \frac{1}{\Lambda^3}\mathrm{Li}_{1/2}(z).$$

Because $\mathrm{Li}_{1/2}(1) = \infty$, $(\partial\rho/\partial\mu)_T$ diverges at $z = 1$, implying that the system becomes unstable.

(ii) Since $\mathrm{Li}_{3/2}(1) = \zeta(3/2) \approx 2.612$, $\rho\Lambda^3 = \mathrm{Li}_{3/2}(z)$ predicts that, at a given temperature, the Bose–Einstein condensation takes place at density

$$\rho_c = \zeta(3/2)/\Lambda^3.$$

(iii) At a given density, $\rho\Lambda^3 = \mathrm{Li}_{3/2}(1)$ predicts that, the Bose–Einstein condensation takes place at temperature

$$T_c = \frac{h^2}{2\pi m k_B}\left[\rho/\zeta(3/2)\right]^{2/3}.$$

(iv) Equation $\rho\Lambda^3 = \mathrm{Li}_{3/2}(z)$ sets a boundary between the Bose gas and the BECs. For each T, there is a unique density corresponding to $z = 1$, which leads to the phase boundary illustrated in Figure 4.4.

(v) At $T < T_c$, the boson particles are distributed into the gas and BEC phases:

$$N = N_0 + \frac{V}{\Lambda^3}\mathrm{Li}_{3/2}(1).$$

Thus

$$N_0/N = 1 - \frac{\mathrm{Li}_{3/2}(1)}{\rho\Lambda^3} = 1 - \frac{\rho\Lambda_c^3}{\rho\Lambda^3} = 1 - (T/T_c)^{3/2},$$

where Λ_c is the thermal wavelength at T_c.

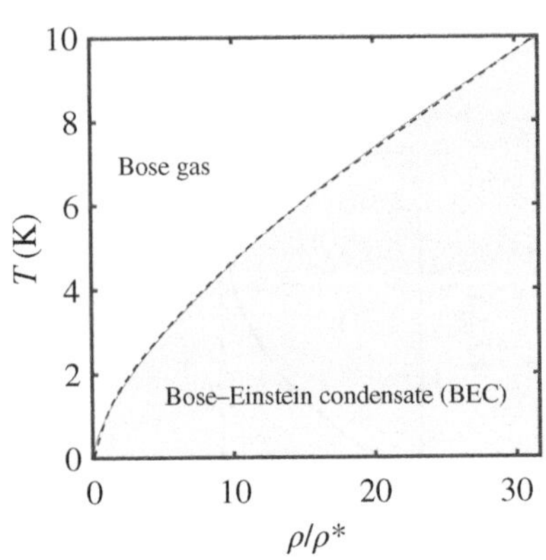

Figure 4.4 The phase boundary between Bose gas and Bose–Einstein condensate (BEC). Here $\rho^* \equiv (h/2\pi m k_B)^{3/2}/\zeta(3/2)$.

Problem 4.13 Estimate the number of photons in an oven at 500 K with the volume of 0.5 m^3. What is the molar internal energy of the photons?

Solution:
At $T = 500$ K, the photon density is

$$\rho \approx 2.0287 \times 10^7 T^3 = 2.0287 \times 10^7 \times 500^3 = 2.5359 \times 10^{15} \text{m}^{-3}.$$

Thus, the number of photons in the oven is

$$N = \rho \times V = 2.5359 \times 10^{15} \times 0.5 = 1.2679 \times 10^{15}.$$

The molar internal energy of the photons is

$$u \approx 2.7012RT = 11.233 \text{ kJ/mol}.$$

Problem 4.14 Consider a photon gas as an open system of volume V, temperature T, and chemical potential μ.

(i) Write an expression for the grand canonical partition function;

(ii) Show that the average number of photons derived from the grand canonical ensemble is the same as that from the canonical ensemble

$$\langle N \rangle = 16\zeta(3)\pi V \left(\frac{k_B T}{hc} \right)^3,$$

where $\zeta(3) \approx 1.2021$ is the Riemann zeta function.

(iii) Show that the relative variance in the number of photons is a constant

$$\frac{\langle N^2 \rangle - \langle N \rangle^2}{\langle N \rangle} = \frac{\pi^2}{6\zeta(3)}.$$

(iv) Explain why the photon gas does not satisfy the fluctuation-compressibility theorem

$$\frac{\langle N^2 \rangle - \langle N \rangle^2}{\langle N \rangle^2} = \frac{k_B T}{V} \kappa_T,$$

where κ_T stands for the isothermal compressibility.

Solution:
This problem and the solution are adapted from an article by H. S. Leff.[1]

(i) The grand canonical partition function is defined as

$$\Xi(T, V, \mu) = \sum_\nu e^{-\beta(E_\nu - N_\nu \mu)}.$$

For a photon gas, the chemical potential is $\mu = 0$. Thus, the grand canonical partition function is the same as the canonical partition function

$$\Xi(T, V, \mu) = \sum_\nu e^{-\beta E_\nu} = \prod_{\alpha=1}^{\infty} \frac{1}{1 - e^{-\beta \epsilon_\alpha}},$$

where ϵ_α stands for energy levels.

1 Leff H. S., "Fluctuations in particle number for a photon gas", *American Journal of Physics*, 83, 362–365 (2015).

(ii) According to the grand canonical ensemble, the average number of photons is given by

$$\langle N\rangle = \frac{\partial \ln \Xi}{\partial \beta\mu} = \frac{1}{\Xi}\sum_{\nu} N_\nu e^{-\beta(E_\nu - N_\nu \mu)} = \frac{1}{Q}\sum_{\nu} N_\nu e^{-\beta E_\nu},$$

for $\mu = 0$. As discussed in the text (Section 4.3), the canonical ensemble average is

$$\langle N\rangle = 16\zeta(3)\pi V\left(\frac{k_B T}{hc}\right)^3.$$

(iii) The variance in the number of photons can be calculated from

$$\langle N^2\rangle - \langle N\rangle^2 = \left\langle \left(\sum_{\alpha=1}^{\infty} n_\alpha\right)^2 \right\rangle - \left(\sum_{\alpha=1}^{\infty}\langle n_\alpha\rangle\right)^2,$$

where n_α is the number of photons at energy level α. As discussed in Section 4.3, we have

$$\langle n_\alpha\rangle = \frac{1}{e^{\beta\epsilon_\alpha} - 1} \qquad \text{and} \qquad \langle n_\alpha n'_\alpha\rangle - \langle n_\alpha\rangle\langle n'_\alpha\rangle = 0 \qquad \text{for} \qquad \alpha \neq \alpha'.$$

Thus

$$\langle N^2\rangle - \langle N\rangle^2 = \sum_{\alpha=1}^{\infty}\sum_{\alpha'=1}^{\infty}\left(\langle n_\alpha n'_\alpha\rangle - \langle n_\alpha\rangle\langle n'_\alpha\rangle\right) = \sum_{\alpha=1}^{\infty}\left(\langle n_\alpha^2\rangle - \langle n_\alpha\rangle^2\right). \tag{A}$$

Note that, similar to the fluctuation of particle numbers in a grand-canonical ensemble,

$$\langle n_\alpha^2\rangle - \langle n_\alpha\rangle^2 = -\frac{\partial\langle n_\alpha\rangle}{\partial\beta\epsilon_\alpha} = \frac{e^{\beta\epsilon_\alpha}}{(e^{\beta\epsilon_\alpha} - 1)^2}.$$

Eq. (A) becomes

$$\langle N^2\rangle - \langle N\rangle^2 = \sum_{\alpha=1}^{\infty}\frac{e^{\beta\epsilon_\alpha}}{(e^{\beta\epsilon_\alpha} - 1)^2} = \int_0^\infty d\upsilon g(\upsilon)\frac{e^{\beta h\upsilon}}{(e^{\beta h\upsilon} - 1)^2} = \frac{8\pi V}{c^3}\int_0^\infty d\upsilon\upsilon^2\frac{e^{\beta h\upsilon}}{(e^{\beta h\upsilon} - 1)^2}$$

$$= \frac{8\pi V}{h^3c^3\beta^3}\int_0^\infty dx\frac{x^2e^x}{(e^x - 1)^2} = \frac{8\pi^3 V}{3h^3c^3\beta^3},$$

where $g(\upsilon) = 8\pi\upsilon^2 V/c^3$ is the density of states for noninteracting photons, $x = \beta h\upsilon$ and

$$\int_0^\infty \frac{x^2e^x}{(e^x - 1)^2}dx = \frac{\pi^2}{3}.$$

Therefore, the relative fluctuation in the number of photons is

$$\frac{\langle N^2\rangle - \langle N\rangle^2}{\langle N\rangle} = \frac{8\pi^3 V}{3h^3c^3\beta^3} \times \frac{(\beta hc)^3}{16\zeta(3)\pi V} = \frac{\pi^2}{6\zeta(3)} \approx 1.3684.$$

(iv) For a photon gas, the chemical potential is fixed at $\mu = 0$ and both the density and pressure depend only on temperature. As a result, the isothermal compressibility

$$\kappa_T = \frac{1}{\rho}\left(\frac{\partial\rho}{\partial P}\right)_T$$

cannot be defined.

Problem 4.15 Suppose that a photon have an effective mass of $m = h\upsilon/c^2$, derive the pressure of a photon gas from the kinetic theory

$$P = \frac{1}{3}\rho m\langle\upsilon^2\rangle.$$

Show that the above result is equivalent to that derived from the thermodynamic route discussed in Section 4.3.

(Hint: For photons, the average velocity is given by the speed of light, i.e., $\langle\upsilon^2\rangle = c^2$, and $mc^2 = h\upsilon$ corresponds to the energy per photon.)

Solution:

The kinetic theory predicts the number of particles per unit area per unit time arriving at a surface perpendicular to the x-axis is ρv_{x+} where v_{x+} is the velocity in the x direction moving toward the surface. The elastic collision leads to a momentum change of $2mv_{x+}$ per particle. Therefore, the surface experiences a pressure of

$$P = \langle (2mv_{x+}) \cdot (\rho v_{x+}) \rangle = 2m\rho \langle v_{x+}^2 \rangle.$$

For a uniform system, $\langle v_{x+}^2 \rangle = \frac{1}{6}\langle v^2 \rangle$ thus

$$P = \frac{1}{3}\rho m\langle v^2 \rangle = \frac{1}{3}\rho\langle mc^2 \rangle = \frac{1}{3}\rho h\langle \upsilon \rangle.$$

Using $m\langle v^2 \rangle = \langle mc^2 \rangle = h\langle \upsilon \rangle$ leads to

$$P = \frac{1}{3}\rho h\langle \upsilon \rangle.$$

Meanwhile, the mean frequency is related to the specific internal energy

$$U/V = Nh\langle \upsilon \rangle / V = \rho h\langle \upsilon \rangle.$$

Therefore, the combination of the above two equations yields

$$P = \frac{U}{3}/V = \frac{1}{3}\sigma T^4,$$

which is the same as that derived from the thermodynamic route discussed in Section 4.3.

Problem 4.16 The entropy analysis plays a major role in understanding the thermodynamic efficiency of solar energy conversion including photosynthesis and biological evolutions.

(i) Verify the following equation for the spectral entropy density for a system of noninteracting photons

$$s_\lambda = \frac{8\pi k_B}{\lambda^4}\left[\frac{\beta hc/\lambda}{e^{\beta hc/\lambda} - 1} - \ln(1 - e^{-\beta hc/\lambda})\right].$$

(ii) Show that the wavelength at the maximum spectral entropy density is given by

$$\lambda_s^{max} \approx 3.0029 \times 10^{-3}/T,$$

where λ_s^{max} has the units of meter and T is in kelvin.

Solution:

(i) We may deduce the frequency dependence of the entropy density for photons following a procedure similar to that used for the spectral internal energy. Specifically, we first express the entropy per unit volume as an integration of the light frequency

$$S/V = \int_0^\infty s_\upsilon d\upsilon = \frac{8\pi k_B}{c^3}\int_0^\infty \left[\frac{\beta h\upsilon}{e^{\beta h\upsilon} - 1} - \ln(1 - e^{-\beta h\upsilon})\right]\upsilon^2 d\upsilon.$$

Hence, the entropy spectral density is given by

$$s_\upsilon = \frac{8\pi k_B}{c^3}\left[\frac{\beta h\upsilon^3}{e^{\beta h\upsilon} - 1} - \upsilon^2\ln(1 - e^{-\beta h\upsilon})\right].$$

(ii) To find the wavelength at the maximum entropy density, we need an expression for s_λ. Because $\upsilon = c/\lambda$ and $d\upsilon = -(c/\lambda^2)d\lambda$, we obtain

$$s_\lambda = \frac{8\pi k_B}{\lambda^4}\left[\frac{\beta hc/\lambda}{e^{\beta hc/\lambda} - 1} - \ln(1 - e^{-\beta hc/\lambda})\right].$$

Using the dimensionless variable $x \equiv \beta hc/\lambda$, we can write the spectral entropy density as

$$s_\lambda = \frac{8\pi k_B}{(\beta hc)^4} f(x),$$

where

$$f(x) = \frac{x^5}{e^x - 1} - x^4 \ln(1 - e^{-x}).$$

The maximum of s_λ takes place when

$$\frac{df(x)}{dx} = \frac{5x^4}{e^x - 1} - 4x^3 \ln(1 - e^{-x}) + \frac{x^4 e^{-x}}{e^{-x} - 1} - \frac{x^5 e^x}{(e^x - 1)^2} = 0,$$

which has a numerical solution $x = 4.7913$. Therefore, the wavelength at maximum s_λ occurs at

$$\lambda_s^{max} = \frac{hc}{k_B x T} \approx 3.0029 \times 10^{-3}/T,$$

where λ_s^{max} has the units of meter and T is in kelvin.

Problem 4.17 Plot the spectral irradiance of solar radiation above Earth's atmosphere based on black-body radiation. What is the intensity of solar energy at the top of the atmosphere? Could you estimate the solar energy intensity at the sea level?

Solution:
Planck's law can be used to predicts the intensity of spectral radiation energy at Sun's surface

$$I(\lambda) = \frac{2\pi hc^2}{\lambda^5} \frac{1}{(e^{\beta hc/\lambda} - 1)}.$$

The radiation intensity is greatly reduced above the atmosphere of Earth because the radiation area increases with distance

$$I_a(\lambda) = I(\lambda) \times \frac{R_S^2}{R_{SE}^2},$$

where $R_S = 6.957 \times 10^8$ m is Sun's radius, and $R_{SE} = 1.4959787 \times 10^{11}$ m is the distance between Sun and Earth. With $T = 5778$ K for the temperature at Sun's surface, Figure 4.5 shows the spectral intensity right above Earth's atmosphere.

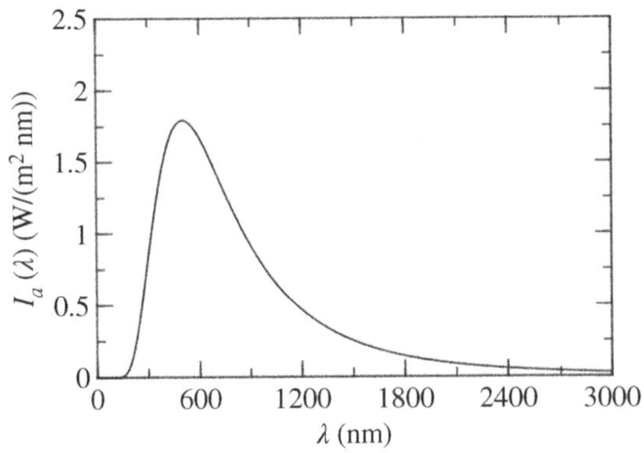

Figure 4.5 Solar irradiance spectrum at the top of the atmosphere predicted by Planck's law (solar.m).

The intensity of solar energy at the top of the atmosphere can be calculated with the Stefan–Boltzmann law:

$$I_a = \Theta \times T^4 \frac{R_S^2}{R_{SE}^2} = 5.6704 \times 10^{-8} \times 5778^4 \times \frac{(6.957 \times 10^8)^2}{(1.4959787 \times 10^{11})^2} = 1.3668\,\text{kW/m}^2.$$

This value, commonly known as the "solar constant," is remarkably close to the result measured by satellite, 1.361 kW/m^2.

Assuming that, on average, 80% of the solar energy reaches the ground, we can predict that the ground-level solar power is 1.0935 kW/m^2. In a cloudless sky when Sun is at the zenith, direct solar radiation is about 1.050 kW/m^2.

Problem 4.18 The photon flux is important in determining the number of electrons and hence the current produced from a solar cell.

(i) Show that the flux of photons from the surface of a blackbody at temperature T is given by

$$J = \frac{4\pi\zeta(3)k_B^3}{c^2h^3}T^3.$$

(ii) Estimate the photon flux at the sea level from the solar radiation.
(iii) Plot the spectral photon flux from the Sun at the sea level.

Solution:

(i) The photon flux is defined as the number of photons per second per unit area. Consider radiation from a differential surface of a backbody, the number of photons emitted is proportional to the speed of light c, the number density of photons ρ, and a coefficient of 1/4 accounting for contributions from all solid angles:

$$J = \frac{1}{4}c\rho.$$

At temperature T, the photon density is given by

$$\rho = \frac{16\pi\zeta(3)k_B^3}{c^3h^3}T^3.$$

Hence, the number of photons per second per unit area is

$$J = \frac{4\pi\zeta(3)k_B^3}{c^2h^3}T^3.$$

(ii) We can estimate the photon flux at the sea level from the solar radiation

$$J_0 = 0.8 \times \frac{4\pi\zeta(3)k_B^3}{c^2h^3}T^3 \times \frac{R_S^2}{R_{SE}^2}.$$

Substituting to above equation with $T = 5778\,\text{K}$ for the temperature at Sun's surface, $R_S = 6.957 \times 10^8$ m for Sun's radius, and $R_{SE} = 1.496 \times 10^{11}$ m for the distance between Sun and Earth, we obtain

$$J_0 = 8.4264 \times 10^{-3}\text{mol/(m}^2\,\text{s)}.$$

(iii) The spectral photon flux can be derived from the spectral irradiance

$$J(\lambda) = \frac{\lambda}{hc} \times I(\lambda) = \frac{\lambda}{hc} \times \frac{2\pi hc^2}{\lambda^5}\frac{1}{(e^{\beta hc/\lambda} - 1)} = \frac{2\pi c}{\lambda^4}\frac{1}{e^{\beta hc/\lambda - 1}}.$$

Alternatively, we may obtain the equation following a procedure similar that used for spectral energy or entropy

$$\rho = \int_0^\infty \rho(\lambda)d\lambda = \frac{8\pi}{c^3\beta^3h^3}\int_0^\infty \frac{x}{e^x-1}dx.$$

Because $x = \beta hc/\lambda$, we have

$$dx = -\frac{\beta hc}{\lambda^2}d\lambda.$$

Plugging it into the above equation for ρ leads to

$$\rho(\lambda) = \frac{8\pi}{\lambda^4}\frac{1}{e^{\beta hc/\lambda}-1}.$$

Therefore, the spectral photon flux is given by

$$J(\lambda) = \frac{c}{4}\rho(\lambda) = \frac{2\pi c}{\lambda^4}\frac{1}{e^{\beta hc/\lambda}-1}.$$

Assuming that, on average, 80% of the solar energy reaches the ground, the photon flux at the ground level is given by

$$J_0(\lambda) = 0.8 \times J(\lambda) \times \frac{R_S^2}{R_{SE}^2}. \tag{B}$$

Figure 4.6 plots the spectral photon flux at the sea level versus the solar wave length predicted by Eq. (B).

Problem 4.19 Verify that, for a single p-n junction solar cell, the Shockley–Queisser (SQ) efficiency (i.e., the ultimate efficiency) is given by

$$\eta_{SQ} = \frac{15}{\pi^4}x_g f_g(x_g),$$

where $x_g = \beta hc/\lambda_g$ with λ_g being the threshold photon wavelength, and

$$f_g(x) = x^2\mathrm{Li}_1(e^{-x}) + 2x\mathrm{Li}_2(e^{-x}) + 2\mathrm{Li}_3(e^{-x}).$$

Solution:

From the blackbody model for the spectral photon flux (Problem 4.18iii),

$$J(\lambda) = \frac{2\pi c}{\lambda^4}\frac{1}{e^{\beta hc/\lambda-1}},$$

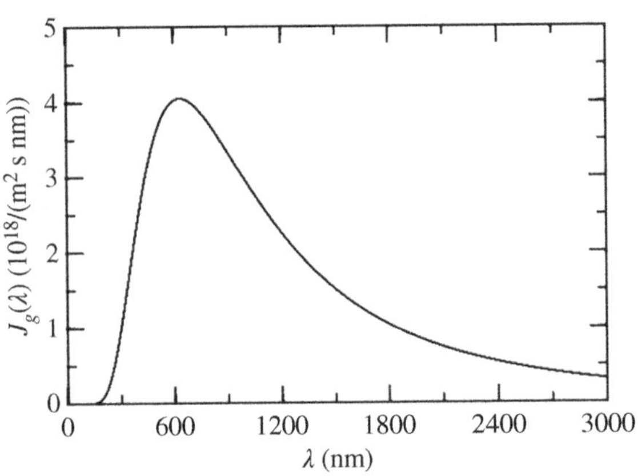

Figure 4.6 The spectral photon flux at the sea level (solar.m).

we can find the flux of photons with the energy larger than the bandgap energy $\epsilon_g = hc/\lambda_g$:

$$J_g = \int_0^{\lambda_g} J(\lambda)d\lambda = \int_0^{\lambda_g} \frac{2\pi c}{\lambda^4} \frac{1}{e^{\beta hc/\lambda - 1}} d\lambda.$$

Let $x = \beta hc/\lambda$, then $d\lambda = -(\beta hc/x^2)dx$. Hence, the above equation becomes

$$J_g = \frac{2\pi c}{(\beta hc)^4} \int_{x_g}^{\infty} \frac{x^4}{e^x - 1} \frac{\beta hc}{x^2} dx = \frac{2\pi}{\beta^3 h^3 c^2} \int_{x_g}^{\infty} \frac{x^2}{e^x - 1} dx$$
$$= \frac{2\pi}{\beta^3 h^3 c^2} f_g(x_g),$$

where $f_g(x) = x^2 \mathrm{Li}_1(e^{-x}) + 2x\mathrm{Li}_2(e^{-x}) + 2\mathrm{Li}_3(e^{-x})$, and the second line is obtained from the integration formula

$$\int_x^{\infty} \frac{x^2}{e^x - 1} dx = x^2 \mathrm{Li}_1(e^{-x}) + 2x\mathrm{Li}_2(e^{-x}) + 2\mathrm{Li}_3(e^{-x}).$$

The maximum electric power per unit area from the solar energy conversion is given by the product of the bandgap energy and J_g:

$$P_{max} = \frac{hc}{\lambda_g} \times \frac{2\pi}{\beta^3 h^3 c^2} f_g(x_g) = \frac{2\pi}{\beta^4 h^3 c^2} x_g \cdot f_g(x_g).$$

The SQ efficiency is defined as

$$\eta_{SQ} \equiv \frac{P_{max}}{I},$$

where the total solar energy flux is I is assumed to be given by the Stephan–Boltzmann law

$$I = \Theta T^4 = \frac{2\pi^5 k_B^4}{15 h^3 c^2}.$$

Substitutions with the expressions for P_{max} and I yields

$$\eta_{SQ} = \frac{2\pi x_g f_g(x_g)}{\beta^4 h^3 c^2} \times \frac{15 h^3 c^2}{2\pi^5 k_B^4 T^4} = \frac{15}{\pi^4} x_g f_g(x_g).$$

Problem 4.20 The global warming may be understood in terms of the enhanced greenhouse effect, i.e., the increase of the concentrations of greenhouse gases such as carbon dioxide, methane, nitrous oxide due to human activities – particularly burning fossil fuels (coal, oil, and natural gas). Because the Earth temperature is regulated by the exchange of energy with the universe mainly through radiations, Planck's law may shed useful light on the problem.

(i) Estimate the Earth temperature if there is no atmosphere; Assume that 30% of the Sun light would be reflected into space.
(ii) Plot the spectral energy density due the Earth's radiation by assuming it as a blackbody with an average temperature of $T = 255$ K.
(iii) Compare the spectrum with that of the sunlight.

Solution:

(i) The Sun emits a power equal to its surface area $4\pi R_S^2$ multiplied by ΘT_S^4. At a distance R_{SE} from the center of the Sun, this power is uniformly distributed over a sphere with surface area $4\pi R_{SE}^2$, and the Earth is only able to "catch" this power over its projected area πR_E^2, where R_E represents the radius of the Earth.

The Earth temperature can be estimated by balancing its radiation energy with the solar energy it receives. Assume that 30% of the Sun light would be reflected into space if there is no atmosphere, we would have

$$I_E \times 4\pi R_E^2 = 0.7 \times I_S \times \frac{R_S^2}{R_{SE}^2} \times \pi R_E^2,$$

where

$$T_E = 0.9147 \times T_S \times \sqrt{\frac{R_S}{2R_{SE}}}.$$

Using $T_S = 5778$ K for Sun's surface temperature, $R_S = 6.957 \times 10^8$ m for its radius and $R_{SE} = 1.496 \times 10^{11}$ m for the distance between Sun and Earth, we obtain $T_E = 254.9$ K. Without the greenhouse effect, the Earth would be very cold!

(ii) Figure 4.7 shows the spectral energy density for the radiation from the Earth by assuming that it can be described in terms of the black-body model with an average surface temperature of $T = 255$ K.

(iii) As shown in FIgure 4.7, the thermal radiation from the Earth is very weak in comparison to that from the Sun. The spectrum is almost all at infrared wavelengths, much longer than that in sunlight. Longer wavelengths in the infrared and beyond (with $\lambda > 700$ nm) are mostly absorbed by vibrational, rotational, and translational modes of gas molecules such as CO_2 and H_2O in the atmosphere. These molecules typically have an uneven distribution of electrons and thus have a permanent dipole moment. They interact with radiation by absorbing and emitting photons, leading to the direct conversion of the radiation into thermal energy without chemical alterations. At $T = 255$ K, Wien's displacement law predicts that the spectral energy density for the Earth exhibits a maximum radiation around wavelength $\lambda_{max} = 3 \times 10^3/T = 11.8$ μm, which is close to a major adsorption wavelength for CO_2 molecules (~ 15 μm). This explains why CO_2 is an important contributor to the greenhouse effect.

Problem 4.21 The chemical potential of photons is nonzero if light is emitted from the excited states of matter. For example, in a light emitting diode (LED), an electron (e), and a hole (h) recombine to produce a photon (γ), which has a chemical potential

$$\mu_\gamma = \mu_e + \mu_h,$$

where μ_e and μ_h are the chemical potentials of the electron and hole, respectively.

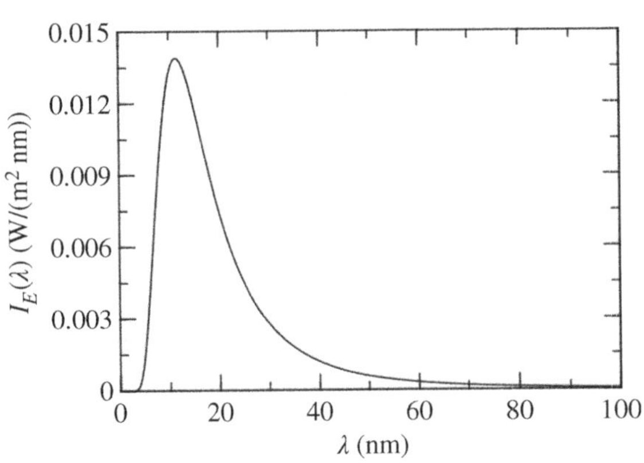

Figure 4.7 The spectral energy density for the radiation reemitted by the Earth (solar.m).

(i) Show that the spectral energy density for the LED light is given by

$$I(\omega) = \frac{\hbar\omega^3}{\pi^2 c^3} \frac{1}{e^{\beta(\hbar\omega - \mu_\gamma)} - 1},$$

where ω stands for the angular frequency of the light.

(ii) Assume that photons cannot be emitted by a LED device if the energy is below the bandgap, i.e., $\hbar\omega < \epsilon_g$, and the emissivity is perfect otherwise. Use $\mu_\gamma = \epsilon_g = 1$ eV and $T = 300$ K, plot the spectral radiation energy density of the LED light and compare it with that from a blackbody radiation at the same temperature.

Solution:

The chemical potential of radiation was discussed in details by Würfel.[2] Herrmann and Würfel gave a pedagogical introduction of nonzero chemical potential for photons.[3]

(i) Photons are bosons thereby following the Bose–Einstein distribution for the single-particle microstates:

$$n_\upsilon = \frac{1}{e^{\beta(\epsilon_\upsilon - \mu_\gamma)} - 1} = \frac{1}{e^{\beta(h\upsilon - \mu_\gamma)} - 1},$$

where $\upsilon = \omega/2\pi$ stands for the photon frequency. The density of states depends only on the photon energy independent of the chemical potential (the same as that for the blackbody radiation):

$$g(\upsilon) = \frac{8\pi\upsilon^2 V}{c^3}.$$

Therefore, the spectral energy density is

$$I(\upsilon) = g(\upsilon) \times n_\upsilon h\upsilon / V = \frac{8\pi\upsilon^2}{c^3} \times \frac{h\upsilon}{e^{\beta(h\upsilon - \mu_\gamma)} - 1}.$$

In terms of angular frequency ω, we can rewrite the above equation using $I(\upsilon)d\upsilon = I(\omega)d\omega$ with $\upsilon = \omega/2\pi$, $d\upsilon = d\omega/2\pi$, and $\hbar = h/2\pi$:

$$I(\omega) = \frac{1}{2\pi} \frac{8\pi(\omega/2\pi)^2}{c^3} \times \frac{h\omega/2\pi}{e^{\beta[(h\omega/2\pi) - \mu_\gamma]} - 1} = \frac{\hbar\omega^3}{\pi^2 c^3} \frac{1}{e^{\beta(\hbar\omega - \mu_\gamma)} - 1}.$$

(ii) Assume that the emissivity of LED is a step function, i.e.,

$$\alpha(\omega) = \begin{cases} 0, & \hbar\omega < \epsilon_g, \\ 1, & \text{otherwise.} \end{cases}$$

Accordingly, the spectral energy density is given by

$$I(\omega) = \alpha(\omega) \frac{\hbar\omega^3}{\pi^2 c^3} \frac{1}{e^{\beta(\hbar\omega - \mu_\gamma)} - 1}.$$

This may be compared with that from a blackbody radiation

$$I_0(\omega) = \frac{\hbar\omega^3}{\pi^2 c^3} \frac{1}{e^{\beta\hbar\omega} - 1}.$$

Figure 4.8 shows the energy spectrum in comparison with that of a blackbody at the same temperature. It shows that LED emits strong luminescent light but with a threshold wavelength controlled by the bandgap energy.

---

2 Würfel P., "The chemical potential of radiation", *J. Phys. C: Solid State Phys.*, 15: 3967–3985 (1982).

3 Herrmann F. and Würfel P., "Light with nonzero chemical potential", *Am. J. Phys.*, 73, 717 (2005).

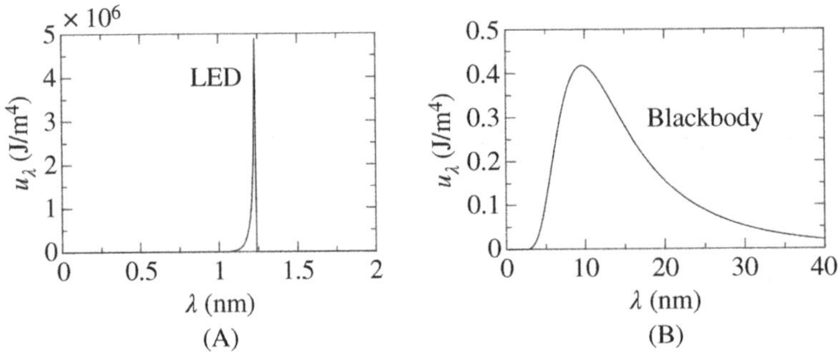

Figure 4.8 The photon emission spectrum of a LED with 1 V applied voltage (A, $\mu_\gamma = 1$ eV) in comparison with the blackbody spectrum (B) at $T = 300$ K. The bandgap energy is $\epsilon_g = 1$ eV (LED.m).

Problem 4.22 Verify Petela's equation for the thermodynamic efficiency for the whole spectrum of the solar energy

$$\eta_T \equiv \frac{\vartheta}{I_{in}} = 1 - \frac{4}{3}\left(\frac{T_0}{T}\right) + \frac{1}{3}\left(\frac{T_0}{T}\right)^4,$$

where ϑ stands for exergy, I_{in} is total solar radiation, $T = 5778$ K, and T_0 is ambient temperature.

Solution:

The exergy of solar radiation over the whole spectrum is given by (Eq. (4.74–76))

$$\begin{aligned}\vartheta &= (I_{in} - I_{out}) - T_0(S_{in} - S_{out}) \\ &= \left(\Theta T^4 - \Theta T_0^4\right) - T_0\left(\frac{4}{3}\Theta T^3 - \frac{4}{3}\Theta T_0^3\right).\end{aligned}$$

Thus

$$\begin{aligned}\eta_T \equiv \frac{\vartheta}{I_{in}} &= \frac{\left(\Theta T^4 - \Theta T_0^4\right) - T_0\left(\frac{4}{3}\Theta T^3 - \frac{4}{3}\Theta T_0^3\right)}{\Theta T^4} \\ &= 1 - \frac{4}{3}\left(\frac{T_0}{T}\right) + \frac{1}{3}\left(\frac{T_0}{T}\right)^4.\end{aligned}$$

Problem 4.23 The photosynthetically active radiation (PAR) region is defined as the spectral range of solar radiation that can be utilized by photosynthetic organisms (or devices) during photosynthesis.

(i) The efficiency of this radiation is usually defined as

$$\eta_{PAR} = \frac{\int_{\lambda_1}^{\lambda_2} I(\lambda)d\lambda}{\int_0^\infty I(\lambda)d\lambda}.$$

What is the value of η_{PAR} if the spectral energy of solar radiation can be represented by that of a blackbody at $T = 5778$ K and that the photosynthesis takes place in the region from $\lambda_1 = 400$ to $\lambda_2 = 700$ nm?

(ii) As a consequence of the entropy increase, not all the radiation reaching the Earth's surface is useful to produce work. The second law PAR efficiency is defined as

$$\eta_{PAR}^{2nd} = \frac{\int_{\lambda_1}^{\lambda_2} \vartheta(\lambda)d\lambda}{\int_0^\infty I(\lambda)d\lambda}.$$

The spectral distribution of the exergy of radiation is defined as

$$\vartheta(\lambda) \equiv I(\lambda, T) - I(\lambda, T_0) - T_0[S(\lambda, T) - S(\lambda, T_0)],$$

where $T_0 = 298.15\,\text{K}$ stands for the ambient temperature, and $S(\lambda, T)$ is the spectral entropy density. What is the value of η_{PAR}^{2nd} under the condition the same as that in part (i)?

Solution:

(i) From the spectral energy radiation

$$I(\lambda) = \frac{2\pi hc^2}{\lambda^5}\frac{1}{(e^{\beta hc/\lambda} - 1)},$$

we can determine the fractional emissive power of solar radiation analytically

$$\begin{aligned} I_g &= \int_0^\lambda I(\lambda)d\lambda = \int_0^\lambda \frac{2\pi hc^2}{\lambda^5}\frac{1}{(e^{\beta hc/\lambda} - 1)}d\lambda \\ &= \frac{2\pi hc^2}{(\beta hc)^5}\int_\infty^x \frac{x^5}{e^x - 1}\frac{(-\beta hc)}{x^2}dx = \frac{2\pi}{\beta^4 h^3 c^2}\int_x^\infty \frac{x^3}{e^x - 1}dx \\ &= \frac{2\pi}{\beta^4 h^3 c^2}[x^3\text{Li}_1(e^{-x}) + 3x^2\text{Li}_2(e^{-x}) + 6x\text{Li}_3(e^{-x}) + 6\text{Li}_4(e^{-x})] \\ &= \frac{15\Theta T^4}{\pi^4}[x^3\text{Li}_1(e^{-x}) + 3x^2\text{Li}_2(e^{-x}) + 6x\text{Li}_3(e^{-x}) + 6\text{Li}_4(e^{-x})], \end{aligned}$$

where $x = \beta hc/\lambda$, $dx = -(\beta hc/\lambda^2)d\lambda$ and $\Theta = \frac{2\pi^5 k_B^4}{15h^3c^2}$. Because $\int_0^\infty I(\lambda)d\lambda = \Theta T^4$, we have

$$\eta_{PAR} = \frac{15}{\pi^4}[x^3\text{Li}_1(e^{-x}) + 3x^2\text{Li}_2(e^{-x}) + 6x\text{Li}_3(e^{-x}) + 6\text{Li}_4(e^{-x})]\Big|_{x_1}^{x_2},$$

where $x_1 = \beta hc/\lambda_1$ and $x_2 = \beta hc/\lambda_2$. Substituting with $T = 5778\,\text{K}$, $\lambda_1 = 400\,\text{nm}$ and $\lambda_2 = 700\,\text{nm}$ yields $\eta_{PAR} = 0.3667$.

(ii) The spectral distribution of the exergy of radiation is defined as

$$\vartheta(\lambda, T) \equiv I(\lambda, T) - I(\lambda, T_0) - T_0[S(\lambda, T) - S(\lambda, T_0)],$$

where $T_0 = 298.15\,\text{K}$ stands for the ambient temperature. From Problem 4.16, we can derive the spectral entropy radiation $S(\lambda, T)$

$$S(\lambda, T) = \frac{c}{4}s(\lambda, T) = \frac{2\pi ck_B}{(\beta hc)^4}\left[\frac{x^5}{e^x - 1} - x^4\ln(1 - e^{-x})\right],$$

where $x = \beta hc/\lambda$.

To derive an analytical expression for the second law PAR efficiency, let

$$\begin{aligned} f(\lambda, T) &\equiv I(\lambda, T) - T_0 S(\lambda, T) \\ &= \frac{2\pi hc^2}{(\beta hc)^5}\frac{x^5}{(e^x - 1)} - \frac{2\pi hc^2 T_0/T}{(\beta hc)^5}\left[\frac{x^5}{e^x - 1} - x^4\ln(1 - e^{-x})\right] \\ &= \frac{2\pi(1 - T_0/T)}{\beta^5 h^4 c^3}\frac{x^5}{e^x - 1} - \frac{2\pi T_0}{\beta^5 h^4 c^3 T}x^4\ln(1 - e^{-x}) \end{aligned}$$

and

$$F(\lambda, T) \equiv \int_0^{\lambda} f(\lambda, T) d\lambda.$$

The integration of the first term in $f(\lambda, T)$ leads to

$$(1 - T_0/T)\frac{15\Theta T^4}{\pi^4}[x^3\mathrm{Li}_1(e^{-x}) + 3x^2\mathrm{Li}_2(e^{-x}) + 6x\mathrm{Li}_3(e^{-x}) + 6\mathrm{Li}_4(e^{-x})],$$

and the integration of the second term gives

$$\begin{aligned}
\int_0^{\lambda} \frac{2\pi T_0}{\beta^5 h^4 c^3 T} x^4 \ln(1 - e^{-x}) d\lambda &= \int_x^{\infty} \frac{2\pi T_0}{\beta^5 h^4 c^3 T} x^4 \ln(1 - e^{-x}) \frac{\beta hc}{x^2} dx \\
&= \frac{2\pi T_0}{\beta^4 h^3 c^2 T} \int_x^{\infty} x^2 \ln(1 - e^{-x}) dx \\
&= \frac{15\Theta T^3 T_0}{\pi^4}[x^3\mathrm{Li}_1(e^{-x}) + 4x^2\mathrm{Li}_2(e^{-x}) + 8x\mathrm{Li}_3(e^{-x}) + 8\mathrm{Li}_4(e^{-x})].
\end{aligned}$$

Hence

$$\begin{aligned}
F(\lambda, T) = \frac{15\Theta T^3}{\pi^4}&[(T - T_0)x^3\mathrm{Li}_1(e^{-x}) + (3T - 7T_0)x^2\mathrm{Li}_2(e^{-x}) \\
&+ (6T - 14T_0)x\mathrm{Li}_3(e^{-x}) + (6T - 14T_0)\mathrm{Li}_4(e^{-x})].
\end{aligned}$$

Finally, with the above expressions, we can derive the second law PAR efficiency

$$\begin{aligned}
\eta_{PAR}^{2nd} &= \frac{\int_{\lambda_1}^{\lambda_2} \vartheta(\lambda) d\lambda}{\int_0^{\infty} I(\lambda) d\lambda} = \frac{\{F(\lambda, T) - F(\lambda, T_0)\}\Big|_{\lambda_1}^{\lambda_2}}{\Theta T^4} \\
&= \Bigg\{ \frac{15}{\pi^4}[(1 - T_0/T)x^3\mathrm{Li}_1(e^{-x}) + (3 - 7T_0/T)x^2\mathrm{Li}_2(e^{-x}) \\
&\quad + (6 - 14T_0/T)x\mathrm{Li}_3(e^{-x}) + (6 - 14T_0/T)\mathrm{Li}_4(e^{-x})] \\
&\quad + \frac{15}{\pi^4}(T_0/T)^4[4x^2\mathrm{Li}_2(e^{-x}) + 8x\mathrm{Li}_3(e^{-x}) + 8\mathrm{Li}_4(e^{-x})] \Bigg\}\Bigg|_{x_1}^{x_2}.
\end{aligned}$$

Plug in the numbers giving $\eta_{PAR}^{2nd} \approx 0.3312$. The second law PAR efficiency is more suitable for the analysis of photosynthesis efficiency because it takes into account the fact that not all the radiation is useful. Furthermore, it accounts for the environmental temperature.

Problem 4.24 It has been shown that the upper limit of the solar energy conversion efficiency is achieved when the energy loss per photon is about $U_{loss} \sim 0.3$–0.4 eV, the exact value depending on the threshold wavelength λ_g. Estimate the threshold wavelength and the efficiency of solar energy conversion for the following fuel generation reactions by photosynthesis for both one-photo and two-photo systems. For simplicity, assume quantum yield $\eta_Q = 1$ and $U_{loss} = 0.37$ eV for all the photosynthesis reactions. Here, n_e is the number of electrons which should be transferred in an electrochemical reaction for the reaction as written.

	Reaction	n_e	ΔG^o (kJ/mol)
(1)	$H_2O(l) \rightarrow H_2(g)+\frac{1}{2}O_2(g)$	2	237
(2)	$CO_2(g) \rightarrow CO(g)+\frac{1}{2}O_2(g)$	2	257
(3)	$CO_2(g) + H_2O(l) \rightarrow HCOOH(l)+\frac{1}{2}O_2(g)$	2	286
(4)	$CO_2(g) + H_2O(l) \rightarrow HCOH(g) + O_2(g)$	4	522
(5)	$CO_2(g) + 2H_2O(l) \rightarrow CH_3OH(l)+\frac{3}{2}O_2(g)$	6	703
(6)	$CO_2(g) + 2H_2O(l) \rightarrow CH_4(g)+2O_2(g)$	8	818
(7)	$N_2 + 3H_2O(l) \rightarrow 2NH_3(g)+\frac{3}{2}O_2(g)$	6	678
(8)	$CO_2(g) + H_2O(l) \rightarrow \frac{1}{6}\ C_6H_{12}O_6(l) + O_2(g)$	4	480

Solution:

The threshold wavelength for the photochemical reactions can be calculated from (Eq. (4.84))

$$\lambda_g = \frac{hc}{\Delta G^0/n + U_{loss}},$$

where ΔG^0 is the Gibbs energy of reaction, and n is the number of photons participating in the reaction. With the assumptions of $U_{loss} = 0.37$ eV and quantum yield $\eta_Q = 1$, the efficiency of solar energy conversion can be calculated from (Eq. (4.85))

$$\eta = \frac{J_g \Delta G^0/n}{I_{in}},$$

where J_g is the flux of photons with wavelength less than λ_g, I_{in} is the total solar energy incidence, and $\Delta G^0/n$ is the free energy per photon. Using J_g and I_{in} from the blackbody radiation model, we can readily calculate λ_g and η for both one-photo ($n = 1$) and two-photo ($n = 2$) systems. It should be noted that n_e is related to the electric voltage but has no effect on the energy conversion efficiency.

Reaction	n_e	ΔG^o kJ/mol	$\lambda_g(1)$ nm	$\lambda_g(2)$ nm	$\eta(1)$	$\eta(2)$
$H_2O(l) \rightarrow H_2(g)+\frac{1}{2}O_2(g)$	2	237	438.7	775.8	0.1188	0.2915
$CO_2(g) \rightarrow CO(g)+\frac{1}{2}O_2(g)$	2	257	408.7	728.5	0.0956	0.2806
$CO_2(g) + H_2O(l) \rightarrow HCOOH(l)+\frac{1}{2}O_2(g)$	2	286	371.9	669.4	0.0683	0.2611
$CO_2(g) + H_2O(l) \rightarrow HCOH(g) + O_2(g)$	4	522	214.5	403.2	0.0024	0.0914
$CO_2(g) + 2H_2O(l) \rightarrow CH_3OH(l)+\frac{3}{2}O_2(g)$	6	703	161.9	309.0	1.3×10^{-4}	0.0296
$CO_2(g) + 2H_2O(l) \rightarrow CH_4(g)+2O_2(g)$	8	818	140.1	269.0	1.8×10^{-5}	0.0133
$N_2 + 3H_2O(l) \rightarrow 2NH_3(g)+\frac{3}{2}O_2(g)$	6	678	167.6	319.3	1.9×10^{-4}	0.0349
$CO_2(g) + H_2O(l) \rightarrow \frac{1}{6}\ C_6H_{12}O_6(l) + O_2(g)$	4	480	232.0	433.9	0.0046	0.1151

Problem 4.25 Verify the following expression for the velocity distribution of free electrons

$$p(\mathbf{v}) = \frac{(m/\hbar)^3}{4\pi^3 \rho} \frac{1}{\exp[\beta(mv^2/2 - \mu)] + 1},$$

and show that it reduces to the Maxwell–Boltzmann distribution at high temperature

$$p(\mathbf{v}) = \left(\frac{m}{2\pi k_B T}\right)^{3/2} \exp[-\beta mv^2/2].$$

Solution:

The number of free electrons in each single-particle energy state ϵ is described by the Fermi–Dirac equation

$$n(\epsilon) = \frac{1}{\exp[\beta(\epsilon - \mu)] + 1}.$$

The electron energy is related to the wavevector by $\epsilon = \hbar^2 k^2/2m$. As discussed in the text, the number of energy states for electrons in a cubic box of length L with the wavevector in the range of $\mathbf{k}$ and $\mathbf{k} + d\mathbf{k}$ is

$$2 \times \left(\frac{L}{2\pi}\right)^3 d\mathbf{k} = \frac{V}{4\pi^3} d\mathbf{k},$$

where prefactor 2 accounts for two spin states. Thus, the number of free electrons in this energy range is

$$\frac{V}{4\pi^3} \frac{d\mathbf{k}}{\exp[\beta(\hbar^2 k^2/2m - \mu)] + 1} \equiv Np(\mathbf{k})d\mathbf{k},$$

where N is the total number of electrons in the system, and $p(\mathbf{k})$ stands for the probability density of *a single electron* in the $\mathbf{k}$ space. Replacing wavevector $\mathbf{k}$ with velocity $\mathbf{v}$, i.e., $\mathbf{k} = m\mathbf{v}/\hbar$ and $d\mathbf{k} = (m/\hbar)^3 d\mathbf{v}$, we have $p(\mathbf{k})d\mathbf{k} = p(\mathbf{v})d\mathbf{v}$ and

$$p(\mathbf{v}) = \frac{V}{N} \frac{(m/\hbar)^3}{4\pi^3} \frac{1}{\exp[\beta(mv^2/2 - \mu)] + 1} = \frac{(m/\hbar)^3}{4\pi^3 \rho} \frac{1}{\exp[\beta(mv^2/2 - \mu)] + 1}.$$

In the classical limit, $\beta(mv^2/2 - \mu)] \gg 1$ and $\exp(\beta\mu) = \rho\Lambda^3/2$ so that

$$p(\mathbf{v}) = \frac{(m/\hbar)^3}{4\pi^3 \rho} \frac{1}{\exp[\beta(mv^2/2 - \mu)] + 1} = \frac{(m/\hbar)^3}{4\pi^3 \rho} (\rho\Lambda^3/2) \exp[-\beta(mv^2/2)]$$

$$= \frac{m^3}{(2\pi\hbar)^3} \left(\frac{h^2}{2\pi m k_B T}\right)^{3/2} \exp[-\beta m v^2/2] = \left(\frac{m}{2\pi k_B T}\right)^{3/2} \exp[-\beta m v^2/2].$$

Note that the velocity of electrons depends on both temperature and density, while the Maxwell–Boltzmann distribution depends only on temperature. As an example, Figure 4.9 shows the velocity distribution of free electrons in copper at two representative temperatures.

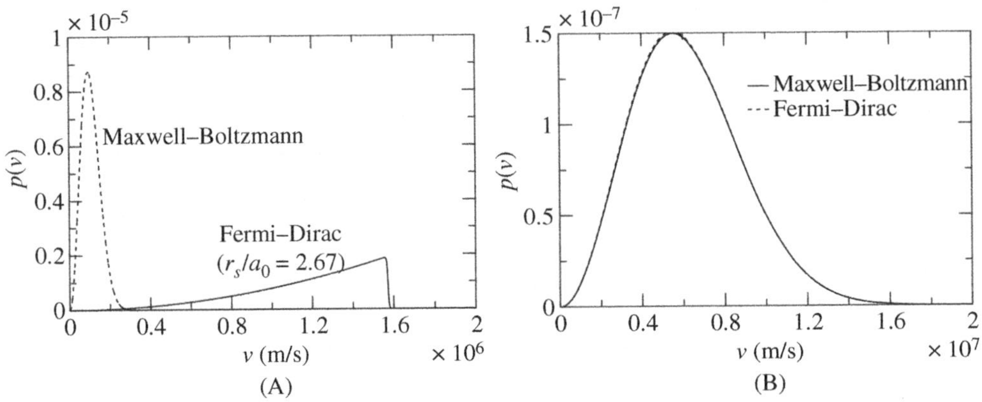

Figure 4.9 The distribution of electron speed $v = |\mathbf{v}|$, $p(v) = 4\pi v^2 p(\mathbf{v})$, in copper at 298.15 K (A) and at 10^6 K (B) (FermiGasvelocity.m).

Problem 4.26 For free electrons at 0 K, show that the internal energy can be expressed as:

$$U = \frac{3N}{5}\epsilon_F,$$

where ϵ_F is the Fermi energy.

Solution:

The free-electron model predicts that the internal energy is given by

$$U = \int d\epsilon \epsilon g(\epsilon) n(\epsilon).$$

At 0 K, $n(\epsilon)$ is a step function, and the average number of electrons is

$$N = \frac{8\pi V}{3}\left(\frac{2m\epsilon_F}{h^2}\right)^{3/2}.$$

Meanwhile $g(\epsilon) = 4\pi V \epsilon^{1/2}\left(\frac{2m}{h^2}\right)^{3/2}$ and thus

$$U = \int_0^{\epsilon_F} d\epsilon \epsilon g(\epsilon) = \int_0^{\epsilon_F} d\epsilon \epsilon 4\pi V \epsilon^{1/2}\left(\frac{2m}{h^2}\right)^{3/2} = 4\pi V\left(\frac{2m}{h^2}\right)^{3/2} \times \frac{2}{5}\epsilon_F^{5/2} = \frac{3N}{5}\epsilon_F.$$

Alternatively, we may use the asymptotic limit of the polylogarithm function

$$U = -\frac{3k_BTV}{\Lambda^3}\mathrm{Li}_{5/2}(-z).$$

As $T \to 0$ K, we have $z \to \infty$ and $\mathrm{Li}_{5/2}(-z) \to (\ln z)^{5/2}/\Gamma(7/2)$ so that

$$U = 3k_BTV\left(\frac{2\pi m k_B T}{h^2}\right)^{3/2}\frac{(\beta\mu)^{2/5}}{\Gamma(7/2)} = 3V\left(\frac{2\pi m}{h^2}\right)^{3/2}\frac{\mu^{2/5}}{15\sqrt{\pi}/8} = \frac{8\pi V}{5}\left(\frac{2m\mu}{h^2}\right)^{3/2}\mu = \frac{3N}{5}\epsilon_F,$$

where $\mu = \epsilon_F$ at 0 K.

Problem 4.27 Show that, for free electrons at low temperature,

(i) the chemical potential can be written as

$$\mu/\epsilon_F = 1 - \frac{\pi^2}{12}\left(\frac{k_BT}{\epsilon_F}\right)^2 + \cdots,$$

(ii) the internal energy is given by

$$\frac{U}{N\epsilon_F} = \frac{3}{5}\left[1 + \frac{5\pi^2}{12}\left(\frac{k_BT}{\epsilon_F}\right) + \cdots\right],$$

where ϵ_F is the Fermi energy.

Solution:

(i) The electron chemical potential is related to the average number of free electrons

$$N = -\frac{2V}{\Lambda^3}\mathrm{Li}_{3/2}(-z).$$

At low temperature, the polylogarithm function can be approximated by its expansion at the large values of fugacity z:

$$-\mathrm{Li}_{3/2}(-z) = \frac{(\ln z)^{3/2}}{\Gamma(5/2)} + \frac{\pi^2}{6}\frac{(\ln z)^{-1/2}}{\Gamma(1/2)} + \cdots$$

$$= \frac{4(\beta\mu)^{3/2}}{3\sqrt{\pi}} + \frac{\pi^2}{6}\frac{(\beta\mu)^{-1/2}}{\sqrt{\pi}} + \cdots = \frac{4(\beta\mu)^{3/2}}{3\sqrt{\pi}}\left[1 + \frac{\pi^2}{8}(\beta\mu)^{-2} + \cdots\right].$$

Using

$$\frac{1}{\Lambda^3} = \left(\frac{2\pi m k_B T}{h^2}\right)^{3/2} = \left(\frac{2\pi m \mu}{h^2}\right)^{3/2} (\beta\mu)^{-3/2},$$

we have

$$N = -\frac{2V}{\Lambda^3}\mathrm{Li}_{3/2}(-z) = \frac{8\pi V}{3}\left(\frac{2m}{h^2}\right)^{3/2} \mu^{3/2}\left[1 + \frac{\pi^2}{8}(k_B T/\mu)^2 + \cdots\right].$$

At $T = 0\,\mathrm{K}$, $\mu = \epsilon_F$ so that, at a fixed electron density and low temperature, the above equation predicts

$$\mu/\epsilon_F = \left[1 + \frac{\pi^2}{8}(k_B T/\mu)^2 + \cdots\right]^{-2/3} = 1 - \frac{\pi^2}{12}\left(\frac{k_B T}{\epsilon_F}\right)^2 + \cdots.$$

(ii) The internal energy of free electrons is given by

$$U = -\frac{3k_B T V}{\Lambda^3}\mathrm{Li}_{5/2}(-z).$$

At low temperature, we have

$$-\mathrm{Li}_{5/2}(-z) = \frac{(\ln z)^{5/2}}{\Gamma(7/2)} + \frac{\pi^2}{6}\frac{(\ln z)^{1/2}}{\Gamma(3/2)} + \cdots$$
$$= \frac{8(\beta\mu)^{5/2}}{15\sqrt{\pi}} + \frac{\pi^2}{3}\frac{(\beta\mu)^{1/2}}{\sqrt{\pi}} + \cdots = \frac{8(\beta\mu)^{5/2}}{15\sqrt{\pi}}\left[1 + \frac{5\pi^2}{8}(\beta\mu)^{-2} + \cdots\right],$$

so that

$$\frac{U}{N} = \frac{3k_B T}{2}\frac{\mathrm{Li}_{5/2}(-z)}{\mathrm{Li}_{3/2}(-z)} = \left(\frac{3k_B T}{2}\cdot\frac{2\beta\mu}{5}\right)\frac{1 + \frac{5\pi^2}{8}(\beta\mu)^{-2} + \cdots}{1 + \frac{\pi^2}{8}(\beta\mu)^{-2} + \cdots}$$
$$= \left(\frac{3\mu}{5}\right)\left[1 + \frac{\pi^2}{2}(\beta\mu)^{-2} + \cdots\right]$$
$$= \left(\frac{3\epsilon_F}{5}\right)\left[1 - \frac{\pi^2}{12}(\beta\mu)^{-2} + \cdots\right]\left[1 + \frac{\pi^2}{2}(\beta\mu)^{-2} + \cdots\right]$$
$$= \left(\frac{3\epsilon_F}{5}\right)\left[1 + \frac{5\pi^2}{12}(\beta\mu)^{-2} + \cdots\right].$$

Using $\mu/\epsilon_F \to 1$ as $T \to 0\,\mathrm{K}$, we obtain from the above equation

$$\frac{U}{N} = \frac{3\epsilon_F}{5}\left[1 + \frac{5\pi^2}{12}\left(\frac{k_B T}{\epsilon_F}\right)^2 + \cdots\right].$$

Problem 4.28 Show that, according to the free-electron model, the reduced entropy per free electron at low temperature is the same as the reduced heat capacity:

$$\frac{S}{Nk_B} = \frac{C_V}{Nk_B} = \frac{\pi^2}{2}\frac{k_B T}{\epsilon_F}.$$

Solution:

According to the free-electron model, the reduced entropy density is given by

$$\frac{S}{k_B V} = \frac{1}{\Lambda^3}\left[-5\mathrm{Li}_{5/2}(-z) + 2\ln z\mathrm{Li}_{3/2}(-z)\right].$$

When $T \to 0\,\text{K}$, $z \to \infty$ and the polylogarithm functions satisfy the asymptotic relations

$$-\text{Li}_{5/2}(-z) = \frac{(\ln z)^{5/2}}{\Gamma(7/2)} + \frac{\pi^2}{6}\frac{(\ln z)^{1/2}}{\Gamma(3/2)},$$

$$\ln z\text{Li}_{3/2}(-z) = -\frac{(\ln z)^{5/2}}{\Gamma(5/2)} - \frac{\pi^2}{6}\frac{(\ln z)^{1/2}}{\Gamma(1/2)},$$

where the higher-order terms disappear because $z \to \infty$. A combination of these two asymptotic terms gives

$$-5\text{Li}_{5/2}(-z) + 2\ln z\text{Li}_{3/2}(-z) = \left[\frac{5}{\Gamma(7/2)} - \frac{2}{\Gamma(5/2)}\right](\ln z)^{5/2} + \frac{\pi^2}{6}\left[\frac{5}{\Gamma(3/2)} - \frac{2}{\Gamma(1/2)}\right](\ln z)^{1/2}.$$

Using the numerical values for the Γ-function, we can evaluate the coefficients

$$\frac{5}{\Gamma(7/2)} - \frac{2}{\Gamma(5/2)} = \frac{5}{15\sqrt{\pi}/8} - \frac{2}{3\sqrt{\pi}/4} = \frac{8}{3\sqrt{\pi}} - \frac{8}{3\sqrt{\pi}} = 0,$$

$$\frac{5}{\Gamma(3/2)} - \frac{2}{\Gamma(1/2)} = \frac{5}{\sqrt{\pi}/2} - \frac{2}{\sqrt{\pi}} = \frac{8}{\sqrt{\pi}}.$$

Hence

$$-5\text{Li}_{5/2}(-z) + 2\ln z\text{Li}_{3/2}(-z) = \frac{\pi^2}{6}\frac{8}{\sqrt{\pi}}(\ln z)^{1/2} = \frac{4\pi^{3/2}}{3}(\ln z)^{1/2}.$$

Substituting the above equation into the expression for the reduced entropy density, along with

$$\frac{1}{\Lambda^3} = \left(\frac{2\pi m k_B T}{h^2}\right)^{3/2} = \left(\frac{2\pi m\mu}{h^2}\right)^{3/2}(\beta\mu)^{-3/2} = \left(\frac{2\pi m\mu}{h^2}\right)^{3/2}(\ln z)^{-3/2},$$

we obtain

$$\frac{S}{k_B V} = \left(\frac{2\pi m\mu}{h^2}\right)^{3/2}(\ln z)^{-3/2}\frac{4\pi^{3/2}}{3}(\ln z)^{1/2} = \frac{\pi^2}{2}\frac{8\pi}{3}\left(\frac{2m\mu}{h^2}\right)^{3/2}(\ln z)^{-1}.$$

Using $\ln z = \beta\mu = \frac{\epsilon_F}{k_B T}$ and $N = \frac{8\pi V}{3}\left(\frac{2m\mu}{h^2}\right)^{3/2}$ as $T \to 0\,\text{K}$, we can rewrite the reduced entropy density as

$$\frac{S}{k_B V} = \frac{\pi^2 N}{2V}\frac{k_B T}{\epsilon_F}.$$

Therefore, at low temperature, the reduced entropy per free electron is

$$\frac{S}{k_B N} = \frac{\pi^2}{2}\frac{k_B T}{\epsilon_F}.$$

Based on the above equation, we can derive the heat capacity

$$\frac{C_V}{T} = \left(\frac{\partial S}{\partial T}\right)_V = \frac{\pi^2}{2}\frac{N k_B^2}{\epsilon_F}.$$

Rearranging the above two equations gives

$$\frac{S}{k_B N} = \frac{C_V}{k_B N} = \frac{\pi^2}{2}\frac{k_B T}{\epsilon_F}.$$

Problem 4.29 Show that, at 0 K, the bulk modulus of a metal can be estimated from

$$B \equiv -V\left(\frac{\partial P}{\partial V}\right)_T = \frac{2\rho\epsilon_F}{3},$$

where is ρ is the number density of free electrons, and ϵ_F is the Fermi energy.

Solution:
Based on the free-electron model for metals, the pressure is related to the internal energy by

$$P = \frac{2U}{3V} = \frac{2N\epsilon_F}{5V} = \frac{2N}{5V}\frac{h^2}{2m}\left(\frac{3\rho}{8\pi}\right)^{2/3} \sim V^{-5/3}.$$

Therefore, we have

$$B \equiv -V\left(\frac{\partial P}{\partial V}\right)_T = \frac{5}{3}P = \frac{5}{3}\frac{2N\epsilon_F}{5V} = \frac{2\rho\epsilon_F}{3}.$$

Problem 4.30 The metallic bonds can be largely attributed to electron delocalization over macroscopic dimensions. To elucidate the microscopic origin of metallic bonds, we may estimate the cohesive energy of a metal from the difference between the internal energy of free electrons at 0 K and that of each free electron confined in confined in a cube of length $a = (1/\rho)^{1/3}$. Show that the free-electron model predicts the cohesive energy per free electron is given by

$$\Delta u = \left[\frac{(3\pi^2)^{2/3}}{5\pi^2} - 1\right] \times \frac{3h^2}{8ma^2} \approx -0.302 \times \frac{h^2}{ma^2}.$$

Compare the theoretical prediction with experimental data for some real metals and discuss the discrepancies.

Solution:
The internal energy of free electrons at 0 K is given by

$$u = U/N = \frac{3}{5}\epsilon_F = \frac{3}{5}\frac{h^2}{2m}\left(\frac{3\rho}{8\pi}\right)^{2/3} = \frac{3}{5}\frac{h^2}{2m}\left(\frac{3}{8\pi a^3}\right)^{2/3} = \frac{(3\pi^2)^{2/3}}{5\pi^2}\frac{3h^2}{8ma^2}.$$

For an electron in a cubic box of length a, the ground-state energy ($k_x = k_y = k_z = \pi/a$) is

$$u_0 = \frac{\hbar^2(k_x^2 + k_y^2 + k_z^2)}{2m} = \frac{3h^2}{8ma^2}.$$

Therefore, the cohesive energy per electron is

$$\Delta u = u - u_0 = \left[\frac{(3\pi^2)^{2/3}}{5\pi^2} - 1\right] \times \frac{3h^2}{8ma^2} \approx -0.302 \times \frac{h^2}{ma^2}.$$

For example, the free-electron density of copper is $\rho = 8.49 \times 10^{28}\,\text{m}^{-3}$, $a = (1/\rho)^{1/3} = 2.2753 \times 10^{-10}$ m, and thus the cohesive energy is

$$\Delta u = u - u_0 = -0.302 \times \frac{(6.626 \times 10^{-34})^2}{9.109 \times 10^{-31}(2.2753 \times 10^{-10})^2}$$
$$\approx -2.8116 \times 10^{-18}\text{J} \approx -1693\ \text{kJ/mol} \approx -17.5\ \text{eV}.$$

The experimental value is about −340 kJ/mol.[4] The free electron model overestimates the cohesive energy by about a factor of 5 because it neglects electron-electron repulsions.

4 Kambe K., "Cohesive energy of noble metals", *Phys. Rev.*, 99, 419 (1955).

Problem 4.31 The free-electron model is often used as a starting point to interpret the electrical conductivity of metallic materials. Image that an external electric field $\mathbf{E}$ is applied to a metal of free electron density ρ, resulting in an electrical current $\mathbf{j}$. The electrical conductivity is defined as

$$\sigma \equiv |\mathbf{j}|/|\mathbf{E}|.$$

The current may be attributed to all free electrons moving with an average velocity $\mathbf{v}$, i.e.,

$$\mathbf{j} = -\rho e \mathbf{v}.$$

Assuming that the directed motion of the free electrons is gained from interacting with the external field over the relaxation time τ (i.e., a duration determined by the consecutive collisions of the electron with other particles in the background), we may estimate the average velocity from Newton's equation

$$\mathbf{v} = -e\mathbf{E}\tau/m.$$

Using these expressions for $\mathbf{v}$ and $\mathbf{j}$, we deduce the Drude equation for the electrical conductivity

$$\sigma = \frac{\rho e^2 \tau}{m}.$$

(i) Estimate the relaxation time of free electrons in copper under ambient condition based on its free electron density and electrical conductivity;
(ii) Based on the relaxation time obtained above, predict the mean free path $\lambda = \tau|\mathbf{v}|$ from $|\mathbf{v}| = \sqrt{3k_BT/m}$, i.e., the electron velocity derived from the classical limit;
(iii) Repeat the above procedure using $|\mathbf{v}| = v_F = \hbar k_F/m$, i.e., the low temperature limit of the electron velocity;
(iv) Discuss the physical significance of the mean free path and why the classical velocity adopted by the classical Drude model may lead to misleading conjectures.

Solution:

(i) At ambient condition, the density of free electrons in coper is about $\rho = 8.46 \times 10^{28}$ m^3 and electrical conductivity is $\sigma = 5.98 \times 10^7$ Siemens. Thus, the relaxation time is

$$\tau = \frac{m\sigma}{\rho e^2} = \frac{9.1 \times 10^{-31} \times 5.98 \times 10^7}{8.46 \times 10^{28} \times (1.602 \times 10^{-19})^2} = 2.51 \times 10^{-14}\text{s}.$$

(ii) In the classical limit, the electron velocity is

$$|\mathbf{v}| = \sqrt{3k_BT/m} = \sqrt{3 \times 1.38 \times 10^{-23} \times 298.15/9.1 \times 10^{-31}} = 1.16 \times 10^5 \text{ m/s}.$$

Thus, the electron mean free path is

$$\lambda = \tau|\mathbf{v}| = 2.51 \times 10^{-14} \times 1.16 \times 10^5 = 2.9 \times 10^{-9} \text{ m}.$$

(iii) In the limit of 0 K, the electron velocity is

$$|\mathbf{v}| = \hbar k_F/m = 1.57 \times 10^6 \text{ m/s},$$

which predicts $\lambda = \tau|\mathbf{v}| = 2.51 \times 10^{-14} \times 1.57 \times 10^6 = 3.95 \times 10^{-8}$ m.
(Strictly speaking, at 0 K, the average velocity of electrons is 0.75 v_F instead of v_F because of the Fermi-Dirac distribution for the electron velocity.)

(iv) The classical model predicts that the electron mean free path is comparable to the inter-atomic spacing, which leads to the erroneous hypothesis in the original Drude model that the collisions are due to electrons bumping into the nuclei of the metal atoms. The quantum model predicts a much larger mean-field path, suggesting that the collisions are most likely affiliated with the crystal defects. Indeed, $\lambda = 3.95 \times 10^{-8}$ m is in excellent agreement with the value (39.9 nm) calculated from more sophisticated theories and that (39 nm) indirectly obtained from experiments.[5]

Problem 4.32 The Wiedemann–Franz law is an empirical correlation between the thermal conductivity (κ) and the electrical conductivity (σ) for metallic materials. It states that the ratio of κ and σ is linearly proportional to the absolute temperature (T), i.e.,

$$\frac{\kappa}{\sigma T} = A,$$

where the proportionality constant A has a numerical value between $2.1 - 3.4 \times 10^{-8}$ WΩK^{-2} for common metals. We may use the free-electron model to justify/understand this empirical relation.

(i) Derive the following expression for the thermal conductivity

$$\kappa = \frac{1}{3}\rho v^2 \tau c_v,$$

where v is the mean-square electron velocity, τ is the electron relaxation time (see Problem 4.31), and c_v represents the molar constant-volume heat capacity of electrons. (Hint: Equating the heat flux in the x-direction, $\mathbf{j}_{x,q} = -\kappa(\partial T/\partial x)$ with the convection of energy in the same direction $-\rho \mathbf{v}_x \cdot (\mathbf{v}_x \tau)(\partial u/\partial x)$, where ρ is the electron density, and u is the molar internal energy.)

(ii) Derive the ratio of κ and σ based on the classical model (i.e., the Maxwell–Boltzmann distribution) for the average velocity

$$\frac{\kappa}{\sigma T} = \frac{3}{2}\left(\frac{k_B}{e}\right)^2.$$

(iii) Derive the ratio of κ and σ based on the results from the Sommerfeld theory at the low temperature limit

$$\frac{\kappa}{\sigma T} = \frac{\pi^2}{3}\left(\frac{k_B}{e}\right)^2.$$

(iv) Discuss the theoretical predictions of the classical and quantum models.

Solution:

(i) The thermal conductivity is defined in terms of the heat flux in the presence of a temperature gradient

$$\mathbf{j}_q = -\kappa \nabla T.$$

Microscopically, the heat flux is affiliated with the electron motion with the flux $\rho\mathbf{v}$ over the duration between consecutive collisions τ, where $\mathbf{v}$ is the average electron velocity. The energy change due to the temperature gradient over a distance corresponding to the mean free path is, $\tau\mathbf{v}\nabla u$. Equating the two expressions of the energy flux in the x-direction leads to

$$-\kappa(\partial T/\partial x) = -\rho \mathbf{v}_x \cdot (\mathbf{v}_x \tau)(\partial u/\partial x) = -\rho v_x^2 \tau c_v (\partial T/\partial x).$$

Using $\langle v_x^2 \rangle = \langle v_y^2 \rangle = \langle v_z^2 \rangle = v^2/3$, we obtain

$$\kappa = \rho v_x^2 \tau c_v = \frac{1}{3}\rho v^2 \tau c_v.$$

5 Gall D., "Electron mean free path in elemental metals", *J. Applied Phys.*, 119, 085101 (2016).

(ii) According to the classical model (viz., the Drude model), we have the average velocity $\langle v^2 \rangle = 3k_BT/m$ predicted by the Maxwell–Boltzmann distribution. Because the electrical conductivity is given by $\sigma = \rho e^2\tau/m$ (Problem 4.31), and molar heat capacity is $c_v = 3k_B/2$, we have

$$\frac{\kappa}{\sigma T} = \frac{\rho v^2 \tau c_v}{3} \times \frac{m}{\rho e^2 \tau T} = \frac{(3k_BT/m)(3k_B/2)}{3} \times \frac{m}{e^2 T}$$
$$= \frac{3}{2}\left(\frac{k_B}{e}\right)^2 \approx 1.11 \times 10^{-8}\ \mathrm{W\Omega K^{-2}}.$$

(iii) According to the Sommerfeld theory of free electrons, at the low temperature limit, we have the average velocity $v^2 = 2\epsilon_F/m$, electrical conductivity $\sigma = \rho e^2\tau/m$, and molar heat capacity $c_v = \pi^2 k_B^2 T/(2\epsilon_F)$. As a result, the thermal conductivity and the electrical conductivity are related by

$$\frac{\kappa}{\sigma T} = \frac{\rho v^2 \tau c_v}{3} \times \frac{m}{\rho e^2 \tau T} = \frac{(2\epsilon_F/m) \cdot \pi^2 k_B^2 T/(2\epsilon_F)}{3} \times \frac{m}{e^2 T}$$
$$= \frac{\pi^2}{3}\left(\frac{k_B}{e}\right)^2 \approx 2.443 \times 10^{-8}\ \mathrm{W\Omega K^{-2}},$$

where the numerical value is known as the Lorenz number.

(iv) The Drude model predicts the correct order of magnitude because of the cancellation of errors in v^2 and c_v. Remarkably, the Sommerfeld theory offers near perfect prediction of the empirical Wiedemann–Franz law!

Problem 4.33 The thermal conductivity of a solid material consists of contributions due to electronic motions and lattice vibration. As discussed in Problem 4.32, the electronic component is well described by the Wiedemann–Franz law. To a first-order approximation, the lattice contribution may be described in terms of phonons as the energy carriers. Accordingly, the thermal conductivity is given by

$$\kappa = \frac{1}{3}\rho v^2 \tau c_v,$$

where ρ and c_v now correspond to the number density and molar heat capacity of phonons, and τ is the phonon relaxation time. The phonon velocity may be simply approximated by the low-frequency speed of sound

$$v = \sqrt{\frac{B + 4C/3}{\hat{\rho}}},$$

where B and C stand for the bulk modulus and shear modulus of the elastic materials, respectively, and $\hat{\rho}$ is the mass density. These equations provide useful guidelines to design novel materials with low thermal conductivity.[6]

(i) Discuss specific strategies that might be taken to reduce the lattice thermal conductivity. Can you name some specific materials that meet the criteria?

(ii) Estimate the low-frequency speed of sound for a typical steel alloy with $B = 170$ GPa, $C = 80$ GPa and $\hat{\rho} = 7{,}700\ \mathrm{kg/m^3}$.

(iii) Assume that the Debye temperature for an alloy is $\theta_D = 470$ K and the molecular weight is 60 g/mol, predict the specific heat capacity of the material at $T = 300$ K based on the Debye model.

6 See, e.g., Toberer E. S. et al, "Phonon engineering through crystal chemistry", *J. Mater. Chem.*, 21, 15843–15852 (2011).

(iv) Using the lattice thermal conductivity $\kappa = 60\,\text{W/(m K)}$, estimate the phonon relaxation time and mean free path and discuss their significance.

Solution:

(i) The kinetic model predicts that low lattice thermal conductivity can be achieved by the reduction of elastic modulus, increase of mass density, or introduction of point defects to reduce the relaxation time of phonons. The article cited shows some successful examples of these approaches in the design of thermoelectric materials.

(ii) We can estimate the low-frequency speed of sound for a typical steel alloy from $B = 170\,\text{GPa}$, $C = 80\,\text{GPa}$ and $\hat{\rho} = 7{,}700\,\text{kg/m}^3$:

$$v = \sqrt{\frac{B + 4C/3}{\hat{\rho}}} = \sqrt{\frac{170 \times 10^9 + 4 \times 80 \times 10^9/3}{7700}} = 5994.2\,\text{m/s}.$$

(iii) At reduced temperature $T/\theta_D = 300/470 = 0.6368$, the molar heat capacity is

$$\begin{aligned} c_v &= 9 \times R \times (T/\theta_D)^3 \int_0^{T/\theta_D} \frac{x^4 e^x}{(e^2 - 1)^2} dx \\ &= 9 \times 8.314 \times 0.6368^3 \times \int_0^{0.6368} \frac{x^4 e^x}{(e^2 - 1)^2} dx = 22.12\,\text{J/(mol K)}. \end{aligned}$$

The specific heat capacity is $\hat{C}_v = c_v/M_w = 0.3687\,\text{J/(g K)}$, where $M_w = 60\,\text{g/mol}$ is the molecular weight of the alloy.

(iv) From $\kappa = 60\,\text{W/(m K)}$, we can estimate the phonon relaxation time from

$$\tau = \frac{3\kappa}{\rho v^2 c_v} = 1.76 \times 10^{-12}\text{s},$$

and the mean free path from

$$\lambda = v \times \tau = 1.06 \times 10^{-8}\,\text{m}.$$

The above prediction shows significant deviation from the experimental results, indicating that 80% of the lattice thermal conductivity arises from phonons with mean free paths between 0.1 and 10 μm. Not surprisingly, more sophisticated theories are needed for quantitative predictions.

Problem 4.34 The Debye function is defined as

$$D(x) \equiv \frac{3}{x^3} \int_0^x \frac{t^3}{e^t - 1} dt.$$

Show that

(i) the molar entropy of a monatomic solid predicted by the Debye model is given by

$$S/R = 4D(\theta_D/T) - 3\ln(1 - e^{-\theta_D/T}),$$

where R is the gas constant, and $\theta_D \equiv h v_D / k_B$.

(ii) at high temperature, the above equation can be approximated by

$$S/R = 4 + 3\ln(T/\theta_D).$$

(iii) at low temperature, the entropy is given by

$$S/R = \frac{4\pi^4}{5}(T/\theta_D)^3.$$

Solution:

(i) According to the Debye model, the Helmholtz energy of a monatomic solid is given by (Section 4.6.2, Eq. (4.159)):

$$F = -\frac{9Nk_BT}{\upsilon_D^3}\int_0^{\upsilon_D} \upsilon^2 \ln\left(\frac{e^{-\beta h\upsilon/2}}{1-e^{-\beta h\upsilon}}\right) d\upsilon.$$

The entropy can be derived from the partial derivative of F with respect to absolute temperature T

$$\begin{aligned}
S &= -\left(\frac{\partial F}{\partial T}\right)_V \\
&= \frac{9Nk_B}{\upsilon_D^3}\int_0^{\upsilon_D} \upsilon^2\left[-\frac{\beta h\upsilon}{2} - \ln(1-e^{-\beta h\upsilon})\right] d\upsilon + \frac{9Nk_BT}{\upsilon_D^3}\int_0^{\upsilon_D} \upsilon^2 \frac{h\upsilon}{2k_BT^2}\left(\frac{e^{\beta h\upsilon}+1}{e^{\beta h\upsilon}-1}\right) d\upsilon \\
&= \frac{9Nk_B}{\upsilon_D^3}\int_0^{\upsilon_D} \upsilon^2\left[-\frac{\beta h\upsilon}{2}\left(1-\frac{e^{\beta h\upsilon}+1}{e^{\beta h\upsilon}-1}\right) - \ln(1-e^{-\beta h\upsilon})\right] d\upsilon \\
&= \frac{9Nk_B}{\upsilon_D^3}\int_0^{\upsilon_D} \upsilon^2\left[\frac{\beta h\upsilon}{e^{\beta h\upsilon}-1} - \ln(1-e^{-\beta h\upsilon})\right] d\upsilon \\
&= \frac{9Nk_B}{x_D^3}\int_0^{x_D} \left[\frac{x^3}{e^x-1} - x^2\ln(1-e^{-x})\right] dx,
\end{aligned}$$

where $x_D = \theta_D/T$. Applying integration by parts to the last term in the above equation gives

$$\begin{aligned}
-\int_0^{x_D} x^2\ln(1-e^{-x})dx &= -\frac{1}{3}\int_0^{x_D} \ln(1-e^{-x})dx^3 \\
&= -\frac{x_D^3}{3}\ln(1-e^{-x_D}) + \frac{1}{3}\int_0^{\frac{\theta_D}{T}} \frac{x^3e^{-x}}{1-e^{-x}}dx.
\end{aligned}$$

A combination of these two equations yields the desired expression

$$\begin{aligned}
\frac{S}{Nk_B} &= \frac{9}{x_D^3}\left[\frac{4}{3}\int_0^{x_D} \frac{x^3}{e^x-1}dx - \frac{x_D^3}{3}\ln(1-e^{-x_D})\right] \\
&= \frac{12}{x_D^3}\int_0^{x_D} \frac{x^3}{e^x-1}dx - 3\ln(1-e^{-x_D}) = 4\mathcal{D}(\theta_D/T) - 3\ln(1-e^{-\theta_D/T}).
\end{aligned}$$

(ii) At high temperature, $x_D = \theta_D/T \to 0$, $\mathcal{D}(\theta_D/T) \to 1$ and $1-e^{-x_D} \to x_D$. Therefore, the reduced entropy can be approximated by

$$S/R = 4 + 3\ln(T/\theta_D).$$

(iii) At low temperature, $x_D = \theta_D/T \to \infty$, the Debye function can be expressed in terms of the gamma function and the Riemann zeta function

$$\mathcal{D}(x) \sim \frac{3}{x^3}\Gamma(4)\zeta(4),$$

where $\Gamma(4) = 6$ and $\zeta(4) = \pi^4/90$. Therefore, the entropy at low temperature is given by

$$S/R = 4\mathcal{D}(\theta_D/T) = 4 \times \frac{3}{x_D^3}\Gamma(4)\zeta(4) = \frac{4\pi^4}{5}(T/\theta_D)^3.$$

Problem 4.35 It has been demonstrated that the Debye model can be used to estimate the entropy at the melting point for a number of monatomic compounds (e.g., Li, Ga, Na, Hg, Al, Mg, In, K,

Zn, Cu, Rb, Ag, Si, Cs, Pb, and Au).[7] The Debye temperature was correlated with the ratio of the densities of the solid and liquid phases

$$\frac{\theta_D^S}{\theta_D^L} \approx \left(\frac{\rho^S}{\rho^L}\right)^\gamma,$$

where γ is known as the Grüneisen parameter. For example, at the melting point ($T = 1356$ K), the densities of copper at the solid and liquid states are 8.2572 g/cm^3 and 7.992 g/cm^3 respectively, and the Grüneisen parameter is $\gamma \approx 1.01$. Estimate the entropy of fusion and compare the result with Richard's melting rule, an empirical observation showing that the normal fusion entropy is approximately a universal constant, $\Delta S_f/R \approx 1.1$.

Solution:
According to Lilley et al., the Debye model predicts that the entropy of fusion is given by

$$\Delta S_f \approx 3\ln\left(\frac{\sqrt{2}\theta_D^S}{\theta_D^L}\right).$$

Substituting the relation between the Debye temperature and density yields

$$\Delta S_f \approx 3\gamma \ln\left(\frac{\rho^S}{\rho^L}\right) = 3 \times 1.01 \times \ln\left(\frac{\sqrt{2} \times 8.2572}{7.992}\right) = 1.149.$$

The number is close to Richard's melting rule, albeit there are significant uncertainties in estimating the Grüneisen parameter.

Problem 4.36 Show that the Debye model for the heat capacity in the low- and high-temperature limits:

$$C_V/(3Nk_B) = \begin{cases} \frac{4\pi^4}{5}(T/\theta_D)^3, & T \to 0\,\text{K}, \\ 1, & T \to \infty. \end{cases}$$

Solution:
At high temperature, $x_D = \theta_D/T \to 0$ and $e^{x_D} \to 1$, thus

$$C_V/(3Nk_B) = \frac{3}{x_D{}^3}\int_0^{x_D} \frac{x^4 e^x}{(e^x-1)^2}dx = \frac{3}{x_D{}^3}\int_0^{x_D} \frac{x^4}{x^2}dx = \frac{3}{x_D^3} \times \frac{x_D^3}{3} = 1.$$

At low temperature, $x_D = \theta_D/T \to \infty$ and $e^{x_D} \to \infty$, we have

$$C_V/(3Nk_B) = \frac{3}{x_D{}^3}\int_0^{\infty} \frac{x^4 e^x}{(e^x-1)^2}dx = \frac{3}{x_D^3} \times \frac{4\pi^4}{15} = \frac{4\pi^4}{5}(T/\theta_D)^3.$$

Problem 4.37 Experimental data suggest that, at low temperature, the heat capacity of many metals can be written in the form

$$C_V/R = AT + BT^3,$$

where R is gas constant, A and B are empirical coefficients that depend on the specific metal. Can you identify the physical meanings of parameters A and B? How are the two contributions to C_V compared with each other for gold at 20 K? Please compare the heat capacity predicted by this

7 Lilley D., Jain A., and Prasher R., "A simple model for the entropy of melting ofmonatomic liquids", *Appl. Phys. Lett.* 118, 083902 (2021).

equation with that predicted by the Debye model. The Debye temperature for gold is 165 K, and the Fermi temperature is about 6.39 $\times 10^4$ K.

Solution:

In $C_V/R = AT + BT^3$, the linear term can be attributed to the electronic degree of freedom

$$C_V^e/R = \frac{\pi^2}{2}(T/T_F),$$

where T_F is the Fermi temperature. Meanwhile, the cubic term may be related to the phonon energy

$$C_V^e/R = \frac{12\pi^4}{5}(T/\theta_D)^3.$$

For gold at $T = 20$ K, $T_F = 6.39 \times 10^4$ K and $\theta_D = 165$ K, a combination of these two contributions predicts

$$C_V/R = \frac{\pi^2}{2}(T/T_F) + \frac{12\pi^4}{5}(T/\theta_D)^3 \approx 0.0015 + 0.4163 = 0.4179.$$

The above equation indicates that the electron contribution to the heat capacity is extremely small. At $x_D = 165/20 = 8.25$, the Debye equation predicts

$$C_V/R = \frac{9}{x_D{}^3}\int_0^{x_D} \frac{x^4 e^x}{(e^x - 1)^2} dx \approx 0.3832.$$

The difference is introduced by the cubic approximation of the Debye function.

5

Cooperative Phenomena and Phase Transitions

Problem 5.1 Find the eigenvalues and eigenvectors of the following matrices

$$(\text{i})\,\hat{\mathbf{A}} = \begin{pmatrix} a & c \\ b & d \end{pmatrix}, \qquad (\text{ii})\,\hat{\mathbf{B}} = \begin{pmatrix} a & c \\ c & d \end{pmatrix}.$$

Here, $a \neq b \neq c \neq d \neq 0$ are real numbers.

Solution:

(i) The eigenvalues for matrix $\hat{\mathbf{A}}$ are solved from

$$\det\begin{pmatrix} a-\lambda & c \\ b & d-\lambda \end{pmatrix} = (a-\lambda)(d-\lambda) - bc = 0.$$

Simplifying the above equation gives

$$\lambda^2 - (a+d)\lambda + (ad-bc) = 0.$$

Therefore, the eigenvalues are

$$\lambda_{1,2} = \frac{(a+d) \pm \sqrt{(a+d)^2 - 4(ad-bc)}}{2} = \frac{(a+d) \pm \sqrt{(a-d)^2 + 4bc}}{2}.$$

The corresponding eigenvalues for $\hat{\mathbf{A}}$ are obtained from

$$\begin{pmatrix} a-\lambda & c \\ b & d-\lambda \end{pmatrix}\begin{pmatrix} v_1 \\ v_2 \end{pmatrix} = \begin{pmatrix} v_1(a-\lambda) + cv_2 \\ bv_1 + (d-\lambda)v_2 \end{pmatrix} = 0.$$

which predicts $v_2/v_1 = (\lambda - a)/c = b/(\lambda - d)$. Hence, the eigenvector satisfies

$$\begin{pmatrix} v_1 \\ v_2 \end{pmatrix} = v_1\begin{pmatrix} 1 \\ (\lambda - a)/c \end{pmatrix},$$

where $v_1 \neq 0$ is an arbitrary nonzero number. Accordingly, we find the two eigenvectors by substituting into the above equation with the two eigenvalues $\lambda_{1,2}$:

$$\mathbf{v}_{1,2} = \begin{pmatrix} v_1 \\ v_2 \end{pmatrix} = v_1\begin{pmatrix} 1 \\ \left[(d-a) \pm \sqrt{(a-d)^2 + 4bc}\right]/2c \end{pmatrix}.$$

(ii) Repeating the above procedure (or replacing b with c) gives

$$\lambda_{1,2} = \frac{(a+d) \pm \sqrt{(a-d)^2 + 4c^2}}{2}, \tag{A}$$

Solutions Manual to Fundamentals and Practice in Statistical Thermodynamics,
First Edition. Jianzhong Wu and John M. Prausnitz.

and

$$v_{1,2} = \left(\begin{matrix} 1 \\ \left[(d-a) \pm \sqrt{(a-d)^2 + 4c^2}\right] / 2c \end{matrix} \right), \tag{B}$$

where $v_1 \neq 0$ has been omitted for simplicity.

Problem 5.2 Show that the eigenvectors for matrix $\hat{\mathbf{B}}$ in Problem 5.1 are orthogonal. How about the eigenvectors for matrix $\hat{\mathbf{A}}$?

Solution:
Matrix $\hat{\mathbf{B}}$ is symmetric and its eigenvectors are orthogonal

$$v_1 \cdot v_2 = 1 + \frac{(d-a)^2 - (a-d)^2 - 4c^2}{4c^2} = 0.$$

For matrix $\hat{\mathbf{A}}$, the eigenvectors are not orthogonal because the product is

$$v_1 \cdot v_2 = 1 + \frac{(d-a)^2 - (a-d)^2 - 4bc^2}{4c^2} = 1 - b/c \neq 0.$$

Problem 5.3 Show that matrix $\hat{\mathbf{B}}$ in Problem 5.1 can be diagonalized with a unitary matrix $\hat{\mathbf{U}}$, i.e.,

$$\hat{\mathbf{U}}^T \cdot \begin{pmatrix} a & c \\ c & d \end{pmatrix} \cdot \hat{\mathbf{U}} = \begin{pmatrix} \lambda_1 & 0 \\ 0 & \lambda_2 \end{pmatrix}.$$

Solution:
Let $\hat{\mathbf{U}} = (v_1^*, v_2^*)$, where $v_1^* = v_1/|v_1|$ and $v_2^* = v_2/|v_2|$ are unit vectors, and $|\cdots|$ represents the vector length. Because the eigenvectors for matrix $\hat{\mathbf{B}}$ are orthogonal, we have

$$v_1^{*T} \cdot v_1^* = v_2^{*T} \cdot v_2^* = 1, \qquad \text{and} \qquad v_1^{*T} \cdot v_2^* = 0.$$

Hence, it is straightforward to show that

$$\begin{aligned}
\hat{\mathbf{U}}^T \cdot \begin{pmatrix} a & c \\ c & d \end{pmatrix} \cdot \hat{\mathbf{U}} &= \hat{\mathbf{U}}^T \cdot \hat{\mathbf{B}} \cdot \hat{\mathbf{U}} = \begin{pmatrix} (v_1^*)^T \\ (v_2^*)^T \end{pmatrix} \cdot \hat{\mathbf{B}} \cdot (v_1^*, v_2^*) = \begin{pmatrix} (\hat{\mathbf{B}} \cdot v_1^*)^T \\ (\hat{\mathbf{B}} \cdot v_2^*)^T \end{pmatrix} \cdot (v_1^*, v_2^*) \\
&= \begin{pmatrix} (\lambda_1 v_1^*)^T \\ (\lambda_2 v_2^*)^T \end{pmatrix} \cdot (v_1^*, v_2^*) = \begin{pmatrix} \lambda_1 v_1^{*T} \cdot v_1^* & \lambda_1 v_1^{*T} \cdot v_2^* \\ \lambda_2 v_2^{*T} \cdot v_1^* & \lambda_2 v_2^{*T} \cdot v_2^* \end{pmatrix} = \begin{pmatrix} \lambda_1 & 0 \\ 0 & \lambda_2 \end{pmatrix}.
\end{aligned}$$

Problem 5.4 Show that the eigenvalues of the transfer matrix for the 1D-Ising model

$$\hat{\mathbf{M}} = \begin{pmatrix} e^{\beta h + \beta \epsilon} & e^{-\beta \epsilon} \\ e^{-\beta \epsilon} & e^{-\beta h + \beta \epsilon} \end{pmatrix},$$

are

$$\lambda_{1,2} = e^{\beta \epsilon} \left[\cosh(\beta h) \pm \sqrt{\sinh^2(\beta h) + e^{-4\beta \epsilon}} \right], \tag{C}$$

and the eigenvectors are

$$v_{1,2} = \left(\begin{matrix} 1 \\ -e^{2\beta \epsilon} \left[\sinh(\beta h) \mp \sqrt{\sinh^2(\beta h) + e^{-4\beta \epsilon}} \right] \end{matrix} \right). \tag{D}$$

Solution:
Substituting

$$a = e^{\beta h+\beta\epsilon}; \qquad c = e^{-\beta\epsilon}; \qquad d = e^{-\beta h+\beta\epsilon},$$

into Eq. (A) derived in Problem 5.1 gives the eigenvalues

$$\lambda_{1,2} = \frac{(a+d) \pm \sqrt{(a-d)^2+4c^2}}{2} = e^{\beta\epsilon}\frac{(e^{\beta h}+e^{-\beta h}) \pm \sqrt{(e^{\beta h}-e^{-\beta h})^2+4e^{-4\beta\epsilon}}}{2}$$

$$= e^{\beta\epsilon}\left[\cosh(\beta h) \pm \sqrt{\sinh^2(\beta h)+e^{-4\beta\epsilon}}\right],$$

and into Eq. (B) gives the eigenvectors

$$\mathbf{v}_{1,2} = \begin{pmatrix} 1 \\ \left[(d-a) \pm \sqrt{(a-d)^2+4c^2}\right]/2c \end{pmatrix}$$

$$= \begin{pmatrix} 1 \\ \left[(e^{-\beta h+\beta\epsilon}-e^{\beta h+\beta\epsilon}) \pm \sqrt{(e^{\beta h+\beta\epsilon}-e^{-\beta h+\beta\epsilon})^2+4e^{-2\beta\epsilon}}\right]/(2e^{-\beta\epsilon}) \end{pmatrix}$$

$$= \begin{pmatrix} 1 \\ -e^{2\beta\epsilon}\left[\sinh(\beta h) \mp \sqrt{\sinh^2(\beta h)+e^{-4\beta\epsilon}}\right] \end{pmatrix}.$$

Problem 5.5 Consider the transfer matrix for the 1D-Ising model

$$\hat{\mathbf{M}} = \begin{pmatrix} e^{\beta h+\beta\epsilon} & e^{-\beta\epsilon} \\ e^{-\beta\epsilon} & e^{-\beta h+\beta\epsilon} \end{pmatrix}.$$

(i) Verify that the eigenvectors are orthogonal;
(ii) Find the corresponding unitary matrix $\hat{\mathbf{U}}$ that diagonalizes the transfer matrix.

Solution:

(i) The eigenvectors are orthogonal because, according to Eq. (D), the product of the two eigenvectors for matrix $\hat{\mathbf{M}}$ is

$$\mathbf{v}_1 \cdot \mathbf{v}_2 = 1 + e^{4\beta\epsilon}\left[\sinh^2(\beta h) - \sinh^2(\beta h) - e^{-4\beta\epsilon}\right] = 1 - e^{4\beta\epsilon}e^{-4\beta\epsilon} = 0.$$

(ii) To find the unitary matrix that diagonalizes the transfer matrix, it is convenient to introduce variables $0 \le \phi \le \pi$ and k such that

$$\cot(2\phi) \equiv e^{2\beta\epsilon}\sinh(\beta h) \equiv k.$$

Using the double-angle identity for tangent

$$\tan(2\phi) = \frac{2\tan\phi}{1-\tan^2(\phi)} = \frac{1}{\cot(2\phi)} = \frac{1}{k},$$

we can obtain $\tan\phi$ from

$$\tan^2(\phi) + 2k\tan\phi - 1 = 0.$$

which leads to

$$\tan\phi = \frac{-2k \pm \sqrt{4k^2+4}}{2} = -e^{2\beta\epsilon}\left[\sinh(\beta h) \mp \sqrt{\sinh^2(\beta h)+e^{-4\beta\epsilon}}\right].$$

According to Eq. (D), the two eigenvectors of matrix $\hat{\mathbf{M}}$ can be written as

$$v_{1,2} = \begin{pmatrix} 1 \\ -e^{2\beta\epsilon}\left[\sinh(\beta h) \mp \sqrt{\sinh^2(\beta h) + e^{-4\beta\epsilon}}\right] \end{pmatrix} = \begin{pmatrix} 1 \\ \tan\phi \end{pmatrix}.$$

For $0 < \phi < \pi/2$, $\tan\phi > 0$ and the eigenvector takes the form of

$$v_1 = \begin{pmatrix} 1 \\ -e^{2\beta\epsilon}\left[\sinh(\beta h) - \sqrt{\sinh^2(\beta h) + e^{-4\beta\epsilon}}\right] \end{pmatrix} = \begin{pmatrix} 1 \\ \tan\phi \end{pmatrix}.$$

In this case, $\cos\phi > 0$ so that the magnitude of this eigenvector can be written as

$$|v_1|^2 = 1 + \tan^2(\phi) = \frac{1}{\cos^2(\phi)} \Rightarrow |v_1| = 1/\cos\phi.$$

Hence, the normalized eigenvector is

$$v_1^* = \frac{1}{|v_1|}\begin{pmatrix} 1 \\ \tan\phi \end{pmatrix} = \begin{pmatrix} \cos\phi \\ \cos\phi\tan\phi \end{pmatrix} = \begin{pmatrix} \cos\phi \\ \sin\phi \end{pmatrix}.$$

Repeating the above procedure but for $\pi/2 < \phi' < \pi$, we have

$$\tan\phi' = -e^{2\beta\epsilon}\left[\sinh(\beta h) + \sqrt{\sinh^2(\beta h) + e^{-4\beta\epsilon}}\right] < 0,$$

which leads to the second normalized eigenvector

$$v_2^* = \frac{1}{|v_2|}\begin{pmatrix} 1 \\ \tan\phi' \end{pmatrix} = \begin{pmatrix} \cos\phi' \\ \cos\phi'\tan\phi' \end{pmatrix} = \begin{pmatrix} \cos\phi' \\ \sin\phi' \end{pmatrix}.$$

Because the two eigenvectors are orthogonal, i.e.,

$$v_1^* \cdot v_2^* = \cos\phi\cos\phi' + \sin\phi\sin\phi' = 0,$$

we have $\cot\phi' = -\tan\phi$, which is satisfied when $\cos\phi' = -\sin\phi$, $\sin\phi' = \cos\phi$. Accordingly, the second normalized eigenvector is

$$v_2^* = \begin{pmatrix} -\sin\phi \\ \cos\phi \end{pmatrix}.$$

Combining the two normalized eigenvectors gives the unitary matrix that diagonalizes transfer matrix $\hat{\mathbf{M}}$:

$$\hat{\mathbf{U}} = (v_1^*, v_2^*) = \begin{pmatrix} \cos\phi & -\sin\phi \\ \sin\phi & \cos\phi \end{pmatrix}.$$

Problem 5.6 Consider a cyclic Ising chain with N spins at zero field. Verify the following equations for the partition function and thermodynamic properties:

(i) Partition function

$$Q = 2^N(\cosh^N K + \sinh^N K),$$

where $K \equiv \beta\epsilon$.

(ii) Helmholtz energy

$$\beta F = -N\ln 2 - \ln[\cosh^N K + \sinh^N K].$$

(iii) Internal energy

$$\frac{U}{N\epsilon} = -\tanh K\left(\frac{1+\tanh^{N-2}K}{1+\tanh^{N}K}\right).$$

(iv) Heat capacity

$$\frac{C_V}{Nk_B} = \frac{K^2}{\cosh^2 K}\frac{1+(N-1)(\tanh^{N-2}K-\tanh^{N}K)-\tanh^{2N-2}K}{(1+\tanh^{N}K)^2}.$$

(v) The average squared magnetization

$$\overline{M^2} \equiv \left\langle\left(\sum_{i=1}^{N} s_i\right)^2\right\rangle = Ne^{2K}\left(\frac{1-\tanh^{N}K}{1+\tanh^{N}K}\right). \quad \text{(E)}$$

Solution:

The derivations below follow the partition function given in Appendix 5A:

$$Q(K,H) = \lambda_+^N + \lambda_-^N, \quad \text{(F)}$$

with

$$\lambda_\pm = e^K\left[\cosh H \pm \sqrt{\sinh^2 H + e^{-4K}}\right],$$

where $K \equiv \beta\epsilon$ and $H \equiv \beta h$.

(i) At $H = 0$, the eigenvalues are

$$\lambda_\pm = e^K\left[1 \pm e^{-2K}\right],$$

so that the partition function becomes

$$Q(K,0) = [e^K(1+e^{-2K})]^N + [e^K(1-e^{-2K})]^N = 2^N(\cosh^N K + \sinh^N K).$$

(ii) The Helmholtz energy at $H = 0$ is given by

$$\beta F = -\ln Q(K,0) = -N\ln 2 - \ln[\cosh^N K + \sinh^N K].$$

(iii) The internal energy at $H = 0$ is

$$\frac{U}{N\epsilon} = \frac{1}{N\epsilon}\frac{\partial \beta F}{\partial \beta} = \frac{1}{N}\frac{\partial \beta F}{\partial K}$$

$$= -\frac{1}{N}\frac{N\cosh^{N-1}K\sinh K + N\sinh^{N-1}K\cosh K}{\cosh^N K + \sinh^N K}$$

$$= -\frac{\cosh^{N-1}K\sinh K}{\cosh^N K}\left(\frac{1+\tanh^{N-2}K}{1+\tanh^{N}K}\right) = -\tanh K\left(\frac{1+\tanh^{N-2}K}{1+\tanh^{N}K}\right).$$

(iv) The heat capacity at $H = 0$ is

$$\frac{C_V}{Nk_B} = \frac{1}{Nk_B}\left(\frac{\partial U}{\partial T}\right) = \frac{\partial U/N\epsilon}{\partial k_B T/\epsilon} = \frac{\partial U/N\epsilon}{\partial \tanh K}\frac{\partial \tanh K}{\partial K}\frac{\partial K}{\partial 1/K}$$

$$= \frac{K^2}{\cosh^2 K}\frac{1+(N-1)(\tanh^{N-2}K-\tanh^{N}K)-\tanh^{2N-2}K}{(1+\tanh^{N}K)^2}.$$

(v) The average squared magnetization is derived from $Q(K,H)$

$$\frac{\partial \ln Q(K,H)}{\partial H} = -\left\langle\sum_{i=1}^{N} s_i\right\rangle$$

and

$$\frac{\partial^2 \ln Q(K,H)}{\partial H^2} = \left\langle \left(\sum_{i=1}^{N} s_i \right)^2 \right\rangle - \left\langle \sum_{i=1}^{N} s_i \right\rangle \left\langle \sum_{i=1}^{N} s_i \right\rangle = \overline{M^2} - (Nm)^2$$

where $m = \left\langle \sum_{i=1}^{N} s_i \right\rangle /N$ is the average magnetization per spin. At $H = 0$, $m = 0$ and thus the above equation becomes

$$\overline{M^2}(H=0) = \left. \frac{\partial^2 \ln Q(K,H)}{\partial H^2} \right|_{H=0}.$$

Substituting Eq. (F) into the above equation, we will find, after some rather straight forward algebra,

$$\overline{M^2}(H=0) = Ne^{2K} \frac{1-\tanh^N K}{1+\tanh^N K}.$$

As $K \to \infty$, $\tanh K \to 1$ and $(1-\tanh^N K) \to N(1-\tanh K)$. Hence, the average squared magnetization becomes

$$\overline{M^2}(H=0) \to N^2 e^{2K} \frac{(1-\tanh K)}{1+\tanh^N K} = N^2 \frac{e^{2K}}{1+\tanh^N K} \frac{2e^{-K}}{e^K + e^{-K}} = N^2. \tag{G}$$

Eq. (G) indicates that the spins are perfectly aligned at $T = 0$ K.

Problem 5.7 A cyclic Ising chain consisting of N spins displays a paramagnetic to ferromagnetic transition at a finite temperature. The transition is not characterized by the mean magnetization per spin m, which is always zero due to the symmetry, but by a drastic increase of the average squared magnetization $\overline{M^2}$.

(i) Show how $\chi^2 \equiv \overline{M^2}/N^2$ varies with temperature for $N = 10, 20$, and 100;

(ii) The transition temperature (viz., the Curie temperature) can be identified from the second-order derivative of the average squared magnetization

$$\frac{\partial^2 \overline{M^2}}{\partial T^2} = 0.$$

Show how the transition temperature varies with the number of spins and compare your numerical results with the prediction of Eq. (5.20), $k_B T/\epsilon = 2/\ln(N-1)$.

Solution:

(i) Based on the average squared magnetization derived in Problem 5.6, we have the following analytic expression for χ^2

$$\chi^2 \equiv \overline{M^2}/N^2 = \frac{e^{2K}}{N} \frac{1-\tanh^N K}{1+\tanh^N K}.$$

Figure 5.1 presents χ^2 versus the reduced temperature $k_B T/\epsilon = 1/K$ for three different chain lengths, $N = 10, 20$, and 100. It shows that a paramagnetic to ferromagnetic transition occurs at low temperature.

(ii) As shown in Figure 5.2, the transition temperature (viz., the inflection point in χ^2 versus T) falls as N increases and approaches absolute zero in the limit $N = \infty$. When $N \gg 1$, the numerical results agree well with that predicted by Eq. (5.20), i.e., $k_B T_m/\epsilon = 2/\ln(N-1)$.

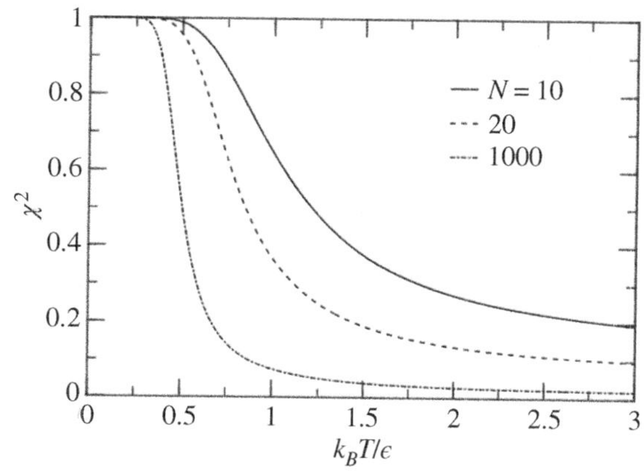

Figure 5.1 The temperature dependences of the average squared magnetization for three cyclic Ising chains (Cyclic1DIsing.m).

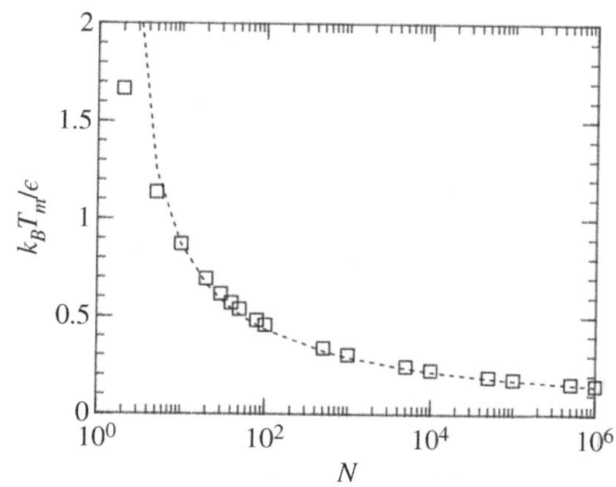

Figure 5.2 The variation of the Curie temperature with the number of spins in the cyclic Ising chain. The symbols are numerical results obtained from the inflection point of χ^2 versus T, and the dashed line is predicted by $2/\ln(N-1)$ (Cyclic1DIsingTm.m).

Problem 5.8 The partition function for an open-ended Ising chain with N spins was reported by Altenberger and Dahler[1]:

$$Q(K,H) = A_+\lambda_+^{N-1} + A_-\lambda_-^{N-1} \tag{H}$$

where

$$\lambda_\pm = e^K\left[\cosh H \pm \sqrt{\sinh^2 H + e^{-4K}}\right]$$

$$A_\pm = \cosh H \pm [\sinh^2 H + e^{-2K}][\sinh^2 H + e^{-4K}]^{-1/2},$$

with $K \equiv \beta\epsilon$ and $H \equiv \beta h$. From $Q(K,H)$, we can derive the average squared magnetization

$$\overline{M^2} \equiv \left\langle \left(\sum_{i=1}^{N} s_i\right)^2 \right\rangle = Ne^{2K}[1 + \sinh(2K)(\tanh^N K - 1)/N].$$

Show how the paramagnetic to ferromagnetic transition temperature varies with the number of spins and compare the numerical results with the prediction of Eq. (5.20), $k_BT/\epsilon = 2/\ln(N-1)$.

Solution:

As discussed in Problem 5.7, we can identify the transition temperature T_m from the inflection point in $\overline{M^2}$ versus T. Similar to the cyclic Ising chain, T_m falls as N grows and approaches the

1 Altenberger A. and Dahler J.,"One-dimensional Ising model for spin systems of finite size", *Adv. Chem. Phys.* 112, 337–356 (2007).

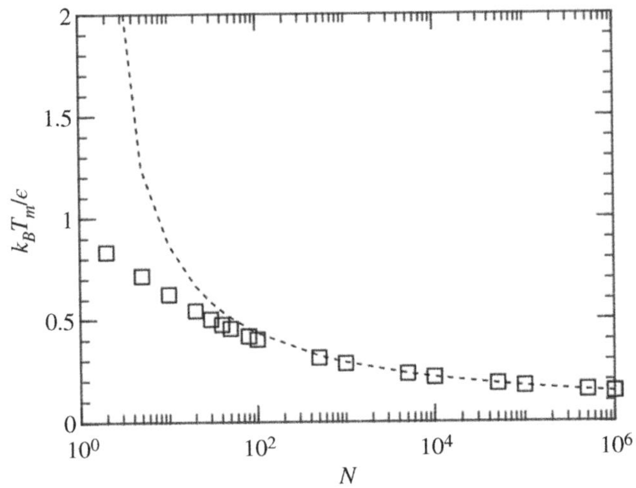

Figure 5.3 The variation of the Curie temperature with the number of spins in the Ising chain with open ends. The symbols are numerical results obtained from the inflection point of $\overline{M^2}$ versus T, and the dashed line is predicted by $2/\ln(N-1)$ (OEIsingChainTm.m).

absolute zero in the limit of $N = \infty$. As shown in Figure 5.3, when $N \gg 1$, the numerical results agree well with the prediction of Eq. (5.20).

Problem 5.9 For an Ising chain consisting of N spins at zero field, the partition function may be expressed in terms of bonding between neighboring spins

$$Q = 2\prod_{i=1}^{N-1}\sum_{b_i=\pm 1} \exp\left(\beta\epsilon\sum_{i=1}^{N-1} b_i\right),$$

where $b_i = s_i s_{i+1}$, and a factor of two accounts for two possible orientations for one of the end spins. Show that

(i) $< b_k > = \tanh(\beta\epsilon)$;

(ii) $< s_k s_{k+n} > = \tanh^n(\beta\epsilon)$.

Solution:

(i)

$$< b_k > = \frac{2\prod_{i=1}^{N-1}\sum_{b_i=\pm 1} b_k \exp(\beta\epsilon\sum_{i=1}^{N-1} b_i)}{2\prod_{i=1}^{N-1}\sum_{b_i=\pm 1} \exp(\beta\epsilon\sum_{i=1}^{N-1} b_i)}$$
$$= \frac{\sum_{b_i=\pm 1} b_k e^{\beta\epsilon b_k}}{\sum_{b_i=\pm 1} e^{\beta\epsilon b_k}} = \frac{e^{\beta\epsilon} - e^{-\beta\epsilon}}{e^{\beta\epsilon} + e^{-\beta\epsilon}} = \tanh(\beta\epsilon);$$

(ii)

$$< s_k s_{k+n} > = < s_k s_{k+1} s_{k+1} \cdots s_{k+n} > = < b_k b_{k+1} \cdots b_{k+n-1} >$$
$$= \frac{2\prod_{i=1}^{N-1}\sum_{b_i=\pm 1} b_k b_{k+1} \cdots b_{k+n-1} \exp(\beta\epsilon\sum_{i=1}^{N-1} b_i)}{2\prod_{i=1}^{N-1}\sum_{b_i=\pm 1} \exp(\beta\epsilon\sum_{i=1}^{N-1} b_i)}$$
$$= \prod_{j=k}^{k+n-1} \langle b_j \rangle = \left[\frac{e^{\beta\epsilon} - e^{-\beta\epsilon}}{e^{\beta\epsilon} + e^{-\beta\epsilon}}\right]^n = \tanh^n(\beta\epsilon).$$

Problem 5.10 Cooperative phenomena are commonplace in chemical and biological systems. The essential concept may be elucidated with a two-state model. For the simplest case, imagine that ligand molecules of finite concentrations are able to bind with a substrate (e.g., a protein or polymer) with two identical binding sites at infinite dilution. When each binding site is occupied by a ligand molecule, the free energy of the system is changed by ΔG_b. Cooperativity is introduced by an additional energy w when both sites are occupied.

(i) Suppose that the chemical reaction for ligand binding of each site may be expressed as

$$\mathrm{L} + \mathrm{S} \rightarrow \mathrm{LS},$$

where L and S represent the ligand and the substrate, respectively. Show that, at infinite dilution of the substrate, ΔG_b can be calculated from the intrinsic binding constant K_b and ligand activity [L]

$$\beta \Delta G_b = -\ln K_b - \ln[\mathrm{L}].$$

(ii) Derive the following expression for the average occupation number of each binding site

$$\alpha = \frac{K_b[\mathrm{L}] + K_b^2[\mathrm{L}]^2 e^{-\beta w}}{1 + 2K_b[\mathrm{L}] + K_b^2[\mathrm{L}]^2 e^{-\beta w}}.$$

(iii) Show that, when $w = 0$, the average occupation number per site is given by the Langmuir binding equation:

$$\alpha = \frac{K_b[\mathrm{L}]}{1 + K_b[\mathrm{L}]}.$$

(iv) Show that the probability of double site occupation is given by

$$\alpha_{12} = \frac{K_b^2[\mathrm{L}]^2 e^{-\beta w}}{1 + 2K_b[\mathrm{L}] + K_b^2[\mathrm{L}]^2 e^{-\beta w}}.$$

(v) Supposed that the intrinsic binding constant for each site is $K_b = 10$, plot the binding curve in terms of the average number of ligand molecules binding with each site as a function of the ligand activity for $\beta w = 0, 1.0, -1.0$.

Solution:

(i) At equilibrium, the binding constant is related to the activities of the underlying chemical species

$$K_b = \frac{[\mathrm{LS}]}{[\mathrm{S}] \cdot [\mathrm{L}]},$$

where $[\cdots]$ stands for activity, which is in the units consistent with that for the reference state (e.g., M). The change in free energy due to the ligand binding is

$$\Delta G_b = \mu_{\mathrm{LS}} - \mu_{\mathrm{S}} - \mu_{\mathrm{L}}. \tag{I}$$

Because the substrate is at infinite dilution, its chemical potential is the same as that in the reference state, i.e.,

$$\mu_{\mathrm{LS}} = \mu_{\mathrm{LS}}^0; \qquad \mu_{\mathrm{S}} = \mu_{\mathrm{S}}^0.$$

Meanwhile, the chemical potential of the ligand depends on its activity

$$\mu_{\mathrm{L}} = \mu_{\mathrm{L}}^0 + k_B T \ln[\mathrm{L}].$$

With the above expressions for the chemical potentials, Eq. (I) becomes

$$\beta \Delta G_b = \beta \Delta G_b^0 - \ln[\mathrm{L}] = -\ln K_b - \ln[\mathrm{L}],$$

where $\Delta G_b^0 \equiv \mu_{\mathrm{LS}}^0 - \mu_{\mathrm{S}}^0 - \mu_{\mathrm{L}}^0$ and $\ln K_b = -\beta \Delta G_b^0$.

(ii) The system of interest consists of two binding sites. Each binding site can exist in one of two energy states, occupied by a ligand molecule or empty. In the context of the two-state model, the system (free) energy may be written as

$$E_\nu = (n_1 + n_2)\Delta G_b + n_1 n_2 w,$$

where $\nu = \{n_1, n_2\}$ stands for a microstate, and $n_i = 1$ or 0 is the occupation number. Hence, the partition function is given by

$$\begin{aligned} Q &= \sum_\nu \exp(-\beta E_\nu) = \sum_{n_1}\sum_{n_2} \exp\{-\beta[(n_1+n_2)\Delta G_b + n_1 n_2 w]\} \\ &= 1 + 2\exp\{-\beta \Delta G_b\} + \exp\{-\beta[2\Delta G_b + w]\} \\ &= 1 + 2K_b[\mathrm{L}] + K_b^2[\mathrm{L}]^2 e^{-\beta w}. \end{aligned}$$

In writing the last line in the above equation, we have used the expression $\exp\{-\beta\Delta G_b\} = K_b[\mathrm{L}]$. In biological sciences, Q is known as the binding polynomial.
From the partition function, we can readily obtain the average occupation number per site

$$\begin{aligned} \alpha &= \frac{< n_1 + n_2 >}{2} = -\frac{1}{2}\cdot\frac{\partial \ln Q}{\partial \beta \Delta G_b} = \frac{\exp\{-\beta\Delta G_b\} + \exp\{-\beta[2\Delta G_b + w]\}}{1 + 2\exp\{-\beta\Delta G_b\} + \exp\{-\beta[2\Delta G_b + w]\}} \\ &= \frac{K_b[\mathrm{L}] + K_b^2[\mathrm{L}]^2 e^{-\beta w}}{1 + 2K_b[\mathrm{L}] + K_b^2[\mathrm{L}]^2 e^{-\beta w}}. \end{aligned} \tag{J}$$

(iii) When $w = 0$, the two binding sites are independent. In that case, Eq. (J) reduces to the Langmuir binding equation

$$\alpha = \frac{K_b[\mathrm{L}] + K_b^2[\mathrm{L}]^2}{1 + 2K_b[\mathrm{L}] + K_b^2[\mathrm{L}]^2} = \frac{K_b[\mathrm{L}]}{1 + K_b[\mathrm{L}]}.$$

(iv) The probability of double site occupation is given by

$$\begin{aligned} \alpha_{12} &= < n_1 n_2 > = -\frac{\partial \ln Q}{\partial \beta w} = \frac{\exp\{-\beta[2\Delta G_b + w]\}}{1 + 2\exp\{-\beta\Delta G_b\} + \exp\{-\beta[2\Delta G_b + w]\}} \\ &= \frac{K_b^2[\mathrm{L}]^2 e^{-\beta w}}{1 + 2K_b[\mathrm{L}] + K_b^2[\mathrm{L}]^2 e^{-\beta w}}. \end{aligned}$$

(v) Figure 5.4 shows the binding curves for $K_b = 10$ and $\beta w = -1.0, 0.0, 1.0$.

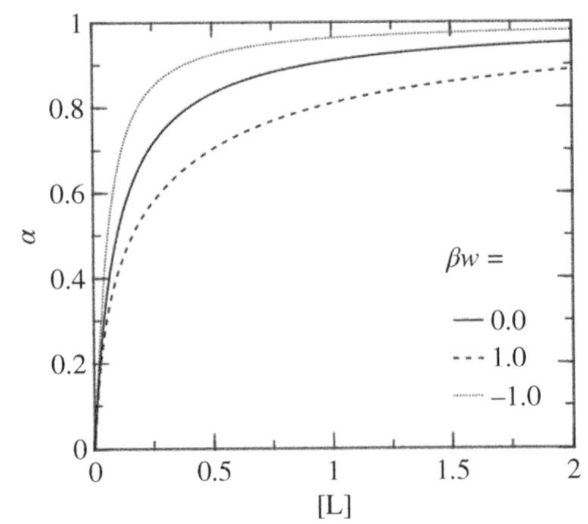

Figure 5.4 Binding curves for a substrate with two identical sites. Here, α is the average occupation number per site, and [L] stands for the ligand activity. $\beta w = 0$ means that the two sites are independent as assumed by the Langmuir model, $\beta w = -1$ represents cooperative binding, and $\beta w = 1$ shows anti-cooperative binding (Bind2site.m).

Problem 5.11 The 1D-Ising model provides a systematic way to represent cooperative phenomena such as ligand binding.

(i) Show that ligand binding with a substrate discussed in Problem 5.10 can be represented by the Ising model.
(ii) Evaluate the average occupation number per site as a function of the ligand activity if the substrate is able to bind a large number of ligand molecules.
(iii) Compare the binding curve for substrates with two and that with a large number of sites using $K_b = 10$ and $\beta w = -3$ and discuss the implications of cooperativity.

Solution:

(i) In the context of 1D-Ising model, $s_i = +1$ means that binding site i is occupied by a ligand molecule, and $s_i = -1$ means that site i is unoccupied. Thus, the system energy is given by

$$E_\nu = (n_1 + n_2)\Delta G_b + n_1 n_2 w,$$

where $\nu = \{s_1, s_2\}$ stands for a microstate, and $n_i = (s_i + 1)/2 = 1$ or 0 represents the occupation number. When $s_i = +1$, $n_i = 1$; when $s_i = -1$, $n_i = 0$. With the spin variables s_1 and s_2, we can rewrite the system energy as

$$\begin{aligned} E_\nu &= \left(\frac{s_1+1}{2} + \frac{s_2+1}{2}\right)\Delta G_b + \left(\frac{s_1+1}{2}\right)\left(\frac{s_1+1}{2}\right) w \\ &= \Delta G_b + \left(\frac{s_1+s_2}{2}\right)\Delta G_b + \frac{s_1+s_2}{4} w + \frac{s_1 s_2}{4} w + w/4 \\ &= \Delta G_b + w/4 + (s_1+s_2)\left(\frac{\Delta G_b}{2} + \frac{w}{4}\right) + \frac{w}{4} s_1 s_2 \\ &\equiv E_0 - h(s_1+s_2) - s_1 s_2 \epsilon, \end{aligned}$$

where $E_0 = \Delta G_b + w/4$, $h = -\left(\Delta G_b/2 + w/4\right)$, and $\epsilon = -w/4$.
For a substrate with two binding sites, the partition function is given by

$$\begin{aligned} Q &= \sum_\nu \exp(-\beta E_\nu) = \sum_{s_1=\pm1} \sum_{s_2=\pm1} \exp\{\beta[h(s_1+s_2) + s_1 s_2 \epsilon]\} \\ &= \exp\{2\beta h + \beta\epsilon\} + 2\exp\{-\beta\epsilon\} + \exp\{-2\beta h + \beta\epsilon\}, \end{aligned}$$

where the contribution from the grand-state energy E_0 has been omitted.

(ii) If the substrate has a large number of binding sites, the average occupation number per site can be calculated from the 1D-Ising model, Eq. (5.16),

$$\alpha = \frac{1}{2}\left[1 + \frac{\sinh(\beta h)}{\sqrt{\sinh^2(\beta h) + e^{-4\beta\epsilon}}}\right],$$

where

$$\beta h = -\left(\frac{\beta \Delta G_b}{2} + \frac{\beta w}{4}\right) = \frac{\ln K_b}{2} + \frac{\ln[\mathrm{L}]}{2} - \frac{\beta w}{4} \qquad \text{and} \qquad 4\beta\epsilon = -\beta w.$$

(iii) Figure 5.5 compares the binding curve for substrates with two and that with a large number of sites using $K_b = 10$ and $\beta w = -3$. The cooperative binding is enhanced as the number of binding sites increases, leading to a sharper transition from the completely unbound state ($\alpha = 0$) to the completely bound state ($\alpha = 1$).

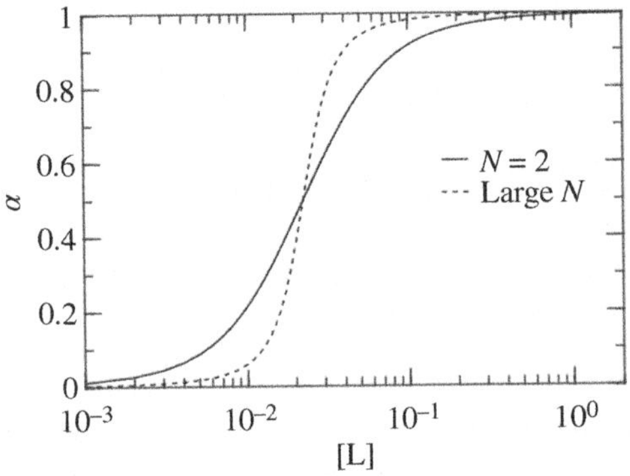

Figure 5.5 Binding curves for substrates with two and a large number of sites (Bind2Large.m).

Problem 5.12 The Ising model has been used to analyze the extent of genetic markers shared by siblings.[2] In the so-called sib pair analysis, each data point for an allele (viz., any of the alternative forms of a gene) being shared and not shared by a sib pair may be represented by $s = +1$ or -1, with the linkage between adjacent genetic markers represented by nearest-neighbor interactions, and the effect of a disease gene by an external energy. For a chromosome with N markers, the total energy is thus given by

$$E_\nu = -\sum_{i=1}^{N-1} \epsilon_i s_i s_{i+1} - \sum_{i}^{N} h_i s_i,$$

where $\nu = \{s_i\}_{i=1,2,\ldots,N}$ stands for an identity-by-descent (IBD) sharing configuration (viz., microstate), "genetic parameters" ϵ_i and h_i are dimensionless, equivalent to the coupling and external energies in a regular Ising model divided by the absolute temperature and the Boltzmann constant. The genetic parameters can be obtained by fitting the Ising model with experimental data for the statistics of gene expression.

(i) Verify that the following iterative relations may be used to evaluate the partition function

$$Q_N = Q_N^+ + Q_N^-,$$

where $Q_N^+ = e^{h_N}[Q_{N-1}^+ e^{\epsilon_{N-1}} + Q_{N-1}^- e^{-\epsilon_{N-1}}]$ is the value of the partition function for a chromosome with N markers, given that the last (Nth) position is positive ($s_N = +1$), and $Q_N^- = e^{-h_N}[Q_{N-1}^+ e^{-\epsilon_{N-1}} + Q_{N-1}^- e^{\epsilon_{N-1}}]$ is the value of the partition function for the same chromosome, given that the last position is negative ($s_N = -1$).

(ii) The genetic parameters, ϵ_i and h_i, can be estimated by using the maximum likelihood method, i.e., by maximizing the probability of observing a particular dataset

$$\mathrm{L} = \prod_{i=1}^{N} p_i,$$

where p_i stands for the probability of observing IBD at marker i. Find an expression for the probability of observing IBD at marker i using the Ising model.

(iii) Under the simplest hypothesis that there is only one marker k affiliated with a disease such that the influence parameter $h_k \neq 0$ and $h_{i\neq k} = 0$. Find an expression for the LOD (logarithm of the odds) score

$$\mathrm{LOD} = \log(L/L_0),$$

where L_0 stands for the likelihood with $h_i = 0$ for all markers.

2 Majewsk J. et al.,"The Ising model in physics and statistical genetics", *Am. J. Hum Genet.* 69(4), 853–862 (2001).

Solution:

(i) The canonical partition function for the Ising chain model is given by

$$Q_N = \sum_v e^{-E_v} = \sum_{s_i=\pm1\}_{i=1,2,\cdots,N}} \exp\left\{\sum_{i=1}^{N-1} \epsilon_i s_i s_{i+1} + \sum_i^N h_i s_i\right\},$$

Because the end position has only two possible states, $s_N = \pm 1$, we have

$$\begin{aligned} Q_N = & \sum_{\{s_i=\pm1\}_{i=1,2,\cdots,N-1}} \exp\left\{\epsilon_{N-1}s_{N-1} + h_N + \sum_{i=1}^{N-2}\epsilon_i s_i s_{i+1} + \sum_i^{N-1} h_i s_i\right\} \\ & + \sum_{\{s_i=\pm1\}_{i=1,2,\cdots,N-1}} \exp\left\{-\epsilon_{N-1}s_{N-1} - h_N + \sum_{i=1}^{N-2}\epsilon_i s_i s_{i+1} + \sum_i^{N-1} h_i s_i\right\} \\ & \equiv Q_N^+ + Q_N^-, \end{aligned}$$

where the last line in the above equation defines the two contributions to the partition function. Q_N^+ is the value of the partition function for N markers, given that the last (Nth) position is positive ($s_N = +1$), and Q_N^- is the value of the partition function, given that the last position is negative ($s_N = -1$). It is straightforward to show that Q_N^+ and Q_N^- follow the iterative relations

$$Q_N^+ = e^{h_N}[Q_{N-1}^+ e^{\epsilon_{N-1}} + Q_{N-1}^- e^{-\epsilon_{N-1}}],$$
$$Q_N^- = e^{-h_N}[Q_{N-1}^+ e^{-\epsilon_{N-1}} + Q_{N-1}^- e^{\epsilon_{N-1}}].$$

(ii) According to the Ising model, the probability of observing IBD at marker k is

$$\begin{aligned} p_k = < s_k > &= \frac{1}{Q_N}\sum_{\{s_i=\pm1\}_{i=1,2,\cdots,N}} s_k \exp\left\{\sum_{i=1}^{N-1}\epsilon_i s_i s_{i+1} + \sum_i^N h_i s_i\right\} \\ &= \frac{Q_k^+ Q_{N-k} - Q_k^- Q_{N-k}}{Q_N}, \end{aligned}$$

where $Q_0 = 1$.

(iii) The likelihood is given by

$$\mathrm{L} = \prod_{i=1}^{N} \frac{Q_k^+ Q_{N-k} - Q_k^- Q_{N-k}}{Q_N},$$

where

$$Q_N^+ = e^{h_N}[Q_{N-1}^+ e^{\epsilon_{N-1}} + Q_{N-1}^- e^{-\epsilon_{N-1}}],$$
$$Q_N^- = e^{-h_N}[Q_{N-1}^+ e^{-\epsilon_{N-1}} + Q_{N-1}^- e^{\epsilon_{N-1}}].$$

When $h_i = 0$ for all markers, the expression for L_0 remains the same but the iterative relation becomes

$$Q_N^+ = Q_{N-1}^+ e^{\epsilon_{N-1}} + Q_{N-1}^- e^{-\epsilon_{N-1}},$$
$$Q_N^- = Q_{N-1}^+ e^{-\epsilon_{N-1}} + Q_{N-1}^- e^{\epsilon_{N-1}}.$$

The LOD score can be calculated recursively from the above expressions.

Problem 5.13 An extended one-dimensional Ising model has been utilized to quantify the folding cooperativity of *de novo*-designed helical repeat proteins (DHRs).[3] In this model, the conformations

3 Barrick et al., "Extreme stability in de novo-designed repeat arrays is determined by unusually stable short-range interactions", *PNAS*, 115(29), 7539–7544 (2018).

of individual repeats in a modular helix–loop–helix–loop protein are represented as either folded or unfolded, resulting in $2n$ possible conformations for an n-repeat superhelical array. For each conformation, the energy is determined by the intrinsic folding energy of each repeat (ΔG_i) as well as the coupling ("interfacial") free energies (ϵ) between adjacent folded repeats. Because the sequences of the N- and C-terminal capping repeats differ from the sequence of central repeats, three intrinsic energies (ΔG_N, ΔG_R, and ΔG_C) were used to represent the intrinsic folding energies of folded repeats at N-terminal, center, and C-terminal, respectively.

(i) Show that the canonical partition function for this system is given by

$$Q = \begin{pmatrix} 0 & 1 \end{pmatrix} \begin{pmatrix} \kappa_N \tau & 1 \\ \kappa_N & 1 \end{pmatrix} \begin{pmatrix} \kappa_R \tau & 1 \\ \kappa_R & 1 \end{pmatrix}^{n-2} \begin{pmatrix} \kappa_C \tau & 1 \\ \kappa_C & 1 \end{pmatrix} \begin{pmatrix} 1 \\ 1 \end{pmatrix},$$

where $\kappa_N = \exp(-\beta \Delta G_N)$, $\kappa_R = \exp(-\beta \Delta G_R)$, $\kappa_C = \exp(-\beta \Delta G_C)$, and $\tau = \exp(-\beta \epsilon)$.

(ii) What is the fraction of folded repeats in the superhelical array?

Solution:

(i) Following the derivation of the Zimm-Bragg model discussed in Appendix 5B, the partition function for a superhelical array with n-repeats can be written as

$$Q = Q_n^C + Q_n^U,$$

where Q_n^C stands for the partition function with the C-terminal repeat in the folded state, and Q_n^U for that with the C-terminal repeat unfolded. In vector form, the above equation can be written as

$$Q = \begin{pmatrix} Q_n^C & Q_n^U \end{pmatrix} \begin{pmatrix} 1 \\ 1 \end{pmatrix},$$

Note that $Q_n^C = Q_{n-1}^C \kappa_C \tau + Q_{n-1}^U \kappa_C$ and $Q_n^U = Q_{n-1}^C + Q_{n-1}^U$, we can write

$$\begin{pmatrix} Q_n^C & Q_n^U \end{pmatrix} = \begin{pmatrix} Q_{n-1}^C & Q_{n-1}^U \end{pmatrix} \begin{pmatrix} \kappa_C \tau & 1 \\ \kappa_C & 1 \end{pmatrix}.$$

Similarly, from $Q_{n-1}^C = Q_{n-2}^C \kappa_R \tau + Q_{n-2}^U \kappa_C$ and $Q_{n-1}^U = Q_{n-2}^C + Q_{n-2}^U$, we have

$$\begin{pmatrix} Q_{n-1}^C & Q_{n-1}^U \end{pmatrix} = \begin{pmatrix} Q_{n-2}^C & Q_{n-2}^U \end{pmatrix} \begin{pmatrix} \kappa_R \tau & 1 \\ \kappa_R & 1 \end{pmatrix}.$$

Repeating the above procedure to the first repeat unit (viz., N-terminal) gives

$$\begin{pmatrix} Q_1^C & Q_1^U \end{pmatrix} = \begin{pmatrix} Q_0^C & Q_0^U \end{pmatrix} \begin{pmatrix} \kappa_N \tau & 1 \\ \kappa_N & 1 \end{pmatrix},$$

where $Q_0^C = 0$ to ensure no interface for the first repeat, and $Q_0^U = 1$ such that the N-terminal in both folded and unfolded states can be included. Thus, the canonical partition function is

$$Q = \begin{pmatrix} 0 & 1 \end{pmatrix} \begin{pmatrix} \kappa_N \tau & 1 \\ \kappa_N & 1 \end{pmatrix} \begin{pmatrix} \kappa_R \tau & 1 \\ \kappa_R & 1 \end{pmatrix}^{n-2} \begin{pmatrix} \kappa_C \tau & 1 \\ \kappa_C & 1 \end{pmatrix} \begin{pmatrix} 1 \\ 1 \end{pmatrix}.$$

(ii) The fraction of folded repeats in the superhelical array can be calculated from the partition function

$$\theta = \frac{1}{nQ} \left(\kappa_N \frac{\partial Q}{\partial \kappa_N} + \kappa_R \frac{\partial Q}{\partial \kappa_R} + \kappa_C \frac{\partial Q}{\partial \kappa_C} \right).$$

Problem 5.14 The one-dimensional Ising model has been applied to understand the cooperativity of histone H1 in binding with a chromatin.[4] The cooperative process results in a long-range correlation of histone H1 distribution along the nucleosome array, which explains several aspects of gene regulation and chromatin structure. In this mode, the chromatin structure is represented by a one-dimensional lattice such that each site represents a nucleosome. Each microstate of the model system (viz., chromatin conformation) is defined by the status of histone binding, i.e., a set of numbers $\{n_1, n_2, \dots, n_N\}$ where $n_i = 0$ or 1 stands for the occupation number for nucleosome i, and N is the total number of binding sites. Accordingly, the system energy is given by

$$E_\nu = -a\sum_i n_i n_{i+1} - (b+\mu)\sum n_i,$$

where a is the free energy associated with the cooperative interaction of histone molecules, b stands for the binding energy, and μ is the chemical potential of histone H1.

(i) Verify that the canonical partition function can be written as

$$Q = \lambda_1^N + \lambda_2^N,$$

where $\lambda_{1,2} = [AB + 1 \pm \sqrt{(AB-1)^2 + 4B}]/2$ with $A = \exp(a/k_BT)$ and $B = \exp[(b+\mu)/k_BT]$.

(ii) What is the probability of nucleosome i occupied by a histone H1 molecule?

(iii) Ishii found that the joint probability for the simultaneous occupancy of nucleosomes i and j is given by

$$< n_i n_j >= \cos^4\phi + e^{|j-i|/\xi_c}\cos^2\phi\sin^2\phi,$$

where $\cot\phi = [AB - 1 + \sqrt{(AB-1)^2 + 4B}]/(2\sqrt{B})$, and $\xi_c = 1/\ln(\lambda_1/\lambda_2) > 0$ stands for the correlation length. Show that the joint probability that nucleosome j is occupied while the nucleosome i is not occupied is given by

$$\frac{< (1-n_i)n_j >}{< 1 - n_i >} = < n > (1 - e^{|j-i|/\xi_c}).$$

(iv) Plot ξ_c as a function of $b + \mu$ for $T = 300$ K and $a \approx 7.5$ kcal/mol, and discuss how gene regulatory DNA sequences can affect the chromatin structure over a long distance.

Solution:

(i) The partition function is given by

$$Q = \sum_\nu \beta E_\nu = \sum_{\{n_i=0,1\}} \exp\left[-\beta a\sum_i n_i n_{i+1} - \beta(b+\mu)\sum n_i\right]$$

In terms of the transfer matrix, Q can be written as

$$Q = \mathrm{Tr}\begin{pmatrix} \exp[\beta(a+b+\mu)] & \exp[\beta(b+\mu)] \\ \exp[\beta(b+\mu)] & 1 \end{pmatrix}^N = \begin{pmatrix} AB & B \\ B & 1 \end{pmatrix}^N$$

The matrix can be diagonalized as follows (Problem 5.3):

$$\begin{pmatrix} AB & B \\ B & 1 \end{pmatrix} = \hat{\mathbf{U}}^T \begin{pmatrix} \lambda_1 & 0 \\ 0 & \lambda_2 \end{pmatrix} \mathbf{U},$$

4 Ishii H., "A statistical-mechanical model for regulation of long-range chromatin structure and gene expression", *J. Theor. Biol.* 203, 215–228 (2000).

where $\lambda_{1,2} = [AB + 1 \pm \sqrt{(AB-1)^2 + 4B}]/2$ with $A = \exp(a/k_BT)$ and $B = \exp[(b+\mu)/k_BT]$ are the eigenvalues, and $\mathbf{U}$ is an orthonormal matrix

$$\mathbf{U} = \begin{pmatrix} \cos\phi & \sin\phi \\ -\sin\phi & \cos\phi \end{pmatrix},$$

with

$$\sin\phi = \frac{2\sqrt{B}}{[4B + (AB - 1 + \sqrt{(AB-1)^2 + 4B})^2]^{1/2}},$$

$$\cos\phi = \frac{AB - 1 + \sqrt{(AB-1)^2 + 4B}}{[4B + (AB - 1 + \sqrt{(AB-1)^2 + 4B})^2]^{1/2}}.$$

Thus, the partition function is

$$Q = \mathrm{Tr}\left[\hat{\mathbf{U}}^T \begin{pmatrix} \lambda_1 & 0 \\ 0 & \lambda_2 \end{pmatrix}^N \mathbf{U}\right] = \lambda_1^N + \lambda_2^N.$$

(ii) The probability of a nucleosome occupied by a histone H1 molecule is given by

$$< n_i > = \frac{1}{Q} \sum_{\{n_i = 0,1\}} n_i \exp\left[-\beta a \sum_i n_i n_{i+1} - \beta(b+\mu) \sum n_i\right]$$

Following the procedure explained in Appendix 5A, the above equation can be evaluated analytically

$$< n_i > = \frac{1}{Q} \begin{pmatrix} AB & B \\ B & 1 \end{pmatrix}^{i-1} \begin{pmatrix} 1 & 0 \\ 0 & 0 \end{pmatrix}^N \begin{pmatrix} AB & B \\ B & 1 \end{pmatrix}^{N-i+1}$$

$$= \frac{\lambda_1^N \cos^2\phi + \lambda_2^N \sin^2\phi}{\lambda_1^N + \lambda_2^N} = \frac{\cos^2\phi + (\lambda_2/\lambda_1)^N \sin^2\phi}{1 + (\lambda_2/\lambda_1)^N}.$$

If the chain is sufficiently long ($N \gg 1$), $< n_i > \approx \cos^2\phi$.

(iii) From the joint probability for the simultaneous occupancy of nucleosomes i and j

$$< n_i n_j > = \cos^4\phi + e^{|j-i|/\xi_c} \cos^2\phi \sin^2\phi,$$

we can readily obtain the joint probability that nucleosome j is occupied while the nucleosome i is not occupied:

$$\frac{< (1-n_i)n_j >}{< 1 - n_i >} = \frac{< n_j > - < n_i n_j >}{< 1 - n_i >}$$

$$= \frac{\cos^2\phi - \cos^4\phi - e^{|j-i|/\xi_c} \cos^2\phi \sin^2\phi}{1 - \cos^2\phi}$$

$$= \cos^2\phi(1 - e^{|j-i|/\xi_c}) = < n > (1 - e^{|j-i|/\xi_c}).$$

(iv) Figure 5.6 shows ξ_c as a function of $b + \mu$ for $T = 300$ K and $a \approx 7.5$ kcal/mol. Note that the correlation length is extremely sensitive to the binding energy and chemical potential of histones. The maximum value is on the order of several hundred nucleosomes, indicating that inhibition of histone H1 binding to only one nucleosome have a large effect, i.e., the lack of histone H1 at one site can destabilize the entire complex.

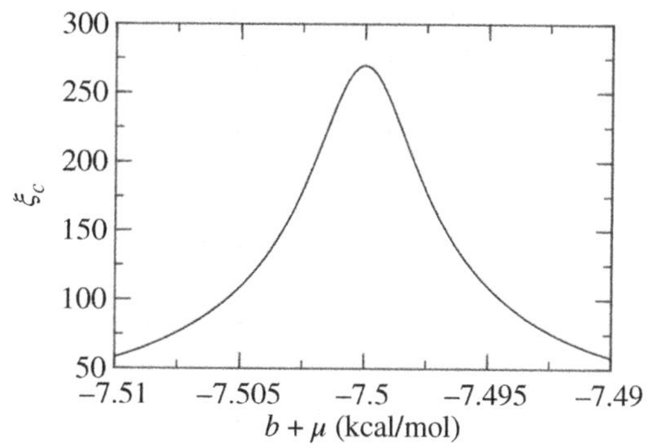

Figure 5.6 Correlation length for histone binding with the chromatin structure (Ishii.m).

Problem 5.15 For the Ising model on the square lattice, the critical point may be identified from the Kramers–Wannier duality, i.e., there is a one-to-one correspondence between terms in the high- and low-temperature series expansions of the partition function. According to the duality relation, the critical temperate satisfies $\tanh(K_C) = e^{-2K_C}$, where $K_C = \beta_C \epsilon$. Show

(i) $\sinh(2K_C) = 1$;

(ii) $K_C = [\ln(1 + \sqrt{2})]/2$.

Solution:

(i) From $\tanh(K) = e^{-2K}$, we have

$$e^{2K} \tanh(K) - 1 = 0.$$

Hence,

$$\frac{e^{2K}(e^K - e^{-K})}{e^K + e^{-K}} - 1 = \frac{e^{2K}(e^{2K} - 1)}{e^{2K} + 1} - 1 = \frac{e^{4K} - 2e^{2K} - 1}{e^{2K} + 1} = 2e^{2K}\frac{(\sinh 2K - 1)}{e^{2K} + 1} = 0,$$

which leads to $\sinh(2K_C) = 1$.

(ii) From $\sinh(2K_C) = 1$, we have $e^{2K_C} - e^{-2K_C} - 2 = 0$. Let $y = e^{2K_C}$, the quadratic equation $y^2 - 2y - 1 = 0$ has the solution of

$$y = \frac{2 \pm \sqrt{8}}{2} = 1 \pm \sqrt{2} \quad \Rightarrow \quad K_C = \frac{1}{2}\ln y = \frac{\ln(1 + \sqrt{2})}{2}.$$

The negative solution is abandoned because $y = e^{2K_C}$ is positive.

Problem 5.16 The Weiss molecular field theory may be alternatively formulated by neglecting the correlated spin fluctuations, i.e., by assuming $(s_i - m)(s_j - m) \approx 0$.

(i) Show that the mean-field energy is given by

$$E_\nu^0 = \sum_i \left[-(h + Zm\epsilon)s_i + Z\epsilon m^2/2\right];$$

(ii) Derive a self-consistent equation for determining the average magnetization;

(iii) Compare the mean-field Helmholtz energy with those obtained from the Gibbs–Bogoliubov variational principle and the Bragg–Williams theory.

Solution:

(i) According to the Ising model, the system energy is given by

$$E_\nu = \sum_i - hs_i - \frac{\epsilon}{2}\sum_{i,j}{}' s_i s_j,$$

where $'$ means that spins i and j are nearest neighbors. Now express the spin variables in terms of the mean value and fluctuation, $s_i = m + (s_i - m)$, the mean-field assumption leads to

$$s_i s_j = [m + (s_i - m)] \cdot [m + (s_j - m)]$$
$$= m^2 + m(s_i - m) + m(s_j - m) + (s_i - m)(s_j - m) \approx m(s_i + s_j) - m^2.$$

Thus, the total energy becomes

$$E_v^0 = \sum_i - hs_i - \epsilon \sum_{i,j}{}' \left[m(s_i + s_j)/2 - m^2/2\right] = \sum_i - hs_i - \epsilon Zm \sum_i s_i + \frac{\epsilon}{2} \sum_i Zm^2 \Bigg]$$
$$= \sum_i \left[-(h + Zm\epsilon)s_i + Z\epsilon m^2/2\right].$$

(ii) According to the mean-field Hamiltonian, the partition function is

$$Q^0 = \sum_i \sum_{s_i=\pm 1} \exp \sum_i \left[\beta(h + Zm\epsilon)s_i - Z\beta\epsilon m^2/2\right] = [2\cosh(\beta h_a)]^N \exp\left[-ZN\beta\epsilon m^2/2\right],$$

where $h_a = h + Zm\epsilon$. From Q^0, we can find a self-consistent equation for determining the average magnetization

$$m = \frac{1}{N} \frac{\partial \ln Q^0}{\partial \beta h_a} = \tanh\left[\beta(h + Zm\epsilon)\right].$$

(iii) The mean-field Helmholtz energy is given by

$$\beta F/N \approx -\frac{1}{N} \ln Q^0 = -\ln\{2\cosh\left[\beta(h + Zm)\right]\} + Z\beta\epsilon m^2/2,$$

which is the same as that obtained from the Gibbs–Bogoliubov variational principle or from the Bragg–Williams theory.

Problem 5.17 Show that the free energy for the Ising model derived from the Bragg–Williams theory

$$\beta F/N = \frac{1+m}{2} \ln \frac{1+m}{2} + \frac{1-m}{2} \ln \frac{1-m}{2} - Z\beta\epsilon m^2/2 - m\beta h,$$

is identical to that from the Gibbs–Bogoliubov variational principle

$$\beta F/N = -\ln\{2\cosh[\beta(h + Zm)]\} + Z\beta\epsilon m^2/2.$$

Solution:

According to the Bragg–Williams theory, the average magnetization can be obtained by minimizing the free energy, i.e.,

$$\frac{\partial \beta F/N}{\partial m} = \frac{1}{2} \ln \frac{1+m}{2} - \frac{1}{2} \ln \frac{1-m}{2} + \frac{1+m}{2} \frac{2}{1+m} - \frac{1-m}{2} \frac{2}{1-m} - Z\beta\epsilon m - \beta h$$
$$= \frac{1}{2} \ln \frac{1+m}{1-m} - Z\beta\epsilon m - \beta h = 0,$$

which is identical to that from the Gibbs–Bogoliubov variational principle

$$m = \tanh\left[\beta(h + Zm\epsilon)\right].$$

Therefore, the free energy can be written as

$$\beta F/N = \frac{1+m}{2} \ln \frac{1+m}{2} - \frac{1+m}{2} \ln \frac{1-m}{2} + \ln \frac{1-m}{2} - Z\beta\epsilon m^2/2 - m\beta h$$

$$
\begin{aligned}
&= \frac{1+m}{2}\ln\frac{1+m}{1-m} + \ln\frac{1-\tanh[\beta(h+Zm\epsilon)}{2} - Z\beta\epsilon m^2/2 - m\beta h \\
&= (1+m)(Z\beta\epsilon m + \beta h) + \ln\frac{e^{-x}}{2\cosh x} - Z\beta\epsilon m^2/2 - m\beta h \\
&= (1+m)(Z\beta\epsilon m + \beta h) - x - \ln 2\cosh x + Z\beta\epsilon m^2/2 \\
&= -\ln\{2\cosh[\beta(Z\epsilon m + h)]\} + Z\beta\epsilon m^2/2,
\end{aligned}
$$

where $x = Z\beta\epsilon m + \beta h$ and $\frac{1}{2}(1 - \frac{e^x - e^{-x}}{e^x - e^{-x}}) = e^{-x}/(2\cosh x)$.

Problem 5.18 The Gibbs–Bogoliubov variational principle may be applied to a generalized Ising model where the external field is dependent on the identity of spins, i.e., the total energy for each microstate is given by

$$E_\nu = \sum_i - h_i s_i - \frac{\epsilon}{2}\sum_{i,j}{}' s_i s_j,$$

where $'$ means that spins i and j are nearest neighbors. Show that the effective mean field for each spin is given by

$$h_i^0 = h_i + \epsilon \sum_j{}' m_j.$$

Solution:

The effective mean field is introduced by defining a noninteracting reference system

$$E_\nu^0 = -\sum_i h_i^0 s_i.$$

In this reference system, the local average of m_i is given by

$$m_i = \tanh[\beta h_i^0].$$

The variational Helmholtz energy is

$$\Phi_0 = \frac{1}{N}\left[\beta F_0 + < \beta \Delta E_\nu >_0\right]$$

where

$$\beta F_0 = -\sum_i \ln\left[2\cosh(\beta h_i^0)\right]$$

$$
\begin{aligned}
< \Delta E_\nu >_0 = \langle E_\nu - E_\nu^0\rangle &= \sum_i - h_i m_i - \frac{\epsilon}{2}\sum_{i,j}{}' m_i m_j + \sum_i h_i^0 m_i \\
&= \sum_i - h_i \tanh[\beta h_i^0] - \frac{\epsilon}{2}\sum_{i,j}{}' \tanh[\beta h_i^0]\tanh[\beta h_j^0] + \sum_i h_i^0 \tanh[\beta h_i^0] \\
&= \sum_i (h_i^0 - h_i)\tanh[\beta h_i^0] - \frac{\epsilon}{2}\sum_{i,j}{}' \tanh[\beta h_i^0]\tanh[\beta h_j^0].
\end{aligned}
$$

Minimizing Φ_0 with respect to h_i^0 leads to

$$\frac{\partial \Phi_0}{\partial \beta h_i^0} = -\frac{\sinh(\beta h_i^0)}{\cosh(\beta h_i^0)} + \frac{\beta(h_i^0 - h_i)}{\cosh^2(\beta h_i^0)} + \tanh(\beta h_i^0) - \frac{\beta\epsilon}{\cosh^2(\beta h_i^0)}\sum_j{}' m_j = 0,$$

which gives

$$h_i^0 = h_i + \epsilon \sum_j{}' m_j.$$

Problem 5.19 Kornyshev[5] proposed a lattice-gas model to describe the capacitance for the electric double layer (EDL) of ionic liquids and molten salts in the presence of an electric field (e.g., introduced by an electrode). Similar to the Bragg–Williams theory, the lattice-gas mode starts with a mean-field free energy for a system of N_+ monovalent cations and N_- monovalent anions placed on a lattice

$$F = e\psi(N_+ - N_-) + B_+ N_+^2 + B_- N_-^2 + B_\pm N_+ N_- - k_B T \ln W,$$

where e stands for unit charge, ψ is the electric potential, B_+, B_-, and $B_\pm$ are binary energy parameters, W is the number of ways to place cations and anions on the lattice, and N is the total number of lattice sites. With the assumption that cations and anions are randomly mixed with each other and with the unoccupied lattice sites, W is given by

$$W = \frac{N!}{(N - N_+ - N_-)!N_+!N_-!}.$$

(i) Supposed that the chemical potentials of cations and anions are given by μ_+ and μ_-, what is the grand potential of the lattice system?

(ii) Assuming that $B_+ = B_- = B_\pm = 0$, show that the ion concentrations can be related to their corresponding values in the bulk (viz., when $\psi = 0$)
(Hint: by minimizing the grand potential)

$$c_+ = c_0 \frac{\exp(-\beta e\psi)}{1 - \gamma + \gamma \cosh(\beta e\psi)}$$
$$c_- = c_0 \frac{\exp(\beta e\psi)}{1 - \gamma + \gamma \cosh(\beta e\psi)},$$

where $c_+ = N_+/v$, $c_- = N_-/v$ are ion concentrations for the lattice system, and v is the lattice volume. $\gamma \equiv 2N_0/N$ is called the lattice-saturation parameter, which provides a measure of the bulk ion concentration. Within the lattice model, N_0 corresponds to the number of cations or anions in the bulk (viz., when $\psi = 0$), and $c_0 = N_0/v$ is the concentration of cations (or anions).

(iii) Consider a planar electrode placed at position $x = 0$ with surface potential ψ_s. The electric potential disappears in the bulk, i.e., $\psi = 0$ at $x = \infty$. Assuming that the above relation between ion concentration and electric potential is valid at any position in space (viz., local density approximation), find an expression for the electric potential as a function of the perpendicular distance x from the surface based on the Poisson equation

$$\frac{\partial^2 \psi}{\partial x^2} = -\frac{4\pi \rho_e}{D},$$

where $D \equiv 4\pi\epsilon_0\epsilon_D$, $\epsilon_0 = 8.854 \times 10^{-12}$ F/m is the vacuum permittivity, ϵ_D stands for the dielectric constant of the ionic fluid, and ρ_e is the local electrical charge density

$$\rho_e = e(c_+ - c_-) = -\frac{2ec_0 \sinh(\beta e\psi)}{1 + 2\gamma \sinh^2(\beta e\psi/2)}.$$

(iv) The differential capacitance per unit surface area of the electrode is defined as

$$C_d \equiv \frac{\partial Q}{\partial \psi_s},$$

where Q is the surface charge density. Derive an expression for capacitance C_d as a function of the surface voltage.

5 Kornyshev A., "Double-layer in ionic liquids: paradigm change?", *J. Phys. Chem. B.*, 111, 5545–5557 (2007).

(v) Plot the capacitance versus the electric potential for several values of γ and discuss how the differential capacitance varies with the ion concentration in the bulk.

Solution:

(i) With the solvent represented as a dielectric continuum, the (semi-) grand potential is given by

$$\begin{aligned}\Omega &= F - N_+\mu_+ - N_-\mu_- \\ &= e\psi(N_+ - N_-) + B_+N_+^2 + B_-N_-^2 + B_\pm N_+N_- \\ &\quad - k_BT\ln\frac{N!}{(N-N_+-N_-)!N_+!N_-!} - N_+\mu_+ - N_-\mu_-.\end{aligned}$$

(ii) The equilibrium ion concentrations can be obtained by minimizing the grand potential with respect to N_+ and N_-:

$$\frac{\partial\Omega}{\partial N_+} = \frac{\partial\Omega}{\partial N_+} = 0,$$

which leads to, noting $\frac{\partial \ln N!}{\partial N} = \ln N$,

$$\mu_+ = \frac{\partial F}{\partial N_+} = e\psi + 2B_+N_+ + B_\pm N_- + k_BT\ln\frac{N_+}{N-N_+-N_-},$$

$$\mu_- = \frac{\partial F}{\partial N_-} = -e\psi + 2B_-N_- + B_\pm N_+ + k_BT\ln\frac{N_-}{N-N_+-N_-}.$$

The chemical potentials of ions for the corresponding bulk ionic system are

$$\mu_+^0 = 2B_+N_+^0 + B_\pm N_-^0 + k_BT\ln\frac{N_+^0}{N-N_+^0-N_-^0},$$

$$\mu_-^0 = 2B_-N_-^0 + B_\pm N_+^0 + k_BT\ln\frac{N_-^0}{N-N_-^0-N_+^0},$$

where N_+^0 and N_-^0 are the number of cations and anions for the bulk phase with $\psi = 0$.

When the ionic system is in equilibrium with the bulk phase, $\mu_+ = \mu_+^0$ and $\mu_- = \mu_-^0$, we have from the above equations, after assuming $B_+ = B_- = B_\pm = 0$,

$$\left(\frac{N_+}{N-N_+-N_-}\right)\left(\frac{N-N_+^0-N_-^0}{N_+^0}\right) = \exp(-\beta e\psi),$$

$$\left(\frac{N_-}{N-N_+-N_-}\right)\left(\frac{N-N_+^0-N_-^0}{N_-^0}\right) = \exp(\beta e\psi). \tag{K}$$

In the bulk phase, $N_+^0 = N_-^0 = N^0$ due to the charge neutrality. With $N_+/N_- = \exp(-2\beta e\psi)$, rearranging Eq. (K) gives

$$N_+ = \frac{N\exp(-\beta e\psi)}{N/N^0 - 2 + 2\cosh(-\beta e\psi)} = \frac{N^0\exp(-\beta e\psi)}{1+\gamma[\cosh(-\beta e\psi)-1]},$$

$$N_- = \frac{N\exp(\beta e\psi)}{N/N^0 - 2 + 2\cosh(-\beta e\psi)} = \frac{N^0\exp(\beta e\psi)}{1+\gamma[\cosh(-\beta e\psi)-1]}, \tag{L}$$

where $\gamma \equiv 2N^0/N$. Using $c_+ = N_+/v$ and $c_- = N_-/v$, we derive the ion concentrations:

$$c_+ = c_0\frac{\exp(-\beta e\psi)}{1-\gamma+\gamma\cosh(\beta e\psi)},$$

$$c_- = c_0\frac{\exp(\beta e\psi)}{1-\gamma+\gamma\cosh(\beta e\psi)}.$$

(iii) Substituting the local charge density to the Poisson equation gives

$$\frac{\partial^2 \psi}{\partial x^2} = \frac{8\pi e c_0}{D} \frac{\sinh(\beta e \psi)}{1 + 2\gamma \sinh^2(\beta e \psi)},$$

Let $\varphi \equiv \beta e \psi$ and $\lambda \equiv x\sqrt{8\pi\beta e^2 c_0/D} = \kappa x$, where $\kappa \equiv \sqrt{8\pi\beta e^2 c_0/D}$ stands for the Debye screening parameter, the Poisson equation becomes

$$\frac{\partial^2 \varphi}{\partial \lambda^2} = \frac{\sinh \varphi}{1 + 2\gamma \sinh^2(\varphi/2)}.$$

Rewrite the right side as $\frac{\partial^2 \varphi}{\partial \lambda^2} = \frac{1}{2}\frac{\partial \varphi'^2}{\partial \varphi}$ and then integrate the above equation with respect to φ:

$$\frac{1}{2}\varphi'^2 = \frac{1}{\gamma} \ln[1 + 2\gamma \sinh^2(\varphi/2)],$$

where the integration constant disappears because $\varphi' = 0$ at $\lambda = \infty$. Hence,

$$\varphi' = \frac{\partial \varphi}{\partial \lambda} = -\text{sgn}(\varphi_s)\sqrt{2/\gamma}\sqrt{\ln[1 + 2\gamma \sinh^2(\varphi/2)]},$$

where $\text{sgn}(\varphi_s)$ is taken to ensure that the variation of the local electric potential is consistent with the electrode charge. With boundary conditions $\varphi = \varphi_s$ at $\lambda = 0$, the above equation can be further integrated to give

$$\lambda = \text{sgn}(\varphi_s)\sqrt{\gamma/2} \int_{\varphi}^{\varphi_s} \frac{d\varphi}{\sqrt{\ln[1 + 2\gamma \sinh^2(\varphi/2)]}}.$$

(iv) The differential capacitance per unit surface area of the electrode is defined as

$$C_d \equiv \frac{\partial Q}{\partial \psi_s}.$$

The surface charge density can be related to the electric potential according to Gauss's law

$$Q = -\frac{D}{4\pi} \frac{\partial \psi}{\partial x}\bigg|_{x=0}.$$

Thus, the specific capacitance becomes

$$\begin{aligned}
C_d &= -\frac{D}{4\pi} \frac{\partial}{\partial \psi_s} \left\{ \frac{\partial \psi}{\partial x}\bigg|_{x=0} \right\} = -\frac{D\kappa}{4\pi} \frac{\partial}{\partial \varphi_s} \left\{ \frac{\partial \varphi}{\partial \lambda}\bigg|_{\lambda=0} \right\} \\
&= \text{sgn}(\varphi_s) C_D \frac{\partial}{\partial \varphi_s} \left\{ \sqrt{2/\gamma}\sqrt{\ln[1 + 2\gamma \sinh^2(\varphi_s/2)]} \right\} \\
&= \text{sgn}(\varphi_s) C_D \sqrt{2/\gamma} \frac{1/2}{\sqrt{\ln[1 + 2\gamma \sinh^2(\varphi_s/2)]}} \frac{2\gamma \sinh(\varphi_s/2)\cosh(\varphi_s/2)}{1 + 2\gamma \sinh^2(\varphi_s/2)} \\
&= \text{sgn}(\varphi_s) C_D \frac{\sinh(\varphi_s)}{1 + 2\gamma \sinh^2(\varphi_s/2)} \sqrt{\frac{\gamma}{2\ln[1 + 2\gamma \sinh^2(\varphi_s/2)]}},
\end{aligned}$$

where $C_D \equiv D\kappa/(4\pi)$ is known as the "Debye" capacitance.

(v) Figure 5.7 shows the capacitance versus the reduced potential for three values of γ. The capacitance varies with the electrical potential in a "bell" shape at high ion concentration but a "bactrian camel" shape at low ion concentration.

Figure 5.7 The relative EDL capacitance (C_d/C_D) versus the reduced surface potential ($\varphi_s = \beta Ze\psi_s$) for three values of the lattice-saturation parameter, γ=0.1, 0.5, and 1.0.

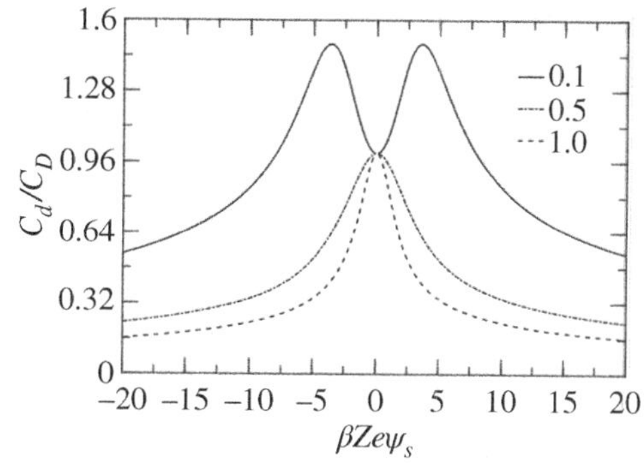

Problem 5.20 Small angle x-ray scattering (SAXS) and small-angle neutron scattering (SANS) are experimental techniques commonly used for quantifying the microscopic structure of matter at atomic dimensions. The intensity and angular distribution of the scattered radiation depend on the fluctuations of electron density in the case of x-rays or nuclear density in the case of neutrons. In both cases, the scattered intensity I can be represented by the Fourier transform of the density–density correlation function $\chi(r)$

$$I(\mathbf{k}) = \langle \eta^2 \rangle \int d\mathbf{r}\chi(r)e^{i\mathbf{k}\cdot\mathbf{r}}, \tag{M}$$

where $\langle \eta^2 \rangle$ is the mean square fluctuation of the scattering density, and the wave vector $\mathbf{q}$ depends on the scattering angle θ and the wave length of the incident beam.

(i) Based on the density–density correlation function from the Teubner–Strey theory of microemulsions

$$\chi(r) = \frac{e^{-r/\xi}}{\kappa r}\sin(\kappa r),$$

show that the scattering intensity can be expressed as

$$I(\mathbf{k}) = \frac{8\pi\langle \eta^2 \rangle c_4}{\xi(a_2 + c_2k^2 + c_4k^4)}, \tag{N}$$

where $k = |\mathbf{k}|$, and

$$\xi = \left[\frac{1}{2}\left(\frac{a_2}{c_4}\right) + \frac{1}{4}\frac{c_2}{c_4}\right]^{-1/2}, \qquad 1/\kappa = \left[\frac{1}{2}\left(\frac{a_2}{c_4}\right) - \frac{1}{4}\frac{c_2}{c_4}\right]^{-1/2}.$$

(ii) Consider an inhomogeneous medium consisting of two types of scattering domains 1 and 2 (e.g., oil and water in a microemulsion). Assume that the fluctuation of the scattering density can be represented by the random distribution of these two scattering regions on a lattice, show that the mean square fluctuation is given by

$$\langle \eta^2 \rangle = \Phi_1\Phi_2\delta^2,$$

where $\Phi_1 = N_1/N$ and $\Phi_2 = N_2/N$ are the fractions of lattice sites occupied by 1 and 2, respectively, $N = N_1 + N_2$ is the total number of lattice sites, and δ is the difference in the amplitude of scatterings from the two scattering regions.
Hint: $\langle \eta^2 \rangle = \frac{\delta^2}{N}\sum_{i,j=1}^{N}[\langle n_in_j\rangle - \langle n_i\rangle\langle n_j\rangle]$ where $n_i = 1$ is the lattice site is occupied by 1 and $n_i = 0$ otherwise.

Solution:

(i) Let $t \equiv \cos\theta$, then $dt = -\sin\theta d\theta$. With Euler's formula $e^{ix} = \cos x + i\sin x$, we may first integrate the angular variables in the three-dimensional Fourier transform

$$\begin{aligned}\int d\mathbf{r}\chi(r)e^{i\mathbf{k}\cdot\mathbf{r}} &= \int_0^{2\pi} d\phi \int_0^{\pi} \sin\theta d\theta \int_0^{\infty} r^2 dr e^{ikr\cos\theta}\chi(r) \\ &= 2\pi \int_0^{\infty} r^2 dr \chi(r) \int_{-1}^{1} dt e^{ikrt} \\ &= 2\pi \int_0^{\infty} r^2 dr \chi(r) \frac{e^{ikr} - e^{-ikr}}{ikr} \\ &= 4\pi \int_0^{\infty} r^2 dr \chi(r) \frac{\sin(kr)}{kr}.\end{aligned}$$

Substituting the above expression and $\chi(r)$ from the Teubner–Strey theory into Eq. (M) gives

$$\begin{aligned}I(\mathbf{q}) &= 4\pi\langle\eta^2\rangle \int_0^{\infty} r^2 dr \frac{e^{-r/\xi}\sin(\kappa r)}{\kappa r}\frac{\sin(kr)}{kr} \\ &= \frac{4\pi\langle\eta^2\rangle}{\kappa k} \int_0^{\infty} dr e^{-r/\xi}\sin(\kappa r)\sin(kr) \\ &= \frac{4\pi\langle\eta^2\rangle}{\kappa k} \frac{2\kappa k/\xi}{[1/\xi^2 + (\kappa - k)^2] \times [1/\xi^2 + (\kappa + k)^2]} \\ &= \frac{8\pi\langle\eta^2\rangle}{\xi} \frac{1}{(\kappa^2 + 1/\xi^2)^2 - 2(\kappa^2 - 1/\xi^2)k^2 + k^4}.\end{aligned}$$

Let $a \equiv (\kappa^2 + 1/\xi^2)$ and $b \equiv \kappa^2 - 1/\xi^2$, the above equation becomes

$$I(\mathbf{k}) = \frac{8\pi\langle\eta^2\rangle/\xi}{a^2 - 2bk^2 + k^4}. \tag{O}$$

Note that $a_2/c_4 = (\kappa^2 + 1/\xi^2)^2$ and $c_2/c_4 = -2(\kappa^2 - 1/\xi^2)$, it is easy to show that Eqs. (O) and (N) are identical.

(ii) For the random mixing of two components 1 and 2 on a lattice, we have $\langle n_i\rangle = \langle n_i n_i\rangle = \Phi_1$ and $\langle n_i n_j\rangle = \langle n_i\rangle\langle n_j\rangle$ for $i \neq j$. Hence,

$$\begin{aligned}\langle\eta^2\rangle &= \frac{\delta^2}{N}\sum_{i,j=1}^{N}[\langle n_i n_j\rangle - \langle n_i\rangle\langle n_j\rangle] \\ &= \frac{\delta^2}{N}\sum_{i=1}^{N}[\langle n_i n_i\rangle - \langle n_i\rangle\langle n_i\rangle] = [\Phi_1 - \Phi_1^2]\delta^2 = \Phi_1\Phi_2\delta^2.\end{aligned}$$

Problem 5.21 The Maier–Saupe theory for predicting phase transition in a nematic liquid crystal is based on the mean-field approximation for van der Waals interaction between rod-like molecules. Specifically, the pair potential between two anisotropic molecules averaged over a random distribution for the center of mass is given by

$$u_{12} = -\epsilon\left(\frac{3}{2}\cos^2\theta_{12} - \frac{1}{2}\right),$$

where $\epsilon > 0$ is an energy parameter, θ_{12} is the angle between the long axes of the two molecules. In the mean-field approximation, the two-body potential is replaced by an effective one-body field

$$u_1 = -\epsilon S\left(\frac{3}{2}\cos^2\theta - \frac{1}{2}\right),$$

where θ is the angle between the long axis of a molecule and the z-axis (i.e., the preferred direction of molecular alignment), and S is the orientational order parameter determined self-consistently from the Boltzmann equation

$$S = <\frac{3}{2}\cos^2\theta - \frac{1}{2}> = \frac{\int_0^\pi \sin\theta d\theta(\frac{3}{2}\cos^2\theta - \frac{1}{2})e^{-\beta u_1}}{\int_0^\pi \sin\theta d\theta e^{-\beta u_1}}. \quad \text{(P)}$$

(i) Show that $S = 0$, which corresponds to the system in an isotropic phase, satisfies the self-consistent equation, Eq. (P);

(ii) Derive the reduced Helmholtz energy of the nematic phase according to the mean-field approximation:

$$\beta F/N = -\frac{1}{N}\ln Q^{IG} - \ln 2 - \frac{3}{2}\ln(\pi) + \frac{\alpha S}{3} - \ln\frac{\text{erfi}(\sqrt{\alpha S})}{\sqrt{\alpha S}}.$$

where $\alpha = 3\beta\epsilon/2$, Q^{IG} stands for the partition function for a system of noninteracting particles, and erfi(x) is the imaginary error function.

(iii) Find nontrivial solution of S numerically and plot it as a function of the reduced temperature k_BT/ϵ;

(iv) Plot the reduced Helmholtz energy per molecule in the nematic phase relative to that in the isotropic phase as a function of reduced temperature k_BT/ϵ;

(v) What are the critical temperature and the orientational order parameter at the isotropic–nematic phase transition? Is it a first-order or second-order phase transition at the critical point?

Solution:

(i) When the system is isotropic, $S = 0$, Eq. (P) is valid because $u_1 = 0$ and

$$S = \frac{\int_0^\pi \sin\theta d\theta(\frac{3}{2}\cos^2\theta - \frac{1}{2})}{\int_0^\pi \sin\theta d\theta} = \frac{\int_0^1 d\cos\theta(\frac{3}{2}\cos^2\theta - \frac{1}{2})}{\int_0^1 d\cos\theta} = \frac{\frac{3}{2}\times\frac{1}{3} - \frac{1}{2}}{2} = 0.$$

(ii) With the mean-field approximation, the canonical partition function can be written as that corresponding to a noninteracting system with an effective one-body potential

$$Q = Q^{IG}\left\{2\pi\int_0^\pi \sin\theta d\theta \exp\left[-\beta\epsilon S\left(\frac{3}{2}\cos^2\theta - \frac{1}{2}\right)\right]\right\}^N,$$

where Q^{IG} stands for the partition function for a system of noninteracting particles, and N is the number of molecules in the system, 2π accounts for the azimuthal angle for the orientation of a single molecule, and θ is the polar angle. For an isotropic phase, $S = 0$, u_1 vanishes thus the Helmholtz energy becomes

$$\beta F^{iso} = -\ln Q^{IG} - N\ln(4\pi).$$

For the nematic phase, the mean-field canonical partition function can be evaluated analytically

$$Q = Q^{IG}\left\{2\pi\int_0^\pi \sin\theta d\theta \exp(-\beta u_1)\right\}^N = Q^{IG}\left\{2\pi\int_{-1}^1 d\xi \exp\left[\alpha S\left(\xi^2 - \frac{1}{3}\right)\right]\right\}^N$$

$$= Q^{IG}\left\{4\pi e^{-\alpha S/3}\frac{\sqrt{\pi}\text{erfi}(\sqrt{\alpha S})}{2\sqrt{\alpha S}}\right\}^N = Q^{IG}\left\{\frac{2\pi^{3/2}e^{-\alpha S/3}}{\sqrt{\alpha S}}\text{erfi}(\sqrt{\alpha S})\right\}^N,$$

where $\alpha \equiv 3\beta\epsilon/2$. Hence, the reduced Helmholtz energy per molecule is

$$\beta F/N = -\frac{1}{N}\ln Q^{IG} - \ln 2 - \frac{3}{2}\ln(\pi) + \frac{\alpha S}{3} - \ln\frac{\text{erfi}(\sqrt{\alpha S})}{\sqrt{\alpha S}}.$$

As $S \to 0$, we have

$$\frac{\text{erfi}(\sqrt{\alpha S})}{\sqrt{\alpha S}} = 2/\sqrt{\pi}.$$

Hence, the expression for the Helmholtz energy of an isotropic phase is recovered.

(iii) Eq. (P) may be simplified as

$$S = \frac{\int_0^\pi \sin\theta d\theta(\frac{3}{2}\cos^2\theta - \frac{1}{2})e^{-\beta u_1}}{\int_0^\pi \sin\theta d\theta e^{-\beta u_1}}$$

$$= \frac{3}{2}\frac{\int_0^\pi \sin\theta d\theta \cos^2\theta \exp[\beta\epsilon S(\frac{3}{2}\cos^2\theta - \frac{1}{2})]}{\int_0^\pi \sin\theta d\theta \exp[\beta\epsilon S(\frac{3}{2}\cos^2\theta - \frac{1}{2})]} - \frac{1}{2}$$

$$= \frac{3}{2}\frac{\int_{-1}^1 d\xi \xi^2 e^{\alpha S\xi^2}}{\int_{-1}^1 d\xi e^{\alpha S\xi^2}} - \frac{1}{2} = \frac{3}{2}\left[\frac{e^{\alpha S}}{2\alpha S} - \frac{\sqrt{\pi}\text{erfi}(\sqrt{\alpha S})}{4(\alpha S)^{3/2}}\right] \times \frac{2\sqrt{\alpha S}}{\sqrt{\pi}\text{erfi}(\sqrt{\alpha S})} - \frac{1}{2}$$

$$= \frac{3}{4}\left[\frac{2e^{\alpha S}}{\sqrt{\pi\alpha S}\times \text{erfi}(\sqrt{\alpha S})} - \frac{1}{\alpha S}\right] - \frac{1}{2},$$

where $\xi = \cos\theta$, and erfi(x) is the imaginary error function. Given $0 < S < 1$, the above equation can be numerically solved for α from which the reduced temperature can be calculated (Figure 5.8).

(iv) Figure 5.9 plots the reduced Helmholtz energy per molecule in the nematic phase relative to that in the isotropic phase as a function of reduced temperature $k_B T/\epsilon$;

(v) By equaling the free energy of the isotropic phase and the nematic phase, we can calculate the reduced critical temperature for the isotropic–nematic phase transition is $k_B T_c/\epsilon \approx 0.215$, and the order parameter at the critical point is $S_c \approx 0.43$. The critical order parameter was found in good agreement with the observed values for many real nematic crystals.[6] The isotropic to nematic transition is a first-order phase transition because the order parameter is discontinuous.

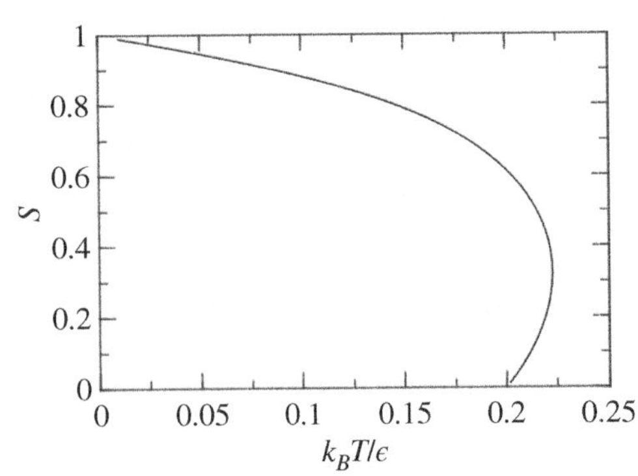

Figure 5.8 The dependence of the order parameter S on the reduced temperature predicted by the Maier–Saupe theory. (nematic.m)

6 See, e.g., Luckhurst G. and Zannoni C., "Why is the Maier–Saupe theory of nematic liquid crystals so successful?" *Nature* 267, 412–414 (1977).

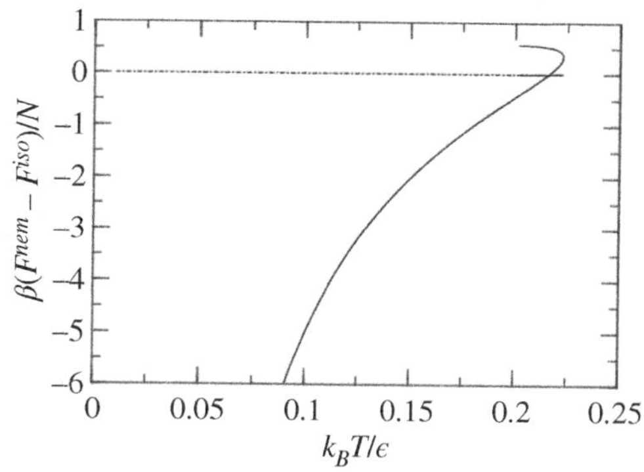

Figure 5.9 The difference in the reduced Helmholtz energy per molecule between the nematic phase and the isotropic phase (nematic.m).

Problem 5.22 When the free energy density (f) of a system does not satisfy the invariance condition $f(m) = f(-m)$ for the order parameter m, a first-order phase transition can be described with the Landau theory by using a fourth-order polynomial that contains both even and odd terms

$$f(m) = f_0 + a_2m^2/2 + a_3m^3/3 + a_4m^4/4,$$

where f_0 is the free energy density for the disordered state ($m = 0$), $a_2 = a_2^0(T - T_0)$, $a_3 < 0$, and $a_4 > 0$. With parameter $a_2^0 > 0$, T_0 controls the sign of a_2 in response to temperature changes.

(i) Show that the disordered phase is always stable when $T > T_A \equiv T_0 + a_3^2/(4a_2^0a_4)$.

(ii) A metastable phase with $m \neq 0$ emerges when the temperature is in the range of $T_C \equiv T_0 + 2a_3^2/(9a_0a_4) < T < T_A$.

(iii) The system undergoes a first-order phase transition at T_C, i.e., the disordered phase is in equilibrium with an ordered phase.

(iv) When $T_0 < T < T_C$, the ordered phase is stable while the disordered phase is metastable.

(v) When $T < T_0$, only the ordered phase is stable, and the disordered phase is unstable.

(vi) Plot the Landau free energy f as a function of m at $T = T_0, T_C$ and T_A and prepare a phase diagram of the system with parameters $f_0 = 0$ and $T_0 = a_2^0 = -a_3 = a_4 = 1$. Label the transition temperatures.

Solution:

This problem is adapted from lecture notes prepared by Dr. An-Chang Shi of McMaster University.

(i) According to the Landau theory of phase transitions, the order parameter of a thermodynamically stable state minimizes the free energy. This condition implies that the first-order derivative of the free energy with respect to the order parameter is zero

$$\frac{\partial f}{\partial m} = a_2m + a_3m^2 + a_4m^3 = 0, \tag{Q}$$

and that the second-order derivative is positive

$$\left.\frac{\partial^2 f}{\partial m^2}\right|_{m=0} = (a_2 + 2a_3m + 3a_4m^2)|_{m=0} = a_2 = a_2^0(T - T_0) > 0. \tag{R}$$

Eq. (Q) has three solutions: $m = 0$ corresponds to the disordered state, and the other two solutions are given by

$$m_\pm = \frac{-a_3 \pm \sqrt{a_3^2 - 4a_2a_4}}{2a_4}. \tag{S}$$

Because m is a real number, the nonzero solutions are physically meaningful only when $a_3^2 > 4a_2a_4$.
Using $a_2 = a_2^0(T - T_0)$, we find from $a_3^2 > 4a_2a_4$ that the nonzero real solution exists only when

$$T < T_A \equiv T_0 + a_3^2/(4a_2^0a_4).$$

In other words, when $T > T_A \equiv T_0 + a_3^2/(4a_2^0a_4)$, there is only one order parameter ($m = 0$) that satisfies $\partial f/\partial m = 0$. In this case, the disordered phase is stable because the second-order derivative at $m = 0$

$$\left.\frac{\partial^2 f}{\partial m^2}\right|_{m=0} = (a_2 + 2a_3m + 3a_4m^2)|_{m=0} = a_2 = a_2^0(T - T_0),$$

is always positive for $T > T_0 + a_3^2/(4a_2^0a_4)$.

(ii) At $T < T_A$, we have two real values of m satisfying $\partial f/\partial m = 0$. To determine whether the solutions correspond to a minimum or a maximum, we need to calculate the second-order derivative of the free energy

$$\left.\frac{\partial^2 f}{\partial m^2}\right|_{m=m_\pm} = (a_2 + 2a_3m + 3a_4m^2)|_{m=m_\pm}.$$

Plugging Eq. (S) into the above equation yields

$$\left.\frac{\partial^2 f}{\partial m^2}\right|_{m=m_\pm} = 2A[(1 - a_2/A) \pm \sqrt{1 - a_2/A}], \tag{T}$$

where $A \equiv a_3^2/4a_4$. For $T < T_A$, $a_3^2 > 4a_2a_4$ or $a_2/A = 4a_2a_4/a_3^2 < 1$, Eq. (T) predicts that the second-order derivative is positive for $m = m_+$, and negative for $m = m_-$. Therefore, the solution $m = m_+$ minimizes the free energy while $m = m_-$ corresponds to a maximum, which is unstable.

While both $m = 0$ and $m = m_+$ minimize the free energy, the thermodynamically stable state corresponds to the one that has the lower free energy, and the other is metastable. The critical temperature is defined by the order-disorder transition, where the disordered and ordered states have equal free energy. From $f(0) = f(m_+)$ and $f'(0) = f'(m_+) = 0$, we have

$$a_2/2 + a_3m/3 + a_4m^2/4 = 0 \quad \text{and} \quad a_2 + a_3m + a_4m^2 = 0.$$

Eliminating the quadratic terms from the above equations gives $m = -3a_2/a_3$. As m satisfies Eq. (S), we obtain

$$m = -3a_2/a_3 = \frac{-a_3 \pm \sqrt{a_3^2 - 4a_2a_4}}{2a_4} \Rightarrow a_3^2 = \frac{9a_2a_4}{2}.$$

Therefore, the critical temperature T_C and the order parameter at the transition point m_c are given by

$$T_C = T_0 + \frac{2a_3^2}{9a_0a_4} \quad \text{and} \quad m_c = \frac{-3a_2}{a_3} = \frac{-3}{a_3} \times \frac{2a_3^2}{9a_4} = \frac{-2a_3}{3a_4}.$$

When the temperature is in the range of $T_C < T < T_A$, it can be shown $f(0) < f(m_+)$, and thus $m = m_+$ corresponds to a metastable sate.

(iii) As discussed above, the disordered phase ($m = 0$) is in equilibrium with an ordered phase with $m_c = \frac{-2a_3}{3a_4}$ at $T = T_C$. This is a first-order phase transition because of the discontinuous change of the order parameter.

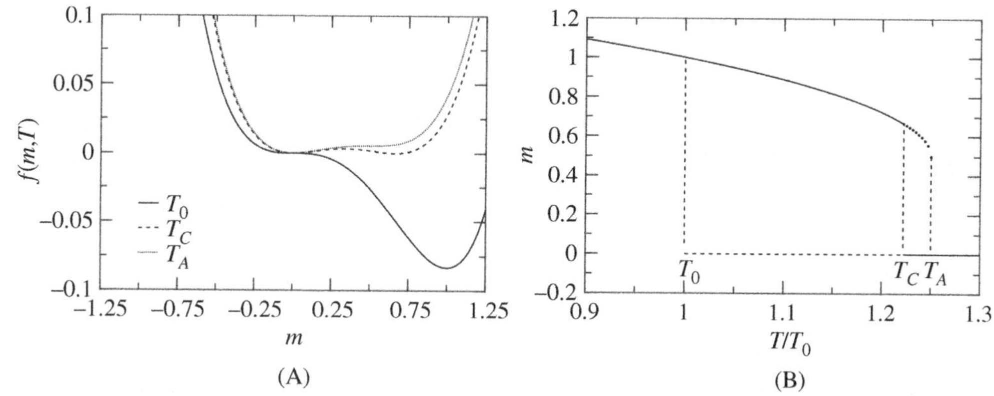

Figure 5.10 A. The Landau free energy represented by a fourth-order polynomial that contains both even and odd terms, $f(m,T) = f_0 + a_2^0(T - T_0)m^2/2 + a_3m^3/3 + a_4m^4/4$ with $f_0 = 0$ and $T_0 = a_2^0 = -a_3 = a_4 = 1$. B. Phase diagram showing the first-order phase transition at T_C (Landau3.m).

(iv) When $T_0 < T < T_C, f(0) > f(m_+)$. Thus, the ordered phase is stable and the disordered phase is metastable.

(v) For $T < T_0$, $a_2 < 0$, we have

$$\left.\frac{\partial^2 f}{\partial m^2}\right|_{m=0} = a_2^0(T - T_0) < 0,$$

which indicates that the disordered phase is unstable.

(vi) Figure 5.10 presents the free energy density (A) and phase diagram (B) predicted by the Landau theory. In part B, the dotted line and the horizontal dashed line show the ranges of temperature where the ordered-phase and the disordered phase are metastable, respectively.

Problem 5.23 The interconnections between different critical exponents were established by B. Widom.[7] The assumption was that, near the critical point, the Gibbs energy per particle, $g \equiv G/N$, includes two contributions, one is referred to as a regular part g_r that has an analytical form, and the other is called a scaling part g_c that has the singular behavior. For a one-component system, the free energy density can be written as

$$g(T, \xi) = g_r(T, \xi) + g_s(\Delta T, \Delta\xi),$$

where T is temperature, ξ (e.g., pressure) is an intensive variable of the system that is conjugated with a dynamic variable x (e.g., particle density or magnetization), and Δ stands for the deviation from the critical condition. The critical behavior implies that the scaling part of the free energy is a generalized homogeneous function, i.e., it satisfies

$$g_s(\lambda^p \Delta T, \lambda^q \Delta\xi) = \lambda g_s(\Delta T, \Delta\xi), \qquad \text{(U)}$$

where $\lambda > 0$ denotes the scaling factor, p and q are parameters.

(i) Using the thermodynamic relation $x = (\partial g/\partial\xi)_T$, show

$$\beta = \frac{1-q}{p},$$

where β is the exponent describing the difference between the two coexisting phases in dynamic variable x, i.e., $\Delta x \sim \Delta T^\beta$.

7 Widom B., "Equation of state in neighborhood of critical point". *J. Chem. Phys.* 43, 3898–3905 (1965).

(ii) For the scaling relation, $\Delta x \sim \Delta\xi^{1/\delta}$, exponent δ satisfies

$$\delta = \frac{q}{1-q}.$$

(iii) According to the linear response theory, the derivative of x with respect to ξ (viz., susceptibility) is related to the fluctuation

$$< \delta x^2 > = -\left(\frac{\partial x}{\partial \xi}\right)_T.$$

Show that the critical exponent for susceptibility is given by

$$\gamma = \frac{2q-1}{p}.$$

(iv) Based on the thermodynamic relation between heat capacity and reduced free energy

$$C_\xi/N = -T\left(\frac{\partial^2 g}{\partial T^2}\right)_\xi,$$

show that the critical exponent for heat capacity, $C_\xi \sim (1/\Delta T)^\alpha$, is given by

$$\alpha = 2 - 1/p.$$

Solution:

The solution for this problem is adapted from[8]

(i) From $x = (\partial g/\partial \xi)_T$, we can obtain by differentiating both sides of Eq. (U) with respect to ξ

$$\lambda^q \Delta x(\lambda^p \Delta T, \lambda^q \Delta\xi) = \lambda \Delta x(\Delta T, \Delta\xi), \quad \text{(V)}$$

Now let $\lambda = (1/\Delta T)^{1/p}$ and $\Delta\xi = 0$, we obtain

$$\Delta x(1,0) = \Delta T^{-(1-q)/p} \Delta x(\Delta T, 0),$$

which leads to

$$\Delta x(\Delta T, 0) = \Delta T^{(1-q)/p} \Delta x(1,0) \sim \Delta T^{(1-q)/p}.$$

Thus

$$\beta = \frac{1-q}{p},$$

where β is the exponent describing the difference between the two coexisting phases in dynamic variable x as the temperature approaches the critical point, i.e., $\Delta x \sim \Delta T^\beta$.

(ii) In Eq. (V) derived above, let $\Delta T = 0$ and $\lambda = (1/\Delta\xi)^{1/q}$ thus

$$\Delta x(0, \Delta\xi) = \Delta\xi^{(1-q)/q} \Delta x(0,1) \sim \Delta\xi^{(1-q)/q}.$$

Hence

$$\delta = \frac{q}{1-q}.$$

(iii) Near the critical point, the susceptibility scales as

$$\chi \sim < \delta x^2 > \sim (1/\Delta T)^\gamma.$$

Differentiating Eq. (V) with respect to ξ gives

$$\lambda^{2q} \chi(\lambda^p \Delta T, \lambda^q \Delta\xi) = \lambda \chi(\Delta T, \Delta\xi),$$

8 Reichl L. R., *A Modern Course in Statistical Physics* (3rd Edition), Wiley-VCH, 2009, pp 149–155.

Let $\lambda = (1/\Delta T)^{1/p}$ and $\Delta\xi = 0$, we obtain

$$\chi(\Delta T, 0) = (1/\Delta T)^{(2q-1)/p}\chi(1, 0).$$

Thus, the critical exponent for susceptibility is given by

$$\gamma = \frac{2q-1}{p}.$$

(iv) Differentiating Eq. (U) with respect to T gives

$$\lambda^p s(\lambda^p \Delta T, \lambda^q \Delta\xi) = \lambda s(\Delta T, \Delta\xi),$$

where $s = S/N$ denotes entropy per particle. Differentiating the above equation with respect to T again gives

$$\lambda^{2p} C_x(\lambda^p \Delta T, \lambda^q \Delta\xi) = \lambda C_x(\Delta T, \Delta\xi).$$

If we let $\lambda = (1/\Delta T)^{1/p}$ and $\Delta\xi = 0$, we obtain

$$C_\xi(\Delta T, 0) = (1/\Delta T)^{(2p-1)/p} C_\xi(1, 0).$$

Hence, we have $\alpha = 2 - 1/p$.

6

Monte Carlo Simulation

Problem 6.1 Generate a sequence of random numbers between 0 and 1 and calculate their mean and standard deviation versus the number of samples. Discuss how your numerical results are compared with the exact values of the mean and standard deviation.

Solution

The probability density of the random numbers between 0 and 1 is $p(x) = 1$. Therefore, its mean and standard deviation are given by

$$M = \int_0^1 xp(x)dx = \int_0^1 xdx = 0.5;$$

$$\sigma^2 = \int_0^1 (x - M)^2 p(x)dx = \int_{-1/2}^{1/2} x^2 dx = \frac{1}{12} \Rightarrow \sigma \approx 0.2887.$$

Figure 6.1 shows how the simulation results approach the exact mean and standard deviation (dashed lines) as the number of samples increases. Here, the numerical results are based on Matlab function *Rand*. The mean absolute percentage error (MAPE), $|\overline{x}_N - M|/M$, scales with the sample size as $\sim 1/\sqrt{N}$.

Problem 6.2 In probability theory, the universality of the uniform refers to the concept that if we have a continuous random variable denoted as x, with a probability density function $p(x)$, then its cumulative distribution function (CDF)

$$F(x) = \int_{-\infty}^{x} f(x')dx'$$

corresponds to a random variable that follows a uniform distribution in the interval [0, 1]. The universality of the uniform is easily proved by the change of variables, $y = F(x)$, which leads to

$$1 = \int_{-\infty}^{\infty} f(x)dx = \int_0^1 f(x)\frac{dx}{dy}dy = \int_0^1 \frac{f(x)}{f(x)}dy = \int_0^1 dy.$$

The above equation indicates that y is a random variable that has a uniform probability distribution in the interval [0,1]. Based on the universality of the uniform, show that the exponential distribution

$$f(x) = \begin{cases} \frac{1}{\langle x \rangle} \exp\left(-\frac{x}{\langle x \rangle}\right), & 0 < x < \infty, \\ 0, & \text{otherwise}, \end{cases}$$

can be generated from $x = -\langle x \rangle \ln y$ where $\langle x \rangle$ represents the mean, and y is a random number with a uniform probability distribution in [0,1].

Solutions Manual to Fundamentals and Practice in Statistical Thermodynamics,
First Edition. Jianzhong Wu and John M. Prausnitz.

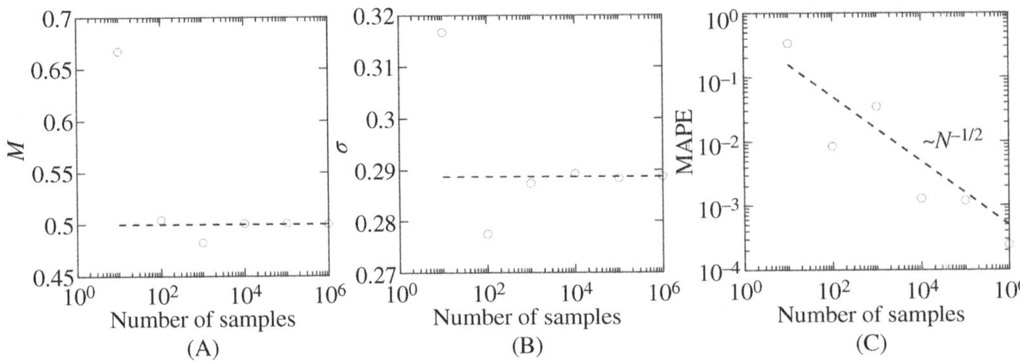

Figure 6.1 Variations of the mean (A), standard deviation (B), and the mean absolute percentage error (C) with the sample size (*N*) (RandomNumbers.m).

Solution:

The CDF for the exponential distribution is given by

$$F(x) = \int_{-\infty}^{x} f(x')dx' = \int_{-\infty}^{x} \frac{1}{\langle x \rangle} \exp\left(-\frac{x'}{\langle x \rangle}\right) dx' = \exp\left(-\frac{x}{\langle x \rangle}\right).$$

Because $y = F(x)$ satisfies uniform probability distribution in the interval [0,1], $x = -\langle x \rangle \ln y$ follows the exponential distribution.

Problem 6.3 Prepare a Monte Carlo (MC) program to estimate the value of π and discuss how the numerical error varies with the sample size.

Solution:

The value of π can be calculated by generating two arrays of random numbers between 0 and 1, one for x and one for y, and counting the number of points with a distance to the coordinate origin less than 1. Figure 6.2 shows how the estimated value varies with the number of samples ($\sim 1/\sqrt{N}$). In order to obtain a highly accurate estimation, it is necessary to perform a lengthy simulation.

Problem 6.4 Suppose that x_1 and x_2 are uniformly and independently distributed random numbers between 0 and 1. The Box–Muller transformation, defined as

$$y_1 = \sqrt{-2\ln x_1}\cos(2\pi x_2), \qquad y_2 = \sqrt{-2\ln x_1}\sin(2\pi x_2),$$

yields two independent standard normal distributions, i.e., variables distributed between $-\infty$ and ∞ with the probability density of $p(y) = e^{-y^2}/\sqrt{\pi}$. Prepare a MC program to generate 1000 sample points following the Gaussian distribution

$$p(x) = \frac{1}{\sqrt{2\pi\sigma^2}} \exp\left[-\frac{(x - x_0)^2}{2\sigma^2}\right],$$

where $x_0 = 2$ and $\sigma = 0.5$. Verify the results by checking the first few moments of the probability distribution:

$$\langle x \rangle = x_0; \qquad \langle x^2 \rangle = x_0^2 + \sigma^2;$$
$$\langle x^3 \rangle = x_0^3 + 3x_0\sigma^2; \qquad \langle x^4 \rangle = x_0^4 + 6x_0^2\sigma^2 + 3\sigma^4.$$

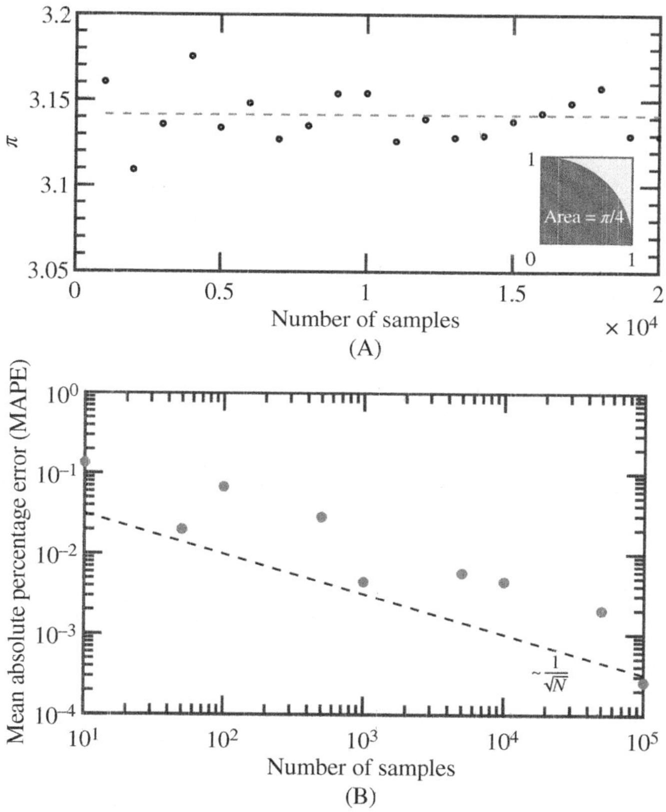

Figure 6.2 The value of π estimated by MC integration (A) and the variation of the mean absolute percentage error with the sample size (B) (pibyMC.m).

Solution:
Figure 6.3 shows the probability densities according to y_1, y_2, and the normally distributed random numbers generated by the builtin Matlab function randn. Table 6.1 compares the numerical results with the exact values for the first four moments of the probability distribution.

Problem 6.5 For a system of classical particles at absolute temperature T, the velocity of translational motions follows the Maxwell–Boltzmann distribution

$$p(\mathbf{v}) = \sqrt{\frac{m}{2\pi k_B T}} \exp\left[-\frac{mv^2}{2k_B T}\right].$$

Prepare a program to assign the velocities of $N = 1000$ particles according to the Maxwell–Boltzmann distribution and verify your result. Assume that all particle have the same mass, $m = 6 \times 10^{-26}$ kg, and that the system temperature is $T = 298.15$ K.

Solution:
The particle velocities can be generated according to the Gaussian distribution as discussed in Problem 6.4. Figure 6.4 compares the probabilities of particle velocities with the Maxwell–Boltzmann distribution.

Problem 6.6 MC simulations often require generating variables within a finite range $[a, b]$ with a given probability density function $p(x)$. This task can be accomplished by rejection sampling,

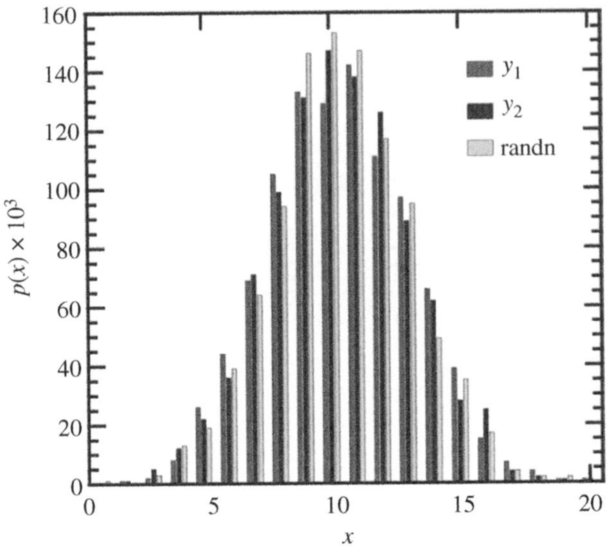

Figure 6.3 Gaussian distributions generated from the Box–Muller transformation. Here, randn means the normally distributed random numbers generated by the builtin Matlab function (Gaussian.m).

Table 6.1 Moments of the Gaussian distribution calculated from different sampling methods.

Sampling method	1st	2nd	3rd	4th
y_1	2.0181	4.3212	9.7067	22.7225
y_2	2.0173	4.3191	9.7244	22.8657
randn	2.0171	4.3209	9.7277	22.8716
exact	2.0000	4.2500	9.5000	22.1875

which is also known as the acceptance–rejection algorithm. The basic idea can be illustrated with Figure 6.5, which demonstrates the generation of variables with probability density $p(x)$ by uniformly sampling the independent variable and retaining the samples falling under the region defined by the graph of the probability density function. Based on the rejection sampling, prepare a program to generate 10^6 numbers from -1 to 1 with a probability satisfying the Gaussian distribution, $p(x) \sim e^{-x^2/2}$.

Solution:
The exact form of the probability density can be found from the normalization condition

$$1 = \int_{-1}^{1} p(x)dx = \frac{1}{C}\int_{-1}^{1} e^{-x^2/2}dx,$$

which gives $C = \sqrt{2\pi}\text{erf}(1/\sqrt{2})$. Figure 6.6 shows the results generated from the rejection sampling algorithm in comparison with the exact distribution function.

Problem 6.7 Consider a shooting game as discussed in Appendix 6A. Suppose that a shooter hits a target with an 80% probability if the previous result is a hit, and 40% probability if the previous result is a miss. Assume that the shooter made the first shot, and consecutive shooting follows a

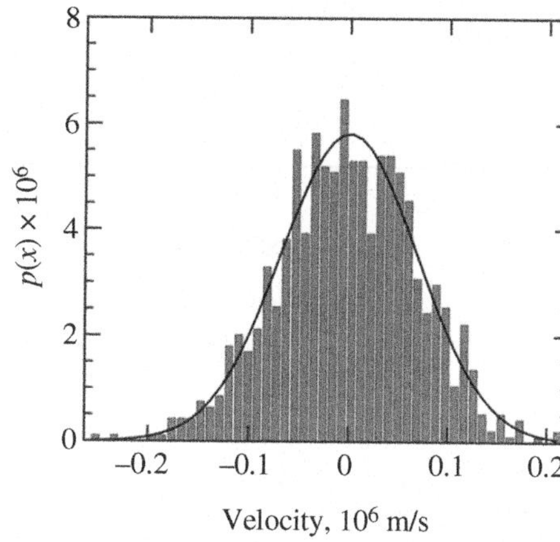

Figure 6.4 The Maxwell–Boltzmann distribution, $p(x)$, for particle velocity in each dimension. The bars are obtained from MC sampling of variables with a Gaussian probability, and the line corresponds to the exact result (Maxwell.m).

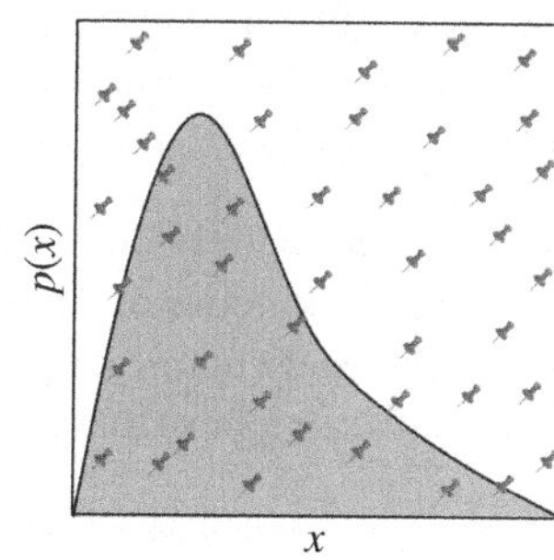

Figure 6.5 Rejection sampling keeps the random samples in the region under the graph of its probability density function, $p(x)$.

Markov chain process. What's the probably that the shooter hit the target in 5th shot? How is it compared with the equilibrium probability.

Solution:

The transition matrix for the Markov process is given by

$$\hat{\mathbf{M}} = \begin{pmatrix} 0.8 & 0.2 \\ 0.4 & 0.6 \end{pmatrix}.$$

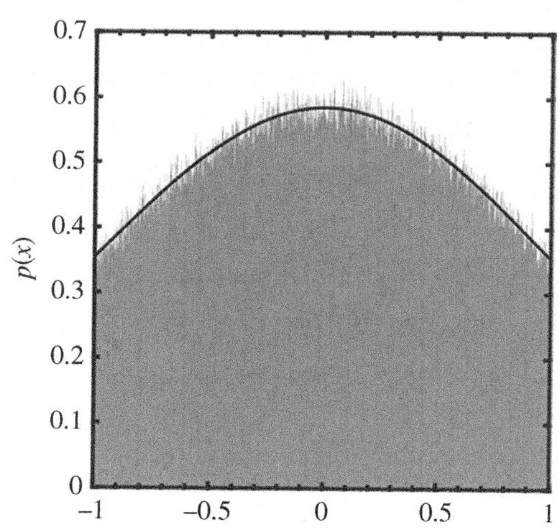

Figure 6.6 Rejection sampling of variables between [−1,1] with a Gaussian probability, $p(x) \sim e^{-x^2/2}$. The solid line represents the exact probability density (Rejection.m).

Table 6.2 The probability distribution in a shooting game (Markov.m).

	Hit	Miss
Initial	1.0000	0
1	0.8000	0.2000
2	0.7200	0.2800
3	0.6880	0.3120
4	0.6752	0.3248
5	0.6701	0.3299
...	...	...

The Markov chain evolves according to $\mathbf{p}_n = \mathbf{p}_0 \times \hat{\mathbf{M}}^n$. Table 6.2 shows the probabilities for the first few steps starting with $\mathbf{p}_0 = (1,0)$. The probably that the shooter hits the target in 5th shot is 0.6701, which is very close to the equilibrium probability, 2/3.

Problem 6.8 Fahidy demonstrated that the Markov chain model is a convenient tool for representing the kinetics of multicomponent linear chemical reactions.[1] In doing so, he related the compositions of chemical species with a probability vector and the rate constants with the transitional matrix. Suppose that three chemical species A, B, C react in a mixture with the rate equations

$$\frac{d[A]}{dt} = -0.3[A] + 0.1[B],$$
$$\frac{d[B]}{dt} = 0.2[A] - 0.3[B] + 0.1[C],$$
$$\frac{d[C]}{dt} = 0.1[A] + 0.2[B] - 0.1[C],$$

where $[\cdots]$ stands for molar concentration. How many independent reactions are there in this system? If the initial composition is $x_A = 0.4$, $x_B = 0.3$, and $x_C = 0.3$, what is the equilibrium composition?

Solution:
This system includes 2 independent reactions, $A \leftrightarrow B$ and $B \leftrightarrow C$. The transition matrix can be constructed from the rate equations, which, in terms of the Markov process, can be expressed as

$$[A]_{n+1} - [A]_n = 0.7[A]_n + 0.1[B]_n,$$
$$[B]_{n+1} - [B]_n = 0.2[A]_n + 0.7[B]_n + 0.1[C]_n,$$
$$[C]_{n+1} - [C]_n = 0.1[A]_n + 0.2[B]_n + 0.9[C]_n.$$

Accordingly, the transfer matrix is given by

$$\hat{\mathbf{M}} = \begin{pmatrix} 0.7 & 0.1 & 0.0 \\ 0.2 & 0.7 & 0.1 \\ 0.1 & 0.2 & 0.9 \end{pmatrix}.$$

The equilibrium composition corresponds to the eigenvector of $\hat{\mathbf{M}}$, $p = (0.0909, 0.2727, 0.6364)$.

1 Fahidy T. Z., "Solving chemical kinetics problems by the Markov-chain approach", *Chem. Eng. Educ.*, Winter, 42–43 (1993).

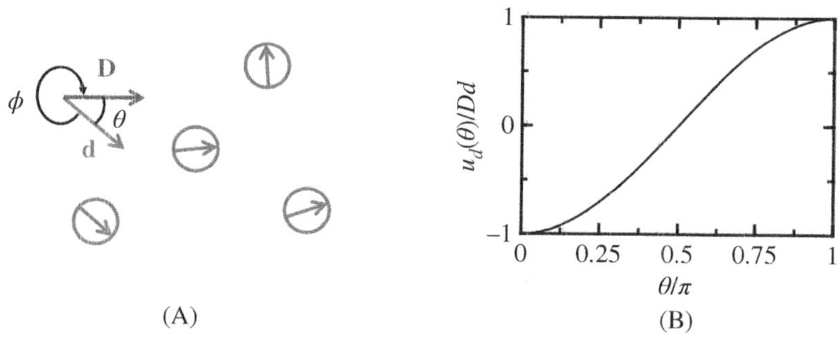

Figure 6.7 A. An ideal gas of point dipole **d** in an electric field **D**. B. The reduced potential energy due to the permanent dipole **d** as a function of the angle θ relative to the direction of the electric field **D**. Here ϕ represents the azimuthal angle around the direction of the electric field.

Problem 6.9 Consider a one-component gas in an external electric field D as depicted schematically in Figure 6.7.

Assume that each gas molecule has polarizability α and permanent dipole $\mathbf{d}$. When the gas density is sufficiently low, the intermolecular interactions can be neglected and, due to the presence of an electric field, each molecule experiences an external energy of

$$u_{ext} = -\frac{1}{2}\alpha D^2 - dD\cos\theta,$$

where $u_{ind} = -\frac{1}{2}\alpha D^2$ represents the energy due to the molecular polarization, and $u_d = -dD\cos\theta$ is the energy due to the permanent dipole $\mathbf{d}$.

(i) Show that the canonical partition function of the system can be written as

$$Q = \frac{1}{N!}\left\{\frac{q_{int}V}{4\pi\Lambda^3}\int_0^{2\pi} d\phi \int_0^{\pi} d\theta \sin\theta e^{-\beta u_{ext}}\right\}^N,$$

where N is the number of gas molecules in the system, Λ is the thermal wave length, q_{int} represents the intrinsic partition function related to the rotational, vibrational, and electronic degrees of freedom of each gas molecule, and V is the system volume.

(ii) Show that the probability density of angle θ is given by

$$p(\theta) = \frac{e^{-\beta u_d(\theta)}\sin\theta}{q_0},$$

where $u_d(\theta) = -dD\cos\theta$, $q_0 = (2/y)\sinh y$, and $y = \beta dD$.

(iii) Plot $p(\theta)$ for $y = 1{,}5$ and 10.

(iv) Why isn't $p(\theta)$ maximized at $\theta = 0$?

(v) Prepare a MC program for sampling the distribution of θ using the Metropolis algorithm and compare the simulation results with the analytic prediction for $y = 1, 5,$ and 10.

Solution:

(i) The system energy includes an ideal gas contribution and the external potential

$$E(\mathbf{x}) = E^{IG}(\mathbf{x}) + Nu_{ext},$$

where $\mathbf{x}$ represents atomic positions and momenta. Accordingly, the canonical partition function is given by

$$Q = \frac{1}{N!C}\int d\mathbf{x}e^{-\beta E(\mathbf{x})},$$

where C is an appropriate normalization constant introduced in the classical representation of microstates. Because the external potential is uncoupled from the atomic positions and momenta, the canonical partition function can be written as that corresponding to a regular ideal gas and an extra term related to the external potential

$$Q = Q^{IG}\left\{\frac{1}{4\pi}\int_0^{2\pi} d\phi \int_0^{\pi} d\theta \sin\theta e^{-\beta u_{ext}}\right\}^N,$$

where $Q^{IG} = (q_{int}V/\Lambda^3)^N/N!$. As expected, $Q = Q^{IG}$ when $u_{ext} = 0$.

(ii) The probability density of angle θ is given by

$$p(\theta) = \frac{e^{-\beta u_d(\theta)}\sin\theta}{\int_0^{\pi} d\theta e^{-\beta u_d(\theta)}\sin\theta} = \frac{e^{-\beta u_d(\theta)}\sin\theta}{\int_{-1}^{1} d\xi e^{y\xi}} = \frac{e^{-\beta u_d(\theta)}\sin\theta}{q_0},$$

where $\xi = \cos\theta$, $y = \beta dD$, and $q_0 = (2/y)\sinh y$.

(iii) Figure 6.8 presents $p(\theta)$ for the reduced electric field $y = 1{,}5$, and 10.

(iv) $p(\theta)$ is not maximized at $\theta = 0$ because of the entropy contributions related to the degree of freedom θ.

(v) The Metropolis algorithm prescribes the probability of acceptance as

$$a_{ij} = \min(1, p_j/p_i) = \min(1, e^{-\beta\Delta E}\sin\theta_j/\sin\theta_i).$$

With random trial moves on $\theta \in [0, \pi]$, the probability density can be readily obtained from the histogram analysis of the MC data. Figure 6.9 compares the simulation and analytical results for $y = 1{,}5$, and 10. As expected, the agreement between MC and the analytical results are perfect.

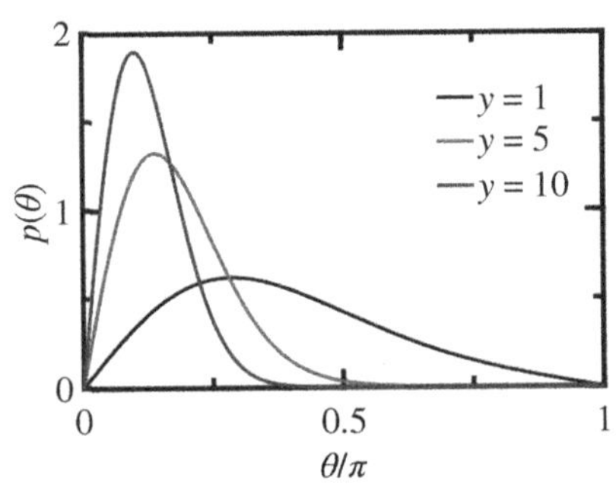

Figure 6.8 The probability density distributions for the dipole angle θ at three reduced electric fields, $y \equiv \beta dD = 1$ (A), 5 (B), and 10 (C). (IdealDipoleE2.m)

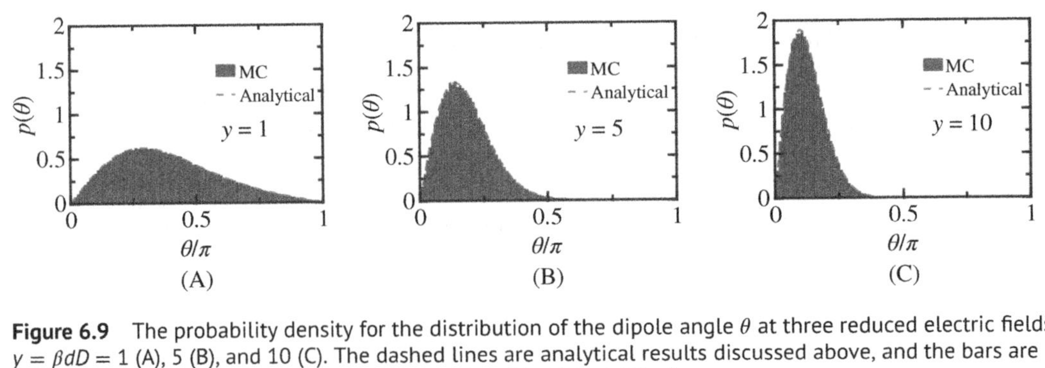

Figure 6.9 The probability density for the distribution of the dipole angle θ at three reduced electric fields $y = \beta dD = 1$ (A), 5 (B), and 10 (C). The dashed lines are analytical results discussed above, and the bars are from MC simulation with the Metropolis algorithm (IdealDipoleE2.m).

Problem 6.10 Implement the Metropolis algorithm for MC simulation of a cyclic Ising chain with $N = 50$ spins and carry out the following analyses:

(i) Consider $\beta\epsilon = 0.5$ and $\beta h = 0$, validate your code by comparing the internal energy and heat capacity with the analytical solutions:

$$\beta U/N = -\beta\epsilon \tanh(\beta\epsilon),$$

$$C_V/Nk_B = \left[\frac{\beta\epsilon}{\cosh(\beta\epsilon)}\right]^2.$$

Why does not the simulation yield an exactly zero magnetization at zero field?

(ii) Consider $\beta\epsilon = 0.5$ and $\beta h = 0.5$, compare your simulation result with the average magnetization per spin predicted by the equation:

$$m = \frac{\sinh(\beta h)}{\sqrt{\sinh^2(\beta h) + e^{-4\beta\epsilon}}}.$$

(iii) Discuss how the simulation results vary with the number of MC steps for $\beta\epsilon = 0.5$ and $\beta h = 0$.

(iv) Can you simulate the reduced Helmholtz energy βF at $\beta\epsilon = 0.5$ and $\beta h = 0.5$?

Solution:

MC simulation for a cyclic Ising chain has been implemented with Matlab program (Ising 1DMC.m).

(i) Comparison of MC and exact results for a cyclic Ising chain with $N = 50$ at $\beta\epsilon = 0.5$ and $\beta h = 0$

	$\beta U/N$	C_V/Nk_B	m
MC	−0.2316	0.1992	−0.0064
Analytical	−0.2311	0.1966	0

The simulation results, including m, contain uncertainties due to thermal fluctuations.

(ii) Further comparison at $\beta\epsilon = 0.5$ and $\beta h = 0.5$

	$\beta U/N$	C_V/Nk_B	m
MC	−0.7892	0.6296	0.8179
Analytic	—	—	0.8169

(iii) As shown in Figure 6.10, the simulation error scales with the number of MC steps as $\sim 1\sqrt{N}$. The trend is the same for all quantities.

(iv) The reduced Helmholtz energy βF may be calculated by integrating the equation

$$\beta F = \int_0^\beta \langle E\rangle d\beta.$$

The thermodynamic integration requires simulation at a number of temperatures.

Problem 6.11 In molecular simulation, the density of states (DOS) is often used both in analyzing the simulation data and in constructing advanced algorithms to speed up microstate sampling. In a canonical ensemble, the DOS is formally related to the partition function through

$$Q = \int dE w(E) e^{-\beta E},$$

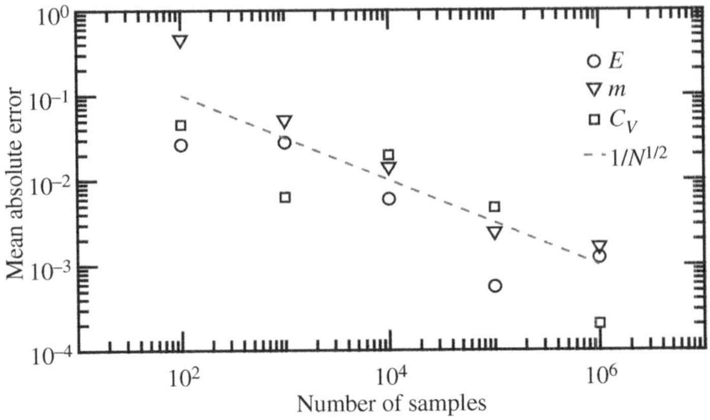

Figure 6.10 Scaling of the mean absolute error with the number of MC steps. (Ising1DMC_NB.m)

where $w(E)dE$ represents the number of microstates with the total energy between E and $E + dE$. Derive an expression for $w(E)$ for a cyclic Ising chain with N spins at zero field and verify the analytical result using MC simulation.

Solution:
For a cyclic Ising chain with N spins at zero field, the total energy is given by

$$E = -\sum_{i=1}^{N} s_i s_{i+1} \epsilon = -\sum_{i=1}^{N} b_i \epsilon,$$

where $s_{N+1} = s_1$, and $b_i \equiv s_i s_{i+1} = \pm 1$. Because $\{b_{i=1,\ldots,N}\}$ consists of N elements, the total energy can be alternatively expressed in terms of the number of elements with $b_i = +1$ and that with $b_i = -1$

$$E(n) = -n\epsilon + (N - n)\epsilon = (N - 2n)\epsilon,$$

where $n = 0, 1, 2, \ldots, N$ denotes the number of occurrences with $b_i = +1$, $(N - n)$ is the number of occurrences with $b_i = -1$. The total energy takes discrete values from $N\epsilon$, $(N - 2)\epsilon$, $(N - 4)\epsilon$, and so on, to $-N\epsilon$, with $dE = 2\epsilon$.
For each value of n, the degeneracy corresponds to the number of ways to assign n bonds (viz., $b_i = s_i s_{i+1}$) out of N with +1. Accordingly, the canonical partition function can be written as

$$Q = \sum_{n=0}^{N} 2 \times \binom{N}{n} \times e^{-\beta E(n)},$$

where a factor of two is added to account for two different ways to assign nearest-neighboring spins in parallel (or anti-parallel) directions. In comparison with $Q = \int dE w(E) e^{-\beta E}$, we find the DOS

$$w(E)dE = 2 \times \binom{N}{n} = \frac{2N!}{n!(N-n)!}$$

or

$$w(E) = \frac{1}{2\epsilon} \times \frac{2N!}{n!(N-n)!}. \tag{A}$$

Figure 6.11 compares the energy histogram calculated from MC simulation and that predicted from Eq. (A). As expected, the agreement is perfect!

Figure 6.11 The density of states for a cyclic Ising chain at zero field. Here, $N = 50$, $\beta\epsilon = 0.5$, and the simulation result is based on 10^6 MC steps (Ising1DMC.m).

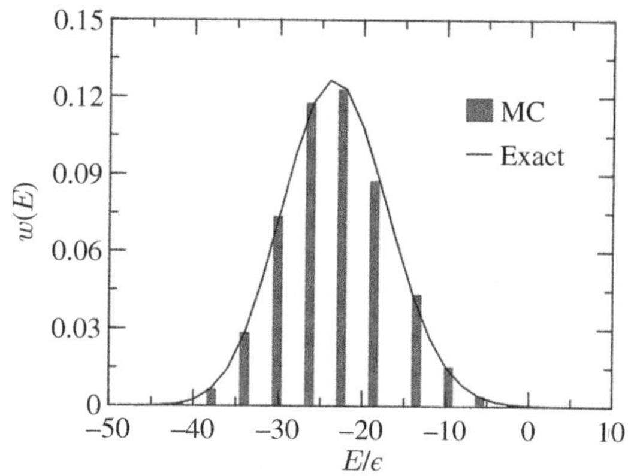

Problem 6.12 Very often molecular simulation is limited by the system size and computer time. Suppose that your Ising chain simulation can run only up to 10^4 MC steps per spin.

(i) Compare your simulation results with the exact values for the internal energy, the mean magnetization, and the heat capacity at zero field and three reduced temperatures, $k_BT/\epsilon = 10,1$, and 0.5.
(ii) Construct the density of states (DOS) at these temperatures and compare with the exact results.
(iii) How does the accuracy of your simulation results change with temperature?

Solution:

(i) Table 6.3 presents the simulation results for a cyclic Ising chain with $N = 50$ spins at zero field with 10^4 MC steps.
(ii) Figure 6.12 compares the density of states at four different temperatures with the exact results.

(iii) The simulation results are less accurate at low temperature because the Metropolis algorithm becomes less efficient, i.e., the simulation could not cover enough microstates due to the low acceptance probability.

Problem 6.13 Implement the Metropolis algorithm for MC simulation of an open-end Ising chain with $N = 50$ spins and carry out the following studies:

(i) Consider $\beta\epsilon = 0.2$ and $\beta h = 0$, validate your results by comparing the internal energy and heat capacity with the exact results

$$\beta U/N = -(1 - 1/N)\beta\epsilon \tanh(\beta\epsilon),$$

$$C_V/Nk_B = (1 - 1/N)\left[\frac{\beta\epsilon}{\cosh(\beta\epsilon)}\right]^2.$$

Table 6.3 MC results for a cyclic Ising chain at different reduced temperatures $T^* \equiv k_BT/\epsilon$.

	$T^* = 10$			$T^* = 1$			$T^* = 0.5$		
	$\beta U/N$	m	C_V/Nk_B	$\beta U/N$	m	C_V/Nk_B	$\beta U/N$	m	C_V/Nk_B
Analytical	−0.0997	0	0.0099	−0.7616	0	0.4200	−0.9640	0	0.2826
MC	−0.1045	−0.0033	0.0093	−0.7520	−0.0260	0.4764	−0.8907	0.1139	1.2405

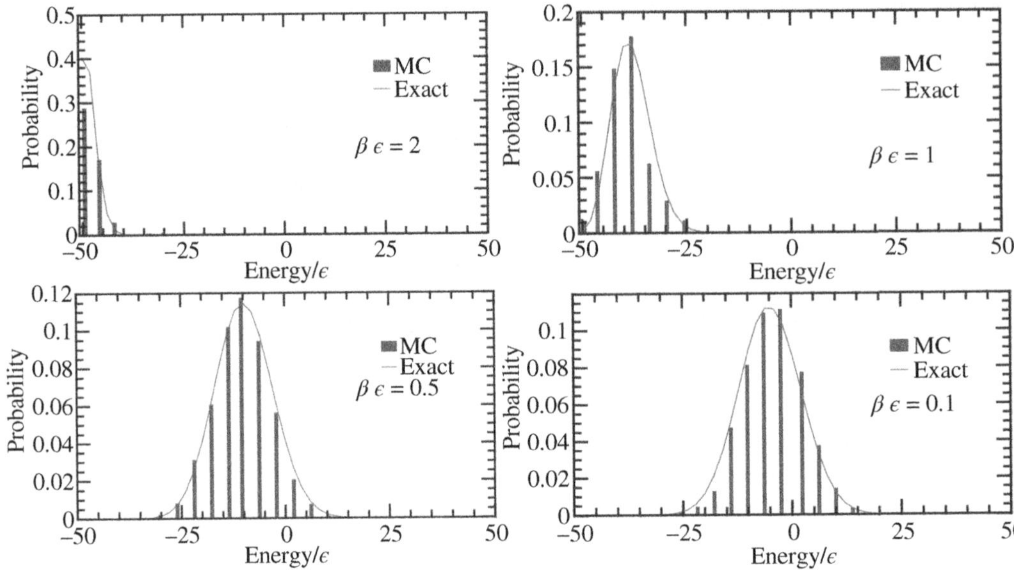

Figure 6.12 The probability density distributions of a cyclic Ising chain at four different reduced temperatures. (A) $\beta\epsilon = 2$; (B) $\beta\epsilon = 1$; (C) $\beta\epsilon = 0.5$, and (D) $\beta\epsilon = 0.1$. In all cases, $N = 50$, and the simulation results are based on 10^4 MC steps per spin (Ising1DMC.m).

(ii) Consider $\beta\epsilon = 0.5$ and $\beta h = 0.5$, compare your simulation result with the average magnetization per spin predicted by the analytical equation:

$$m = \frac{\sinh(\beta h)}{\sqrt{\sinh^2(\beta h) + e^{-4\beta\epsilon}}}.$$

(iii) Show how the simulation results vary with the number of MC steps for $\beta\epsilon = 0.2$ and $\beta h = 0$.
(iv) How would you calculate the Helmholtz energy?

Solution:
MC simulation for both cyclic and open-end Ising chains can be implemented with Matlab program (Ising1DMC.m).

(i) Comparison of MC and exact results for an open Ising chain at $\beta\epsilon = 0.2$ and $\beta h = 0$. Here, $N = 50$ and the simulation results were obtained after 10^6 MC steps.

	$U/(N\epsilon)$	C_V/Nk_B	m
MC	−0.1934	0.0377	−0.0011
Analytical	−0.1934	0.0377	0

(ii) Further comparison at $\beta\epsilon = 0.5$ and $\beta h = 0.5$

	$U/(N\epsilon)$	C_V/Nk_B	m
MC	−0.8560	0.5014	0.5585
Analytical	—	—	0.5661

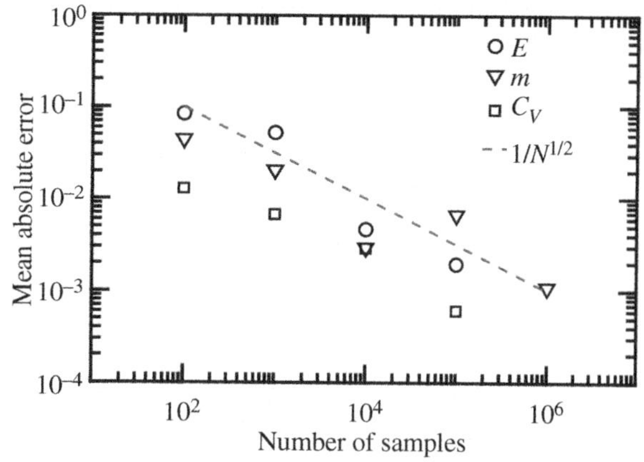

Figure 6.13 Scaling of the mean absolute error with the number of MC steps. (Ising1DMC.m)

(iii) As shown in Figure 6.13, the simulation error scales with the number of MC steps as $\sim 1\sqrt{N}$. The trend is the same for all quantities.

(iv) The reduced Helmholtz energy βF may be calculated by integrating the Gibbs-Helmholtz equation

$$\beta F = \int_0^\beta \langle E \rangle d\beta.$$

The thermodynamic integration requires simulation at a number of temperatures.

Problem 6.14 Carry out MC simulation for the following thermodynamic properties versus temperature for a cyclic Ising chain with $N = 50$ spins without external field (i.e., $h = 0$). Compare the simulation results with the analytic solutions:

(i) The reduced internal energy

$$\frac{U}{N\epsilon} = -\tanh K \left(\frac{1 + \tanh^{N-2} K}{1 + \tanh^N K} \right).$$

(ii) The mean magnetization per spin

$$m = 0.$$

(iii) The reduced heat capacity

$$\frac{C_V}{Nk_B} = \frac{K^2}{\cosh^2(K)} \frac{1 + (N-1)(\tanh^{N-2}K - \tanh^N K) - \tanh^{2N-2}K}{(1 + \tanh^N K)^2}.$$

(iv) The normalized average squared magnetization

$$\chi^2 \equiv \left\langle \left(\sum_{i=1}^N s_i \right)^2 \right\rangle \Big/ N^2 = \frac{e^{2K}}{N} \frac{1 - \tanh^N K}{1 + \tanh^N K}.$$

Solution:

Figure 6.14 compares the simulation and analytical results for the thermodynamic properties of the Ising chain near the temperature when the free energy attains a minimum value, $k_B T_m/\epsilon = 2/\ln(N-1)$.

Approximately, the paramagnetic to ferromagnetic transition takes place at $k_B T_m/\epsilon \approx 2/\ln(N-1) \approx 0.5139$. Below this temperature, the agreement is poor for the heat capacity and mean magnetization per spin. However, the analytical and simulation results agree well with each other

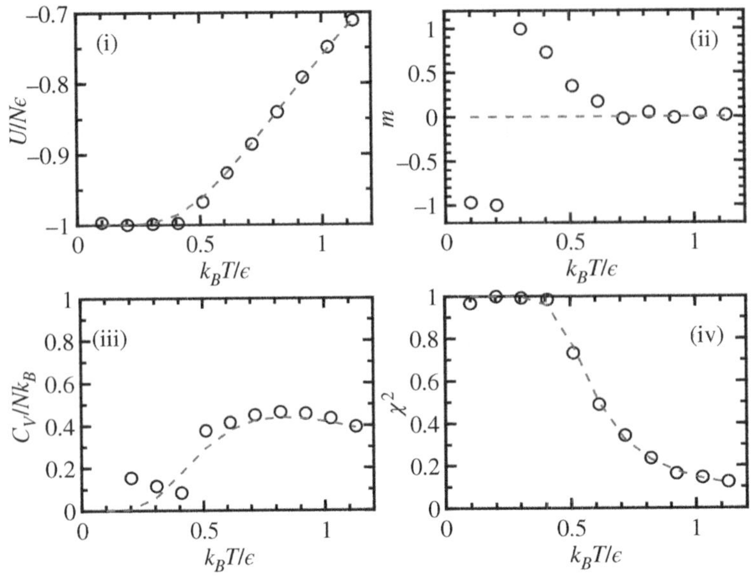

Figure 6.14 Comparison of MC and analytical results for a cyclic Ising chain with $N = 50$ spins (Ising1DMCnT.m).

for the internal energy and average squared magnetization because these quantities are the same for the spin up and spin down phases.

Problem 6.15 Carry out MC simulation for the following thermodynamic properties versus temperature for an open-end Ising chain with $N = 50$ spins without external field (i.e., $h = 0$). Compare the simulation results with the analytic solutions:

(i) The reduced internal energy

$$\frac{U}{N\epsilon} = -(1 - 1/N)\tanh K.$$

(ii) The mean magnetization per spin

$$m = 0.$$

(iii) The reduced heat capacity

$$\frac{C_V}{Nk_B} = (1 - 1/N)\frac{K^2}{\cosh^2(K)}.$$

(iv) The normalized average squared magnetization

$$\chi^2 == \left\langle \left(\sum_{i=1}^{N} s_i \right)^2 \right\rangle \Big/ N^2 = \frac{e^{2K}}{N}[1 + \sinh(2K)(\tanh^N K - 1)/N].$$

Solution:

Figure 6.15 compares the simulation and analytical results for the Ising chain.

Approximately, the paramagnetic to ferromagnetic transition takes place at $k_BT_m/\epsilon \approx 2/\ln(N-1) \approx 0.5139$. Below this temperature, the agreement is poor for the heat capacity and mean magnetization per spin. However, the analytical and simulation results agree well with each other for the internal energy and average squared magnetization because these quantities are the same for the spin up and spin down phases.

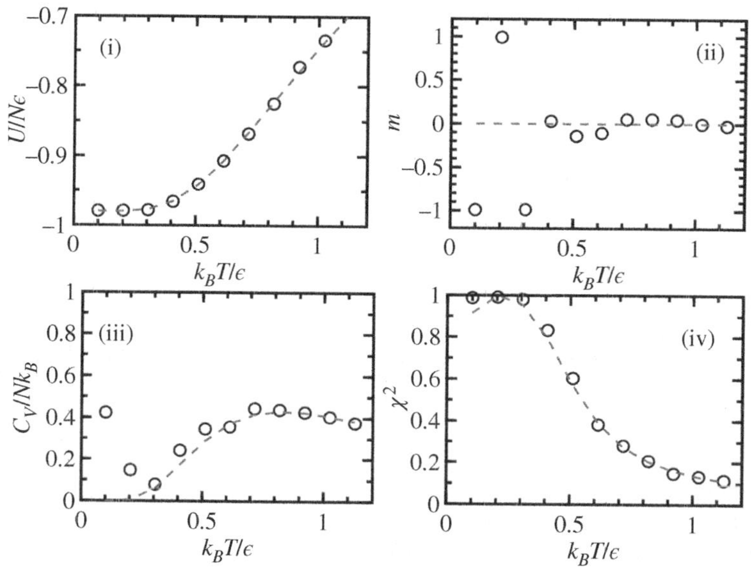

Figure 6.15 Comparison of MC and analytical results for an open-end Ising chain with $N = 50$ spins (Ising1DMCnT.m).

Problem 6.16 An Ising chain of finite length exhibits an order-disorder transition at temperature $k_BT_c/\epsilon \approx 2/\ln(N-1)$. To obtain the degree of magnetization below T_c, we may perform MC simulation in the presence of an external field and calculate the desired result through non-Boltzmann sampling

$$m = \frac{1}{N}\frac{\langle \sum_\nu \sum_i s_i \exp(\beta \Delta E_\nu)\rangle}{\langle \exp(\beta \Delta E_\nu)\rangle},$$

where $\Delta E_\nu \equiv E_\nu - E_{0,\nu}$, subscript "0" denotes the system at zero field, and $\langle \cdots \rangle$ stands for the ensemble average with total energy E_ν. Implement the non-Boltzmann sampling with the non-zero field, calculate the magnetization of a cyclic Ising chain with $N = 50$ spins below and above T_c, and compare the simulation results with those obtained from Problem 6.14.

Solution:

The essential idea of non-Boltzmann sampling is to calculate the simulation results based on microstates generated according to a reference system. For this problem, the reference system is a cyclic Ising chain in the presence of a field, and the system of interest has zero field. Accordingly, the difference in the total energy is

$$\Delta E_\nu = -\sum_{i=1}^{N} s_i h.$$

The ensemble averages $\langle \sum_\nu \sum_i s_i \exp(\beta \Delta E_\nu)\rangle$ and $\langle \exp(\beta \Delta E_\nu)\rangle$ can be calculated from the MC simulation with a nonzero field.

Different from the results obtained in Problem 6.14, Figure 6.16 shows a sharp disorder-to-order transition near the Curie temperature. The magnetization prefers a specific direction due to the biased sampling. With an appropriate selection of the external field, the magnetization obtained from non-Boltzmann sampling is more reliable because it avoids the coexistence of two phases in a small system.

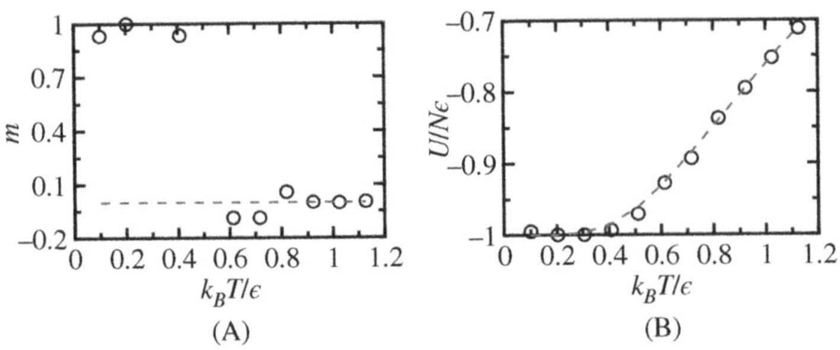

Figure 6.16 The mean magnetization (A) and internal energy (B) calculated from non-Boltzmann sampling ($\beta h = 0.05$) for a cyclic Ising chain with $N = 50$ spins. The dashed lines correspond to asymptotic results at $N \to \infty$ (Ising1DMC_NB.m).

Problem 6.17 Prepare a computer program for the MC simulation of the 2D-Ising model with the square lattice and zero field using the Metropolis method. Investigate the dependence of the total energy, heat capacity, and magnetization on temperature near the Curie temperature. Compare the simulation results with Onsager's theory discussed in Section 5.5 and explore how the accuracy of numerical results vary with temperature.

Solution:
Figure 6.17 compares the simulation and analytical results for the 2D-Ising model.
One can improve the simulation results by taking more MC steps and using a larger system size. At low temperature, the agreement between simulation and analytical results deteriorates because the Metropolis method becomes inefficient. Besides, the direct simulation of the two coexisting phases is problematic because they have different spin orientations.

Problem 6.18 Investigate the size effects on the simulation results using a MC simulation program for the 2D-Ising model. Report the simulation results at two reduced temperatures, $T/T_c = 2.0$ and 0.5, for the internal energy per spin, heat capacity, magnetization, and susceptibility with different lattice sites in each dimension, $L = 5, 10, 20, 30, 50$. Discuss the accuracy of the simulation results versus the system size.

Solution:
Figure 6.18 shows the size effects on the simulation results for the 2D-Ising model at $T/T_c = 2.0$ (4 panels on the left) and 0.5 (4 panels on the right).

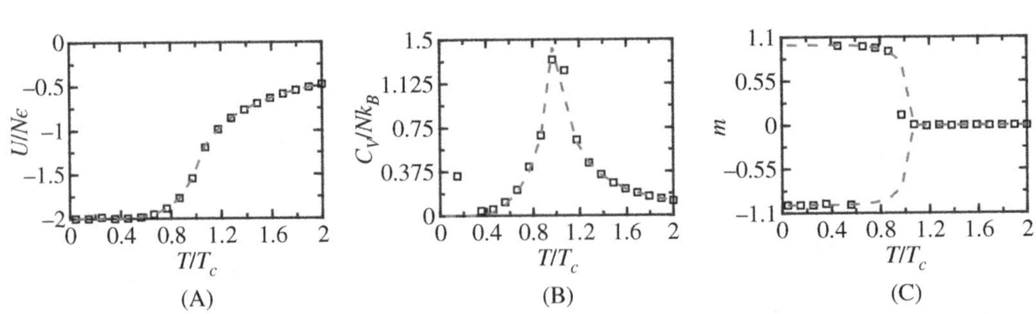

Figure 6.17 Comparison of the analytical results for the 2D-Ising model with MC with a 20×20 square lattice and periodic boundary conditions. The simulation results for the internal energy (A), heat capacity (B), and magnetization (C) are based on 10^5 MC cycles per spin after 10^3 cycles for equilibrium. The Curie temperature is $k_BT_c/\epsilon \approx 2.2692$ (Ising2DMC_T.m).

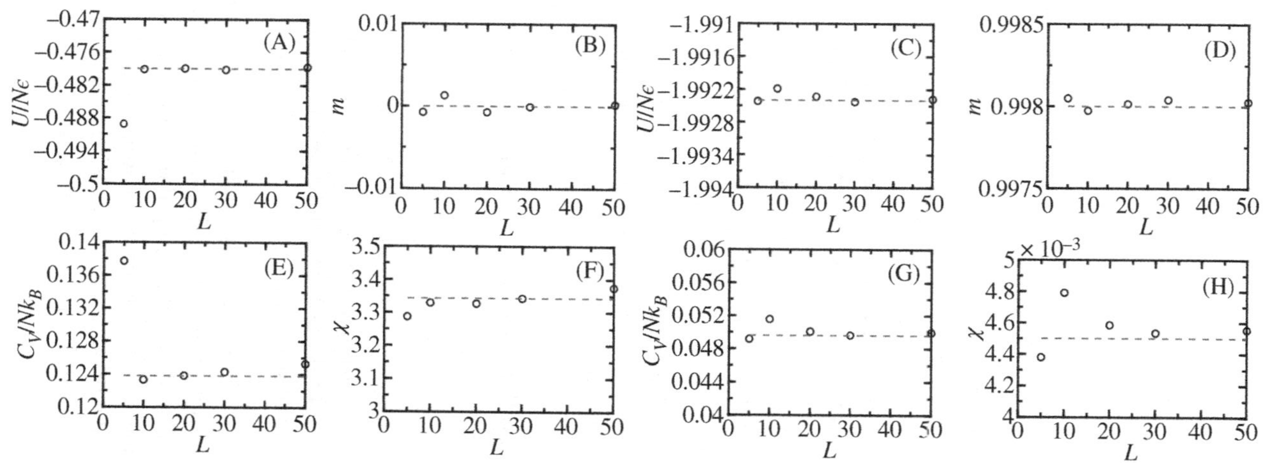

Figure 6.18 The dependence of the simulation results on the number of lattice sites in each dimension (L) at two reduced temperatures ($T/T_c = 2.0$ for 4 left panels and 0.5 for 4 right panels). The simulation results (symbols) for internal energy (A, C), magnetization (B, D), heat capacity (C, G) and susceptibility (F, H) are based on 10^5 MC moves per spin after 10^3 cycles of equilibrium with the random initial spin orientations. The dashed lines are exact results (Ising2DMC_L.m).

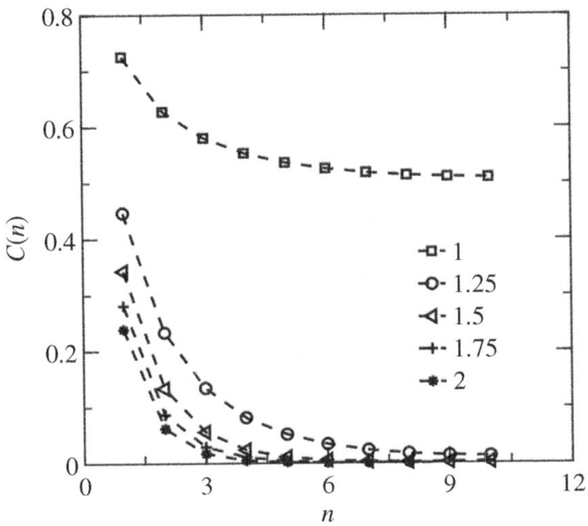

Figure 6.19 The spin–spin correlation function at different reduced temperatures (T/T_c). The simulation results are based on 10^5 MC moves per spin with random initial spin orientations (Critical2Ddata.m).

As expected, the simulation results become more accurate for larger systems. In order to get reliable simulation results, the system size must be much larger than the correlation length.

Problem 6.19 For the square-lattice Ising model, the correlation function is defined by neighboring spins either within the same row or the same column

$$C(n) = \langle s_{i,j}s_{i+n,j}\rangle - \langle s_{i,j}\rangle\langle s_{i+n,j}\rangle = \langle s_{i,j}s_{i,j+n}\rangle - \langle s_{i,j}\rangle\langle s_{i,j+n}\rangle,$$

where (i,j) are spin indexes. Perform MC simulation to calculate $c(n)$ using a 20×20 lattice with the periodic boundary conditions at a few different temperatures, and demonstrate the divergence of the correlation length at the critical point.

Solution:
At zero field, there is no net magnetization, i.e., $\langle s_{i,j}\rangle = 0$. Thus, the correlation function can be calculated from

$$C(n) = \langle s_{i,j}s_{i+n,j} + s_{i,j}s_{i-n,j} + s_{i,j}s_{i,j+n} + s_{i,j}s_{i,j-n}\rangle/4.$$

Figure 6.19 shows the results at five reduced temperatures (T/T_c). At a condition remote from the critical temperature, the spin–spin correlation approaches zero beyond a few lattice sites. However, the correlation becomes stronger closer to the critical point and diverges at $T = T_c$ (not correctly shown in Figure 6.19 because of the finite size effect).

Problem 6.20 Due to the divergence of the correlation length, thermodynamic properties calculated from MC simulation become inaccurate near the critical point. Nevertheless, the simulation data can be utilized to extrapolate accurate results using the scaling relations derived from the renormalization group theory (Section 5.12). To demonstrate this, suppose that the heat capacity and susceptibility have been simulated with a 20×20 lattice Ising model shown in Figure 6.20. Extrapolate the simulation data to the critical point using appropriate scaling relations (Critical2Ddata.m).

Solution:
The renormalization group theory predicts that the magnetic susceptibility satisfies the following asymptotic result (Section 5.11)

$$\chi(T) \approx 0.963(1 - T_c/T)^{-7/4},$$

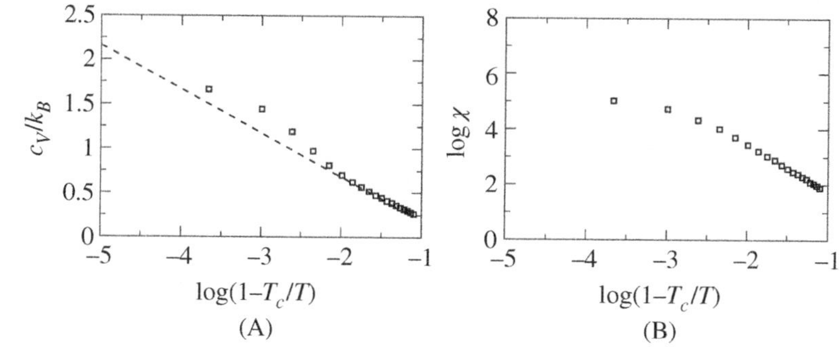

Figure 6.20 The heat capacity (A) and susceptibility (B) of the 2D-Ising model obtained from finite size MC simulation (Critical2Ddata.m).

and that the heat capacity scales as

$$c_V/k_B \approx -\frac{8}{\pi}\left(\frac{\epsilon}{k_B T_c}\right)^2 \ln(1 - T_c/T) + \text{constant}.$$

With the help of the scaling relations, we may extrapolate the critical properties from the simulation data near the critical temperature, which is shown in Figure 6.21.

Problem 6.21 Consider a binary mixture containing components 1 and 2 at constant pressure. Show that the Clapeyron equation for equilibrium between two phases α and γ can be written as

$$\left(\frac{\partial \beta}{\partial z_2}\right)_P = \frac{x_2^\alpha - x_2^\gamma}{z_1 z_2 (h^\alpha - h^\gamma)},$$

where $x_{i=1,2}$ represents the mole fractions, $z_{i=1,2} = f_i/f_T$ is the fugacity fraction of component i, and h stands for the molar enthalpy. Here, the fugacity is defined in terms of chemical potential

$$\mu_i = \mu_i^0 + k_B T \ln(f_i/f_i^0),$$

where superscript "0" denotes a reference state. The total fugacity is $f_T = f_1 + f_2$. Hint: For a binary system, $d(\beta G) = H d\beta + \beta V dP + \beta\mu_1 dN_1 + \beta\mu_2 dN_2$.

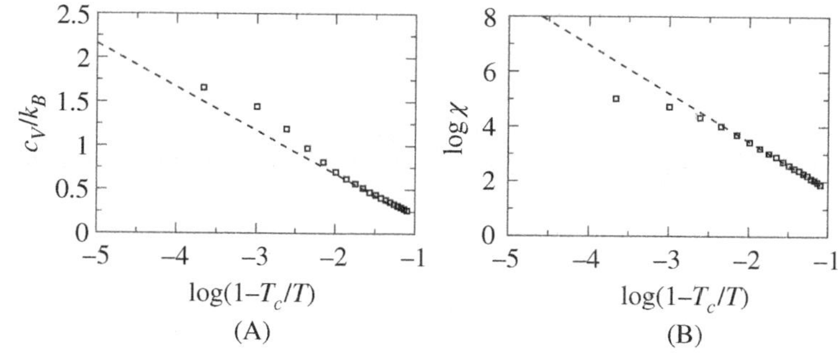

Figure 6.21 The heat capacity (A) and susceptibility (B) of the 2D-Ising model obtained from finite size MC simulation. The dashed lines are the predictions of scaling relations (Critical2Ddata.m).

Solution:
For the binary system, the reduced Gibbs energy can be written as

$$d(\beta G) = \left(\frac{\partial \beta G}{\partial \beta}\right)_{P,N_i} d\beta + \left(\frac{\partial \beta G}{\partial P}\right)_{\beta,N_i} dP + \sum_{i=1}^{2}\left(\frac{\partial \beta G}{\partial N_i}\right)_{\beta,P,N_{j\neq i}} dN_i$$

$$= Hd\beta + \beta VdP + \sum_{i=1}^{2}\beta\mu_i dN_i.$$

Because $G = N_1\mu_1 + N_2\mu_2$, at constant pressure P, the Gibbs–Duhem equation becomes

$$Hd\beta - \sum_{i=1}^{2} N_i d(\beta\mu_i) = 0.$$

We can rewrite the above equation using $d(\beta\mu_i) = d\ln f_i$ and dividing both sides by $N_T = N_1 + N_2$

$$hd\beta - \sum_{i=1}^{2} x_i d\ln f_i = hd\beta - \sum_{i=1}^{2} x_i d\ln z_i - d\ln f_T,$$

where $h \equiv H/N_T$, and $x_i = N_i/N_T$. Noting that

$$\sum_{i=1}^{2} x_i d\ln z_i = \frac{x_1}{z_1}dz_1 + \frac{x_2}{z_2}dz_2 = \left(-\frac{x_1}{z_1} + \frac{x_2}{z_2}\right)dz_2 = \frac{x_2 z_1 - x_1 z_2}{z_1 z_2}dz_2 = \frac{x_2 - z_2}{z_1 z_2}dz_2,$$

and that f_T and z_2 are the same for the coexisting phases, we obtain the desired equation by applying the Gibbs–Duhem equation to both α and γ phases

$$\left(\frac{\partial \beta}{\partial z_2}\right)_P = \frac{x_2^\alpha - x_2^\gamma}{z_1 z_2 (h^\alpha - h^\gamma)}.$$

Problem 6.22 Perform the Wang–Landau sampling for the 2-dimensional Ising model using a 20×20 square lattice with the periodic boundary conditions. Use the density of states (DOS) from MC simulation, predict the Helmholtz energy, internal energy and heat capacity per spin over the entire range of temperature.

Solution:
The Wang–Landau algorithm can be implemented in Matlab (wangLand_Matlab.m). Figures 6.22 and 6.23 show some representative results by running the program in comparison with Onsager's theory.

Problem 6.23 The torsional potential for n-butane can be expressed as a cosine-function series[2]:

$$u(\phi) = \frac{1}{2}\sum_{i=1}^{6} A_i(1 - \cos i\phi),$$

where ϕ is the dihedral angle (or torsion angle), the coefficients from A_1 to A_6 have the units of kJ/mol and are given by

$$A = [2.8948, 0.5144, 13.7323, 0.4785, -0.0718, -0.4306].$$

The torsional potential predicts three energy minima corresponding to trans ($\phi = 180°$), gauche– ($\phi = 60°$) and gauche+ ($\phi = 300°$) conformers of n-butane.

2 Herrebout W. A., et al., "Enthalpy difference between conformers of n-butane and the potential function governing conformational interchange", *J. Phys. Chem.*, 99, 578–585 (1995).

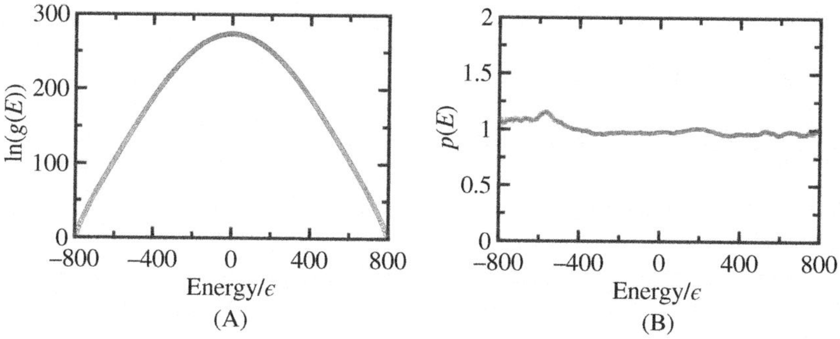

Figure 6.22 A. Logarithm of the density of states for a 20×20 square lattice with the periodic boundary conditions. B. The probability density of the system energy after 10^7 MC moves for each value of f the multiplying factor (WangLandMatlab.m).

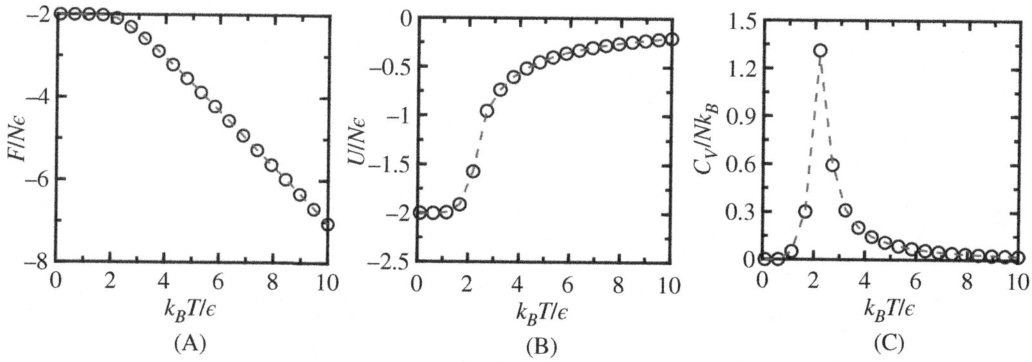

Figure 6.23 Thermodynamic quantities for the 2D-Ising model calculated from the density of states (symbols). A. Helmholtz free energy; B. Internal energy; and C. Constant-volume heat capacity. The dashed lines are predictions of Onsager's theory (WangLandMatlab.m).

(i) Derive an analytical expression for predicting the probability distribution of the dihedral angle as a function of temperature. Assume that the kinetic energy of the n-butane atoms has no effect on the distribution of the dihedral angle.

(ii) Develop a MC scheme for sampling the probability distribution of the dihedral angle using the Metropolis method.

(iii) Compare the analytical and simulation results at T=100 and 300 K and discuss how the efficiency of the sampling varies with temperature.

(iv) Estimate the equilibrium constant for the transformation between the trans ($120° < \phi < 240°$) and gauche states ($0° < \phi < 120°$ or $240° < \phi < 360°$) of n-butane at 300 K. What is the equilibrium composition of different conformers of n-butane in the gas phase?

Solution:

(i) Without considering the kinetic energy effects, the probability density for the dihedral angle can be calculated from the Boltzmann distribution

$$p(\phi) = \frac{\exp[-\beta u(\phi)]}{Q},$$

where

$$Q = \int_{0^\circ}^{360^\circ} d\phi \exp[-\beta u(\phi)].$$

The effect of kinetic energy on the dihedral angle distribution was discussed by van Gunsteren.[3]

(ii) See Matlab code (Mnbutane.m) for sampling the probability distribution of the dihedral angle with the Metropolis method.

(iii) Figure 6.24 compares the analytical and simulation results for the probability density of the dihedral angle at T=100 and 300 K. The MC data are taken after Niter = 10^6 steps starting from the minimum energy state ($\phi = 180°$).
Figure 6.24 shows that, after 10^6 MC steps, the simulation results are reasonably accurate at 300 K. At 100 K, however, the Metropolis sampling is easily trapped at the trans state, indicating the decline of the sampling efficiency at low temperature. In Figure 6.24A, the maximum step length is set as $\Delta\phi_{max} = 10°$. As shown in Figure 6.25, the energy barrier can be overcome

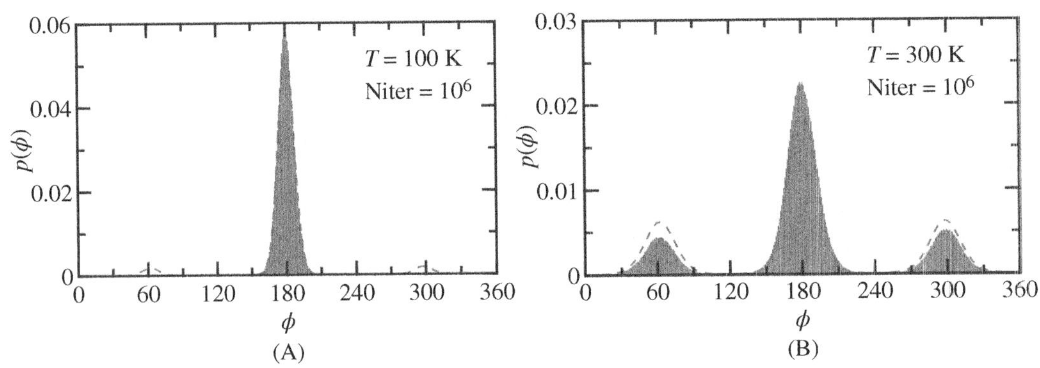

Figure 6.24 The probability density for the dihedral angle of *n*-butane at 100 K (A) and 300 K (B). The bars are from MC simulation based on the histogram analysis, and the dashed lines are the Boltzmann distributions (Mnbutane.m).

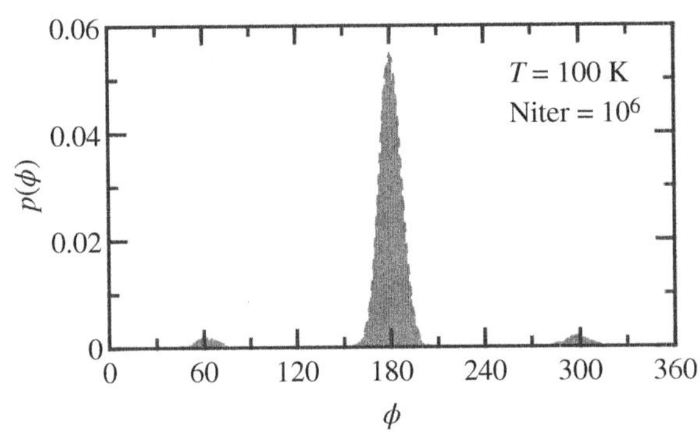

Figure 6.25 The same as Figure 6.24 but the simulation results were obtained at a larger step length, $\Delta\phi = 180°$ (USC4.m).

3 van Gunsteren W. F., "Constrained dynamics of flexible molecules", *Mol. Phys.* 40, 1015–1019 (1980).

by using a larger step, $\Delta\phi_{max} = 180°$, but with less sampling efficiency (the acceptance rate drops from 0.86 to 0.12).

(iv) The equilibrium constant for the transformation between the trans and gauche− of n-butane can be calculated from

$$K_{eq}^{-} = \frac{y_{g^-}}{y_t} = e^{-\beta\Delta F} = \frac{e^{-\beta F_{gauche-}}}{e^{-\beta F_{trans}}} = \frac{\int_{0°}^{120°} d\phi \exp[-\beta u(\phi)]}{\int_{120°}^{240°} d\phi \exp[-\beta u(\phi)]} = 0.3467.$$

Similarly, the equilibrium constant for the transformation between the trans and gauche+ of n-butane is calculated from

$$K_{eq}^{+} = \frac{y_{g^+}}{y_t} = \frac{\int_{240°}^{360°} d\phi \exp[-\beta u(\phi)]}{\int_{120°}^{240°} d\phi \exp[-\beta u(\phi)} = 0.3467.$$

Because $y_{g^-} + y_{g^+} + y_t = 1$, we can find the equilibrium compositions of different conformers

$$y_{g^-} = \frac{K_{eq}^{-}}{1 + K_{eq}^{-} + K_{eq}^{+}} = 0.2047,$$

$$y_{g_+} = \frac{K_{eq}^{+}}{1 + K_{eq}^{-} + K_{eq}^{+}} = 0.2047,$$

$$y_t = \frac{1}{1 + K_{eq}^{-} + K_{eq}^{+}} = 0.5905.$$

These numbers agree reasonably well with more sophisticated calculations of the distribution of n-butane conformers in the gas and liquid phases.[4]

Problem 6.24 Develop a parallel tempering scheme for sampling the probability distribution of the dihedral angle of n-butane using three replicas at temperature $T = 100$, 200, and 300 K. Show that the probability density conforms to the Boltzmann distribution for each replica.

Solution:

See Matlab code (TemperC4.m) for sampling the probability distribution of the dihedral angle with the parallel tempering method. The generalized ensemble ensures that each replica conforms to the usual canonical distribution.

Figure 6.26 shows the probability density of the dihedral angle at the three temperatures used in the parallel tempering simulation. Here, the initial state for each replica is selected to be one of the conformer states, i.e., trans, gauche−, and gauche+. Unlike the Metropolis algorithm, the replica exchange enables more efficient sampling of all minimum energy states. The discrepancy between the simulation and exact results is due to the duration of simulation.

---

4 Wielopolski P. A. and Smith E. R., "Dihedral angle distribution in liquid n-butane: Molecular dynamics simulations", *J. Chem. Phys.* 84, 6940–6942 (1986).

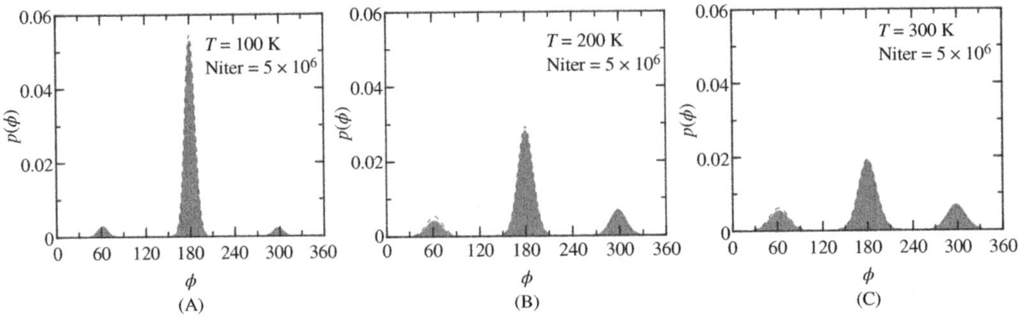

Figure 6.26 The probability distribution for the dihedral angle of n-butane at 100, 200, and 300 K calculated from parallel tempering with three replicas. The bars are from parallel tempering simulation and the dashed lines are from the exact results (TemperC4.m).

7

Simple Fluids and Colloidal Dispersions

Problem 7.1 Extend the Gibbs entropy to systems containing m types of classical particles of spherical shape. Show that, if the particles are not interacting with each other, the entropy of mixing is given by

$$\Delta S = -k_B \sum_{i=1}^{m} N_i \ln(N_i/N),$$

where k_B is the Boltzmann constant, and $N = \sum_{i=1}^{m} N_i$ is the total number of particles in the system.

Solution:

Similar to that for a one-component system of spherical particles (Eq. (7.10)), the probability of microstate ν for an m-component mixture of classical particles can be written as

$$p_\nu = p(\nu) \prod_{i=1}^{m} N_i! h^{3N_i},$$

where $p(\nu)$ is the probability density in the phase space, i.e., it is a function of continuous variable ν representing the positions and momenta of all classical particles in the system. Therefore, the Gibbs entropy is

$$S/k_B = -\sum_\nu p_\nu \ln p_\nu = -\int d\nu p(\nu) \ln \left[p(\nu) \prod_{i=1}^{m} N_i! h^{3N_i} \right].$$

Because the degrees of freedom for individual particles are independent of each other, we have $d\nu = \prod_{i=1}^{m} d\nu_i$ and $p(\nu) = \prod_{i=1}^{m} p_i(\nu_i)$ if the particles do not interact with each other, where ν_i denotes the degrees of freedom for all particles of type i. Accordingly, the entropy of the mixture is

$$\begin{aligned} S/k_B &= -\int d\nu p(\nu) \sum_{i=1}^{m} \ln[p_i(\nu_i) N_i! h^{3N_i}] = -\sum_{i=1}^{m} \int d\nu p(\nu) \ln[p_i(\nu_i) N_i! h^{3N_i}] \\ &= -\sum_{i=1}^{m} \int d\nu_i p_i(\nu_i) \ln[p_i(\nu_i) N_i! h^{3N_i}] = \sum_{i=1}^{m} S_i/k_B, \end{aligned}$$

where S_i stands for the entropy of the pure species i at system temperature T, volume V, and particle number N_i.

Solutions Manual to Fundamentals and Practice in Statistical Thermodynamics, First Edition. Jianzhong Wu and John M. Prausnitz.

The entropy of mixing is defined as the difference between the entropy of the mixture and those of the pure species at the system temperature T and pressure P

$$\Delta S = S(T,P,N_{i=1,2,\ldots,m}) - \sum_{i=1}^{m} S(T,P,N_i)$$
$$= \sum_{i=1}^{m}[S_i(T,V,N_i) - S_i(T,P,N_i)] = \sum_{i=1}^{m}[S_i(T,V,N_i) - S_i(T,N_ik_BT/P,N_i)]$$
$$= \sum_{i=1}^{m} N_ik_B \ln(PV/N_ik_BT) = -k_B\sum_{i=1}^{m} N_i \ln(N_i/N),$$

where $V = Nk_BT/P$ follows the ideal gas law.

Problem 7.2 For a system of spherical particles at low density, the radial distribution function can be approximated by the Boltzmann distribution

$$g(r) \approx e^{-\beta u(r)},$$

where $\beta = 1/(k_BT)$, and $u(r)$ stands for the pair potential. Based on the above expression, derive equations for

(i) internal energy
(ii) Helmholtz energy
(iii) compressibility factor
(iv) the second virial coefficient.

Solution:

(i) The internal energy can be directly derived from the energy equation (Eq. (7.50))

$$U = U^{IG} + 2\pi\rho\int drr^2g(r)u(r) = U^{IG} + 2\pi\rho\int drr^2u(r)e^{-\beta u(r)},$$

where $U^{IG} = \frac{3}{2}Nk_BT$ is the internal energy of non-interacting particles, and $\rho = N/V$ is the particle density.

(ii) By integrating the Gibbs–Helmholtz equation $(\partial\beta F/\partial\beta)_V = U$, we can obtain the reduced Helmholtz energy

$$\beta F = \beta F^{IG} + \int_0^\beta d\beta U^{ex} = \int_0^\beta d\beta\left[2\pi\rho\int drr^2u(r)e^{-\beta u(r)}\right]$$
$$= \beta F^{IG} + 2\pi\rho\int drr^2\int_0^\beta d\beta u(r)e^{-\beta u(r)}$$
$$= \beta F^{IG} - 2\pi\rho\int drr^2\left[e^{-\beta u(r)} - 1\right],$$

where $\beta F^{IG} = N(\ln\rho\Lambda^3 - 1)$ is the reduced Helmholtz energy of noninteracting particles.

(iii) The pressure is given by

$$P = -\left(\frac{\partial F}{\partial V}\right)_T = P^{IG} + \left(\frac{\partial}{\partial V}\frac{2\pi k_BTN}{V}\int drr^2\left[e^{-\beta u(r)} - 1\right]\right)_T$$
$$= P^{IG} - \frac{2\pi k_BTN}{V^2}\int drr^2\left[e^{-\beta u(r)} - 1\right].$$

Hence, the compressibility factor is

$$Z = \frac{PV}{Nk_BT} = 1 - \frac{2\pi N}{V}\int drr^2\left[e^{-\beta u(r)} - 1\right].$$

(iv) The second virial coefficient is obtained by a comparison of the above expression with the virial equation $Z = 1 + \rho B + \cdots$:

$$B = -2\pi \int drr^2 \left[e^{-\beta u(r)} - 1\right].$$

Problem 7.3 Consider the grand canonical partition function for a uniform system of spherical particles at temperature T and chemical potential μ

$$\Xi = \sum_{N=0}^{\infty} \frac{1}{N!\Lambda^{3N}} \int d\mathbf{r}^N \exp[-\beta\Phi(\mathbf{r}^N) + N\beta\mu] = \sum_{N=0}^{\infty} \frac{z^N}{N!} \int d\mathbf{r}^N \exp[-\beta\Phi(\mathbf{r}^N)],$$

where $\Phi(\mathbf{r}^N)$ stands for the total potential energy, $\beta = 1/(k_BT)$, $z \equiv e^{\beta\mu}/\Lambda^3$ represents the fugacity, and Λ is the thermal wavelength. Show that

(i) the grand potential can be expressed as

$$\Omega = -k_BT \ln \Xi = -PV,$$

where P is the pressure.

(ii) the average number density of particles $\rho \equiv \langle N\rangle/V$ is related to fugacity z by

$$\rho = \frac{z}{V}\left(\frac{\partial \ln \Xi}{\partial z}\right)_{T,V},$$

(iii) assuming pairwise additivity for the total potential energy $\Phi(\mathbf{r}^N)$, the grand canonical partition function can be written as a power series of z, i.e.,

$$\Xi = 1 + zV + \frac{1}{2}z^2V^2(1 + I_2) + \frac{1}{3!}z^3V^3(1 + 3I_2 + 3I_2^2 + I_3) + \cdots,$$

where $I_{n=1,2,\ldots}$ are constants defined by the Mayer function $f_{ij} \equiv \exp[-\beta u(|\mathbf{r}_i - \mathbf{r}_j|)] - 1$

$$I_2 \equiv \frac{1}{V^2}\int d\mathbf{r}_1 \int d\mathbf{r}_2 f_{12}, \quad I_3 \equiv \frac{1}{V^3}\int d\mathbf{r}_1 \int d\mathbf{r}_2 \int d\mathbf{r}_3 f_{12}f_{23}f_{13}, \ldots,$$

(iv) the reduced pressure can be expressed in terms of a power series of z, i.e.,

$$\frac{PV}{k_BT} = \ln \Xi = zV + \frac{z^2V^2}{2}I_2 + \frac{1}{6}z^3V^3(3I_2^2 + I_3) + \cdots,$$

(v) the average particle density can also be expressed in a power series of z,

$$\rho = z + z^2VI_2 + \frac{z^3V^2}{2}(3I_2^2 + I_3) + \cdots,$$

(vi) the compressibility factor Z in the power series of ρ is given by

$$Z \equiv \frac{PV}{\langle N\rangle k_BT} = 1 + B_2\rho + B_3\rho^3 + \cdots.$$

The above equation is known as the virial expansion, and B_2, B_3 and … are virial coefficients

$$B_2 = -\frac{1}{2}\int d\mathbf{r}_{12} f_{12},$$

$$B_3 = -\frac{1}{3}\int d\mathbf{r}_{12} \int d\mathbf{r}_{13} f_{12}f_{23}f_{13},$$

…

Solution:

(i) For a uniform system, the grand potential is

$$\Omega = U - \langle N\rangle\mu - TS = U - (U - TS + PV) - TS = -PV.$$

(ii) The ensemble average for the number of particles is

$$\langle N\rangle = \frac{1}{\Xi}\sum_N \frac{1}{N!\Lambda^{3N}}\int d\mathbf{r}^N N\exp[-\beta\Phi(\mathbf{r}^N) + N\beta\mu] = \frac{\partial \ln \Xi}{\partial \beta\mu}.$$

Because $(\partial z/\partial \beta\mu)_T = z$, the number density of particles ρ in the system is then given by

$$\rho = \frac{z}{V}\left(\frac{\partial \ln \Xi}{\partial z}\right)_{T,V}.$$

(iii) The pairwise additivity for the potential energy $\Phi(\mathbf{r}^N)$ indicates

$$\exp[-\beta\Phi(\mathbf{r}^N)] = \exp\left[-\sum_{i<j}\beta u(|\mathbf{r}_i - \mathbf{r}_j|)\right] = \prod_{i<j}(f_{ij} + 1).$$

Hence,

$$\begin{aligned}
\Xi &= \sum_{N=0}^{\infty}\frac{z^N}{N!}\int d\mathbf{r}^N \exp[-\beta\Phi(\mathbf{r}^N)] \\
&= 1 + z\int d\mathbf{r} + \frac{z^2}{2}\int d\mathbf{r}_1\int d\mathbf{r}_2(f_{12} + 1) \\
&\quad + \frac{z^3}{3!}\int d\mathbf{r}_1\int d\mathbf{r}_2\int d\mathbf{r}_3(f_{12} + 1)(f_{23} + 1)(f_{13} + 1) + \cdots \\
&= 1 + zV + \frac{z^2V^2}{2}(1 + I_2) + \frac{z^3V^3}{3!}(1 + 3I_2 + 3I_2^2 + I_3) + \cdots,
\end{aligned}$$

where the first term corresponds to $N = 0$, the second term for $N = 1$, the third term for $N = 2$, and so on for higher order terms.

(iv) Using the power series $\ln(1 + x) = x - x^2/2 + x^3/3 + \cdots$, we can write $\ln\Xi = \ln(1 + \cdots)$ as

$$\begin{aligned}
\frac{PV}{k_BT} &= \ln\Xi = zV + \frac{z^2V^2}{2}(1 + I_2) + \frac{1}{3!}z^3V^3(1 + 3I_2 + 3I_2^2 + I_3) \\
&\quad - \frac{z^2V^2}{2} - \frac{z^3V^3}{2}(1 + I_2) + \frac{z^3V^3}{3} - \frac{z^4V^4}{8}(1 + I_2)^2 \\
&\quad - \frac{z^4V^4}{3!}(1 + 3I_2 + 3I_2^2 + I_3) + \cdots \\
&= zV + \frac{z^2V^2}{2}I_2 + \frac{1}{6}z^3V^3(3I_2^2 + I_3) + \cdots,
\end{aligned}$$

where the second and third lines come from the lowest terms in $-x^2/2 + x^3/3$.

(v) Following part (iv), we can derive the particle density in the power series of z

$$\rho = \frac{z}{V}\left(\frac{\partial \ln \Xi}{\partial z}\right)_{T,V} = z + z^2VI_2 + \frac{z^3V^2}{2}(3I_2^2 + I_3) + \cdots.$$

(vi) To express pressure as a function of ρ, consider first the coefficient relations between the power series $y = a_1x + a_2x^2 + a_3x^3 + \cdots$ and the reverse series $x = b_1y + b_2y^2 + b_3y^3 + \cdots$,

$$\begin{aligned}
x &= b_1(a_1x + a_2x^2 + a_3x^3 + \cdots) + b_2(a_1x + a_2x^2 + a_3x^3 + \cdots)^2 \\
&\quad + b_3(a_1x + a_2x^2 + a_3x^3 + \cdots)^3 \cdots.
\end{aligned}$$

A comparison of the two sides of the above equation indicates

$$b_1 = 1/a_1;$$
$$b_2 = -(b_1 a_2/a_1^2) = -a_2/a_1^3;$$
$$b_3 = -(b_1 a_3 + 2b_2 a_1 a_2)/a_1^3 = (2a_2^2 - a_1 a_3)a_1^5; \quad \cdots.$$

From $\rho = z + VI_2 z^2 + \frac{V^2}{2}(3I_2^2 + I_3)z^3 + \cdots$ derived in part (v), we can express z in a power series of ρ

$$z = b_1\rho + b_2\rho^2 + b_3\rho^3 + \cdots$$

with

$$b_1 = 1/a_1 = 1;$$
$$b_2 = -a_2/a_1^3 = -VI_2;$$
$$b_3 = (2a_2^2 - a_1 a_3)a_1^5 = [2V^2 I_2^2 - V^2(3I_2^2 + I_3)/2] = V^2(I_2^2 - I_3)/2; \quad \cdots.$$

Substituting $z = \rho - VI_2\rho^2 + \frac{1}{2}V^2(I_2^2 - I_3)\rho^3 + \cdots$ into the expression for the reduced pressure leads to

$$\begin{aligned}
\frac{PV}{k_BT} &= \left[\rho - VI_2\rho^2 + \frac{1}{2}V^2(I_2^2 - I_3)\rho^3 + \cdots\right]V \\
&+ \frac{V^2}{2}I_2\left[\rho - VI_2\rho^2 + \frac{1}{2}V^2(I_2^2 - I_3)\rho^3 + \cdots\right]^2 \\
&+ \frac{1}{6}V^3(3I_2^2 + I_3)\left[\rho - VI_2\rho^2 + \frac{1}{2}V^2(I_2^2 - I_3)\rho^3 + \cdots\right]^3 + \cdots \\
&= \rho V + \left[-V^2 I_2 + \frac{1}{2}I_2 V^2\right]\rho^2 + \left[\frac{1}{2}V^3(I_2^2 - I_3) - V^3 I_2^2 + \frac{1}{6}V^3(3I_2^2 + I_3)\right]\rho^3 + \cdots \\
&= \rho V - \frac{1}{2}I_2 V^2\rho^2 - \frac{1}{3}I_3 V^3\rho^3 + \cdots.
\end{aligned}$$

In terms of the compressibility factor, the above equation becomes

$$Z \equiv \frac{PV}{\langle N\rangle k_BT} = 1 - \frac{I_2V}{2}\rho - \frac{I_3V^2}{3}\rho^2 + \cdots,$$

where $\langle N\rangle = \rho V$. Comparing this equation with $Z = 1 + B_2\rho + B_3\rho^3 + \cdots$ indicates

$$B_2 \equiv -I_2V/2 = -\frac{1}{2V}\int d\mathbf{r}_1 \int d\mathbf{r}_2 f_{12} = -\frac{1}{2}\int d\mathbf{r}_{12} f_{12},$$
$$B_3 \equiv -I_3V^2/3 = -\frac{1}{3V}\int d\mathbf{r}_1 \int d\mathbf{r}_2 \int d\mathbf{r}_3 f_{12}f_{23}f_{13} = -\frac{1}{3}\int d\mathbf{r}_{12}\int d\mathbf{r}_{13} f_{12}f_{23}f_{13},$$

$\cdots.$

Problem 7.4 Evaluate the second and third virial coefficients for a one-component system of hard spheres with diameter σ. Compare the results with the virial coefficients predicted by the Carnahan–Starling (CS) equation

$$Z = \frac{1 + \eta + \eta^2 - \eta^3}{(1 - \eta)^3}.$$

Solution:

For a hard-sphere system, the Mayer function is a step function, i.e.,

$$f_{ij}(r) = \begin{cases} -1 & r < \sigma; \\ 0 & r \geq \sigma. \end{cases}$$

Therefore, the second virial coefficient is

$$B_2 = -\frac{1}{2}\int d\mathbf{r}_{12} f_{12} = \frac{1}{2}\int_0^\sigma dr 4\pi r^2 = \frac{2\pi\sigma^3}{3}.$$

The third virial coefficient for a hard-sphere potential was first solved by Boltzmann in 1909.[1] First, we express B_3 as an integration over the Fourier transform of the Mayer function

$$\begin{aligned}
\hat{f}(k) &\equiv \int d\mathbf{r} f(r) e^{i\mathbf{r}\cdot\mathbf{k}} = \int_0^{2\pi} d\phi \int_0^\pi d\theta \int_0^\infty dr r^2 \sin\theta f(r) e^{iqr\cos\theta} \\
&= -2\pi\sigma^3 \int_{-1}^1 d\xi \int_0^1 dx x^2 e^{ik\sigma x\xi} = -4\pi\sigma^3 \int_0^1 dx x^2 \left(\frac{\sin k\sigma x}{k\sigma x}\right) \\
&= -\frac{4\pi\sigma^3}{(k\sigma)^3}\int_0^{k\sigma} dt t \sin t = -\frac{4\pi\sigma^3}{(k\sigma)^3}[\sin t - t\cos t]\Big|_0^{k\sigma} = -4\pi\sigma^3\left[\frac{\sin k\sigma}{(k\sigma)^3} - \frac{\cos k\sigma}{(k\sigma)^2}\right].
\end{aligned}$$

Using the relations $\mathbf{r}_{23} = \mathbf{r}_{13} - \mathbf{r}_{12}$, $\int d\mathbf{r} e^{i\mathbf{r}\cdot\mathbf{k}} = (2\pi)^3\delta(\mathbf{k})$ and $\hat{f}(-k) = \hat{f}(k)$, we can write

$$\begin{aligned}
B_3 &= -\frac{1}{3}\int d\mathbf{r}_{12}\int d\mathbf{r}_{13} f_{12} f_{23} f_{13} \\
&= -\frac{1}{3}\int d\mathbf{r}_{12}\int d\mathbf{r}_{13}\int \frac{d\mathbf{k}}{(2\pi)^3}\hat{f}(k)e^{-i\mathbf{k}\cdot\mathbf{r}_{12}}\int \frac{d\mathbf{k}'}{(2\pi)^3}\hat{f}(k')e^{-i\mathbf{k}'\cdot\mathbf{r}_{23}}\int \frac{d\mathbf{k}''}{(2\pi)^3}\hat{f}(k'')e^{-i\mathbf{k}''\cdot\mathbf{r}_{13}} \\
&= -\frac{1}{3}\int \frac{d\mathbf{k}}{(2\pi)^3}\hat{f}(k)\int \frac{d\mathbf{k}'}{(2\pi)^3}\hat{f}(k')\int \frac{d\mathbf{k}''}{(2\pi)^3}\hat{f}(k'')\int d\mathbf{r}_{12}\int d\mathbf{r}_{13} e^{-i(\mathbf{k}-\mathbf{k}')\cdot\mathbf{r}_{12}} e^{-i(\mathbf{k}'+\mathbf{k}'')\cdot\mathbf{r}_{13}} \\
&= -\frac{1}{3}\int \frac{d\mathbf{k}}{(2\pi)^3}\hat{f}(k)\int \frac{d\mathbf{k}'}{(2\pi)^3}\hat{f}(k')\int \frac{d\mathbf{k}''}{(2\pi)^3}\hat{f}(k'')(2\pi)^6\delta(\mathbf{k}-\mathbf{k}'))\delta(\mathbf{k}'+\mathbf{k}'') \\
&= -\frac{1}{3(2\pi)^3}\int d\mathbf{k}\hat{f}(k)\hat{f}(-k)\hat{f}(k) = -\frac{1}{3(2\pi)^3}\int d\mathbf{k}\hat{f}^3(k).
\end{aligned}$$

Integration of $\hat{f}^3(k)$ gives

$$\begin{aligned}
B_3 &= \frac{1}{3(2\pi)^3}\int dk 4\pi k^2\left\{4\pi\sigma^3\left[\frac{\sin k\sigma}{(k\sigma)^3} - \frac{\cos k\sigma}{(k\sigma)^2}\right]\right\}^3 \\
&= \frac{(4\pi)^4\sigma^6}{3(2\pi)^3}\int_0^\infty dt t^2\left[\frac{\sin t}{t^3} - \frac{\cos t}{t^2}\right]^3 = \frac{(4\pi)^4\sigma^6}{3(2\pi)^3}\times\frac{5\pi}{192} = \frac{5\pi^2\sigma^6}{18}.
\end{aligned}$$

The Carnahan–Starling equation can be expanded into a power series of η

$$Z = \frac{1+\eta+\eta^2-\eta^3}{(1-\eta)^3} = 1 + 4\eta + 10\eta^2 + 18\eta^3 + \cdots.$$

As $\eta = \pi\rho\sigma^3/6$, the power series predicts

$$B_2 = 2\pi\sigma^3/3, \quad B_3 = 10\times(\pi\sigma^3/6)^2 = \frac{5\pi^2\sigma^6}{18}.$$

Both B_2 and B_3 agree with the exact results.

Problem 7.5 Calculate the second virial coefficient using the Lennard–Jones (LJ) potential

$$u(r) = 4\epsilon\left[\left(\frac{\sigma}{r}\right)^{12} - \left(\frac{\sigma}{r}\right)^6\right].$$

---

1 See, e.g., Getter N. J., Calculation of Virial Coefficients and Their Uses, PhD thesis, Stony Brook University (2010).

Plot B_2/B_2^0 versus the reduced temperature k_BT/ϵ and estimate the Boyle temperature, i.e., the temperature when $B_2 = 0$. Here, $B_2^0 = 2\pi\sigma^3/3$ is the second virial coefficient according to the hard-sphere potential.

Solution:
The second virial coefficient is

$$B_2 = -\frac{1}{2}\int_0^\infty dr 4\pi r^2[e^{-\beta u(r)} - 1] = -2\pi\sigma^3 \int_0^\infty dx x^2[e^{-\beta u(x)} - 1],$$

where $x \equiv r/\sigma$. Thus

$$B_2/B_2^0 = -3\int_0^\infty dx x^2[e^{-\beta u(x)} - 1],$$

where $B_2^0 = 2\pi\sigma^3/3$. Figure 7.1 shows B_2/B_2^0 versus the reduced temperature k_BT/ϵ. It predicts that the Boyle temperature for the LJ fluid is $k_BT/\epsilon \approx 3.42$.

Problem 7.6 At cryogenic conditions, gas molecules such as hydrogen and helium may exhibit non-negligible quantum effects that cannot be reliably captured with classical models. Approximately, such effects may be accounted for by modifying the Lennard–Jones (LJ) model with the 4th-order approximation to the Feynman–Hibbs (FH) equation[2]

$$\frac{u(r)}{4\epsilon} = \frac{u^{LJ}(r)}{4\epsilon} + \frac{\beta\hbar^2}{24\mu}\left[u''^{LJ}(r) + \frac{2u'^{LJ}(r)}{r}\right] + \frac{\beta^2\hbar^4}{1152\mu^2}\left[u''''^{LJ}(r) + \frac{4u'''^{LJ}(r)}{r} + \frac{15u'^{LJ}(r)}{r^3}\right].$$

In the above equation, r denotes the center-to-center distance, $u^{LJ}(r)$ represents the 12-6 LJ potential, $\beta = 1/k_BT$, k_B is the Boltzmann constant, T is the absolute temperature, $\hbar = h/2\pi$ is the reduced Planck constant, $\mu = m_im_j/(m_i + m_j)$ is the reduced mass for particles i and j with mass m_i and m_j, respectively, and the number of primes represents the order of derivatives of the LJ potential with respect to distance r.

(i) Compare the quantum-corrected pair potential for H_2 at 50 K with the LJ potential;
(ii) Plot the second virial coefficients of H_2 gas in the temperature range from 50 to 300 K.

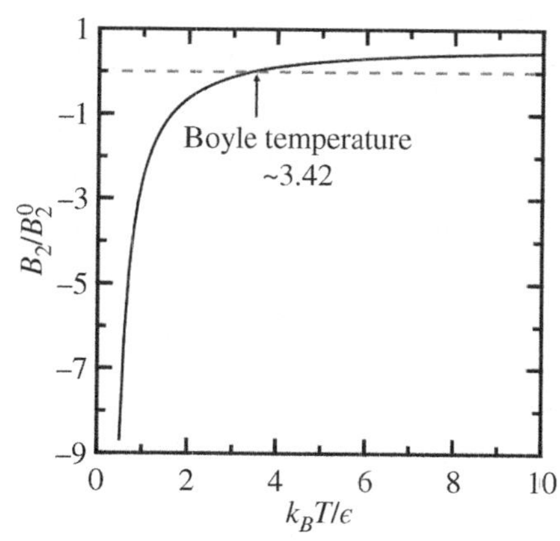

Figure 7.1 The second virial coefficient calculated from the Lennard–Jones potential (B2LJ.m).

2 Feynman R. P., Hibbs A. R. and Styer D. F., Quantum Mechanics and Path Integrals, Emended ed. McGraw-Hill, New York, 2005, p. 384.

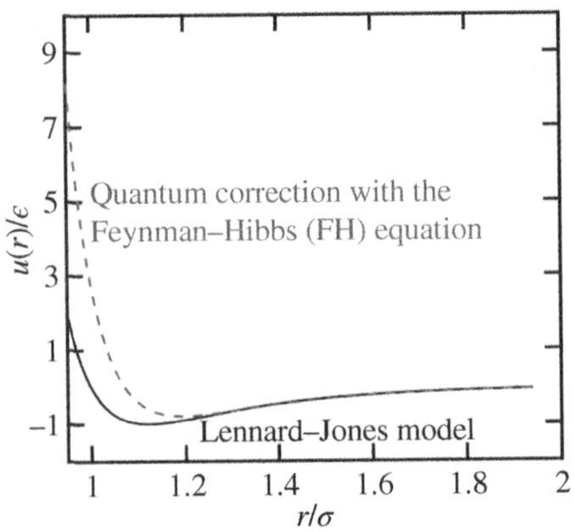

Figure 7.2 The quantum effects on the pair potential between hydrogen molecules (B2LJFH4.m).

Assume that the LJ parameters for different isotopic forms of hydrogen molecules are independent of the molecular weight, with $\sigma = 0.296$ nm and ϵ/k_B=34.2 K.

Solution:

(i) Figure 7.2 compares the quantum-corrected pair potentials for H_2 at 50 K with the LJ potential. In comparison with the classical model, the FH correction increases the effective diameter while reduces the strength of attraction.

(ii) Figure 7.3 shows the second virial coefficients of H_2 gas in the temperature range from 50 to 300 K. For H_2 gas, the quantum effects becomes negligible near the room temperature.

Problem 7.7 Derive the second-order virial coefficient from the van der Waals equation of state

$$P = \frac{\rho k_B T}{1 - \rho b} - a\rho^2,$$

where ρ is the number density, and compare the result with that predicted by the Lennard–Jones potential. What is the Boyle temperature predicted by the van der Waals equation? Comment on the performance at low density for various cubic equations of state (e.g., the Peng–Robinson equation) derived from the empirical modifications of the van der Waals equation.

Solution:

Expanding the van der Waals equation in a power series of ρ

$$\frac{P}{\rho k_B T} = \frac{1}{1 - \rho b} - \frac{a\rho}{k_B T} = 1 + b\rho + b^2\rho^2 + \cdots - \frac{a\rho}{k_B T}$$

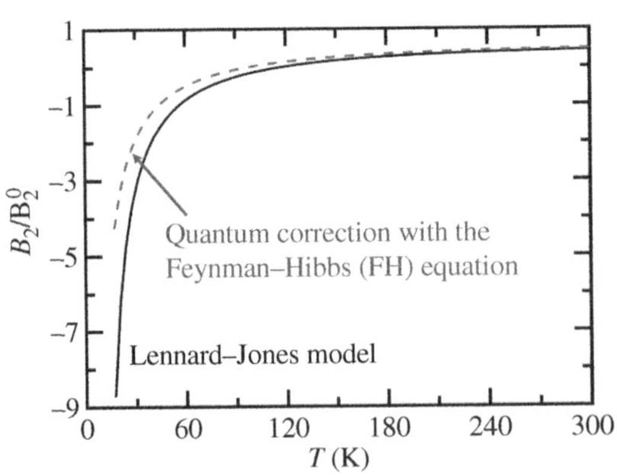

Figure 7.3 The second virial coefficient calculated from the Lennard–Jones potential with and without the quantum corrections (B2LJFH4.m).

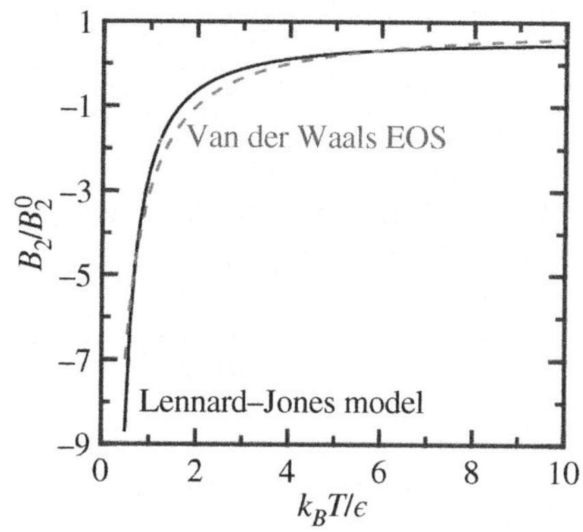

Figure 7.4 The second virial coefficient calculated from the van der Waals equation of state (EOS) for the Lennard–Jones potential (B2LJ.m).

yields

$$B_2 = b - \frac{a}{k_B T}.$$

Hence, $B_2 = 0$ predicts that the Boyle temperature is $T = a/(bk_B)$. The parameters of the van der Waals equation of state can be calculated from the Lennard–Jones model (Eqs. (7.140) and (7.141)):

$$a = 2\pi \int_\sigma^\infty dr r^2 4\epsilon \left(\frac{\sigma}{r}\right)^6 = 8\pi\epsilon\sigma^3 \int_1^\infty dx \frac{1}{x^4} = \frac{8\pi\epsilon\sigma^3}{3}.$$
$$b = 2\pi\sigma^3/3 = B_2^0.$$

Accordingly, the reduced second virial coefficient can be written as

$$B_2/B_2^0 = 1 - \frac{4\epsilon}{k_B T}.$$

Figure 7.4 compares B_2/B_2^0 predicted by the van der Waals theory with the exact result calculated from the LJ potential. The agreement is fair over the entire range of the temperature. If a cubic equation of state cannot reproduce the second virial coefficient, its performance at low density will be problematic.

Problem 7.8 Show that for an open system of spherical particles defined by the chemical potential μ, absolute temperature T and volume V, the isothermal compressibility is related to the fluctuation of particle density

$$\kappa_T = \frac{V}{k_B T} \frac{\langle \Delta\rho^2 \rangle}{\rho^2}.$$

Solution:

In a grand canonical ensemble specified by temperature T, particle chemical potential μ, and volume V, the fluctuation of the number of particles in the system is given by

$$\langle \Delta N^2 \rangle = \left(\frac{\partial \langle N \rangle}{\partial \beta\mu}\right)_{V,T} = k_B T V \left(\frac{\partial \rho}{\partial \mu}\right)_T,$$

where $\rho = \langle N \rangle / V$ is the average density. Meanwhile, the Gibbs–Duhem equation predicts

$$\left(\frac{\partial \rho}{\partial \mu}\right)_T = \rho \left(\frac{\partial \rho}{\partial P}\right)_T.$$

Hence, the isothermal compressibility is related to the average particle density by

$$\kappa_T = \frac{1}{\rho}\left(\frac{\partial \rho}{\partial P}\right)_T = \frac{1}{\rho^2}\left(\frac{\partial \rho}{\partial \mu}\right)_T = \frac{1}{\rho^2}\frac{\langle \Delta N^2\rangle}{k_B T V} = \frac{V}{k_B T}\frac{\langle \Delta \rho^2\rangle}{\rho^2}.$$

Problem 7.9 Based on the following expression for the radial distribution function of spherical particles discussed in Section 7.2

$$g(r) = e^{-\beta u(r)} \frac{V \int d\mathbf{r}_3 d\mathbf{r}_4 \cdots d\mathbf{r}_N \exp\{-\beta \Phi'(\mathbf{r}^{N-2})\}}{\int d\mathbf{r}_2 d\mathbf{r}_3 \cdots d\mathbf{r}_N \exp\{-\beta \Phi(\mathbf{r}^N)\}}. \tag{A}$$

(i) Show that

$$-k_B T \ln y(r) = \mu_2^{ex}(r) - 2\mu^{ex},$$

where $y(r) = g(r)e^{\beta u(r)}$ is the cavity correlation function, μ^{ex} is the excess chemical potential, and $\mu_2^{ex}(r)$ represents the excess chemical potential for a "dimer" of two particles with the center-to-center distance fixed at distance r.

(ii) Prove the zero-separation theorem, i.e., for a system of hard spheres,

$$\mu^{ex} = k_B T \ln y(0).$$

Solution:

(i) First, we write Eq. (A) in terms of the ensemble average

$$y(r) = \frac{V \int d\mathbf{r}_3 d\mathbf{r}_4 \cdots d\mathbf{r}_N \exp\{-\beta \Phi'(\mathbf{r}^{N-2})\}}{\int d\mathbf{r}_2 d\mathbf{r}_3 \cdots d\mathbf{r}_N \exp\{-\beta \Phi(\mathbf{r}^N)\}}$$

$$= \frac{V \int d\mathbf{r}_3 d\mathbf{r}_4 \cdots d\mathbf{r}_N \exp\{-\beta \Phi'(\mathbf{r}^{N-2})\}}{\int d\mathbf{r}^N \delta(\mathbf{r}_1) \exp\{-\beta \Phi(\mathbf{r}^N)\}} \times \frac{Q'_{N-2}}{Q'_{N-2}} \times \frac{Q'_N}{Q'_N}$$

$$= \frac{\langle e^{-\beta \Delta\Phi(\mathbf{r}_1,\mathbf{r}_2)}\rangle_{N-2}}{\langle \delta(\mathbf{r}_1)\rangle^2} \times \frac{Q'_{N-2}}{Q'_N},$$

where $\langle \delta(\mathbf{r}_1)\rangle = 1/V$. As discussed in Section 7.2, $\Delta\Phi(\mathbf{r}_1, \mathbf{r}_2)$ represents the potential energy due to the interaction of a pair of tagged particles, one at origin and the other at position $\mathbf{r}$, interacting with other $N-2$ particles in the system, $\Phi'(\mathbf{r}^{N-2})$ is defined as the total potential energy of the N-particle system subtracting the pair potential between tagged particles, $\langle \cdots \rangle$ denotes the canonical ensemble average with respect to the configurations of the N particles (or $N-2$ particles) when associated with subscript $N-2$), and Q'_N stands for the configurational integral with respect to the positions of N particles.

In a uniform system, the single particle density can be written as

$$\langle \delta(\mathbf{r}_1)\rangle = \frac{1}{V} = \frac{\int d\mathbf{r}^N \delta(\mathbf{r}_1) \exp\{-\beta \Phi(\mathbf{r}^N)\}}{\int d\mathbf{r}^N \exp\{-\beta \Phi(\mathbf{r}^N)\}} = \frac{\langle e^{-\beta \Delta\Phi(\mathbf{r})}\rangle_{N-1} Q'_{N-1}}{Q'_N},$$

where $\Delta\Phi(\mathbf{r})$ corresponds to the potential energy change due to the insertion of a particle at position $\mathbf{r}$ in a system of $N-1$ particles. Combing the above two equations leads to

$$y(r) = \frac{\langle e^{-\beta \Delta\Phi(\mathbf{r}_1,\mathbf{r}_2)}\rangle_{N-2}}{\langle e^{-\beta \Delta\Phi(\mathbf{r})}\rangle^2_{N-1}} \times \frac{Q'_{N-2} Q'_N}{Q'^2_{N-1}}.$$

In the thermodynamic limit, $Q'_N/Q'_{N-1} = Q'_{N-1}/Q'_{N-2}$. Meanwhile, Widom's insertion theorem predicts

$$\beta\mu^{ex} = -\ln \langle e^{-\beta \Delta\Phi(\mathbf{r})}\rangle_{N-1} \quad \text{and} \quad \beta\mu_2^{ex} = -\ln \langle e^{-\beta \Delta\Phi(\mathbf{r}_1,\mathbf{r}_2)}\rangle_{N-2}.$$

Hence, we have, by combining the above equations,

$$-\ln y(r) = \beta\mu_2^{ex}(r) - 2\beta\mu^{ex}.$$

Note that $y(r)$ depends only on the radial distance because the properties of the isotropic system is angle independent.

(ii) For a system of hard spheres, $\mu_2^{ex} = \mu^{ex}$ when $r = 0$. Therefore, $\mu^{ex} = k_B T \ln y(0)$.

Problem 7.10 Widom's insertion theorem, $\beta\mu^{ex} = -\ln\langle e^{-\beta\Delta\Phi(\mathbf{r})}\rangle$, can be extended to inhomogeneous systems, i.e., systems with a nonuniform distribution of particles due to the presence of an external potential. To understand this extension, consider a system of spherical particles subject to a one-body potential $V^{ext}(\mathbf{r})$. The grand canonical partition function of the system can be written as

$$\Xi = \sum_{N=0}^{\infty} \frac{z^N}{N!} \int d\mathbf{r}^N \exp\left[-\beta\Phi(\mathbf{r}^N) - \sum_{i=1}^{N} \beta V^{ext}(\mathbf{r}_i)\right],$$

where $\Phi(\mathbf{r}^N)$ stands for the potential energy due to inter-particle interactions, $\beta = 1/(k_B T)$, and $z \equiv e^{\beta\mu}/\Lambda^3$ with Λ being the thermal wavelength.

(i) Derive the potential distribution theorem, i.e., the one-body density profile can be written as

$$\rho(\mathbf{r}) \equiv \sum_{i=1}^{N} \langle \delta(\mathbf{r} - \mathbf{r}_i)\rangle = z e^{-\beta V^{ext}(\mathbf{r})} \langle e^{-\beta\Delta\Phi(\mathbf{r})}\rangle,$$

where $\Delta\Phi(\mathbf{r})$ stands for the change in potential energy upon inserting a particle at position $\mathbf{r}$ due to its interaction with all particles in the system.

(ii) In the low density limit, the one-body density profile is reduced to the Boltzmann distribution

$$\rho(\mathbf{r}) = \rho_0 e^{-\beta V^{ext}(\mathbf{r})},$$

where ρ_0 stands for the particle density in the bulk (viz., $V^{ext}(\mathbf{r}) = 0$).

(iii) The local excess chemical potential can be expressed as

$$\mu^{ex}(\mathbf{r}) = -k_B T \ln\langle e^{-\beta\Delta\Phi(\mathbf{r})}\rangle,$$

where $\mu^{ex}(\mathbf{r}) \equiv \mu - \mu^{IG} - V^{ext}(\mathbf{r})$, and μ^{IG} stands for the chemical potential of the noninteracting reference system with local density $\rho(\mathbf{r})$.

(iv) Suppose that the external potential is introduced by fixing a particle at the origin, the cavity correlation function $y(r)$ satisfies

$$-\ln y(r) = \beta\mu^{ex}(r) - \beta\mu^{ex},$$

where μ^{ex} stands for the excess chemical potential of the bulk system, and $\beta\mu^{ex}(\mathbf{r}) = -\ln\langle e^{-\beta\Delta\Phi(\mathbf{r})}\rangle$.

Solution:

(i) The one-body density profile can be written as

$$\begin{aligned}
\rho(\mathbf{r}) &= \frac{1}{\Xi}\sum_{N=0}^{\infty} \frac{z^N}{N!} \int d\mathbf{r}^N \sum_{i=1}^{N} \delta(\mathbf{r} - \mathbf{r}_i) \exp\left[-\beta\Phi(\mathbf{r}^N) - \sum_{i=1}^{N} \beta V^{ext}(\mathbf{r}_i)\right] \\
&= z e^{-\beta V^{ext}(\mathbf{r})} \frac{1}{\Xi} \sum_{N=1}^{\infty} \frac{z^{N-1}}{(N-1)!} e^{-\beta\Delta\Phi(\mathbf{r},N-1)} \int d\mathbf{r}^{N-1} \exp\left[-\beta\Phi(\mathbf{r}^{N-1}) - \sum_{i=1}^{N-1} \beta V^{ext}(\mathbf{r}_i)\right]
\end{aligned}$$

$$= ze^{-\beta V^{ext}(\mathbf{r})}\frac{1}{\Xi}\sum_{N=0}^{\infty}\frac{z^N}{N!}e^{-\beta\Delta\Phi(\mathbf{r},N)}\int d\mathbf{r}^N \exp\left[-\beta\Phi(\mathbf{r}^N)-\sum_{i=1}^{N}\beta V^{ext}(\mathbf{r}_i)\right]$$

$$= ze^{-\beta V^{ext}(\mathbf{r})}\langle e^{-\beta\Delta\Phi(\mathbf{r})}\rangle,$$

where $\Delta\Phi(\mathbf{r}) = \Delta\Phi(\mathbf{r}, N)$ stands for the potential energy of inserting a particle at position $\mathbf{r}$ due to its interactions with all particles in the system, and $\langle\cdots\rangle$ represents the ensemble average.

(ii) In the low density limit, the inter-particle interactions are negligible. Thus, the one-body density becomes

$$\rho(\mathbf{r}) = ze^{-\beta V^{ext}(\mathbf{r})} = \rho_0 e^{-\beta V^{ext}(\mathbf{r})},$$

where we have used the ideal-gas chemical potential for the low-density system, i.e., $\beta\mu = \ln(\rho_0\Lambda^3)$ or $z = \rho_0$.

(iii) As the chemical potential is uniform throughout the system, the local excess chemical potential can be expressed as

$$\mu^{ex}(\mathbf{r}) \equiv \mu - \mu^{IG} - V^{ext}(\mathbf{r}) = \mu - k_BT\ln[\rho(\mathbf{r})\Lambda^3] - V^{ext}(\mathbf{r}) = -k_BT\ln\langle e^{-\beta\Delta\Phi(\mathbf{r})}\rangle.$$

(iv) With a particle placed at the origin, the external potential is the same as the pair potential, $V^{ext}(\mathbf{r}) = u(r)$. Thus, $g(r) = \rho(r)/\rho_0$ and the cavity correlation function is

$$y(r) = g(r)e^{\beta u(r)} = \frac{z}{\rho_0}\langle e^{-\beta\Delta\Phi(\mathbf{r})}\rangle = \exp[-\beta\mu^{ex}(\mathbf{r}) + \beta\mu^{ex}].$$

Using $\beta\mu = \ln(\rho_0\Lambda^3) + \beta\mu^{ex}$, we can rearrange the above equation to obtain

$$-\ln y(r) = \beta\mu^{ex}(r) - \beta\mu^{ex},$$

where μ^{ex} stands for the excess chemical potential of the bulk system, and $\beta\mu^{ex}(r) = -\ln\langle e^{-\beta\Delta\Phi(\mathbf{r})}\rangle$.

Problem 7.11 With the pairwise additive approximation for the inter-particle potentials, the virial theorem predicts that the pressure equation for a binary mixture of spherical particles is

$$P = \rho k_BT - \frac{1}{6}\sum_{i=1}^{2}\sum_{j=1}^{2}\rho_i\rho_j\int d\mathbf{r}\frac{\partial u_{ij}(r)}{\partial r}rg_{ij}(r),$$

where $\rho = \rho_1 + \rho_2$, $\rho_{i=1,2}$ is the number density for type i particles, and $g_{ij}(r) = g_{ji}(r)$ is the pair distribution function between particles i and j, which has a pair potential $u_{ij}(r)$.

(i) Verify the above equation using the virial theorem

$$P = \rho k_BT + \frac{1}{3V}\left[\sum_{n_1=1}^{N_1}\langle\mathbf{r}_{n_1}\cdot\mathbf{F}_{n_1}\rangle + \sum_{n_2=1}^{N_2}\langle\mathbf{r}_{n_2}\cdot\mathbf{F}_{n_2}\rangle\right],$$

where N_1 and N_2 are the number of particles of type 1 and 2, respectively; $\mathbf{r}_{n_i}$ and $\mathbf{F}_{n_i}$ stand for the position and force on a particle of type i; and $\langle\cdots\rangle$ represents the ensemble average.

(ii) Show that, at low particle density, the radial distribution function can be approximated by the Boltzmann distribution

$$g_{ij}(r) = e^{-\beta u_{ij}(r)}.$$

(iii) Derive the second virial coefficient for the binary mixture.

(iv) Extend the above procedure to a multicomponent system with m types of spherical particles.

Solution:

(i) According to the pairwise additive approximation, the force on each particle includes contributions due to its pair interaction with both types of particles

$$\mathbf{F}_{n_1} = \sum_{m_1=1}^{N_1}{}' \mathbf{F}_{n_1,m_1} + \sum_{m_2=1}^{N_2} \mathbf{F}_{n_1,m_2},$$

$$\mathbf{F}_{n_2} = \sum_{m_1=1}^{N_1} \mathbf{F}_{n_2,m_1} + \sum_{m_2=1}^{N_2}{}' \mathbf{F}_{n_2,m_2},$$

where the prime sign means that the self-interaction has been excluded, and $\mathbf{F}_{i,j}$ stands for the force on particle i due to its interaction with particle j. Newton's third law predicts $\mathbf{F}_{i,j} = -\mathbf{F}_{j,i}$. The ensemble average of each virial is

$$\langle \mathbf{r}_{n_1} \cdot \mathbf{F}_{n_1} \rangle = \sum_{m_1=1}^{N_1}{}' \langle \mathbf{r}_{n_1} \cdot \mathbf{F}_{n_1,m_1} \rangle + \sum_{m_2=1}^{N_2} \langle \mathbf{r}_{n_1} \cdot \mathbf{F}_{n_1,m_2} \rangle = \sum_{i=1}^{2} \sum_{m_i}^{N_i}{}' \langle \mathbf{r}_{n_1} \cdot \mathbf{F}_{n_1,m_i} \rangle$$

Thus, the summation in the virial theorem becomes

$$\sum_{n_1=1}^{N_1} \langle \mathbf{r}_{n_1} \cdot \mathbf{F}_{n_1} \rangle + \sum_{n_2=1}^{N_2} \langle \mathbf{r}_{n_2} \cdot \mathbf{F}_{n_2} \rangle = \sum_{i=1}^{2} \sum_{n_i=1}^{N_i} \langle \mathbf{r}_{n_i} \cdot \mathbf{F}_{n_i} \rangle = \sum_{i=1}^{2} \sum_{j=1}^{2} \sum_{n_i=1}^{N_i} \sum_{n_j=1}^{N_j}{}' \langle \mathbf{r}_{n_i} \cdot \mathbf{F}_{n_i,n_j} \rangle$$

$$= \frac{1}{2} \sum_{i=1}^{2} \sum_{j=1}^{2} \sum_{n_i=1}^{N_i} \sum_{n_j=1}^{N_j}{}' \langle \mathbf{r}_{n_i} \cdot \mathbf{F}_{n_i,n_j} + \mathbf{r}_{n_j} \cdot \mathbf{F}_{n_j,n_i} \rangle$$

$$= \frac{1}{2} \sum_{i=1}^{2} \sum_{j=1}^{2} \sum_{n_i=1}^{N_i} \sum_{n_j=1}^{N_j}{}' \langle (\mathbf{r}_{n_i} - \mathbf{r}_{n_j}) \cdot \mathbf{F}_{n_i,n_j} \rangle$$

$$= -\frac{1}{2} \sum_{i=1}^{2} \sum_{j=1}^{2} \rho_i \rho_j \int d\mathbf{r} \frac{\partial u_{ij}(r)}{\partial r} r g_{ij}(r),$$

where we have applied $\mathbf{F}_{i,j} = -\mathbf{F}_{j,i}$, $(\mathbf{r}_i - \mathbf{r}_j) \cdot \mathbf{F}_{i,j} = -r_{ij} (\partial u / \partial r)|_{r=r_{ij}}$ and

$$\left\langle \sum_{n_i=1}^{N_i} \sum_{n_j=1}^{N_j}{}' \delta(\mathbf{r} - \mathbf{r}_i) \delta(\mathbf{r}' - \mathbf{r}_j) \right\rangle = \rho_i \rho_j g_{ij}(r)$$

with $r = |\mathbf{r} - \mathbf{r}'|$. Hence, the virial equation for the binary mixture becomes

$$P = \rho k_B T - \frac{1}{6} \sum_{i=1}^{2} \sum_{j=1}^{2} \rho_i \rho_j \int d\mathbf{r} \frac{\partial u_{ij}(r)}{\partial r} r g_{ij}(r).$$

(ii) We may write the local density near a particle of type i fixed at the origin of the coordinate as (See Problem 7.10)

$$\rho_{i,j}(\mathbf{r}) = \rho_j^0 e^{-\beta u_{i,j}(\mathbf{r}) - \beta \Delta \mu^{ex}(\mathbf{r})},$$

where ρ_j^0 is the number density remote from the fixed particle, $\Delta \mu^{ex}(\mathbf{r})$ represents the difference between the local excess chemical potential and the excess chemical potential remote from the fixed particle (viz., the excess chemical potential of the bulk system). At low density, the excess chemical potential vanishes thus the radial distribution function becomes

$$g_{ij}(r) = \frac{\rho_{i,j}(\mathbf{r})}{\rho_j^0} = e^{-\beta u_{ij}(r)}.$$

(iii) With $g_{ij}(r)$ approximated by the Boltzmann distribution, we can simplify the virial equation

$$P = \rho k_B T - \frac{1}{6}\sum_{i=1}^{2}\sum_{j=1}^{2}\rho_i\rho_j \int d\mathbf{r}\frac{\partial u_{ij}(r)}{\partial r} r e^{-\beta u_{ij}(r)}$$

$$= \rho k_B T + \frac{k_B T}{6}\sum_{i=1}^{2}\sum_{j=1}^{2}\rho_i\rho_j \int dr 4\pi r^3 \frac{\partial[e^{-\beta u_{ij}(r)} - 1]}{\partial r}$$

$$= \rho k_B T - \frac{2k_B T\pi}{3}\sum_{i=1}^{2}\sum_{j=1}^{2}\rho_i\rho_j \int dr 3r^2[e^{-\beta u_{ij}(r)} - 1]$$

$$= \rho k_B T - 2k_B T\pi\sum_{i=1}^{2}\sum_{j=1}^{2}\rho_i\rho_j \int dr r^2[e^{-\beta u_{ij}(r)} - 1].$$

Thus, the compressibility factor can be expressed as

$$Z = \frac{P}{\rho k_B T} = 1 - \left\{2\pi\sum_{i=1}^{2}\sum_{j=1}^{2}x_i x_j \int dr r^2[e^{-\beta u_{ij}(r)} - 1]\right\}\rho = 1 + B\rho,$$

where $x_i = \rho_i/\rho$ is the mole fraction. The above equation indicates that the second virial coefficient for the binary mixture is

$$B = -\sum_{i=1}^{2}\sum_{j=1}^{2}x_i x_j 2\pi \int dr r^2[e^{-\beta u_{ij}(r)} - 1] = \sum_{i=1}^{2}\sum_{j=1}^{2}x_i x_j B_{ij}$$

with $B_{ij} = -2\pi \int dr r^2[e^{-\beta u_{ij}(r)} - 1]$. B_{ii} and B_{jj} are the same as the virial coefficients of pure species.

(iv) For a multicomponent system with m types of spherical particles, it is straightforward to show that the second virial coefficient is given by

$$B = \sum_{i=1}^{m}\sum_{j=1}^{m}x_i x_j B_{ij}$$

with $B_{ij} = -2\pi \int dr r^2[e^{-\beta u_{ij}(r)} - 1]$.

Problem 7.12 The Ornstein–Zernike (OZ) equation may serve as a definition of the direct correlation function $c(r)$. To get this point, we first write the OZ equation through the Fourier transformation:

$$\hat{h}(k) = \hat{c}(k) + \rho\hat{h}(k)\hat{c}(k).$$

Rearranging the above equation gives

$$\hat{c}(k) = \frac{\hat{h}(k)}{1 + \rho\hat{h}(k)}.$$

Thus $c(r)$ is completely determined by the total correlation function $h(r)$ provided that the denominator $1 + \rho\hat{h}(k)$ is nonzero. Show that $1 + \rho\hat{h}(k) \neq 0$ for any thermodynamic system containing spherical particles.

Solution:

As discussed in Section 7.2, $1 + \rho\hat{h}(k)$ is related to the static structure factor, $\hat{S}(k)$, which is defined as

$$\hat{S}(k) = \frac{1}{N}\langle|\hat{\rho}(k)|^2\rangle,$$

where $\hat{\rho}(k) = \sum_i^N e^{i\mathbf{k}\cdot\mathbf{r}}$. Since $|\hat{\rho}(k)|^2 > 0$, we must have $\hat{S}(k) > 0$, which justifies the definition of the direct correlation function $c(r)$.

Problem 7.13 Establish a numerical procedure for solving the OZ equation for one-component Lennard–Jones fluids with the Percus–Yevick (PY) and the hypernetted chain (HNC) closures. Compare the results from these two closures for $g(r)$, $c(r)$, $\gamma(r) \equiv h(r) - c(r)$, and $B(r)$ at $k_B T = 1.5$ and $\rho\sigma^3 = 0.5$.

Hint: The OZ equation can be solved by using the Picard iteration:

(1) initial guess for $c(r) = \exp[-\beta u(r)] - 1$;
(2) fast Fourier transform (FFT) to calculate $\hat{c}(k)$;
(3) calculate $\hat{h}(k) = \hat{c}(k)/[1 - \rho\hat{c}(k)]$;
(4) inverse FFT to obtain $\gamma(r) \equiv h(r) - c(r)$;
(5) calculate new $c(r)$ based on a closure and iterate the procedure until convergence.

Solution:

The Matlab program prepared by Tsogbayar Tsednee (PhD), California State University Northridge (OZE.m) was used to generate the correlation functions of the Lennard-Jones fluid shown in Figure 7.5.

Comments:

(i) At this particular condition, the results from PY and HNC closures are virtually identical for $g(r)$, $c(r)$ and $\gamma(r)$ even though the bridge functions $B(r)$ are very different;
(ii) The solutions to the OZ equation are relatively insensitive to the choice of the bridge function because it is mostly screened by the pair potential $u(r)$, i.e., the closure is given by

$$\gamma(r) \equiv h(r) - c(r) = \ln[h(r) + 1] + \beta u(r) + B(r).$$

$B(r)$ is significant in the range of r where $\beta u(r)$ is virtually infinite.

(iii) The Picard iteration fails to converge at low temperature and high density.

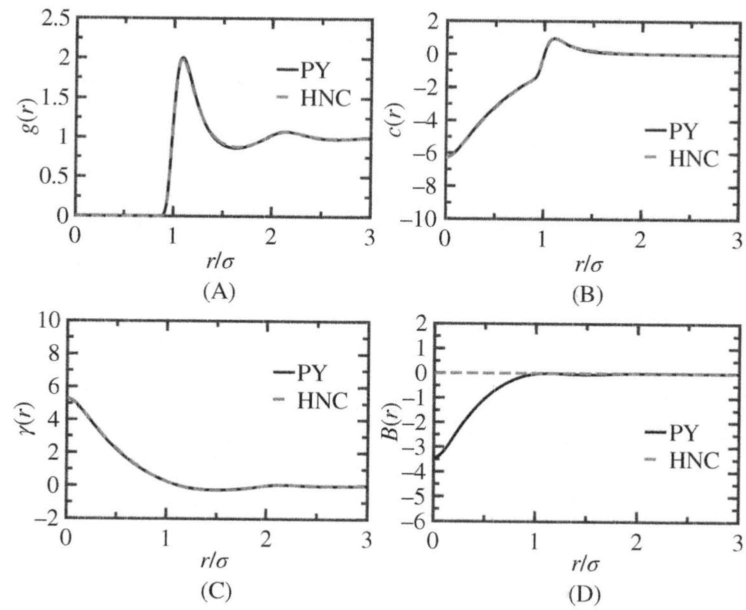

Figure 7.5 Radial distribution function (A), direct correlation function (B), gamma correlation function (C), and bridge function (D) for a Lennard–Jones fluid at $k_B T = 1.5$ and $\rho\sigma^3 = 0.5$ predicted by the PY and HNC closures (OZE.m).

Problem 7.14 Show that the second virial coefficient predicted by the sticky hard-sphere potential is given by

$$B_2 = B_2^{HS}\left(1 - \frac{1}{4\tau}\right),$$

where τ is a dimensionless parameter. Based on the van der Waals theory of simple fluids, discuss how parameter τ may vary with temperature.

Solution:

The second virial coefficient is defined as

$$B_2 = -\frac{1}{2}\int_0^\infty dr 4\pi r^2[e^{-\beta u(r)} - 1]. \tag{B}$$

Plugging into Eq. (B) with the sticky hard-sphere potential

$$e^{-\beta u(r)} = \begin{cases} 0 & r < \sigma \\ \frac{\sigma+d}{12\tau d} & \sigma \le r \le \sigma + d \\ 1 & r > \sigma + d \end{cases}$$

gives

$$\begin{aligned} B_2 &= \frac{1}{2}\int_0^\sigma dr 4\pi r^2 - \frac{1}{2}\lim_{d\to 0}\int_\sigma^{\sigma+d} dr 4\pi r^2\left(\frac{\sigma+d}{12\tau d} - 1\right) \\ &= B_2^{HS} - \frac{1}{2}\lim_{d\to 0}\left(\frac{\sigma+d}{12\tau d}\right)\left(\frac{4\pi \times 3d\sigma^2}{3}\right) = B_2^{HS} - \frac{2\pi\sigma^3}{12\tau} = B_2^{HS}\left(1 - \frac{1}{4\tau}\right), \end{aligned}$$

where $B_2^{HS} = 2\pi\sigma^3/3$. In comparison of the above equation with the reduced second virial coefficient from van der Waals theory (Problem 7.7)

$$B_2/B_2^0 = 1 - \frac{4\epsilon}{k_BT},$$

where ϵ is a constant, we may assume that τ is approximately a linear function of temperature.

Problem 7.15 Consider a uniform hard-sphere fluid with bulk density ρ and diameter σ.

(i) Based on the virial equation

$$\frac{P}{\rho k_BT} = 1 + 4\eta y(\sigma),$$

where $\eta = \pi\sigma^3/6$ is the packing fraction, derive the contact value of the cavity correlation function $y(\sigma)$ based on the equation of state predicted by the PY theory through the compressibility route, and compare the result with that derived from the Carnahan–Starling equation;

(ii) Compare the compressibility factor predicted by the PY virial equation, by the PY compressibility equation, and by the Carnahan–Starling equation for the packing fraction up to the freezing point;

(iii) Compare the contact value of the cavity correlation function predicted by the PY virial equation, by the PY compressibility equation, and by the Carnahan–Starling equation over the entire range of the packing fraction of the hard-sphere fluid.

Solution:

(i) The virial equation predicts that the contact value of the cavity correlation function is related to the compressibility factor $Z = P/(\rho k_BT)$

$$y(\sigma) = \frac{Z - 1}{4\eta}.$$

Plugging into the above equation with the equation of state predicted by the PY theory through the compressibility route (C) yields

$$y^c(\sigma) = \frac{1}{4\eta}\left[\frac{1+\eta+\eta^2}{(1-\eta)^3} - 1\right] = \frac{1}{4\eta}\frac{4-2\eta+\eta^2}{(1-\eta)^3} = \frac{1-\eta/2+\eta^2/4}{(1-\eta)^3}.$$

Based on the compressibility factor from the Carnahan–Starling (CS) equation, we have

$$y^{CS}(\sigma) = \frac{1}{4\eta}\left[\frac{1+\eta+\eta^2-\eta^3}{(1-\eta)^3} - 1\right] = \frac{1}{4\eta}\frac{4\eta-2\eta^2}{(1-\eta)^3} = \frac{1-\eta/2}{(1-\eta)^3}.$$

For comparison, the contact value of the cavity correlation function predicted the PY virial equation (V) is

$$y^v(\sigma) = \frac{1+\eta/2}{(1-\eta)^2}.$$

(ii) Figure 7.6A shows the compressibility factor and the contact value of the cavity correlation function predicted by various equations. For both quantities, we observe substantial discrepancy between these methods at high densities (the packing fraction at the freezing point is $\eta_f \approx 0.494$).

(iii) Figure 7.6B compares the contact value of the cavity correlation function predicted by different theoretical methods.

Problem 7.16 Analytical expressions for the radial distribution functions of hard-sphere fluids have been reported for $x \equiv r/\sigma < 1.5$ by de Souza and Ben-Amotz[3]

$$\ln g_0(x) = \frac{3-\eta}{(1-\eta)^3} - 3 + \frac{-3\eta(2-\eta)}{(1-\eta)^3}x + \left\{\ln\left[\frac{2-\eta}{2(1-\eta)^3}\right] - \frac{\eta(2-6\eta+3\eta^2)}{(1-\eta)^3}\right\}x^2,$$

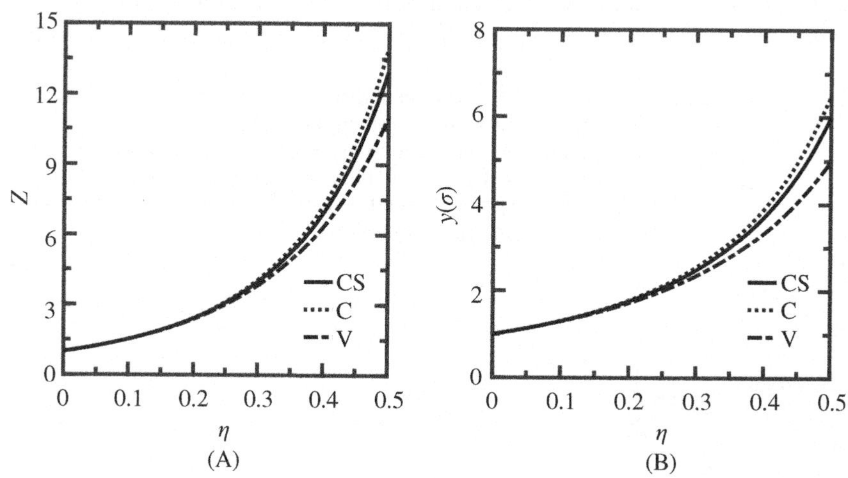

Figure 7.6 The compressibility factor (A) and the contact value of the cavity correlation function (B) predicted from various hard-sphere equations (PYHS.m).

3 de Souza L. E. S. and Ben-Amotz D., "Optimized perturbed hard sphere expressions for the structure and thermodynamics of Lennard-Jones fluids", *Mol. Phys.* 78, 137 (1993).

Table 7.1 The radial distribution function near contact at two reduced densities $\rho\sigma^3 = 0.3$ and 0.860.

x	0.3	0.860	x	0.3	0.860
1	1.535	4.645	1.025	1.504	3.966
1.05	1.471	3.34	1.075	1.436	2.846
1.1	1.402	2.441	1.125	1.367	2.111
1.15	1.334	1.845	1.175	1.303	1.625
1.2	1.274	1.442	1.225	1.249	1.289
1.25	1.225	1.162	1.275	1.204	1.056
1.3	1.183	0.969	1.325	1.163	0.896
1.35	1.145	0.836	1.375	1.126	0.787
1.4	1.107	0.747	1.425	1.09	0.716
1.45	1.075	0.691	1.475	1.06	0.673
1.5	1.046	0.662			

and, for $x \lessapprox 1.2$, by Tang and Lu[4]

$$g_0(x) = \frac{1+\eta/2}{(1-\eta)^2}\frac{1}{x} + \frac{1-5\eta-5\eta^2}{(1-\eta)^3}(1-1/x) + \frac{-3\eta+6\eta^2+21\eta^3/2}{(1-\eta)^4}\frac{(x-1)^2}{x}.$$

Plots these two approximations of $g_0(x)$ for $1 < x < 1.5$ at reduced densities $\rho\sigma^3 = 0.3$ and 0.860, and compare the theoretical predictions with the numerical results obtained from Monte Carlo simulation (Table 7.1).[5]

Solution:

Figure 7.7 compares the two approximations for the radial distribution functions of hard-sphere fluids with the simulation data. The two theoretical methods yield similar results at low density.

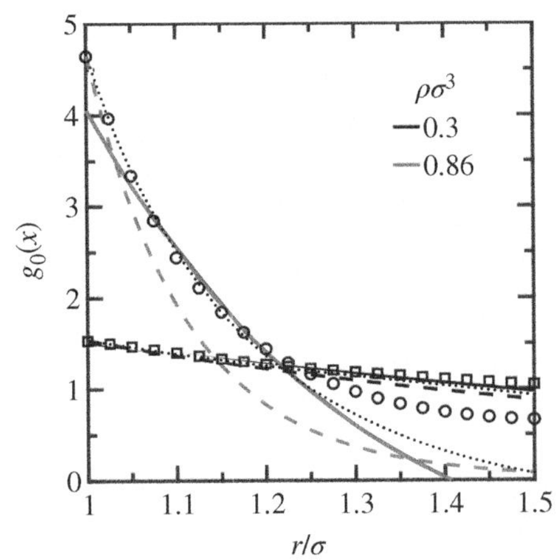

Figure 7.7 The radial distribution functions of two hard-sphere fluids predicted from the Tang–Lu approximation (solid lines), the deSouza–Ben-Amotz approximation (dashed lines), and Monte Carlo simulation (symbols). The dotted lines are obtained by a combination of the two approximations (PYHS.m).

4 Tang Y. and Lu B. C.-Y., "A study of associating Lennard–Jones chains by a new reference radial distribution function", *Fluid Phase Equilibria* 171, 27–44 (2000).

5 Labík S. and Malijevský A., "An effective Monte Carlo calculation of the radial distribution function of hard spheres on a minicomputer", *Czech. J. Phys.* 33, 128–134 (1983).

However, overall the Tang–Lu approximation is more accurate at high density. Better results can be obtained by replacing the coefficient in the first term of the Tang–Lu approximation with the contact value predicted by the Carnahan–Starling (CS) equation and averaging it with the prediction of the de-Souza–Ben-Amotz approximation.

Problem 7.17 The PY theory for a one-component system of sticky hard spheres predicts that the thermodynamic properties of the system are dependent on the sticky parameter τ and packing fraction η through parameter λ defined as

$$\lambda \equiv 6\left\{1-\tau+\tau/\eta-\sqrt{(1-\tau+\tau/\eta)^2-(1+2/\eta)/6}\right\}. \tag{C}$$

Show that

(i) $\lambda \to 0$ as $\tau \to \infty$ for any reasonable value of η;
(ii) As λ must be a real number, the PY theory yields a real solution for any reasonable value of η when $\tau \geq \tau_c \equiv (2-\sqrt{2})/6$;
(iii) At $\tau = \tau_c$, $\lambda = 1/\tau_c$ for any $\eta \leq \eta_c \equiv (3\sqrt{(2)}-4)/2$. For $\eta > \eta_c$, λ is given by

$$\lambda = 12(1-\tau_c+\tau_c/\eta)-1/\tau_c.$$

Solution:

(i) As $\tau \to \infty$, $(1-\tau+\tau/\eta)^2 \gg (1+2/\eta)/6$ for any reasonable value of η that satisfies $(1-\tau+\tau/\eta) > 0$ and $0 < \eta < \sqrt{2}/2$. Thus, as $\tau \to \infty$, we have

$$\lambda = 6\left\{1-\tau+\tau/\eta-\sqrt{(1-\tau+\tau/\eta)^2}\right\} \to 0.$$

(ii) As λ must be a real number, the term under the square root in Eq. (C) must be nonnegative, i.e.,

$$(1-\tau+\tau/\eta)^2-(1+2/\eta)/6 \geq 0.$$

The above quadratic inequality is satisfied except in the range $\eta_- < \eta < \eta_+$ with

$$\eta_\pm = \frac{\tau^2-\tau+1/6 \pm \sqrt{\tau^2/2-\tau/3+1/36}}{\tau^2-2\tau+5/6}.$$

When $\tau^2/2-\tau/3+1/36 = 0$ or $\tau \equiv \tau_c = (2-\sqrt{2})/6, \eta_- = \eta_+ = (3\sqrt{2}-4)/2$. In this case, the inequality is always satisfied and thus we have a real value for λ for any η.

(iii) At $\tau = \tau_c$, we have

$$\begin{aligned}(1-\tau_c+\tau_c/\eta)^2-(1+2/\eta)/6 &= (1-\tau_c+\tau_c/\eta)^2-1/6-1/3\eta \\ &= (1-\tau_c+\tau_c/\eta)^2-1/2+1/3-1/3\eta \\ &= (1-\tau_c+\tau_c/\eta)^2-1/3\tau+1/36\tau^2+1/3-1/3\eta \\ &= (1-\tau_c+\tau_c/\eta-1/6\tau_c)^2,\end{aligned}$$

where $1/2 = 1/3\tau - 1/36\tau^2 = 0$ is used in above equation. Hence, for $(1-\tau_c+\tau_c/\eta-1/6\tau_c)^2 \geq 0$ or equivalently, $\eta \leq \eta_c = (3\sqrt{(2)}-4)/2$, we have

$$\begin{aligned}\lambda &= 6\left\{1-\tau_c+\tau_c/\eta-\sqrt{(1-\tau_c+\tau_c/\eta)^2-(1+2/\eta)/6}\right\} \\ &= 6\left\{1-\tau_c+\tau_c/\eta-\sqrt{(1-\tau_c+\tau_c/\eta-1/6\tau_c)^2}\right\} = 1/\tau_c.\end{aligned}$$

For $\eta > \eta_c$, $(1-\tau_c+\tau_c/\eta-1/6\tau_c)^2 < 0$. Because the square-root term must be positive, we get

$$\lambda = 12(1-\tau_c+\tau_c/\eta)-1/\tau_c.$$

λ declines monotonically with η for $\eta > \eta_c$.

Problem 7.18 The osmotic pressure for an aqueous solution of globular proteins can be measured with dynamic light scattering (DLS) techniques. The key output from the DLS experiment is the excess scattered intensity with respect to the solvent, which can be written as

$$I_s = A\left(\frac{dn}{dc}\right) k_B Tc\left(\frac{\partial \Pi}{\partial c}\right)^{-1},$$

where A is an instrumental constant, dn/dc is related to the refractive index increment with respect to the solvent, Π stands for the osmotic pressure, and c is the protein concentration in the units of g/L. The osmotic pressure of a protein solution may be described by the sticky hard-sphere model in the PY approximation

$$\frac{\Pi}{\rho k_B T} = \frac{1+\eta+\eta^2}{(1-\eta)^3} - \eta\lambda\frac{18(2+\eta)-\eta\lambda^2}{36(1-\eta)^3},$$

where ρ is the protein number density, $\eta = \pi\sigma^3/6$ is the packing fraction of globular proteins with hard-sphere diameter σ, and λ is defined by the sticky parameter τ and packing fraction η

$$\lambda \equiv 6\left\{1-\tau+\tau/\eta-\sqrt{(1-\tau+\tau/\eta)^2-(1+2/\eta)/6}\right\}.$$

(i) Show that the derivative of the osmotic pressure with respect of the protein concentration is given by

$$\frac{\partial \Pi}{\partial c} = \frac{RT}{M}\frac{(1+2\eta-\lambda\eta)^2}{(1-\eta)^4},$$

where $R = k_B N_A$ is the gas constant, $N_A = 6.02214 \times 10^{23}$ is Avogadro's number, and M is the molecular weight of the globular protein.

(ii) For a given protein solution, it was found by fitting with experimental data that the sticky parameter can be calculated from $\tau = -1.6826 + 0.0065T/K$. Plot $\frac{M}{RT}\frac{\partial \Pi}{\partial c}$ versus η at temperatures 12.2, 17.2, 24.5, 30.2, 35.0 °C and compare the theoretical results with experimental data from the literature.[6]

Solution:

(i) See Barboy B., *J. Chem. Phys.* 61, 3194 (1974) and references cited in this article.

(ii) Figure 7.8 shows the theoretical results. In comparison with the experimental data, we find excellent agreement.

Problem 7.19 Consider a simple fluid with the pairwise-additive potential described by the Sutherland model:

$$u(r) = \begin{cases} \infty, & r < \sigma, \\ -\epsilon\left(\frac{\sigma}{r}\right)^n, & r \geq \sigma, \end{cases}$$

where σ represents the hard-sphere diameter, and r is the center-to-center distance, ϵ is an energy parameter, and $n > 3$ is a positive number.

(i) Show that the van der Waals parameter is given by

$$a = \frac{2\pi}{n-3}\epsilon\sigma^3.$$

6 Fig. 3, Piazza R., "Interactions in protein solutions near crystallisation: a colloid physics approach", *J. Cryst. Growth* 196, 415–423 (1999).

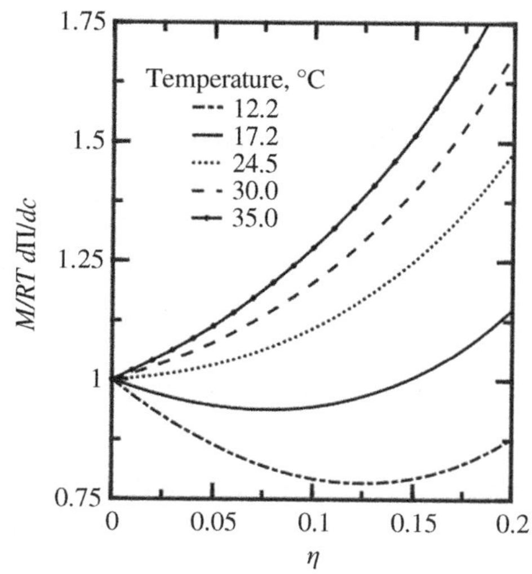

Figure 7.8 Reduced osmotic compressibility of hen egg-white lysozyme in NaAcO buffer and 0.2 M NaC1 at pH = 4.7 (SHS.m).

(ii) Find the critical temperature and critical density of the system from the modified van der Waals equation of state

$$\frac{P}{\rho k_B T} = \frac{1+\eta+\eta^2}{(1-\eta)^3} - \frac{a\rho}{k_B T},$$

where $\eta = \pi\rho\sigma^3/6$.

(iii) What is the reduced critical temperature $k_B T_c/\epsilon$ for $n = 6$?

Solution:

(i) The van der Waals parameter is given by

$$a = -\frac{1}{2}\int d\mathbf{r}u_A(r) = \frac{1}{2}\int_\sigma^\infty 4\pi r^2 dr\epsilon\left(\frac{\sigma}{r}\right)^n$$
$$= 2\pi\epsilon\int_\sigma^\infty drr^2\left(\frac{\sigma}{r}\right)^n = 2\pi\epsilon\sigma^3\int_1^\infty dx\frac{1}{x^{n-2}} = \frac{2\pi}{n-3}\epsilon\sigma^3.$$

(ii) At the critical point, the first and second order derivatives of the pressure with respect to density disappears

$$(\partial P/\partial\rho)_T = 0 \quad \text{and} \quad (\partial^2 P/\partial\rho^2)_T = 0.$$

Rewrite the equation of state in the dimensionless form

$$P^* = \frac{\eta+\eta^2+\eta^3}{(1-\eta)^3} - a^*\eta^2,$$

where $P^* = \frac{P\pi\sigma^3}{6\rho k_B T}$, $a^* = \frac{6a}{\pi\sigma^3 k_B T}$ and $\eta = \pi\rho\sigma^3/6$. Thus

$$(\partial P^*/\partial\eta)_T = -2a^*\eta + \frac{(1+2\eta)^2}{(1-\eta)^4} = 0,$$
$$(\partial^2 P^*/\partial\eta^2)_T = -2a^* + \frac{4(2+\eta)(1+2\eta)}{(1-\eta)^5} = 0.$$

Simplifying the above equations gives

$$a^* = \frac{(1+2\eta)^2}{2\eta(1-\eta)^4} \quad \text{and} \quad a^* = \frac{2(2+\eta)(1+2\eta)}{(1-\eta)^5}.$$

Therefore

$$a^* = \frac{(1+2\eta)^2}{2\eta(1-\eta)^4} \quad \text{and} \quad 1 = \frac{4\eta(1+2\eta)}{(1-\eta)(2+\eta)}.$$

Solving the above equations leads to $\eta_c \approx 0.128667$ and $a^* \approx 10.6578$. Recall

$$a^* = \frac{6a}{\pi\sigma^3 k_B T} = \frac{6}{\pi\sigma^3 k_B T} \times \frac{2\pi}{n-3}\epsilon\sigma^3 = \frac{12\epsilon}{(n-3)k_B T_c}.$$

so that

$$k_B T_c/\epsilon = \frac{12}{(n-3)a^*} \approx \frac{1.12594}{n-3}.$$

(iii) When $n = 6$, the reduced temperature is $k_B T_c/\epsilon \approx 0.375312$.

Problem 7.20 In combination with the Carnahan–Starling equation for hard spheres, the modified van der Waals equation of state for the Sutherland model is given by

$$\frac{P}{\rho k_B T} = \frac{1+\eta+\eta^2-\eta^3}{(1-\eta)^3} - \frac{a\rho}{k_B T},$$

where $\eta = \pi\rho\sigma^3/6$ and $a = 2\pi\epsilon\sigma^3/3$.

(i) Show that the corresponding chemical potential is given by

$$\frac{\mu}{k_B T} = \ln(\rho\Lambda^3) + \eta\frac{8-9\eta+3\eta^2}{(1-\eta)^3} - \frac{2\rho a}{k_B T}.$$

(ii) What are the critical temperature and the critical density predicted by the modified van der Waals equation of state?

(iii) Construct a vapor–liquid phase diagram (viz., the vapor–liquid coexistence curve) predicted by the above equations.

Solution:

(i) The chemical potential can be calculated from the pressure using the Gibbs–Duhenm equation $\rho d\mu = dP$. In terms of excess properties, we have

$$\beta\mu^{ex} = \int_0^\rho d\rho \frac{1}{\rho}\frac{\partial \beta P^{ex}}{\partial \rho} = \int_0^\rho d\rho \frac{1}{\rho}\frac{\partial}{\partial \rho}\left[\frac{1+\eta+\eta^2-\eta^3}{(1-\eta)^3} - 1\right]\rho - \int_0^\rho d\rho \frac{1}{\rho}\frac{1}{\partial \rho}\frac{a\rho^2}{k_B T}$$

$$= \int_0^\eta d\eta \frac{1}{\eta}\frac{\partial}{\partial \eta}\left[\frac{2\eta(2-\eta)}{(1-\eta)^3}\right]\eta - \int_0^\rho d\rho \frac{2a}{k_B T} = \eta\frac{8-9\eta+3\eta^2}{(1-\eta)^3} - \frac{2\rho a}{k_B T}.$$

Hence, the chemical potential is

$$\frac{\mu}{k_B T} = \frac{\mu^{IG}}{k_B T} + \frac{\mu^{ex}}{k_B T} = \ln(\rho\Lambda^3) + \eta\frac{8-9\eta+3\eta^2}{(1-\eta)^3} - \frac{2\rho a}{k_B T},$$

where $\beta\mu^{IG} = \ln(\rho\Lambda^3)$.

(ii) The critical temperature and density can be obtained by solving the equations

$$(\partial P/\partial \rho)_T = 0 \quad \text{and} \quad (\partial^2 P/\partial \rho^2)_T = 0,$$

which leads to $k_B T_c/\epsilon \approx 0.377315$ and $\eta_c \approx 0.130444$.

(iii) The vapor–liquid coexistence curve is obtained by solving the thermodynamic equations for phase equilibrium:

$$P^L = P^V$$
$$\mu^L = \mu^V.$$

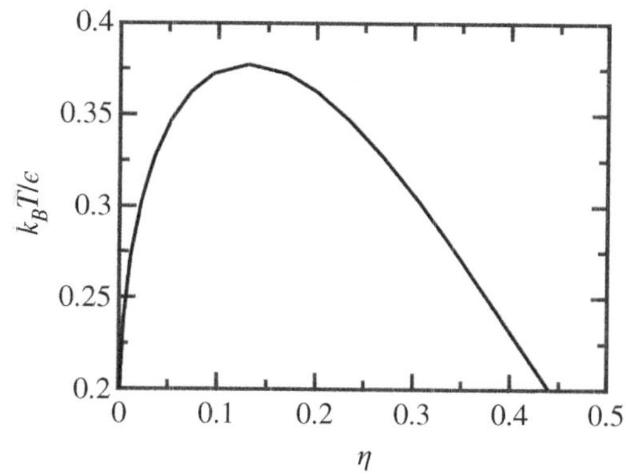

Figure 7.9 The vapor–liquid coexistence calculated from the modified van der Waals equation of state for the Sutherland model (vle.m).

For each temperature $T < T_c$, we can find two densities satisfying the above two equations, one corresponds to the vapor and the other for the liquid. Figure 7.9 shows the phase diagram calculated from the modified van der Waals equation of state for the Sutherland model. In calculating the vapor–liquid coexistence curve, one may start with the critical point and identify the two densities by reducing the temperature.

Problem 7.21 Predict the coexisting densities at the fluid-solid equilibrium (FSE) for a system of uniform hard spheres using the Carnahan–Starling equation for the fluid phase and the cell model for the solid with the face-centered-cubic (FCC) structure:

$$\frac{P_f}{\rho k_B T} = \frac{1+\eta+\eta^2-\eta^3}{(1-\eta)^3}, \qquad \frac{P_s}{\rho k_B T} = \frac{1}{1-(\rho/\rho_c)^{1/3}},$$

where $\rho_c\sigma^3 = \sqrt{2}$ is the reduced density at the close packing of uniform hard spheres. Compare the molar entropy of the hard-sphere system as a fluid and as a solid at packing fraction $\eta = 0.6$ and comment on the relationship between entropy and the degree of randomness.

Solution:

The coexisting densities at the fluid–solid transition can be identified by solving the equilibrium equations:

$$P_f = P_s, \qquad \mu_f = \mu_s.$$

From Section 7.9, we have the expressions for the chemical potentials of the fluid and solid phases

$$\beta\mu_f = \ln(\rho\Lambda^3) + \eta\frac{8-9\eta+3\eta^2}{(1-\eta)^3}.$$

$$\beta\mu_s = \ln(\rho\Lambda^3) + \frac{1}{1-(\rho/\rho_c)^{1/3}} - 3\ln[1-(\rho/\rho_c)^{1/3}] - \ln 8.$$

The equations for FSE can be solved with a math tool (e.g., MatLab) to give $\rho_f\sigma^3 = 0.9397$ and $\rho_s\sigma^3 = 1.0602$.

At $\eta = 0.6$, the molar entropy of the hard-sphere system as a fluid is -48.9379 J/(mol K) and as a solid with the FCC structure is -38.5257 J/(mol K), respectively. The solid phase has a higher entropy because, at this packing fraction, each particle has more free volume on average.

Problem 7.22 Prepare a computer program to construct the fluid–solid phase diagram for the Sutherland model with $n = 5, 6$ and 7 using the mean-field theory for the fluid phase and the cell model for the solid as discussed in Sections 7.7 and 7.8, respectively. Discuss whether the vapor–liquid transition is stable or metastable for these systems.

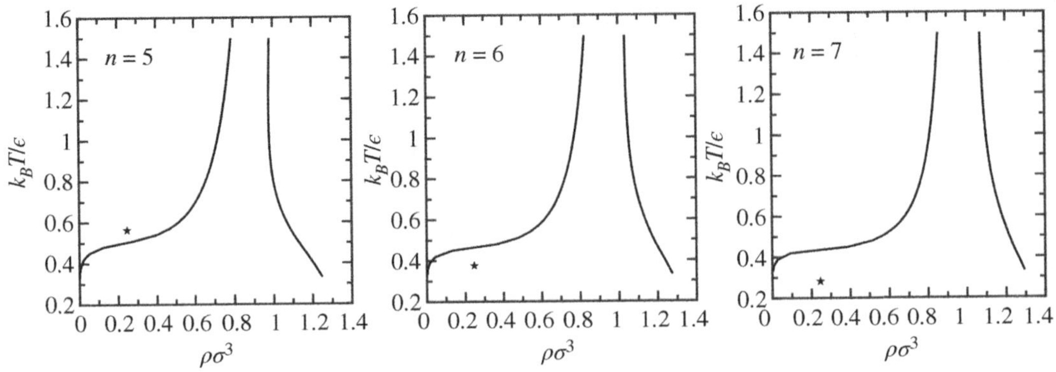

Figure 7.10 The fluid–solid phase diagram predicted by the mean-field theory for the Sutherland model with $n = 5, 6$, and 7. In each panel, the star symbol represents the critical point of the vapor–liquid equilibrium (vlse.m).

Solution:
Figure 7.10 shows the phase diagram calculated from the mean-field theory and cell model based on the Sutherland model. The critical point of the vapor–liquid equilibrium can be calculated from $\eta_c \approx 0.128667$ and $k_BT_c/\epsilon \approx 1.12594/(n-3)$ (Problem 7.19). The mean-field theory predicts that the vapor-liquid equilibrium (VLE) is unstable for $n = 6$ and 7 but stable for $n = 5$.

Problem 7.23 In the self-consistent Ornstein–Zernike approximation (SCOZA),[7] a state-dependent parameter is introduced into the long-range component of the direct correlation function. This parameter is determinated self-consistently from the thermodynamic consistency between the energy and compressibility equations:

$$\omega = \frac{\rho}{2V}\int d\mathbf{r}u(r)g(r) \quad \text{and} \quad \left(\frac{\partial \beta P}{\partial \rho}\right)_T = 1 - \rho\hat{c}(k=0),$$

where $\beta = 1/(k_BT)$, and $\omega = \langle\Phi\rangle/V$ is the potential energy density. Show that the thermodynamic consistency is satisfied by imposing

$$-\frac{\partial \hat{c}(k=0)}{\partial \beta} = \frac{\partial^2 \omega}{\partial \rho^2}.$$

Solution:
Starting from the Gibbs–Duhem equation for a one-component system at fixed temperature T

$$\left(\frac{\partial \beta P}{\partial \rho}\right) = \rho\left(\frac{\partial \beta \mu}{\partial \rho}\right),$$

we apply the compressibility equation and differentiate both sides with respect to β leading to

$$-\frac{\partial \hat{c}(k=0)}{\partial \beta} = \frac{\partial^2 \beta\mu}{\partial \rho \partial \beta}.$$

Meanwhile, we can write the chemical potential in terms of the Helmholtz energy density

$$\frac{\partial \beta\mu}{\partial \rho} = \frac{\partial}{\partial \rho}\frac{\partial \beta f}{\partial \rho} = \frac{\partial^2 \beta f}{\partial \rho^2},$$

7 Dickman R. and Stell G., "Self-consistent Ornstein-Zernike approximation for lattice gases", *Phys. Rev. Lett.* 77, 996 (1996).

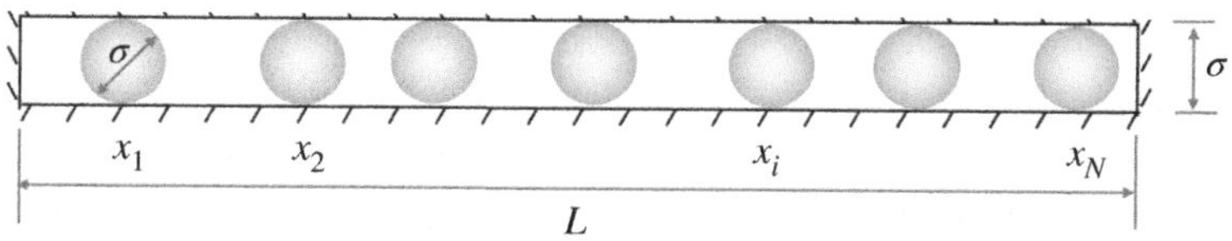

Figure 7.11 Schematic of a Tonks–Takahashi fluid confined between two hard walls.

where $\beta\mu = (\partial \beta f/\partial \rho)_T$ and $f = F/V$ is the Helmholtz energy density. Thus

$$\frac{\partial^2 \beta\mu}{\partial\rho\partial\beta} = \frac{\partial^2}{\partial\rho^2}\frac{\partial\beta f}{\partial\beta} = \frac{\partial^2(U/V)}{\partial\rho^2},$$

where we have used the Gibbs–Helmholtz equation to relate U/V, the internal energy density, to the reduced Helmholtz energy density βf. Because $U/V = 3k_BT\rho/2 + \omega$, we have

$$-\frac{\partial\hat{c}(k=0)}{\partial\beta} = \frac{\partial^2\omega}{\partial\rho^2}.$$

Problem 7.24 A Tonks gas, also known as a Tonks–Takahashi fluid, refers to a one-dimensional system of particles with hard-sphere interactions.

As shown schematically in Figure 7.11, the hard-sphere potential is described as

$$u(r) = \begin{cases} \infty, & r < \sigma, \\ 0, & r \geq \sigma, \end{cases}$$

where $r = |x_i - x_j|$ is the center-to-center distance between any pair of particles at positions x_i and x_j, and σ is the particle diameter. Because of the one-dimensional confinement, the particle positions are restricted to $x_1 \geq \sigma/2$ and $x_N \leq L - \sigma/2$. As the particles cannot pass each other, the pair potential applies only to the nearest neighbors. The canonical partition function of the system is given by

$$Q(N, L) = \frac{1}{\Lambda^N}\int_0^L dx_1 \int_0^L dx_2 \cdots \int_0^L dx_N \exp[-\beta\Phi(x_1, x_2, \ldots, x_N)],$$

where Λ stands for the thermal wave length, and $\Phi(x_1, x_2, \ldots, x_n)$ is the total potential energy. Here $N!$ is not applied because the particles are identifiable by their relative positions or sequences.

(i) Show

$$\exp[-\beta\Phi(x_1, x_2, \ldots, x_N)] = \Theta(x_1 - \sigma/2)\Theta(L - \sigma/2 - x_N)\prod_{i=2}^{N}\Theta(x_i - x_{i-1} - \sigma),$$

where $\Theta(x)$ stands for the Heaviside step function, i.e., $\Theta(x) = 1$ for $x \geq 0$ and $\theta(x) = 0$ for $x < 0$.

(ii) Develop expressions for $Q(1, L)$ and $Q(2, L)$.

(iii) Verify

$$Q(N, L) = \frac{(L - N\sigma)^N}{N!\Lambda^N}.$$

(iv) Derive the equation of state of the system based on $\beta P = (\partial \ln Q/\partial L)_T$. What are the internal energy and chemical potential of this system predicted by the canonical partition function?

(v) Show that the grand canonical partition function is given by

$$\Xi(\mu, L) = \sum_{i=1}^{\infty}\frac{e^{N\beta\mu}}{N!\Lambda^N}(L - N\sigma)^N\Theta(L - N\sigma).$$

(vi) Show for any real number $x > 0$,

$$\lim_{L\to\infty} \frac{\Xi(\mu, L-x)}{\Xi(\mu, L)} = e^{-Px}.$$

(vii) In the bulk phase, the radial distribution function of a Tonks gas may be understood as the normalized local density $g(r) = \rho(x = r - \sigma/2)/\rho$ for an open system with one particle fixed at $x = -\sigma/2$ and another at $L + \sigma/2 \to \infty$. Show that

$$g(r) = e^{\beta\mu^{ex}} e^{-\beta Pr} \Xi(\mu, r - \sigma),$$

where $\mu^{ex} = \mu - k_B T \ln(\rho\Lambda)$.

(viii) Plot $g(r)$ at reduced densities $\rho\sigma = 0.1, 0.5$ and 0.9.

Solution:

This problem and the corresponding solutions were adopted from Davis H. T.,[8] and from Salsburg Z. W. et al.[9]

(i) The total potential energy includes a contribution due to the pairwise inter-particle potential and an external energy for each end particle

$$\Phi(x_1, x_2, \dots, x_N) = V^{ext}(x_1) + V^{ext}(x_N) + \sum_{i=2}^{N} u(x_i - x_{i-1})$$

where $x_1 < x_2 < \cdots < x_N$ stand for the particle positions. Because the external energy limits the positions of the terminal particles to $x_1 \geq \sigma/2$ and $x_N \leq L - \sigma/2$, we have

$$\exp[-\beta V^{ext}(x_1)] = \Theta(x_1 - \sigma/2) \quad \text{and} \quad \exp[-\beta V^{ext}(x_N)] = \Theta(L - \sigma/2 - x_N).$$

Similarly, the hard-sphere potential predicts

$$\exp[-\beta u(x_i - x_{i-1})] = \Theta(x_i - x_{i-1} - \sigma).$$

Thus,

$$\exp[-\beta\Phi(x_1, x_2, \dots, x_N)] = \Theta(x_1 - \sigma/2)\Theta(L - \sigma/2 - x_N)\prod_{i=2}^{N}\Theta(x_i - x_{i-1} - \sigma).$$

(ii) If the system contains a single particle, there is no particle–particle interaction. Thus

$$Q(1, L) = \frac{1}{\Lambda}\int_0^L dx_1 \Theta(x_1 - \sigma/2)\Theta(L - \sigma/2 - x_1) = \frac{1}{\Lambda}\int_{\sigma/2}^{L-\sigma/2} dx_1 = \frac{L - \sigma}{\Lambda}.$$

For $N = 2$, we have

$$\begin{aligned}
Q(2, L) &= \frac{1}{\Lambda^2}\int_0^L dx_1 \int_0^L dx_2 \Theta(x_1 - \sigma/2)\Theta(x_2 - x_1 - \sigma)\Theta(L - \sigma/2 - x_2) \\
&= \frac{1}{\Lambda^2}\int_0^{L-3\sigma/2} dx_1 \Theta(x_1 - \sigma/2)\int_{x_1+\sigma}^{L-\sigma/2} dx_2 \\
&= \frac{1}{\Lambda^2}\int_0^{L-3\sigma/2} dx_1 \Theta(x_1 - \sigma/2)(L - 3\sigma/2 - x_1) \\
&= \frac{1}{\Lambda^2}\int_{\sigma/2}^{L-3\sigma/2} dx_1 (L - 3\sigma/2 - x_1).
\end{aligned}$$

8 Davis H. T., "Density distribution functions of confined Tonks–Takahashi fluids", *J. Chem. Phys.* 93, 4339–4344 (1990).

9 Salsburg Z. W., Zwanzig R. W. and Kirkwood J. G., "Molecular distribution functions in a one-dimensional fluid", *J. Chem. Phys.* 21, 1098–1107 (1953).

The above integral can be easily evaluated by variable change $t \equiv (L - 3\sigma/2 - x_1)$,

$$Q(2, L) = -\frac{1}{\Lambda^2}\int_{L-2\sigma}^{0} t dt = \frac{(L-2\sigma)^2}{2\Lambda^2}.$$

(iii) For a system with N particles, the canonical partition function is

$$Q(N, L) = \frac{1}{\Lambda^N}\int_0^L dx_1 \ldots \int_0^L dx_N \Theta(x_1 - \sigma/2)\Theta(L - \sigma/2 - x_N)\prod_{i=2}^{N}\Theta(x_i - x_{i-1} - \sigma)$$
$$= \frac{1}{\Lambda^N}\int_{\sigma/2}^{L+\sigma/2-N\sigma} dx_1 \cdots \int_{x_{i-1}+\sigma}^{L+\sigma/2-(N-i+1)\sigma} dx_i \cdots \int_{x_{N-1}+\sigma}^{L+\sigma/2-(N-N+1)\sigma} dx_N.$$

Let $t_i \equiv x_i - (i-1/2)\sigma$ and note $x_{i-1} + \sigma - (i-1/2)\sigma = t_{i-1}$, we can rewrite the above equation as

$$Q(N, L) = \frac{1}{\Lambda^N}\int_0^{L-N\sigma} dt_1 \int_{t_1}^{L-N\sigma} dt_2 \cdots \int_{t_{i-1}}^{L-N\sigma} dt_i \cdots \int_{t_{N-1}}^{L-N\sigma} dt_N.$$

An iterative evaluation of the integrals leads to

$$Q(N, L) = \frac{1}{\Lambda^N}\int_0^{L-N\sigma} dt_1 \cdots \int_{t_{N-2}}^{L-N\sigma} dt_{N-1}(L - N\sigma - t_{N-1})$$
$$= \frac{1}{\Lambda^N}\int_0^{L-N\sigma} dt_1 \cdots \int_{t_{N-3}}^{L-N\sigma} dt_{N-2}(L - N\sigma - t_{N-2})^2/2!$$
$$= \frac{1}{\Lambda^N}\int_0^{L-N\sigma} dt_1 \cdots \int_{t_{N-4}}^{L-N\sigma} dt_{N-3}(L - N\sigma - t_{N-3})^3/3!$$
$$= \frac{1}{\Lambda^N}\int_0^{L-N\sigma} dt_1 \cdots \int_{t_{N-i-1}}^{L-N\sigma} dt_{N-i}(L - N\sigma - t_i)^i/i!$$
$$= \frac{1}{\Lambda^N}\int_0^{L-N\sigma} dt_1(L - N\sigma - t_1)^{N-1}/(N-1)! = \frac{(L-N\sigma)^N}{N!\Lambda^N}.$$

(iv) The Helmholtz energy is

$$F = -k_BT \ln Q = -Nk_BT[\ln(L - N\sigma)/\Lambda - \ln N + 1]$$
$$= Nk_BT[\ln(\rho\Lambda) - 1 - \ln(1 - \rho\sigma)] = k_BT\left[\ln\frac{\rho\Lambda}{1-\rho\sigma} - 1\right],$$

where $\rho = N/L$ is the number density. The system pressure is

$$\beta P = (\partial \ln Q/\partial L)_T = \frac{N}{L - N\sigma} = \frac{\rho}{1-\rho\sigma}.$$

The internal energy of the system can be derived from the Gibbs–Helmholtz equation

$$U = (\partial\beta F/\partial\beta)_{L,N} = N(\partial \ln \Lambda/\partial\beta)_{L,N} = Nk_BT/2.$$

The chemical potential can be derived from F and PV (or from $(\partial F/\partial N)_{L,T}$):

$$\mu = (F + PV)/N = k_BT\left[\ln\frac{\rho\Lambda}{1-\rho\sigma} - 1 + \frac{1}{1-\rho\sigma}\right]$$
$$= k_BT\left[\ln\frac{\rho\Lambda}{1-\rho\sigma} + \frac{\rho\sigma}{1-\rho\sigma}\right].$$

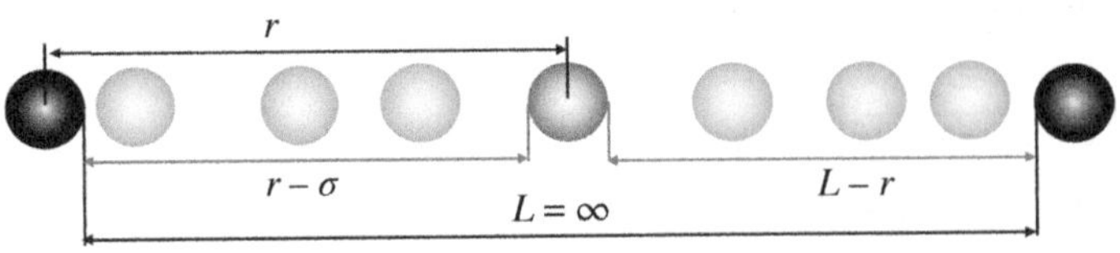

Figure 7.12 A Tonks–Takahashi fluid confined between two particles that are macroscopically separated. The probability of finding a particle (darker colored) at position $x = r - \sigma/2$ is related to the radial distribution function $g(r)$ of a bulk fluid.

(v) The grand partition function is given by

$$\Xi(\mu, L) = \sum_{N=0}^{\infty} \frac{1}{\Lambda^N} \int_0^L dx_1 \int_0^L dx_2 \cdots \int_0^L dx_N \exp[-\beta\Phi(x_1, x_2, \ldots, x_N) + N\beta\mu]$$

$$= \sum_{N=0}^{\infty} e^{N\beta\mu} Q(N, L) = \sum_{N=0}^{\infty} \frac{e^{N\beta\mu}}{N!\Lambda^N} (L - N\sigma)^N \Theta(L - N\sigma),$$

where the Heaviside function $\Theta(L - N\sigma)$ sets the maximum number of particles in the system.

(vi) In the thermodynamic limit ($L \to \infty$), $\beta PL = \ln \Xi(\mu, L)$. Hence,

$$\lim_{L\to\infty} \frac{\Xi(\mu, L - x)}{\Xi(\mu, L)} = e^{-Px}.$$

(vii) As shown in Figure 7.12, we may relate the radial distribution function $g(r)$ of a bulk fluid to an open system with one particle fixed at $x = -\sigma/2$ and another at $L + \sigma/2 \to \infty$. The semi-infinite system is equivalent to a bulk Tonks–Takahashi fluid.
The local density at position $x = r - \sigma/2$ can be expressed in terms of an average of the instantaneous density in the grand canonical ensemble

$$\rho(x = r - \sigma/2) = \left\langle \sum_{i=1}^{N} \delta(r - \sigma/2 - x_i) \right\rangle$$

$$= \frac{1}{\Xi(\mu, L)} \sum_{N=0}^{\infty} \frac{e^{N\beta\mu}}{\Lambda^N} \int_0^L dx_1 \cdots \int_0^L dx_N$$

$$\times \sum_{i=1}^{N} \delta(r - \sigma/2 - x_i) \exp[-\beta\Phi(x_1, x_2, \ldots, x_N)]$$

$$= \frac{1}{\Xi(\mu, L)} \sum_{N=0}^{\infty} \sum_{i=1}^{N} e^{N\beta\mu} Q(i - 1, r - \sigma) Q(N - i, L - r)$$

$$= \frac{e^{\beta\mu}}{\Xi(\mu, L)} \sum_{N=0}^{\infty} \sum_{i=1}^{N} Q(i - 1, r - \sigma) e^{(i-1)\beta\mu} Q(N - i, L - r) e^{(N-i)\beta\mu}$$

$$= \frac{e^{\beta\mu}}{\Xi(\mu, L)} \sum_{i'=0}^{\infty} Q(i', r - \sigma) e^{i'\beta\mu} \sum_{N'=0}^{\infty} Q(N', L - r) e^{N'\beta\mu}$$

$$= \frac{e^{\beta\mu}}{\Xi(\mu, L)} \Xi(\mu, r - \sigma) \Xi(\mu, L - r).$$

Using relations $\frac{\Xi(\mu, L-r)}{\Xi(\mu, L)} = e^{-Pr}$ and $\beta\mu = \ln \rho\Lambda + \beta\mu^{ex}$, we get

$$g(r) = \rho(x = r - \sigma/2)/\rho = \frac{e^{\beta\mu}}{\rho\Lambda} e^{-\beta Pr} \Xi(\mu, r - \sigma) = e^{\beta\mu^{ex}} e^{-\beta Pr} \Xi(\mu, r - \sigma).$$

(viii) From the above derivations, we have

$$g(r) = e^{\beta\mu^{ex}} e^{-\beta P r} \Xi(\mu, r - \sigma),$$

where

$$\beta\mu^{ex} = \ln \frac{1}{1-\rho\sigma} + \frac{\rho\sigma}{1-\rho\sigma}; \qquad \beta P = \frac{\rho}{1-\rho\sigma},$$

and the grand partition function is given by

$$\Xi(\mu, r - \sigma) = \sum_{0 \le N \le r/\sigma - 1} \frac{e^{N\beta\mu^{ex}}}{N!} [\rho \times (r - \sigma - N\sigma)]^N.$$

Figure 7.13 shows the radial distribution function of the Tonk gas $g(r)$ at three values of reduced densities.

Problem 7.25 The Kornberg–Stryer model for the distributions of nucleosomes in chromatin is analogous to the statistical thermodynamic model of the Tonks gas. According to this model, nucleosome arrangements in a chromatin fiber can be appropriately described by the statistical packing principle, i.e., in terms of steric exclusion between nucleosome as represented by the one-dimensional hard-sphere interaction (see Figure 7.14). To rationalize the experimental results, the average nucleosome density ρ and the number of base pairs (bp) per nucleosome (a.k.a., nucleosome width) are typically used as adjustable parameters.
Using parameters $b = 149$ bp for the nucleosome length and $\rho = 1/182$ bp^{-1} for the number density of nucleosomes in *Candida albicans*, plot nucleosome density versus the distance from the "nucleosome-free region" (NFR) up to 1500 bp and compare the result with that reported in Möbius et al.[10] Can the Kornberg–Stryer model reproduce the experimental data shown in Figure 7.14?

Solution:

The statistical distribution of nucleosomes is equivalent to the radial distribution function for the Tonks gas. The only difference is that the particle density ρ and the hard-sphere diameter σ are expressed in the units of base pairs (bp). For $b = 149$ bp and $\rho = 1/182$ bp^{-1}, $\rho\sigma = 149/182 \approx 0.8187$. Figure 7.15 reproduces the results reported in the PNAS article by Möbius et al.
Figure 7.16 reproduces the nucleosome distribution shown in Figure 7.14 with $\rho = 1/170$ bp^{-1}. The agreement is almost perfect!

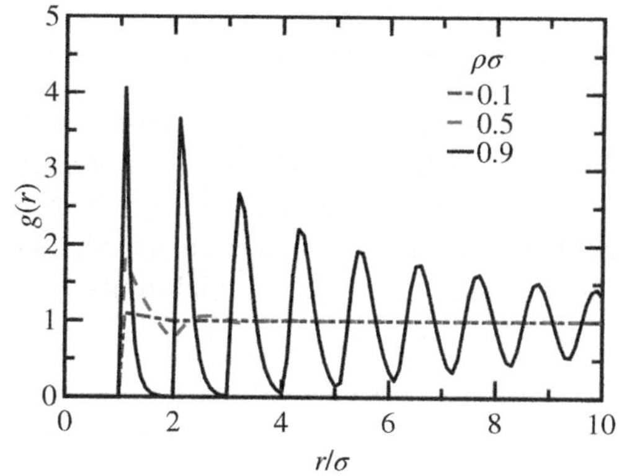

Figure 7.13 The radial distribution function $g(r)$ of Tonks–Takahashi fluids reduced densities $\rho\sigma = 0.1$, 0.5, and 0.9.

10 Möbius W., et al. "Toward a unified physical model of nucleosome patterns flanking transcription start sites", *PNAS* 110, 5719–5724 (2013).

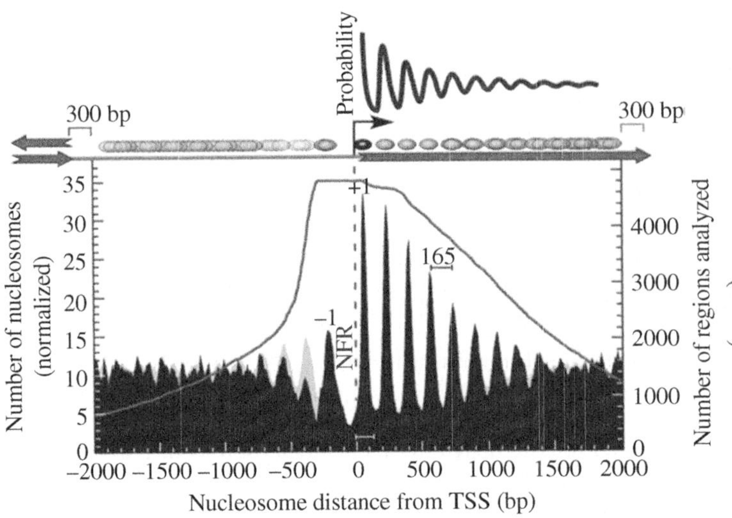

Figure 7.14 Distribution of nucleosome locations relative to transcriptional start sites (TSS), i.e., the location where the first DNA nucleotide is transcribed into RNA. The left side of TSS is referred to as the "nucleosome-free region" (NFR). Source: Mavrich T. N. et al.[11]

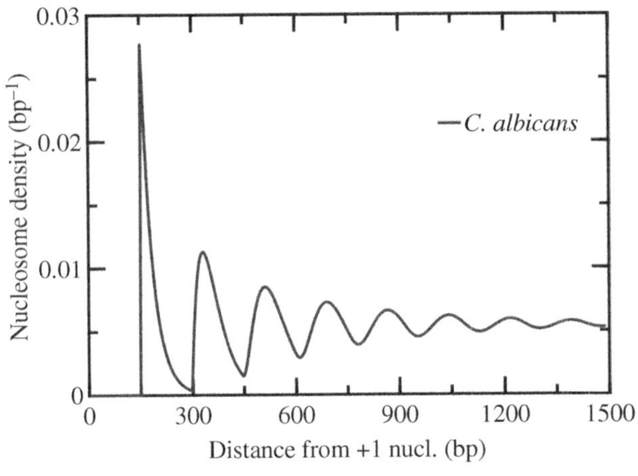

Figure 7.15 Distribution of nucleosome locations in *C. albicans*.

Problem 7.26 Waisman and Lebowitz derived an analytical solution for the Ornstein–Zernike (OZ) equation with the mean-spherical approximation (MSA) for a thermodynamic system consisting of two types of parallel plates, one positively charged and one negatively charged.[12] In this so-called one-dimensional analog of electrolyte solutions, the plate motions are restricted to the perpendicular direction with a pair potential similar to that given by the primitive model, i.e.,

$$u_{ij}(r) = \begin{cases} \infty, & r < \sigma_{ij}; \\ -q_i q_j r/(8\pi\epsilon_0\epsilon), & r \geq \sigma_{ij}. \end{cases}$$

Here r refers to the center-to-center distance between plates i and j, $\sigma_{ij} = (\sigma_i + \sigma_j)/2$ is their closest separation, σ_i and q_i represent the thickness and electrical charge density of plate i, respectively,

11 Mavrich T. N., et al. "A barrier nucleosome model for statistical positioning of nucleosomes throughout the yeast genome", *Genome Res.* 18: 1073–1083 (2008).

12 Waisman E. and Lebowitz J. L., "Mean spherical model integral equation for charged hard spheres", *J. Chem. Phys.* 56, 3086 (1972); *ibid.* 56, 3093 (1972).

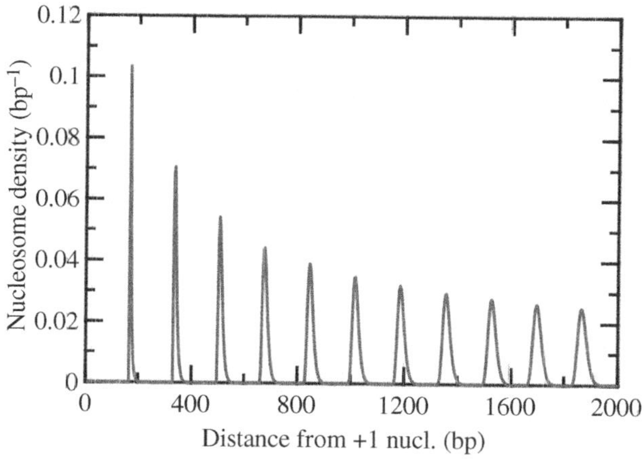

Figure 7.16 The best fit of the nucleosome locations relative to transcriptional start sites (TSS).

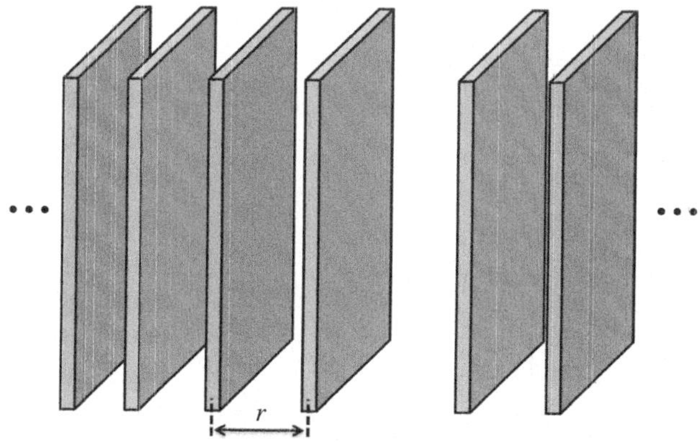

Figure 7.17 Charged parallel plates aligned in one dimension.

ϵ_0 is the permittivity of free space, and ϵ is the dielectric constant of the background (viz., the solvent). Figure 7.17 shows a schematic representation of the model system.

It has been shown that the average reduced potential energy per unit volume (viz., the reduced excess internal energy per volume) is

$$\beta\omega = \frac{\kappa}{4} + \frac{\kappa^2}{8}\left(\frac{q_+\sigma_+ - q_-\sigma_-}{q_+ - q_-}\right), \tag{D}$$

where q_+ and q_- stands for the charge densities of positively and negatively charged plates, respectively, and κ is a screening parameter defined by

$$\kappa^2 = \frac{\beta(\rho_+q_+^2 + \rho_-q_-^2)}{4\pi\epsilon_0\epsilon}.$$

Similar to that in a bulk electrolyte solution, the number densities of positively and negatively charged plates, ρ_+ and ρ_-, satisfy the condition of charge neutrality $\rho_+q_+ + \rho_-q_- = 0$. Based on the potential energy density given by Eq. (D), show that MSA for the one-dimensional system predicts:

(i) the reduced excess Helmholtz energy per unit volume

$$\beta f^{ex} = \beta f^{ex}_{HR} + \frac{\kappa}{2} + \frac{\kappa^2}{8}\left(\frac{q_+\sigma_+ - q_-\sigma_-}{q_+ - q_-}\right),$$

where $f^{ex} \equiv f(T, \rho_i) - f^{id}(\rho_i)$, f^{id} represents the Helmholtz energy density of the ideal system, and βf^{ex}_{HR} is the reduced excess Helmholtz energy of the uncharged system. The latter is the same as that of uncharged one-dimensional hard rods (HR) (viz., the Tonks gas).

$$\beta f^{ex}_{HR} = -\rho \ln(1-\eta)$$

with $\rho = \rho_+ + \rho_-$ and $\eta \equiv \rho_+\sigma_+ + \rho_-\sigma_-$.

(ii) the osmotic pressure Π is given by

$$\frac{\beta\Pi}{\rho} = \frac{1}{1-\eta} - \frac{\kappa}{4\rho}$$

(iii) the system does not exhibit phase transition.

Solution:

(i) The reduced excess Helmholtz energy per unit volume can be obtained by integrating the Gibbs–Helmholtz equation

$$\begin{aligned}\beta f^{ex} &= \beta f^{ex}_{HR} + \int_0^{\beta} \omega d\beta \\ &= \beta f^{ex}_{HR} + \int_0^{\kappa} d\kappa \frac{2}{\kappa}\left[\frac{\kappa}{4} + \frac{\kappa^2}{8}\left(\frac{q_+\sigma_+ - q_-\sigma_-}{q_+ - q_-}\right)\right] \\ &= \beta f^{ex}_{HR} + \frac{\kappa}{2} + \frac{\kappa^2}{8}\left(\frac{q_+\sigma_+ - q_-\sigma_-}{q_+ - q_-}\right),\end{aligned}$$

where we have used $d\ln\beta = 2d\ln\kappa$, and $\beta f^{ex}_{HR} = \beta f^{ex}(\beta = 0)$.

(ii) We can obtain the osmotic pressure Π from the Helmholtz energy

$$\begin{aligned}\beta\Pi &= -\left(\frac{\partial \beta F}{\partial V}\right)_{T,N_+,N_-} = \beta\Pi^{id} - \left[\frac{\partial(\beta f^{ex} V)}{\partial V}\right]_{T,N_+,N_-} \\ &= \rho + \left\{\frac{\partial[N\ln(1-\eta)]}{\partial V}\right\}_{T,N_+,N_-} - \left\{\frac{\partial}{\partial V}\left[\frac{\kappa V}{2} + \frac{\kappa^2 V}{8}\left(\frac{q_+\sigma_+ - q_-\sigma_-}{q_+ - q_-}\right)\right]\right\}_{T,N_+,N_-} \\ &= \rho + \frac{\rho\eta}{1-\eta} - \frac{\kappa}{4} = \frac{\rho}{1-\eta} - \frac{\kappa}{4},\end{aligned}$$

where $\Pi^{id} = \rho k_B T$ represents the osmotic pressure of the ideal system. In deriving the above equation, we have used $(\partial\kappa V/\partial V) = \kappa/2$ and $(\partial\kappa^2 V/\partial V) = 0$. Hence, the osmotic coefficient is

$$\frac{\beta\Pi}{\rho} = \frac{1}{1-\eta} - \frac{\kappa}{4\rho}.$$

(iii) MSA predicts no phase transition for the one-dimensional system of charged plates. The negative pressure at low density is simply unphysical.

Problem 7.27 In the same work discussed in Problem 7.26, Waisman and Lebowitz obtained the MSA equations for the restricted primitive model (RPM) of electrolyte solutions, i.e., a model system of cations and anions with same size and valence but opposite charges. In this model, the pairwise additive potential is given by

$$u_{ij}(r) = \begin{cases} \infty, & r < \sigma; \\ q_i q_j/(4\pi\epsilon_0\epsilon r), & r \geq \sigma. \end{cases}$$

where r refers to the center-to-center distance between ions i and j, σ is their hard-sphere diameter, q_i is the electrostatic charge, ϵ_0 is the permittivity of free space, and ϵ is the solvent dielectric constant. MSA predicts the average potential energy per unit volume of the electrolyte solution (viz., the excess internal energy per volume)

$$\omega = -\frac{x^2 + x - x(1+2x)^{1/2}}{4\pi\beta\sigma^3}, \tag{E}$$

where $x^2 \equiv (\rho_+ q_+^2 + \rho_- q_-^2)\beta\sigma^2/\epsilon_0\epsilon$, ρ_+ and ρ_- are the number densities of positively and negatively ions. Show that

(i) the reduced excess Helmholtz energy per unit volume is

$$\beta f^{ex} = -\frac{2 + 6x + 3x^2 - 2(1+2x)^{3/2}}{12\pi\sigma^3},$$

(ii) the osmotic pressure Π satisfies

$$\frac{\beta\Pi}{\rho} = \frac{\beta\Pi_0}{\rho} + \frac{2 + 3x + 3x(1+2x)^{1/2} - 2(1+2x)^{3/2}}{72\eta},$$

where $\rho \equiv (\rho_+ + \rho_-)$, $\eta \equiv \pi\sigma^3\rho/6$, and Π_0 is the osmotic pressure of the system without the charge (viz., the reference hard-sphere system). According to the Carnahan–Starling equation, the hard-sphere contribution is

$$\frac{\beta\Pi_0}{\rho} = \frac{1 + \eta + \eta^2 - \eta^3}{(1-\eta)^3}.$$

(iii) the logarithm of the mean activity coefficient is given by

$$\ln\gamma_\pm = \ln\gamma_0 + \frac{x(1+2x)^{1/2} - x - x^2}{24\eta},$$

where the hard-sphere contribution is

$$\ln\gamma_0 = \eta\frac{8 - 9\eta + 3\eta^2}{(1-\eta)^3}.$$

(iv) the system exhibits a phase transition with the critical temperature and the packing fraction of ions $T_c^* \approx 0.0785773$ and $\eta_c \approx 0.00758476$, respectively. Here the reduced temperature is defined as $T^* \equiv \sigma/l_B$ where $l_B = \beta q^2/(4\pi\epsilon_0\epsilon)$ is the Bjerrum length, $q = q_+ = -q_-$ is the net charge of a cation or an anion.

Solution:

(i) The reduced excess Helmholtz energy per unit volume can be obtained by integrating the Gibbs–Helmholtz equation

$$\beta f^{ex} = \beta f_{HR}^{ex} + \int_0^\beta \omega d\beta = \beta f_{HR}^{ex} - 2\int_0^x dx\left[\frac{x^2 + x - x(1+2x)^{1/2}}{4\pi x\sigma^3}\right]$$

$$= \beta f_{HR}^{ex} - \frac{1}{2\pi\sigma^3}\int_0^x dx\left[x + 1 - (1+2x)^{1/2}\right]$$

$$= \beta f_{HR}^{ex} - \frac{2 + 6x + 3x^2 - 2(1+2x)^{3/2}}{12\pi\sigma^3},$$

where we have used $d\ln\beta = 2d\ln x$ and $\beta f_{HR}^{ex} = \beta f^{ex}(\beta = 0)$.

(ii) We can obtain the osmotic pressure Π from the Helmholtz energy

$$\beta\Pi = -\left(\frac{\partial\beta F}{\partial V}\right)_{T,N_+,N_-} = \beta\Pi^{id} - \left[\frac{\partial(\beta f^{ex}V)}{\partial V}\right]_{T,N_+,N_-}$$

$$= \beta\Pi_0 + \left\{ \frac{\partial}{\partial V} \left[\frac{2V + 6xV + 3x^2V - 2V(1+2x)^{3/2}}{12\pi\sigma^3} \right] \right\}_{T,N_+,N_-}$$

$$= \beta\Pi_0 + \frac{2 + 3x + 3x(1+2x)^{1/2} - 2(1+2x)^{3/2}}{12\pi\sigma^3},$$

where $\Pi^{id} = \rho k_B T$ represents the osmotic pressure of the ideal system. In deriving the above equation, we have used $(\partial xV/\partial V) = x/2$, $V(\partial x/\partial V) = -x/2$ and $(\partial x^2 V/\partial V) = 0$. Hence, the osmotic coefficient is

$$\phi \equiv \frac{\beta\Pi}{\rho} = \frac{\beta\Pi_0}{\rho} + \frac{2 + 3x + 3x(1+2x)^{1/2} - 2(1+2x)^{3/2}}{72\eta},$$

where $\rho \equiv (\rho_+ + \rho_-)$, $\eta \equiv \pi\sigma^3\rho/6$, and Π_0 is the osmotic pressure of the system without the charge (viz., the reference hard-sphere system).

(iii) the logarithm of the activity coefficient is related to the excess Helmholtz energy density by

$$\ln\gamma_+ = \beta\mu_+^{ex} = \frac{\partial\beta f^{ex}}{\partial\rho_+}$$

$$= \ln\gamma_0 - \frac{\partial}{\partial\rho}\left[\frac{2 + 6x + 3x^2 - 2(1+2x)^{3/2}}{12\pi\sigma^3}\right]$$

$$= \ln\gamma_0 - \frac{\partial x}{\partial\rho}\frac{\partial}{\partial x}\left[\frac{2 + 6x + 3x^2 - 2(1+2x)^{3/2}}{12\pi\sigma^3}\right]$$

$$= \ln\gamma_0 - \times\frac{x}{2\rho} \times \frac{1}{12\pi\sigma^3} \times \left[6 + 6x - 6(1+2x)^{1/2}\right]$$

$$= \ln\gamma_0 - \frac{x}{24\eta}\left[1 + x - (1+2x)^{1/2}\right],$$

where we have used $(\partial\rho/\partial\rho_+) = 1$, $\eta = \pi\sigma^3\rho/6$ and $(\partial x/\partial\rho) = x/2\rho$. The same expression can be obtained for the activity coefficient of anions. Hence, the logarithm of the mean activity coefficient is

$$\ln\gamma_\pm = (\ln\gamma_+ + \ln\gamma_-)/2 = \ln\gamma_0 + \frac{x(1+2x)^{1/2} - x - x^2}{24\eta}.$$

Alternatively, the same equation can be obtained from

$$\ln\gamma_\pm = \frac{\partial\beta f^{ex}}{\partial\rho} = \frac{\partial\beta f^{ex}}{\partial\beta}\frac{\partial\beta}{\partial x}\frac{\partial x}{\partial\rho} = \ln\gamma_0 + \omega\frac{2\beta}{x}\frac{x}{2\rho}$$

$$= \ln\gamma_0 - \frac{x^2 + x - x(1+2x)^{1/2}}{4\pi\beta\sigma^3}\frac{\beta}{\rho} = \ln\gamma_0 + \frac{x(1+2x)^{1/2} - x - x^2}{24\eta},$$

where we have used $(\partial x/\partial\beta) = x/2\beta$ and $(\partial x/\partial\rho) = x/2\rho$.

(iv) The reduced osmotic pressure Π varies with the packing fraction of ions η and the reduced temperature T^* according to

$$\Pi^* \equiv \frac{\pi\beta\Pi\sigma^3}{6} = \frac{\eta(1 + \eta + \eta^2 - \eta^3)}{(1-\eta)^3} + \frac{2 + 3x + 3x(1+2x)^{1/2} - 2(1+2x)^{3/2}}{72},$$

where

$$x^2 = \rho q^2\beta\sigma^2/\epsilon_0\epsilon = 24\eta l_B/\sigma = 24\eta/T^*,$$

and $l_B = \beta q^2/(4\pi\epsilon_0\epsilon)$ is the Bjerrum length with $q \equiv |q_\pm|$ being the net charge of cations and anions and $T^* = \sigma/l_B$.

To identify the critical point, we solve the equations

$$\frac{\partial \Pi^*}{\partial \eta} = 0 = \frac{1+4\eta+4\eta^2-4\eta^3+\eta^4}{(1-\eta)^4} + \frac{a\left(\sqrt{1+2a\sqrt{\eta}}-1-a\sqrt{\eta}\right)}{48\sqrt{\eta}+2a\eta^{3/2}}$$

$$\frac{\partial^2 \Pi^*}{\partial \eta^2} = 0 = \frac{4(2+5\eta-\eta^2)}{(1-\eta)^5} + y\frac{\frac{a^2\eta^2}{192}+a\eta^{3/2}\left(\frac{1}{64}-\frac{1}{96}\sqrt{1+2a\sqrt{\eta}}\right)+\frac{\eta}{192}\left(1-\sqrt{1+2a\sqrt{\eta}}\right)}{\eta^{5/2}(1/2+a\sqrt{\eta})\sqrt{1+2a\sqrt{\eta}}},$$

where $a \equiv \sqrt{24/T^*}$. Solving the above equations yield $\eta_c \approx 0.00758476$ and $a \approx 17.4766$ which amounts to $T_c^* \approx 0.0785773$.

Figure 7.18 shows the isothermal curves at three representative temperatures near the critical point. The system exhibits a phase transition (viz., van der Waals loop) when the reduced temperature is less than T_c^*. The MSA predictions are in fair agreement with the simulation result $T_c^* = 0.0496$ and $\eta = 0.041469$.[13]

Problem 7.28 Predict the osmotic and mean activity coefficients for aqueous solutions of LiCI using the MSA equations from Problem 7.27. Compare the theoretical results with the experimental data shown in Table 7.2 and with the predictions of the Debye–Hückel (DH) theory (see Section 9.6):

$$\phi = \frac{\beta \Pi}{\rho} = 1 - \frac{x^2+2x-2(1+x)\ln(1+x)}{48\eta(1+x)},$$

$$\ln \gamma_\pm = -\frac{x}{2(1+x)T^*},$$

where $\eta \equiv \pi\sigma^3\rho/6$ is the packing fraction of ions, $x \equiv \sqrt{24\eta/T^*}$, and $T^* \equiv \sigma/l_B$ is the reduced temperature as defined in Problem 7.27. The average hard-sphere diameter of hydrated ions is about $\sigma \approx 3.6$ Å. Discuss why MSA fails at high salt concentration and how the theoretical results can be improved.

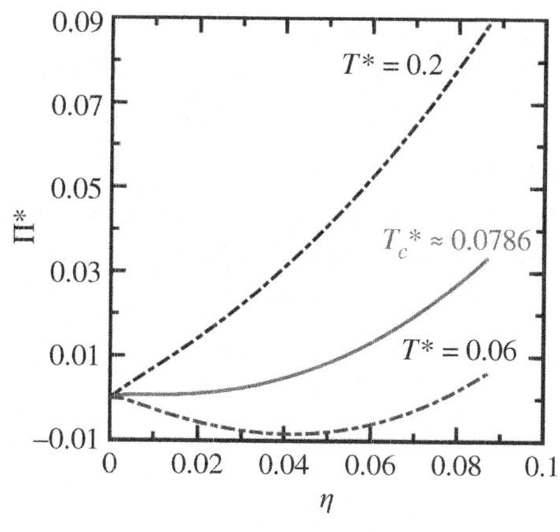

Figure 7.18 The reduced osmotic pressure versus packing density at three temperatures predicted by MSA (RPM.m).

13 Panagiotopoulos A. Z., "Critical parameters of the restricted primitive model", *J. Chem. Phys.* 116, 3007 (2002).

Table 7.2 Osmotic and mean activity coefficients of LiCl aqueous solutions.[a)]

m	ϕ	γ_m	m	ϕ	γ_m	m	ϕ	γ_m
0.001	0.988	0.965	1.000	1.020	0.775	8.000	2.159	5.190
0.002	0.984	0.952	1.200	1.044	0.798	9.000	2.328	7.110
0.005	0.976	0.928	1.400	1.068	0.825	10.000	2.480	9.600
0.010	0.969	0.904	1.600	1.093	0.855	11.000	2.612	12.700
0.020	0.960	0.874	1.800	1.119	0.888	12.000	2.723	16.400
0.050	0.947	0.827	2.000	1.145	0.924	13.000	2.814	20.700
0.100	0.940	0.789	2.500	1.213	1.029	14.000	2.887	25.500
0.200	0.940	0.756	3.000	1.284	1.157	15.000	2.947	30.900
0.300	0.946	0.743	3.500	1.361	1.312	16.000	2.995	36.900
0.400	0.954	0.739	4.000	1.441	1.499	17.000	3.036	43.400
0.500	0.964	0.739	4.500	1.525	1.726	18.000	3.066	50.300
0.600	0.974	0.742	5.000	1.613	2.000	19.000	3.079	57.000
0.700	0.985	0.748	5.500	1.703	2.330	19.219	3.079	58.400
0.800	0.996	0.756	6.000	1.794	2.730			
0.900	1.008	0.765	7.000	1.979	3.760			

a) Source: Hamer W. J. and Wu Y.-C., "Osmotic coefficients and mean activity coefficients of uni-univalent electrolytes in water at 25 °C", *J. Phys. Chem. Ref. Data* 1, 1047 (1972).

Hints:
(1) In Table 7.2, m represents molality, i.e., moles of solute per 1000 grams of solvent; γ_m is the activity coefficient defined in terms of molality, which can be related to the mean activity coefficient by $\gamma_m = C\gamma_\pm/md_0$, where C represents molarity, i.e., moles of solute (electrolyte) per liter of solution, and $d_0 = 0.997$ is the solvent density in units of kg/L.
(2) The molality and molarity can be converted to each other by using experimental data for the densities of LiCI solutions[14]:

$$d(\mathrm{kg/m^3}) = d_1 + d_2 t + d_3 t^2 + d_4\zeta + d_5\zeta^2 + d_6\zeta t + d_7\zeta^2 t + d_8\zeta t^2 + d_9\zeta^3 + d_{10}\zeta^4,$$

where the coefficients $d_{i=1,2,\ldots,10}$ are as follows: 1002.8 -0.15582 -2.88385e-3 6.1379 -5.8452e-2 6.0650e-4 -1.2546e-4 5.8029e-5 2.6623e-3 -2.5941e-5, t is temperature in °C, $\zeta \equiv 100\,w$, and w is the mass fraction of the solute.

Solution:
The numerical values of the activity coefficient depend on the definition of the reference state, which differs in theoretical derivation and in practical use. In derivation of the MSA equations, the mean activity coefficient is defined by

$$\ln\gamma_\pm \equiv \beta\mu^{ex} \equiv \beta\mu - \beta\mu^{id}(T,\rho) = \eta\frac{8-9\eta+3\eta^2}{(1-\eta)^3} + \frac{x(1+2x)^{1/2} - x - x^2}{24\eta},$$

where μ^{id} represents chemical potential of the salt in an ideal solution (id) at system temperature T and number density of ions ρ. On the other hand, the activity coefficient in molality is defined as

$$\beta\mu \equiv \beta\mu_0 + \ln a \equiv \beta\mu_0 + \ln(m\gamma_m/m_0),$$

14 Wimby J. M. and Berntsson T. S., "Viscosity and density of aqueous solutions of LiBr, LiCl, $ZnBr_2$, $CaCl_2$, and $LiNO_3$. I: Single salt solutions", *J. Chem. Eng. Data* 39(1), 68–72 (1994).

where μ_0 is the chemical potential at the reference state, i.e., an ideal solution at system temperature and m_0=1 mol/kg solvent. With μ_0 expressed as $\beta\mu_0 = \beta\mu^{id}(T, \rho_0)$ where ρ_0 is the number density corresponding to m_0, we find

$$\ln \gamma_m = \beta\mu^{id}(T,\rho) + \ln \gamma_\pm - \beta\mu^{id}(T,\rho_0) - \ln(m\gamma_m/m_0) = \ln \gamma_\pm + \ln\left(\frac{\rho m_0}{\rho_0 m}\right).$$

Rearranging the above equation leads to

$$\gamma_m = \gamma_\pm\left(\frac{\rho m_0}{\rho_0 m}\right) = \gamma_\pm\left(\frac{C m_0}{C_0 m}\right),$$

where $C = md/(1 + 0.001\,\text{mM})$ stands for molar concentration (a.k.a, molarity), d is the solution density in kg/L, M is the molecular weight of the solute (g/mol), and $C_0 = m_0 d_0$ is the molar concentration at m_0 but with the properties of infinity dilution (viz., ideal solution). Thus, we obtain $\gamma_m = \gamma_\pm C/(md_0)$.

Figure 7.19 compares the theoretical and experimental results. Despite its simplicity, MSA is able to reproduce the experimental data up to about m = 8 mol/kg, while the DH theory is clearly way too simplistic. The poor performance at high salt concentration should be expected because the continuum representation of the solvent breaks down. To improve the theoretical performance at high density, we need to use a nonprimitive model of electrolytes that accounts for the solvent molecules explicitly.[15] Alternatively, we may empirically modify the PM model by assuming that the hard-sphere diameter varies with the salt concentration.[16] A subtle way to improve the prediction is converting the mean activity coefficient from the McMillan–Mayer model (i.e., implicit solvent) to the Lewis–Randel framework (explicit solvent)[17]:

$$\ln \gamma_\pm^{LR} = \ln \gamma_\pm^{MM} - \phi\rho v_{salt},$$

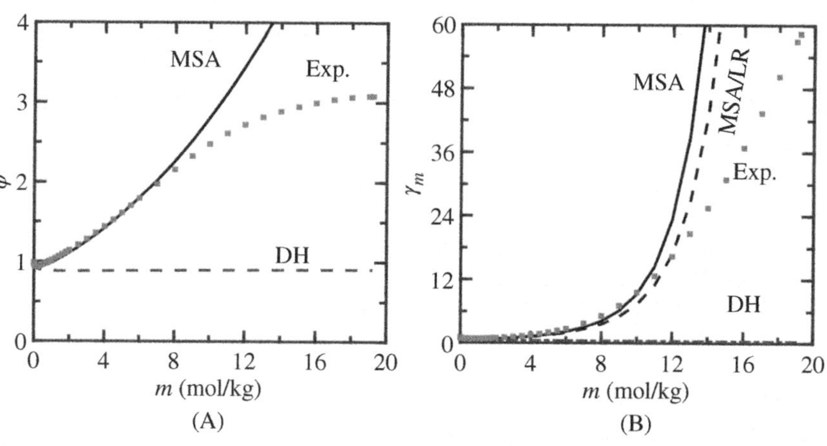

Figure 7.19 Osmotic (A) and activity coefficients (B) of LiCl aqueous solutions at 25 °C predicted by MSA and by the Debye–Hückel (DH) theory in comparison with experimental data (Exp). MSA/LR represents MSA with the Lewis–Randel correction.

15 e.g., Mester Z. and Panagiotopoulos A. Z., "Mean ionic activity coefficients in aqueous NaCl solutions from molecular dynamics simulations", *J. Chem. Phys.* 142, 044507 (2015).

16 e.g., Lu J.-F., Yu Y.-X., and Li Y.-G., "Modification and application of the mean spherical approximation method", *Fluid Phase Equilibria* 85, 81–100 (1993).

17 Section 9.1.4. For more in-depth discussions, see, e.g., Simonin J. P., "Study of experimental-to-McMillan–Mayer conversion of thermodynamic excess functions", *J. Chem. Soc., Faraday Trans.* 92, 3519–3523 (1996).

where v_{salt} represents the partial molar volume of the salt. With $v_{salt} \approx 0.01$ l/mol used as an adjustable parameter, the results from MSA are clearly improved (Figure 7.19B).

Problem 7.29 Consider the Barker–Henderson (BH) perturbation theory for a square-well (SW) fluid with the pair potential

$$u(r) = \begin{cases} \infty, & r < \sigma, \\ -\epsilon, & \sigma \leq r \leq \lambda\sigma, \\ 0, & r > \lambda\sigma, \end{cases}$$

where r is the center-to-center distance between two spherical particles, σ is the hard-sphere (HS) diameter, $\lambda > 1$ specifies the range of attraction, and ϵ is an energy parameter. The reduced Helmholtz energy can be written as

$$f \equiv \beta F/N \approx f_0 + \beta f_1 + \beta^2 f_2,$$

where $\beta = 1/(k_B T), f_0 = (4\eta - 3\eta^2)/(1-\eta)^2$ corresponds to that of the HS reference system as predicted by the Carnahan–Starling (CS) equation, f_1 and f_2 are the first and second perturbation terms associated with the attractive energy.

(i) Show that the first-order perturbation term can be written as

$$f_1 = -12\epsilon\eta \int_1^\lambda dx x^2 g_0(x),$$

where $x = r/\sigma$.

(ii) Show that, with the local compressibility assumption (LCA), the second-order perturbation term can be written as

$$f_2 = \frac{1}{2}\epsilon\chi_0\eta\frac{\partial f_1}{\partial \eta},$$

where $\chi_0 \equiv k_B T(\partial\rho/\partial P) = (1-\eta)^4/(1+4\eta+4\eta^2)$ is calculated from the isothermal compressibility of the hard-sphere fluid. What is f_2 according to the macroscopic compressibility approximation (MCA)?

(iii) An analytical expression for $g_0(x)$ with $x \lesssim 1.2$ has been reported by Tang and Lu (Problem 7.16)[18]

$$g_0(x) = \frac{1+\eta/2}{(1-\eta)^2}\frac{1}{x} + \frac{1-5\eta-5\eta^2}{(1-\eta)^3}(1-1/x) + \frac{-3\eta+6\eta^2+21\eta^3/2}{(1-\eta)^4}\frac{(x-1)^2}{x}$$

show that the integration term can be calculated from

$$\int_1^\lambda dx x^2 g_0(x) = \frac{a_4\lambda^4 + a_3\lambda^3 + a_2\lambda^2 + a_1}{(1-\eta)^4}$$

with

$$a_1 = -(8+12\eta^2+7\eta^3)/24; \; a_2 = 3\eta(1+2\eta)^2/4; \\ a_3 = (1+2\eta)^2(1-4\eta)/3; \quad a_4 = -3\eta(2-4\eta-7\eta^2)/8. \tag{F}$$

(iv) Compare Eq. (F) with the correlation used in the statistical associating fluid theory for potentials of variable attractive range (SAFT-VR) equation[19]:

$$\int_1^\lambda dx x^2 g_0(x) = (\lambda^3 - 1)\frac{1-\eta_{eff}/2}{(1-\eta_{eff})^3}$$

18 Tang Y.-P. and Lu B. C.-Y., "A study of associating Lennard–Jones chains by a new reference radial distribution function", *Fluid Phase Equilibria* 171, 27–44 (2000).
19 Gil-Villegas A. et al., "Statistical associating fluid theory for chain molecules with attractive potentials of variable range", *J. Chem. Phys.* 106, 4168 (1997).

with $\eta_{eff} = c_1\eta + c_2\eta^2 + c_3\eta^3$ and

$$\begin{bmatrix} c_1 \\ c_2 \\ c_3 \end{bmatrix} = \begin{bmatrix} 2.25855 & -1.50349 & 0.249434 \\ -0.669270 & 1.40049 & -0.827739 \\ 10.1576 & -15.0427 & 5.30827 \end{bmatrix} \begin{bmatrix} 1 \\ \lambda \\ \lambda^2 \end{bmatrix}.$$

Solution:

(i) The first-order perturbation term is defined as

$$f_1 = \frac{\rho}{2}\int d\mathbf{r}u_A(r)g_0(r) = \frac{\rho}{2}\int_\sigma^{\lambda\sigma} dr4\pi r^2(-\epsilon)g_0(r)$$
$$= -2\pi\rho\sigma^3\epsilon\int_1^\lambda dxx^2g_0(x) = -12\epsilon\eta\int_1^\lambda dxx^2g_0(x),$$

where $x = r/\sigma$.

(ii) The LCA predicts that the second-order perturbation term is given by

$$f_2 = -\frac{\rho k_BT}{4}\int d\mathbf{r}u_A^2(r)\left[\frac{\partial\rho g_0(r)}{\partial P}\right]_T = -\frac{\rho k_BT}{4}\frac{\partial}{\partial P}\int d\mathbf{r}u_A^2(r)\rho g_0(r)$$
$$= -\frac{\rho k_BT}{4}\frac{\partial}{\partial\rho}\frac{\partial\rho}{\partial P}\int d\mathbf{r}u_A^2(r)\rho g_0(r) = -\frac{\rho k_BT}{4}\frac{\partial}{\partial\rho}\frac{\partial\rho}{\partial P}\int_\sigma^{\lambda\sigma} dr4\pi r^2(-\epsilon)^2\rho g_0(r)$$
$$\approx \frac{\epsilon k_BT\rho}{2}\left(\frac{\partial\rho}{\partial P}\right)_T\frac{\partial f_1}{\partial\rho} = \frac{1}{2}\epsilon\chi_0\eta\frac{\partial f_1}{\partial\eta},$$

where $\chi_0 \equiv k_BT(\partial\rho/\partial P)_T$, the isothermal condition and the reference hard-sphere fluid are applied to all partial derivatives. In the MCA, the second-order term is given by

$$f_2 = -\frac{\rho k_BT}{4}\left(\frac{\partial\rho}{\partial P}\right)_T\int d\mathbf{r}u_A^2(r)g_0(r) \approx \frac{1}{2}\epsilon\chi_0 f_1.$$

(iii) Let $g_0(x) = b_1/x + b_2(1 - 1/x) + b_3(x-1)^2/x$, the integration of $g_0(x)$ with respect to x^2dx leads to

$$\int_1^\lambda dxx^2g_0(x) = \frac{-6b_1 + 2b_2 - b_3}{12} + \frac{b_1 - b_2 + b_3}{2}\lambda^2 + \frac{b_2 - 2b_3}{3}\lambda^3 + \frac{b_3}{4}\lambda^4.$$

Substituting with the relations between $b_i = 1, 2, 3$, and η into the above equation yields

$$\int_1^\lambda dxx^2g_0(x) = \frac{a_4\lambda^4 + a_3\lambda^3 + a_2\lambda^2 + a_1}{(1-\eta)^4}.$$

(iv) Figure 7.20 compares the numerical correlation in the SAFT-VR equation with the analytical result based on an approximation for $g_0(x)$. The discrepancy at large densities is most likely due to numerical errors.

Problem 7.30 The hard-sphere diameter plays an important role in determining both the structure and thermodynamic properties of condensed systems with perturbation theories. To gain an understanding, calculate the hard-sphere diameter for a Lennard–Jones fluid near the triple point (e.g., $k_BT/\epsilon = 0.7$ and $\rho\sigma^3 = 0.85$) according to the Weeks–Chandler–Anderson (WCA) and the Barker–Henderson (BH) theories and discuss their implications in predicting the compressibility factor from the first-order perturbation theory

$$Z \equiv \beta P/\rho \approx Z_0 - \frac{\rho^2}{6k_BT}\int d\mathbf{r}ru_A'(r)g_0(r),$$

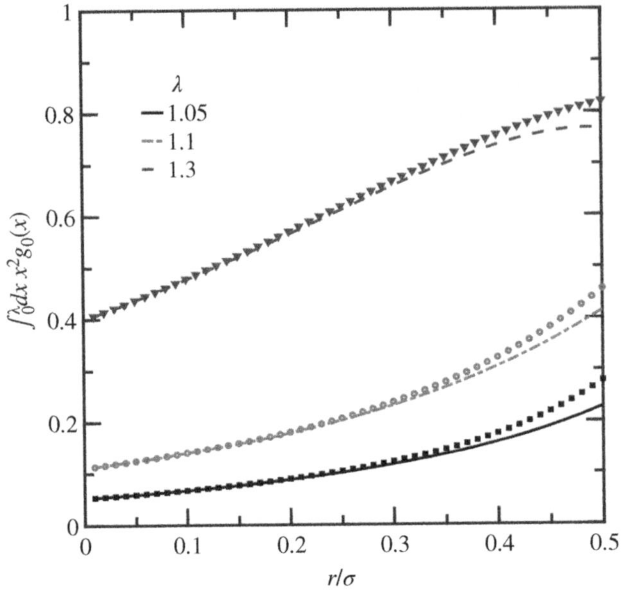

Figure 7.20 A comparison of the numerical correlation (markers) and an analytical approximation (lines) of the integral $\int_0^\lambda x^2 g_0(x)$ in the BH theory.

where Z_0 and $g_0(r)$ are the compressibility factor and radial distribution function of the hard-sphere reference system, and u_A' is the derivative of the perturbation potential.

Solution:

In the original formulation of the BH theory, the perturbation potential is defined as the attractive part of the pair potential and the hard-sphere diameter depends only on temperature. By contract, in the WCA theory the perturbation potential refers to the part of the pair potential where the force is negative, and the hard-sphere diameter depends on both the number density and temperature. At $k_BT/\epsilon = 0.7$ and $\rho\sigma^3 = 0.85$, the numerical fittings predict $\sigma_{BH} = 0.9796\sigma$ and $\sigma_{WCA} = 1.0335\sigma$. Figure 7.21 shows the force between two LJ particles near contact. For the WCA theory, the force is

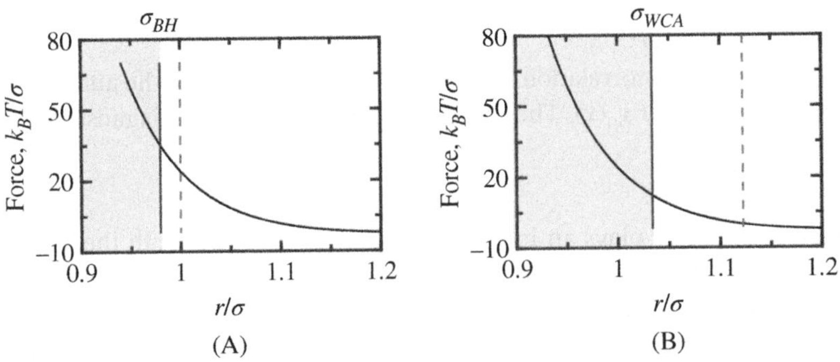

Figure 7.21 (A) In the BH theory, the inter-particle force is quite large near contact ($r = \sigma$, dashed line). (B) In the WCA theory, the inter-particle force vanishes when the separation is smaller than $r_c = 2^{1/6}\sigma$ (dashed line). The perpendicular solid lines and shaded areas represent the positions of the hard-sphere diameter at $k_BT/\epsilon = 0.7$ and $\rho\sigma^3 = 0.85$.

nearly zero for a large part in r where the pair-distribution function is large. Conversely, in the BH theory the force is quite significant when the separation between particles is near the hard-sphere diameter. As a result, a small error in $g_0(r)$ may lead to large errors.

Problem 7.31 The following exercise extends the Gibbs–Bogoliubov variational principle discussed in Section 5.6 to a system of spherical particles.

(i) Consider two systems with identical temperature T and particle density ρ but one with the total potential energy Φ and the other with the total potential energy Φ_0. Show the two Helmholtz energies satisfy the Gibbs–Bogoliubov inequalities

$$F_0 + \langle \Delta\Phi \rangle \leq F \leq F_0 + \langle \Delta\Phi \rangle_0,$$

where $\Delta\Phi = \Phi - \Phi_0$, $\langle \cdots \rangle_0$ and $\langle \cdots \rangle$ represent canonical ensemble averages under the potentials Φ and Φ_0, respectively.

(ii) Assume that the total potential energy is pairwise additive and that $(\Delta\beta u)e^{-\beta u_0} \approx e^{-\beta u} - e^{-\beta u_0}$, show

$$\langle \Delta\Phi \rangle_0 \approx \frac{N^2 k_B T}{2V} \int d\mathbf{r} y_0(r)[e^{-\beta u(r)} - e^{-\beta u_0(r)}],$$

where $y_0(r) = g_0(r)e^{\beta u_0(r)}$ is the cavity correlation function.

(iii) Based on the Weeks–Chandler–Anderson (WCA) potential for $u(r)$ and its corresponding hard-sphere potential for $u_0(r)$, plot the so-called blip function $\Delta e(r) \equiv e^{-\beta u(r)} - e^{-\beta u_0(r)}$ in the neighborhood of $r = \sigma$ for a Lennard–Jones fluid near the triple point (e.g., $k_B T/\epsilon = 0.7$ and $\rho\sigma^3 = 0.85$, which defines a WCA hard-sphere diameter $\sigma_{\text{WCA}} = 1.0335\sigma$).

(iv) In the WCA theory, the hard-sphere diameter is defined by $\langle \Delta\Phi \rangle_0 = 0$. Show that the approximation $F \approx F_0$ is accurate to the order of ξ^4 where ξ is the range of r where $\Delta e(r)$ is non-zero (i.e., the width of the blip function).

Solution:

(i) As discussed in Section 5.6, the Gibbs–Bogoliubov inequality

$$F \leq F_0 + \langle \Delta\Phi \rangle_0$$

follows directly from the Jensen inequality, i.e.,

$$\frac{Q}{Q_0} = \langle \exp(-\beta\Delta\Phi) \rangle_0 \geq \exp(\langle -\beta\Delta\Phi \rangle_0).$$

The Jensen inequality can be derived from $e^x \geq 1 + x$ for any real value x. Specifically, $\langle e^x \rangle = e^{\langle x \rangle}\langle e^{x-\langle x \rangle} \rangle \geq e^{\langle x \rangle}\langle 1 + (x - \langle x \rangle) \rangle$. If we interchange the definitions of Φ and Φ_0 leads to

$$F_0 + \langle \Delta\Phi \rangle \leq F \leq F_0 + \langle \Delta\Phi \rangle_0.$$

(ii) If the total potential energy is pairwise additive

$$\langle \Delta\Phi \rangle_0 = \left\langle \sum_i \sum_{j>i} \Delta u(|\mathbf{r}_i - \mathbf{r}_j|) \right\rangle_0$$

$$= \sum_i \sum_{j>i} \int d\mathbf{r} \int d\mathbf{r}' \langle \delta(\mathbf{r} - \mathbf{r}_i)\delta(\mathbf{r} - \mathbf{r}_j) \rangle_0 \Delta u(|\mathbf{r}_i - \mathbf{r}_j|) = \frac{N^2}{2V} \int d\mathbf{r} g_0(r)\Delta u(r).$$

With $y_0(r) = g_0(r)e^{\beta u_0(r)}$ and the assumption of $(\Delta\beta u)e^{-\beta u_0} \approx e^{-\beta u} - e^{-\beta u_0}$, we have

$$\langle \Delta\Phi \rangle_0 \approx \frac{N^2 k_B T}{2V} \int d\mathbf{r} y_0(r)[e^{-\beta u(r)} - e^{-\beta u_0(r)}].$$

(iii) Figure 7.22 shows the "blip" function of the WCA theory at $k_B T/\epsilon = 0.7$ and $\rho\sigma^3 = 0.85$.

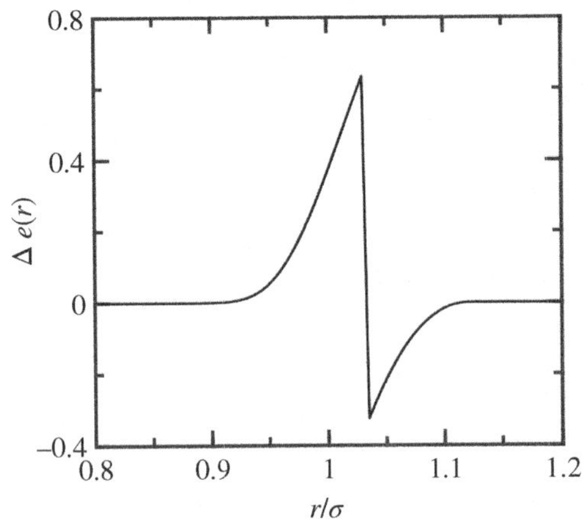

Figure 7.22 The "blip" function at $k_BT/\epsilon = 0.7$ and $\rho\sigma^3 = 0.85$ (WCA.m).

(iv) The Helmholtz energy corresponding to the WCA potential can be formally expanded relative to that of a hard-sphere system

$$F = F_0 + \int d\mathbf{r}\delta F_0 \Delta e(r) + \frac{1}{2}\int d\mathbf{r} \int d\mathbf{r}'\delta^2 F_0 \Delta e(r)\Delta e(r') + \cdots,$$

where δF_0 and $\delta^2 F_0$ represent the functional derivatives. Because $\Delta e(r)$ is non-zero in the range of ξ, the quadratic expansion is expected to be accurate on the order of ξ^2. When the hard-sphere diameter is defined by $\langle\Delta\Phi\rangle_0 = 0$, the first-order term in the above expansion disappears, and the second-order term will be accurate on the order of ξ^4 because in this case $\delta^2 F_0$ is also on the order of ξ^2. A more rigorous proof is given in Appendix B of the original article by Andersen H. C. et al.[20]

Problem 7.32 According to the Ornstein–Zernike (OZ) theory, the turbidity of a thermodynamic system, τ, is determined by integrating the scattering intensity over all angles

$$\tau = A\kappa_T \left[\frac{2\alpha^2 + 2\alpha + 1}{\alpha^3}\ln(1 + 2\alpha) - \frac{2(1 + \alpha)}{\alpha^2}\right],$$

where A is an instrumental constant, κ_T is isothermal compressibility, and $\alpha = 2(k_0\xi)^2$ with k_0 being the wave vector of incident light in the scattering medium and ξ the correlation length. The above equation allows to characterize the critical behavior of simple fluids as well as colloidal dispersions with turbidity measurements.[21]

(i) Show that the critical exponent for the isothermal compressibility can be obtained from turbidity in the small angle limit τ_0 (i.e., at $\alpha \to 0$).
(ii) Suppose that experimental data are available for τ/τ_0 in the vicinity of the critical point, how are the data used to determine the correlation length ξ.
(iii) Check the experimental results from the literature.

20 Andersen H. C., Weeks J. D. and Chandler D., "Relationship between the hard-sphere fluid and fluids with realistic repulsive forces", *Phys. Rev. A* 4, 1597 (1971).
21 e.g., Chen B. H. et al., "Turbidity and critical behavior of a colloid-polymer system", *Phys. Rev. E* 64, 042401 (2001).

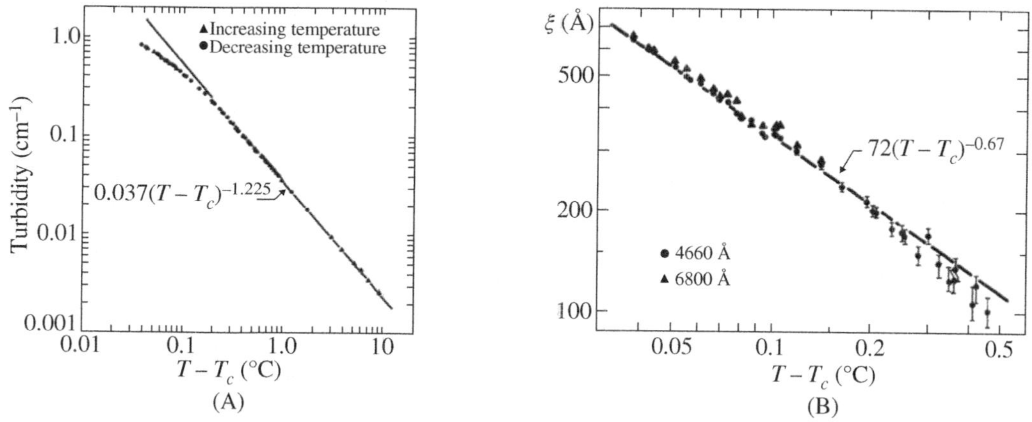

Figure 7.23 The temperature dependence of the turbidity (A) and the correlation length (B) of SF_6 near the vapor–liquid critical point. In panel B, the experimental data were obtained at two wavelengths of the laser light, as noted in the figure legend. Reproduced from Puglielli and Ford[16].

Solution:

(i) At small angle θ, the scattering wave vector k_0 thus $\alpha \to 0$. Accordingly, the turbidity becomes

$$\tau_0 = \lim_{\alpha \to 0} = \tau = \frac{8}{3} A \kappa_T.$$

With the numerical prefactor determined from conditions remote from the critical point, a log–log plot of the turbidity τ_0 versus the relative temperature $T - T_c$ would result in the critical exponent for the isothermal compressibility, $\kappa_T \sim (T - T_c)^{-\gamma}$ with $\gamma \approx 1.24$.

(ii) Fitting the experimental data for τ/τ_0 with

$$f(\alpha) = \frac{3}{8} \left[\frac{2\alpha^2 + 2\alpha + 1}{\alpha^3} \ln(1 + 2\alpha) - \frac{2(1 + \alpha)}{\alpha^2} \right]$$

would yield the correlation length ξ at different temperatures. The critical exponent for ξ can be obtained from a log–log plot of ξ with respect to $T - T_c$.

(iii) Figure 7.23 shows the temperature dependence of both the isothermal compressibility and the correlation length from turbidity measurements.[22]

22 Puglielli V. G. and Ford Jr. N. C., "Turbidity measurements in SF_6 near its critical point", *Phys. Rev. Lett.* 25, 143 (1970).

8

Polymer Solutions, Blends, and Complex Fluids

Problem 8.1 Based on the number of ways to place polymer chains on a lattice predicted by the Flory–Huggins theory (viz., Eq. 8.13), verify that the entropy of mixing is given by

$$\Delta S/k_B = -n_1 \ln \phi_1 - n_2 \ln \phi_2,$$

where n_1 is the number of solvent molecules, n_2 is the number of polymer chains, and $\phi_{i=1,2}$ are the corresponding volume fractions.

Solution:
According to Eq. (8.14), the entropy of mixing is

$$\begin{aligned}
\Delta S/k_B &= \ln\left\{\frac{(n_1+mn_2)!}{n_1!n_2!(n_1+mn_2)^{mn_2}}\left[(n_1+mn_2)z^{m-1}\right]^{n_2}\right\} - \ln\left\{\frac{(mn_2)!}{n_2!(mn_2)^{mn_2}}\left(mn_2 z^{m-1}\right)^{n_2}\right\}\\
&= \ln n_T! - \ln n_1! - (m-1)n_2 \ln n_T - \ln(mn_2)! + (m-1)n_2\ln(mn_2)\\
&= n_T \ln n_T - n_T - n_1 \ln n_1 + n_1 - mn_2\ln(mn_2) + mn_2 + (m-1)n_2 \ln(mn_2/n_T)\\
&= -n_1\ln(n_1/n_T) - mn_2\ln(mn_2/n_T) + (m-1)n_2\ln(mn_2/n_T)\\
&= -n_1\ln(n_1/n_T) - n_2\ln(mn_2/n_T),
\end{aligned}$$

where $n_T = n_1 + mn_2$. Plugging into the above equation with $\phi_1 = n_1/n_T$ and $\phi_2 = mn_2/n_T$ leads to

$$\Delta S/k_B = -n_1 \ln \phi_1 - n_2 \ln \phi_2.$$

Problem 8.2 Using the Flory–Huggins theory for a binary mixture of a monodisperse homopolymer and a monomeric solvent, show that the spinodal line can be described by

$$\phi = \alpha \pm \sqrt{\alpha^2 - 1/(2m\chi_F)},$$

where ϕ is the polymer volume fraction, m and χ_F have their usual meanings, and $\alpha \equiv 1/2 - (1-m)/(4\chi_F)$. Derive the Flory parameter and the polymer volume fraction at the critical point of the liquid-liquid equilibrium.

Solution:
The spinodal line specifies the boundary of thermodynamic stability, which can be expressed in terms of the second-order derivative of the free energy density

$$\left(\frac{\partial^2 f}{\partial \phi^2}\right)_{n_T,T} = 0.$$

Solutions Manual to Fundamentals and Practice in Statistical Thermodynamics,
First Edition. Jianzhong Wu and John M. Prausnitz.

According to the Flory–Huggins theory, the free energy density of mixing is

$$\Delta f \equiv \frac{\Delta F}{n_T v_0} = \frac{k_B T}{v_0}\left[\frac{\phi}{m}\ln\phi + (1-\phi)\ln(1-\phi) + (1-\phi)\phi\chi_F\right], \tag{A}$$

where v_0 represents the volume of each lattice site. Note that the free energy density of the polymer solution is given by

$$f = \frac{F}{n_T v_0} = \frac{\Delta F + n_1\mu_1^0 + n_2\mu_2^0}{n_T v_0} = \Delta f + (1-\phi)f_1^0 + \frac{\phi}{m}f_2^0,$$

where $f_{i=1,2}^0 = \mu_i^0/v_0$, and μ_i^0 represents the chemical potentials of pure species. Taking the derivative of Eq. (A) with respect to ϕ gives

$$\frac{\partial(\beta f v_0)}{\partial\phi} = \frac{\beta\mu_2^0}{m} - \beta\mu_1^0 + \frac{\ln\phi}{m} + \frac{1}{m} - \ln(1-\phi) - 1 + (1-2\phi)\chi_F,$$

and applying the second-order derivative to Eq. (A) yields

$$\frac{\partial^2(\beta f v_0)}{\partial\phi^2} = \frac{1}{m\phi} + \frac{1}{1-\phi} - 2\chi_F = 0.$$

Therefore, the spinodal line can be obtained from

$$2m\chi_F\phi(1-\phi) - (1-\phi) - m\phi = 0 \quad \rightarrow \quad -2m\chi_F\phi^2 + (2m\chi_F + 1 - m)\phi - 1 = 0.$$

Solving the quadratic equation yields

$$\phi = \frac{(2m\chi_F + 1 - m) \pm \sqrt{(2m\chi_F + 1 - m)^2 - 8m\chi_F}}{4m\chi_F} = \alpha \pm \sqrt{\alpha^2 - 1/(2m\chi_F)}$$

with $\alpha \equiv 1/2 - (1-1/m)/4\chi_F$.

At the critical condition, we have only one root for the above quadratic equation, which means

$$\alpha^2 = [1/2 - (1-1/m)/4\chi_F]^2 = 1/(2m\chi_F) \quad \rightarrow \quad \frac{1-1/m}{4\chi_F} + \frac{1}{\sqrt{2m\chi_F}} - \frac{1}{2} = 0.$$

Thus, the Flory parameter and the polymer volume fraction at the critical point are given by

$$\chi_{Fc} = \frac{1}{2}\left(1 + \frac{1}{\sqrt{m}}\right)^2 \quad \text{and} \quad \phi_c = \alpha = 1/\sqrt{2m\chi_{Fc}} = \frac{1}{1+\sqrt{m}}.$$

Problem 8.3 To understand the critical behavior of the liquid–liquid phase separation (LLPS) in a polymer solution as discussed in Problem 8.2, it is helpful to define the Landau free energy using $\delta\phi = \phi - \phi_c$ as the order parameter (see Section 5.9)

$$f_L = \frac{a_2}{2}\delta\phi^2 + \frac{a_4}{4}\delta\phi^4 + \cdots,$$

where ϕ is the polymer volume fraction, ϕ_c is the polymer volume fraction at the critical point, and a_2 and a_4 are phenomenological parameters.

(i) Show that, near the critical point, the Landau free energy can be expressed as

$$f_L = -\delta\chi_F\delta\phi^2 + \frac{\sqrt{m}\chi_{Fc}^2}{3}\delta\phi^4 + \mathcal{O}(\delta\phi^5),$$

where $\delta\chi_F = \chi_F - \chi_{Fc}$, and m is the number of segments per polymer chain.

(ii) Verify the following scaling relation predicted by the Landau theory

$$\delta\phi \sim (\delta\chi_F/\sqrt{m})^{1/2},$$

where $\delta\phi$ represents the difference between the volume fractions of the coexisting phases.

(iii) The renormalization-group correction to the mean-field theory indicates that the difference between the coexisting densities follows the scaling relation[1]

$$\delta\phi \sim (\delta\chi_F/m)^{1/3}.$$

Compare the mean-field prediction of $\delta\phi$ versus $\delta\chi_F$ with the scaling law above for $m = 10^3, 10^4$, and 10^5.

Hint: The Landau free energy is defined as the mean-field free energy relative to the "background" contribution.

Solution:

This problem is adapted from Qian et al.[2]

(i) As shown in the solution for Problem 8.2, the Flory–Huggins theory predicts that the Helmholtz energy density of a homopolymer dissolved in a monomeric solvent is given by

$$f(\phi,\chi_F) = \frac{k_BT}{v_0}\left[\frac{\phi}{m}\ln\phi + (1-\phi)\ln(1-\phi) + (1-\phi)\phi\chi_F\right] + (1-\phi)f_1^0 + \frac{\phi}{m}f_2^0.$$

Near the critical point, $f(\phi,\chi_F)$ can be expanded in terms of $\delta\phi = \phi - \phi_c$

$$\begin{aligned} f &= f(\phi_c,\chi_F) + \left(\frac{\partial f}{\partial\phi}\right)\delta\phi + \tfrac{1}{2}\left(\frac{\partial^2 f}{\partial\phi^2}\right)\delta\phi^2 + \tfrac{1}{3!}\left(\frac{\partial^3 f}{\partial\phi^3}\right)\delta\phi^3 + \tfrac{1}{4!}\left(\frac{\partial^4 f}{\partial\phi^4}\right)\delta\phi^4 + \cdots \\ &= f(\phi_c,\chi_F) + \left[\frac{f_2^0}{m} - f_1^0 + \frac{\partial\Delta f}{\partial\phi}\right]\delta\phi + \frac{1}{2}\frac{\partial^2\Delta f}{\partial\phi^2}\delta\phi^2 + \frac{1}{3!}\frac{\partial^3\Delta f}{\partial\phi^3}\delta\phi^3 + \frac{1}{4!}\frac{\partial^4\Delta f}{\partial\phi^4}\delta\phi^4 + \cdots, \end{aligned} \tag{B}$$

where all partial derivatives are evaluated at ϕ_c and χ_F. The mixing free energy density is

$$\Delta f = \frac{k_BT}{v_0}\left[\frac{\phi}{m}\ln\phi + (1-\phi)\ln(1-\phi) + (1-\phi)\phi\chi_F\right].$$

In the Taylor series given by Eq. (B), the zero-order term depends only on temperature (equivalently, χ_F). Meanwhile, the linear term is related to the difference between the chemical potentials of the polymer segments and the solvent

$$\begin{aligned} \left.\frac{\partial(\beta f v_0)}{\partial\phi}\right|_{\phi_c,\chi_F} &= \beta v_0\left[\frac{f_2^0}{m} - f_1^0 + \frac{\partial\Delta f}{\partial\phi}\right] \\ &= \frac{\beta\mu_2^0}{m} - \beta\mu_1^0 + \frac{\ln\phi}{m} + \frac{1}{m} - \ln(1-\phi) - 1 + (1-2\phi)\chi_F \\ &= [\beta\mu_2(\phi_c,\chi_F)/m - \beta\mu_1(\phi_c,\chi_F)]. \end{aligned}$$

The higher-order terms are given by

$$\frac{1}{2}\left(\frac{\partial^2\beta v_0\Delta f}{\partial\phi^2}\right)\bigg|_{\phi_c} = \frac{(1+\sqrt{m})^2 - 2\chi_F m}{2m} = \chi_{Fc} - \chi_F$$

1 de Gennes P. G., *Scaling Concepts in Polymer Physics*. Cornell University Press, 1979, Page 121.

2 Qian D., Michaels T. C. T., and Knowles T. P. J., "Analytical solution to the Flory–Huggins Model", *J. Phys. Chem. Lett.* 13(33), 7853–7860 (2022).

$$\frac{1}{3!}\left(\frac{\partial^3 \beta v_0 \Delta f}{\partial \phi^3}\right)\bigg|_{\phi_c} = 0$$

$$\frac{1}{4!}\left(\frac{\partial^4 \beta v_0 \Delta f}{\partial \phi^4}\right)\bigg|_{\phi_c} = \frac{(\sqrt{m}+1)^4}{12m^{3/2}} = \frac{\sqrt{m}}{3}\chi_{Fc}^2,$$

The Landau free energy is defined as the mean-field free energy minus the "background" contribution:

$$f_L \equiv \beta v_0[f - f_0(\phi, \chi_F)],$$

where

$$f_0(\phi, \chi_F) \equiv f(\phi_c, \chi_F) + \frac{\phi}{v_0}\left[\frac{\mu_2(\phi_c, \chi_F)}{m} - \mu_1(\phi_c, \chi_F)\right].$$

A comparison of the Landau free energy with Eq. (B) leads to

$$f_L = -\delta\chi_F\delta\phi^2 + \frac{\sqrt{m}\chi_{Fc}^2}{3}\delta\phi^4 + \mathcal{O}(\delta\phi^5),$$

where $\delta\chi_F = \chi_F - \chi_{Fc} > 0$.

(ii) The Landau theory predicts that, at equilibrium, the order parameters of coexisting phases satisfy (Eq. 5.160)

$$\delta\phi = \pm\sqrt{-a_2/a_4}.$$

Substituting into the above equation with $a_2 = -2\delta\chi_F$ and $a_4 = 4\sqrt{m}\chi_{Fc}^2/3$, we obtain the difference between the volume fractions of the coexisting phases

$$\delta\phi = 2\sqrt{(2\delta\chi_F)\frac{3}{4\sqrt{m}\chi_{Fc}^2}} \approx \sqrt{\frac{24\delta\chi_F}{\sqrt{m}}} \sim (\delta\chi_F/\sqrt{m})^{1/2}.$$

(iii) Figure 8.1 compares the mean-field predictions with de Gennes' scaling law for the critical behavior of homopolymer solutions at the three chain lengths. The difference near the vicinity of coexistence highlights the importance of fluctuation effects on the critical behavior.

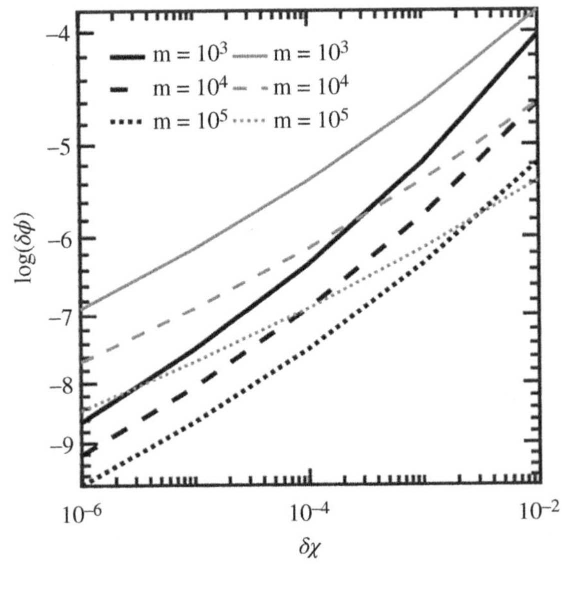

Figure 8.1 Comparison of the Landau theory (black lines) with de Gennes' scaling law (gray lines) for the critical behavior of homopolymer solutions with $m = 10^3, 10^4$, and 10^5. In this plot, the proportionality constants are assumed to be 1.

Problem 8.4 The following exercise extends the Flory–Huggins theory to a ternary liquid mixture containing homopolymer B and a co-solvent of small-molecule liquids A and C.

(i) Show that the entropy of mixing is given by

$$\Delta S/k_B = -n_A \ln \phi_A - n_B \ln \phi_B - n_C \ln \phi_C,$$

where $n_{i=A,B,C}$ are the numbers of molecules for each chemical species in the system, and $\phi_{i=A,B,C}$ refer to the volume fractions of the three chemical species.

(ii) Show that the internal energy of mixing is given by

$$\Delta\beta U/n_T = \phi_A\phi_B\chi_{AB} + \phi_B\phi_C\chi_{BC} + \phi_A\phi_C\chi_{AC},$$

where $\chi_{AB} \equiv Z(\epsilon_{A,A} + \epsilon_{B,B} - 2\epsilon_{AB})/2k_BT$ is the Flory parameter for the binary mixture of A and B, χ_{AC} and χ_{BC} have similar definitions.

Solution:

(i) According to the Flory–Huggins theory, the entropy of a polymer system is related to the number of ways to place polymer segments and solvent molecules on a lattice

$$S/k_B = \ln(W_A \times W_B \times W_C).$$

For pure species, the entropy is

$$S_{\alpha=A,B,C}/k_B = \ln(W_\alpha^0),$$

with $W_A^0 = W_C^0 = 1$. When the lattice is first filled with the polymer segments, W_B depends only on the total number of lattice sites $n_T = n_A + m_B n_B + n_C$ and the number of polymer chains n_B. Thus, W_B/W_B^0 is the same as that for the polymer in a pure solvent that occupies $n_0 = n_T - m_B n_B$ lattice sites

$$\ln(W_B/W_B^0) = -n_B \ln \phi_B - n_0 \ln \phi_0,$$

where $\phi_B = n_B m_B/n_T$, and $\phi_0 = n_0/n_T$. Meanwhile, $W_A \times W_C$ is related to the number of ways to fill the rest sites without overlap

$$W_A \times W_C = \frac{n_0!}{n_A! n_B!}.$$

Thus, the entropy of mixing is

$$\begin{aligned}\Delta S/k_B &= \ln(W_B/W_B^0) + \ln(W_A \times W_C) = -n_B \ln \phi_B - n_0 \ln \phi_0 - n_A \ln \frac{n_A}{n_0} - n_C \ln \frac{n_C}{n_0} \\ &= -n_B \ln \phi_B - n_A \ln\left(\frac{n_A}{n_0} \times \frac{n_0}{n_T}\right) - n_C \ln\left(\frac{n_C}{n_0} \times \frac{n_0}{n_T}\right) \\ &= -n_B \ln \phi_B - n_A \ln \phi_A - n_C \ln \phi_C.\end{aligned}$$

(ii) The internal energy of mixing is the same as that for a ternary mixture of monomers with random mixing

$$\begin{aligned}\Delta U/n_T &= \frac{Z}{2} \sum_{\alpha,\beta=A,B,C} \phi_\alpha\phi_\beta\epsilon_{\alpha,\beta} - \frac{Z}{2} \sum_{\alpha=A,B,C} \epsilon_{\alpha\alpha} \\ &= \frac{Z}{4} \sum_{\alpha\beta=A,B,C} \phi_\alpha\phi_\beta(2\epsilon_{\alpha\beta} - \epsilon_{\alpha\alpha} - \epsilon_{\beta\beta}) = \frac{k_BT}{2} \sum_{\alpha,\beta=A,B,C} \phi_\alpha\phi_\beta\chi_{\alpha\beta} \\ &= k_BT(\phi_A\phi_B\chi_{AB} + \phi_B\phi_C\chi_{BC} + \phi_A\phi_C\chi_{AC}),\end{aligned}$$

where $\chi_{AB} \equiv Z(\epsilon_{A,A} + \epsilon_{B,B} - 2\epsilon_{AB})/(2k_BT)$, and Z is the coordination number of the lattice.

Problem 8.5 At given temperature and pressure, the stability of a ternary mixture is determined by the second derivatives of the free energy density

$$D = \det \begin{pmatrix} \frac{\partial^2 f}{\partial \phi_A^2} & \frac{\partial^2 f}{\partial \phi_A \partial \phi_B} \\ \frac{\partial^2 f}{\partial \phi_A \partial \phi_B} & \frac{\partial^2 f}{\partial \phi_B^2} \end{pmatrix} > 0.$$

Based on the energy and entropy of mixing derived in Problem 8.4, show that the spinodal curve of homopolymer B in a co-solvent of small-molecule liquids A and C is described by

$$\left(\frac{1}{m_B\phi_B} + \frac{1}{\phi_C} - 2\chi_{BC}\right) \times \left(\frac{1}{\phi_A} + \frac{1}{\phi_C} - 2\chi_{AC}\right) = \left(\frac{1}{\phi_C} + \chi_{AB} - \chi_{BC} - \chi_{AC}\right)^2.$$

Solution:
This problem is adapted from Dudowicz et al.[3]
Based on the energy and entropy of mixing derived in Problem 8.4, we can obtain the Helmholtz energy density

$$\Delta f \equiv \frac{\Delta F}{n_T v_0} = \frac{1}{n_T v_0}[\Delta U - T\Delta S] = \frac{k_B T}{v_0}\left[\phi_A\phi_B\chi_{AB} + \phi_B\phi_C\chi_{BC} + \phi_A\phi_C\chi_{AC} + \frac{\phi_B}{m_B}\ln\phi_B + \phi_A\ln\phi_A + \phi_C\ln\phi_C\right],$$

where v_0 represents the volume of each lattice site. Because the free energy of each pure species is linearly dependent on the volume fraction, the second derivatives of the free energy density for the ternary mixture are the same as those for the mixing free energy density

$$\frac{\partial^2 \beta f v_0}{\partial \phi_A^2} = \left(\frac{\partial^2 \Delta\beta f v_0}{\partial \phi_A^2}\right)_{\phi_B} = \frac{\partial}{\partial \phi_A}(\phi_B\chi_{AB} - \phi_B\chi_{BC} + \phi_C\chi_{AC} - \phi_A\chi_{AC} + \ln\phi_A - \ln\phi_C)|_{\phi_B}$$
$$= -2\chi_{AC} + \frac{1}{\phi_A} + \frac{1}{\phi_C}.$$

Similarly, we have

$$\frac{\partial^2 \beta f v_0}{\partial \phi_B^2} = \left(\frac{\partial^2 \Delta\beta f v_0}{\partial \phi_B^2}\right)_{\phi_A} = \frac{1}{m_B\phi_B} + \frac{1}{\phi_C} - 2\chi_{BC},$$

and

$$\frac{\partial^2 \beta f v_0}{\partial \phi_A \partial \phi_B} = \frac{\partial^2 \Delta\beta f v_0}{\partial \phi_A \partial \phi_B} = \frac{1}{\phi_C} + \chi_{AB} - \chi_{BC} - \chi_{AC}.$$

Therefore, the spinodal curve of the ternary mixture, determined by $D = 0$, leads to

$$\left(\frac{1}{m_B\phi_B} + \frac{1}{\phi_C} - 2\chi_{BC}\right) \times \left(\frac{1}{\phi_A} + \frac{1}{\phi_C} - 2\chi_{AC}\right) = \left(\frac{1}{\phi_C} + \chi_{AB} - \chi_{BC} - \chi_{AC}\right)^2.$$

Problem 8.6 The Flory–Huggins theory for polymer solutions is similarly applicable to polymer blends. To elucidate, consider a binary mixture of homopolymers A and B in a rubbery state (i.e., the system may be considered as a binary liquid mixture). Assume that each chain of polymer A takes m_A lattice sites, and for polymer B each chain takes m_B sites. Based on the assumptions

---

3 Dudowicz J., Freed K. F. and Douglas J. F., "Cosolvency and cononsolvency explained in terms of a Flory-Huggins type theory", *J. Chem. Phys.*, 143, 131101 (2015).

used in deriving the Flory–Huggins theory for polymer solutions, show the following expressions for

(i) the entropy of mixing

$$\Delta S/k_B = -n_A \ln \phi_A - n_B \ln \phi_B,$$

where n_A and n_B are the numbers of polymer chains, and ϕ_A and ϕ_B are the volume fractions of polymers A and B;

(ii) the internal energy of mixing

$$\Delta\beta U = n_T \phi_A \phi_B \chi_F,$$

where $n_T = n_A m_A + n_B m_B$ is the total number of lattice sites, and χ_F is the Flory parameter;

(iii) the Helmholtz energy of mixing

$$\Delta\beta F = n_A \ln \phi_A + n_B \ln \phi_B + n_T \phi_A \phi_B \chi_F;$$

(iv) the chemical potentials of polymer species

$$\beta\mu_A = \beta\mu_A^0 + \ln \phi_A + \phi_B(1 - m_A/m_B) + m_A \phi_B^2 \chi_F,$$
$$\beta\mu_B = \beta\mu_B^0 + \ln \phi_B + \phi_A(1 - m_B/m_A) + m_B \phi_A^2 \chi_F.$$

Solution:

(i) First, consider the number of ways to place polymer A on an empty lattice. With the same number of lattice sites n_T, the number is identical to that for a polymer solution

$$W_A = \frac{n_T!}{n_A!(n_T - m_A n_A)! n_T^{m_A n_A}} \left[n_T Z^{m_A - 1}\right]^{n_A} = \frac{n_T!}{n_A!(m_B n_B)! n_T^{m_A n_A}} \left[n_T Z^{m_A - 1}\right]^{n_A}.$$

Within the assumptions of the Flory–Huggins theory, the number of ways to place polymer B chains on the remaining $(n_T - m_A n_A)$ sites is

$$W_B = \frac{\left[n_T Z^{m_B - 1}\right]^{n_B}}{n_B!} \prod_{i=1}^{m_B n_B} \left(\frac{n_T - m_A n_A - i + 1}{n_T}\right)$$
$$= \frac{(n_T - m_A n_A)!}{n_B! n_T^{m_B n_B}} \left[n_T Z^{m_B - 1}\right]^{n_B} = \frac{(m_B n_B)!}{n_B! n_T^{m_B n_B}} \left[n_T Z^{m_B - 1}\right]^{n_B}.$$

The number of ways for placing the pure polymer chains is the same as that considered for polymer solutions:

$$W_A^0 = \frac{(m_A n_A)!}{n_A!(m_A n_A)^{m_A n_A}} \left(m_A n_A Z^{m_A - 1}\right)^{n_A},$$

$$W_B^0 = \frac{(m_B n_B)!}{n_B!(m_B n_B)^{m_B n_B}} \left(m_B n_B Z^{m_B - 1}\right)^{n_B}.$$

Hence, the entropy of mixing is

$$\Delta S/k_B = \ln(W_A W_B) - \ln(W_A^0) - \ln(W_B^0)$$
$$= \ln\left\{\frac{n_T!\left(n_T Z^{m_A - 1}\right)^{n_A}}{n_A!(m_B n_B)! n_T^{m_A n_A}}\right\} - \ln\left\{\frac{(m_A n_A)!\left(m_A n_A Z^{m_A - 1}\right)^{n_A}}{n_A!(m_A n_A)^{m_A n_A}}\right\}$$
$$+ \ln\left\{\frac{(m_B n_B)!}{n_B! n_T^{m_B n_B}} \left[n_T Z^{m_B - 1}\right]^{n_B}\right\} - \ln\left\{\frac{(m_B n_B)!\left(m_B n_B Z^{m_B - 1}\right)^{n_B}}{n_B!(m_B n_B)^{m_B n_B}}\right\}$$

$$= \ln \frac{n_T!}{(m_A n_A)!(m_B n_B)!} + (m_A - 1)n_A \ln(m_A n_A/n_T) + (m_B - 1)n_B \ln(m_B n_B/n_T)$$

$$= n_T \ln n_T - n_T - m_B n_B \ln(m_B n_B) + m_B n_B - (m_A n_A)\ln(m_A n_A)$$

$$+ m_A n_A + (m_A - 1)n_A \ln(m_A n_A/n_T) + (m_B - 1)n_B \ln(m_B n_B/n_T)$$

$$= -(m_A n_A)\ln(m_A n_A/n_T) - (m_B n_B)\ln(m_B n_B/n_T)$$

$$+ (m_A - 1)n_A \ln(m_A n_A/n_T) + (m_B - 1)n_B \ln(m_B n_B/n_T)$$

$$= -n_A \ln \phi_A - n_B \ln \phi_B,$$

where $\phi_A = m_A n_A/n_T$ and $\phi_B = m_B n_B/n_T$ are the volume fractions of A and B.

(ii) The internal energy of mixing is assumed to be the same as that for the corresponding monomeric system

$$U = U - U_A - U_B$$

$$= -\frac{n_T Z}{2}(\phi_A^2 \epsilon_{AA} + \phi_B^2 \epsilon_{BB} + 2\phi_A \phi_B \epsilon_{AB}) + \frac{n_A m_A Z \epsilon_{AA}}{2} + \frac{n_B m_B Z \epsilon_{BB}}{2}$$

$$= \frac{n_T Z \phi_A \phi_B}{2}(\epsilon_{AA} + \epsilon_{BB} - 2\epsilon_{AB}) = n_T \phi_A \phi_B k_B T \chi_F,$$

where $\chi_F = Z(\epsilon_{AA} + \epsilon_{BB} - 2\epsilon_{AB})/2k_B T$ is the Flory parameter.

(iii) The Helmholtz energy of mixing is obtained by the combination of ΔU and ΔS:

$$\Delta \beta F = \beta(\Delta U - T\Delta S) = n_A \ln \phi_A + n_B \ln \phi_B + n_T \phi_A \phi_B \chi_F.$$

(iv) We obtain the chemical potential of each species from the partial derivative of the Helmholtz energy with respect to the number of polymer chains

$$\beta\mu_A - \beta\mu_A^0 = \left(\frac{\partial \Delta F}{\partial n_A}\right)_{T,n_B}$$

$$= \ln \phi_A + \frac{n_A}{\phi_A}\frac{\partial \phi_A}{\partial n_A} + \frac{n_B}{\phi_B}\frac{\partial \phi_B}{\partial n_A} + \chi_F \frac{\partial}{\partial n_A}\left(\frac{m_A n_A m_B n_B}{m_A n_A + m_B n_B}\right)$$

$$= \ln \phi_A + \left(\frac{n_A}{\phi_A} - \frac{n_B}{\phi_B}\right)\frac{m_A}{n_T}\left(1 - \frac{m_A n_A}{n_T}\right) + \chi_F \frac{m_A m_B n_B}{n_T}\left(1 - \frac{m_A n_A}{n_T}\right)$$

$$= \ln \phi_A + \left(\frac{n_T}{m_A} - \frac{n_T}{m_B}\right)\frac{m_A}{n_T}\phi_B + m_A \phi_B^2 \chi_F$$

$$= \ln \phi_A + (1 - m_A/m_B)\phi_B + m_A \phi_B^2 \chi_F.$$

Similarly, we have

$$\beta\mu_B - \beta\mu_B^0 = \left(\frac{\partial \Delta F}{\partial n_B}\right)_{T,n_A} = \ln \phi_B + (1 - m_B/m_A)\phi_A + m_B \phi_A^2 \chi_F.$$

Problem 8.7 Consider of a binary mixture of homopolymers A and B. Show that, at the critical point of demixing, the polymer volume fraction and the Flory parameter are given by, respectively,

$$\phi_c = \frac{\sqrt{m_B}}{\sqrt{m_A} + \sqrt{m_B}} \quad \text{and} \quad \chi_{Fc} = \frac{1}{2}\left[\frac{1}{\sqrt{m_A}} + \frac{1}{\sqrt{m_B}}\right]^2,$$

where ϕ stands for the volume fraction of polymer A, subscript "c" stands for the critical condition.

Solution:

The critical volume fraction and critical temperature can be determined from the stability condition:

$$\left(\frac{\partial^2 \Delta F}{\partial \phi^2}\right)\bigg|_T = 0 \quad \text{and} \quad \left(\frac{\partial^3 \Delta F}{\partial \phi^3}\right)\bigg|_T = 0.$$

Substituting the above equations with the free energy derived from Problem 8.6

$$\Delta\beta F/n_T = \frac{\phi}{m_A}\ln\phi + \frac{(1-\phi)}{m_B}\ln(1-\phi) + \phi(1-\phi)\chi_F$$

leads to

$$\left(\frac{\partial^2 \Delta\beta F/n_T}{\partial \phi^2}\right)\bigg|_T = 0 = \frac{1}{m_A\phi} + \frac{1}{m_B(1-\phi)} - 2\chi_F,$$

$$\left(\frac{\partial^3 \Delta\beta F/n_T}{\partial \phi^3}\right)\bigg|_T = 0 = \frac{1}{m_B(1-\phi)^2} - \frac{1}{m_A\phi^2}.$$

Solving the above two equations gives

$$\phi_c = \frac{\sqrt{m_B}}{\sqrt{m_A}+\sqrt{m_B}} \quad \text{and} \quad \chi_{Fc} = \frac{1}{2}\left[\frac{1}{\sqrt{m_A}} + \frac{1}{\sqrt{m_B}}\right]^2.$$

Problem 8.8 The Sanchez–Lacombe (SL) equation of state can be considered as an extension of the Flory–Huggins theory that allows for the description of compressibility effects.[4] Like the Flory–Huggins theory, the SL equation combines the configurational entropy estimated from a lattice model and the mean-field approximation for the internal energy. However, the procedure applies only to pure species and empty sites are allowed in filling the lattice. The thermodynamic properties of polymer solutions and blends are predicted with mixing rules similar to those used in the van der Waals equation of state.

(i) Consider the number of ways to place n polymer chains onto an empty lattice such that each polymer chain occupies m sites without overlap. The total number of lattice sites is $n_T = nm + n_0$ where n_0 represents the number of empty sites. Following the Flory–Huggins theory, show that the number of ways to place the polymer chains can be approximated by

$$W = \frac{(n_0+mn)!}{n_0!n!(n_0+mn)^{mn}}\left[(n_0+mn)Z^{m-1}\right]^n,$$

where Z is the coordination number.

(ii) Show that the mean-field approximation results in an internal energy of

$$U/n_T = -\frac{Z\epsilon}{2}\left(\frac{mn}{mn+n_0}\right)^2,$$

where ϵ is the energy due to the attraction of nearest-neighboring segments.

(iii) Derive the SL equation of state

$$\frac{P^*v^*}{T^*} = \frac{1}{m} - \left[1 + v^*\ln(1-1/v^*) + \frac{1}{v^*T^*}\right],$$

where $P^* = 2Pv_0/(Z\epsilon)$, $T^* = 2k_BT/(Z\epsilon)$, $v^* = V/(nmv_0)$, and v_0 is the volume of each lattice site.

4 Sanchez I. C. and Lacombe R. H., "An elementary molecular theory of classical fluids", *J. Phys. Chem.* 80, 2352 (1976).

Solution:

(i) The number of ways to place the polymer chains on the lattice is identical to that derived in the original Flory–Huggins theory but with n_0 empty sites

$$W = \frac{(n_0 + mn)!}{n_0!n!(n_0 + mn)^{mn}} \left[(n_0 + mn)Z^{m-1}\right]^n.$$

(ii) The mean-field internal energy is also the same as that adopted in the original Flory–Huggins theory but without solvent-solvent and solvent-polymer interactions

$$U/n_T = -\frac{Z\epsilon}{2}\left(\frac{mn}{mn + n_0}\right)^2.$$

(iii) According to the SL theory, the Helmholtz energy for a pure polymeric specie is

$$F = -\frac{Z\epsilon(mn + n_0)}{2}\left(\frac{mn}{mn + n_0}\right)^2 - k_BT\ln\left\{\frac{(n_0 + mn)!}{n_0!n!(n_0 + mn)^{mn}}\left[(n_0 + mn)Z^{m-1}\right]^n\right\}.$$

Using the Stirling equation $\ln n! = n\ln n - n$, we can obtain the entropy

$$\begin{aligned} S/k_B = {} & (n_0 + mn)\ln(n_0 + mn) - (n_0 + mn) - n_0\ln n_0 + n_0 - n\ln n + n \\ & - (mn - n)\ln(n_0 + mn) + n(m - 1)\ln Z. \end{aligned}$$

The pressure can be derived from the partial derivative of the Helmholtz energy with respect to the system volume, $V = (n_0 + mn)v_0$,

$$\begin{aligned} P &= -\left(\frac{\partial F}{\partial V}\right)_{T,n} = -\frac{1}{v_0}\left(\frac{\partial F}{\partial n_0}\right)_{T,n} \\ &= -\frac{Z\epsilon(mn)^2}{2v_0(n_0 + mn)^2} + \frac{k_BT}{v_0}\left\{\ln\left(\frac{n_0 + mn}{n_0}\right) - \frac{mn - n}{n_0 + mn}\right\} \\ &= -\frac{Z\epsilon v_0(mn)^2}{2V^2} + \frac{k_BT}{v_0}\left\{-\ln\left(1 - \frac{mn}{n_0 + mn}\right) + \frac{1}{m}\frac{mn}{n_0 + mn} - \frac{mn}{n_0 + mn}\right\}. \end{aligned}$$

Let $P^* = 2Pv_0/(Z\epsilon)$, $T^* = 2k_BT/(Z\epsilon)$ and $v^* = V/(nmv_0) = (nm + n_0)/(nm)$. Multiplying both sides of the above equation with $V/(nmk_BT)$ leads to

$$\frac{P^*v^*}{T^*} = -\frac{1}{v^*T^*} + \frac{1}{m} - 1 - v^*\ln(1 - 1/v^*) = \frac{1}{m} - \left[1 + v^*\ln(1 - 1/v^*) + \frac{1}{v^*T^*}\right].$$

In the SL equation, the thermodynamic properties of the pure components can be characterized by three dimensionless variables, namely, the reduced temperature (T^*), reduced pressure (P^*) and reduced volume (v^*). For multicomponent systems, the parameters in the SL equation are calculated from the mixing rules

$$\frac{1}{v_0} = \sum_i \frac{\phi_i}{v_{0,i}} \quad \text{and} \quad \frac{1}{m} = \sum_i \frac{\phi_i v_0}{m_i v_{0,i}}$$

$$\epsilon = \sum_i\sum_j \phi_i\phi_j\epsilon_{ij} = \sum_i \phi_i\epsilon_{ii} - \frac{2k_BT}{Z}\sum_i\sum_{j>i}\phi_i\phi_j\chi_{ij},$$

where $\phi_i = m_in_i/\sum_j m_jn_j$ is the volume fraction of species i, and χ_{ij} is the Flory parameter. These mixing rules are empirical, and other forms have also been proposed.

Problem 8.9 Ruzette and Mayes proposed a modification of the Flory–Huggins theory that is able to capture the phase behavior of many weakly interacting polymer blends with the pure component data for the mass density, solubility parameter, and thermal expansion coefficient as the input.[5] One essential idea follows earlier developments by Hildebrand and by Flory who suggested that, upon mixing two *compressible* species A and B, the change in entropy should scale with the logarithm of the ratio of the "free volume" available in the mixture (blend), $V_{f,b}$, to those corresponding to the pure components, $V_{f,A}$ and $V_{f,B}$:

$$\Delta S/k_B = n_A \ln\left(\frac{V_{f,b}}{V_{f,A}}\right) + n_B \ln\left(\frac{V_{f,b}}{V_{f,B}}\right).$$

Ruzette and Mayes assumed further that the free volume can be estimated from the difference between the actual volume of polymeric species V_i and the lattice volume, i.e.,

$$V_{f,i} = V_i - m_i n_i v_i \quad \text{and} \quad V_{f,b} = V_b - (m_A n_A + m_B n_B) v_b,$$

where subscript $i = A, B$ represents pure polymer A and B, respectively; parameters m_i and n_i are the same as those defined in the Flory–Huggins theory, $v_{i=A,B}$ is the "segment" volume of polymers A and B, and $v_b = \sqrt{v_A v_B}$ is the cell volume for the lattice representing the polymer blend.

(i) Show that the modified Flory–Huggins theory predicts the entropy of mixing

$$\Delta S/k_B = -n_A \ln \phi_A - n_B \ln \phi_B + n_A \ln\left(\frac{1-\eta_b}{1-\eta_A}\right) + n_B \ln\left(\frac{1-\eta_b}{1-\eta_B}\right),$$

where $\eta_{i=A,B} = m_i n_i v_i / V_i$ and $\eta_b = (m_A n_A + m_B n_B) v_b / V_b$ are the packing fractions of the pure polymers and the blend, respectively, $\phi_{i=A,B} = V_i/V_b$ stands for the volume fraction of each polymer in the blend.

(ii) Assume that the internal energy of each pure polymer is linearly proportional to its packing fraction

$$U_{i=A,B} = -\frac{n_i m_i Z \epsilon_{ii}}{2} \eta_i = -V_i \delta_{i,0}^2 \eta_i^2,$$

where $\delta_{i,0} = \sqrt{Z\epsilon_{ii}/2v_i}$ is the solubility parameter, and assume further that the internal energy of the mixture can be approximated by the solubility model

$$U_b/V_b = \phi_A^2 C_{AA} + \phi_B^2 C_{BB} + 2\phi_A \phi_B C_{AB},$$

where $C_{ii} = -U_i/V_i$ represents the cohesive energy density, and $C_{AB} = \sqrt{C_{AA} C_{BB}}$. Show that the energy of mixing satisfies

$$\Delta U/V_b = \phi_A \phi_B (\eta_A \delta_{A,0} - \eta_B \delta_{B,0})^2.$$

Solution:

(i) Note that the free volume is defined as $V_{f,i} = V_i(1-\eta_i)$. The entropy of mixing can be rewritten as

$$\begin{aligned}\Delta S/k_B &= n_A \ln\left(\frac{V_{f,b}}{V_{f,A}}\right) + n_B \ln\left(\frac{V_{f,b}}{V_{f,B}}\right) \\ &= n_A \ln\left(\frac{V_b}{V_A}\right) + n_B \ln\left(\frac{V_b}{V_B}\right) + n_A \ln\left(\frac{1-\eta_b}{1-\eta_A}\right) + n_B \ln\left(\frac{1-\eta_b}{1-\eta_B}\right) \\ &= -n_A \ln \phi_A - n_B \ln \phi_B + n_A \ln\left(\frac{1-\eta_b}{1-\eta_A}\right) + n_B \ln\left(\frac{1-\eta_b}{1-\eta_B}\right),\end{aligned}$$

where $\phi_{i=A,B} = V_i/V_b$ stands for the volume fraction of polymer i.

5 Ruzette A. V. G. and Mayes A. M., "A simple free energy model for weakly interacting polymer blends", *Macromolecules* 34, 1894–1907 (2001).

(ii) The internal energy of mixing per total volume is given by

$$\begin{aligned}\frac{\Delta U}{V_b} &= \frac{U - U_A - U_B}{V_b} \\ &= \phi_A^2 C_{AA} + \phi_B^2 C_{BB} + 2\phi_A \phi_B C_{AB} + \phi_A \delta_{A,0}^2 \eta_A^2 + \phi_B \delta_{B,0}^2 \eta_B^2 \\ &= \phi_A(1-\phi_A)\delta_{A,0}^2\eta_A^2 + \phi_B(1-\phi_B)\delta_{B,0}^2\eta_B^2 + 2\phi_A\phi_B\delta_{A,0}\eta_A\delta_{B,0}\eta_B \\ &= \phi_A\phi_B(\eta_A\delta_{A,0} - \eta_B\delta_{B,0})^2,\end{aligned}$$

where we have used the relations

$$C_{ii} = -U_i/V_i = \delta_{i,0}^2\eta_i^2 \quad \text{and} \quad U_{i=A,B}/V_b = -V_i\delta_{i,0}^2\eta_i^2/V_b = -\phi_i\delta_{i,0}^2\eta_i^2.$$

Problem 8.10 Overbeek and Voorn extended the Flory–Huggins theory to describe complex coacervation in polyelectrolyte solutions.[6] To understand the liquid–liquid phase separation (LLPS) in such systems, consider an aqueous solution of polycations and polyanions of the same chain length m and degree of ionization α. Each polymer chain contains m segments, and each charged segment is monovalent, i.e., it has the valence of $|Z_\pm| = 1$. When the polycation and polyanion have the same concentration, the polyelectrolyte solution may be considered as an *effective* binary system, i.e., a polyelectrolyte complex (viz., "ploysalt") with volume fraction $\phi = \phi_+ + \phi_-$ dissolved in a solvent of volume fraction $1 - \phi$. Overbeek and Voorn assumed that the polyion segments and solvent molecules have the same molecular volume v_0 and that the free energy density due to electrostatic interactions can be represented by the Debye–Hückel theory (see Section 9.4)

$$\frac{f_{DH}}{k_B T} = -\frac{\kappa^3}{12\pi}.$$

The Debye screening parameter is defined as

$$\kappa = \sqrt{4\pi l_B(\alpha\rho_+ Z_+^2 + \alpha\rho_- Z_-^2)} = \sqrt{4\pi l_B \alpha\phi/v_0},$$

where $l_B \equiv e^2/(4\pi\epsilon\epsilon_0 k_B T)$ stands for the Bjerrum length, α is the degree of ionization (i.e., the fraction of charged polymer segments), $\rho_+ = \rho_-$ are the number densities of polymer segments. Because the van der Waals interactions are much weaker than electrostatic interactions, Overbeek and Voorn neglected the internal energy term related to the Flory parameter χ_F.

(i) Show that the Overbeek–Voorn theory predicts the free energy of mixing per lattice site

$$\frac{\Delta f v_0}{k_B T} = \frac{\phi}{m}\ln(\phi/2) + (1-\phi)\ln(1-\phi) - \lambda(\alpha\phi)^{3/2},$$

where $\lambda \equiv 2\sqrt{\pi/v_0}\, l_B^{3/2}/3$, and v_0 is the molecular volume of the solvent.[7]

(ii) Show that the chemical potentials of the polymer complex and solvent in the effective binary mixture are given by

$$\beta\mu_p = \beta\mu_p^0 + \ln(\phi/2) + (1-m)(1-\phi) - \frac{\lambda\alpha^{3/2} m\sqrt{\phi}}{2}(3-\phi),$$

$$\beta\mu_s = \beta\mu_s^0 + \ln(1-\phi) + (1-1/m)\phi + \frac{\lambda\alpha^{3/2}\phi^{3/2}}{2},$$

where subscript 0 denotes the reference system of pure species.

---

6 Overbeek J. T. G. and Voorn M. J., "Phase separation in polyelectrolyte solutions. Theory of complex coacervation", *J. Cell. Comp. Physiol.* 49, 7–26 (1957).

7 For water at room temperature ($T = 298.15$ K), $l_B \approx 7.14$ Å, the molecular volume for water is $v_0 \approx 10\,\text{Å}^3$, and thus $\lambda \approx 7.13$.

(iii) At the critical point of LLPS, the degree of ionization α_c and the polymer volume fraction ϕ_c can be approximated by the following scaling relations

$$\phi_c \approx 1/(m+3) \quad \text{and} \quad \alpha_c \approx 0.52/m^{1/3}.$$

(iv) Plot the spinodal line of the system for a few representative values of the polymer chain length m and discuss the phase behavior.

(v) Show that, at $m = \infty$, the coacervation boundary can be determined from

$$\alpha = \frac{1}{\phi}\left[\frac{-2\ln(1-\phi) - 2\phi}{\lambda}\right]^{3/2}.$$

For small ϕ, the above equation can be approximated by $\alpha \approx 0.27\phi^{1/3}$.

(vi) Plot the phase diagram for the LLPS predicted by the Overbeek–Voorn theory for polymer chain length $m = 2, 5, 10, 1000, 3000$, and ∞.

Solution:

This problem is adapted from Voorn.[8]

(i) The free energy of mixing includes an entropic contribution the same as that predicted by the Flory–Huggins theory and an electrostatic term given by the Debye–Hückel theory:

$$\frac{\Delta f v_0}{k_B T} = \frac{\phi_+}{m}\ln(\phi_+) + \frac{\phi_-}{m}\ln(\phi_-) + (1-\phi_+-\phi_-)\ln(1-\phi_+-\phi_-)$$
$$-\frac{v_0}{12\pi}\left(\frac{4\pi l_B \alpha\phi}{v_0}\right)^{3/2} = \frac{\phi}{m}\ln(\phi/2) + (1-\phi)\ln(1-\phi) - \lambda(\alpha\phi)^{3/2},$$

where $\lambda \equiv 2\sqrt{\pi/v_0}\, l_B^{3/2}/3$.

(ii) The chemical potential of the polymer complex is given by

$$\beta\mu_p = \beta\mu_p^0 + \left(\frac{\partial \Delta\beta F}{\partial n_p}\right)_{T,n_0} = \beta\mu_p^0 + \ln(\phi/2) + (1-m)(1-\phi) - \frac{\lambda\alpha^{3/2} m\sqrt{\phi}}{2}(3-\phi),$$

where μ_p^0 denotes the reference chemical potential, $n_p = n_+ + n_-$ represents the number of polymer chains, and n_0 is the number of solvent molecules. Similarly, the solvent chemical potential is

$$\beta\mu_s = \beta\mu_s^0 + \left(\frac{\partial \Delta\beta F}{\partial n_s}\right)_{T,n_0} = \beta\mu_s^0 + \ln(1-\phi) + (1-1/m)\phi + \frac{\lambda\alpha^{3/2}\phi^{3/2}}{2}.$$

(iii) At the critical point of LLPS, the second and third-order derivatives of the free energy vanish, i.e.,

$$\frac{\partial^2 \Delta\beta f v_0}{\partial\phi^2} = 0 = \frac{1}{1-\phi} + \frac{1}{m\phi} - \frac{3\alpha^{3/2}\lambda}{4\phi^{1/2}},$$

$$\frac{\partial^3 \Delta\beta f v_0}{\partial\phi^3} = 0 = \frac{1}{(1-\phi)^2} - \frac{1}{m\phi^2} + \frac{3\alpha^{3/2}\lambda}{8\phi^{3/2}}.$$

Solving the above equations gives the critical degree of ionization α_c and the critical polymer volume fraction ϕ_c

$$\phi_c = \frac{-m-2+\sqrt{m(m+8)}}{2(m-1)} = \frac{1}{m}(1-\frac{3}{m}+\cdots) \approx \frac{1}{m+3},$$

$$\frac{3\alpha_c^{3/2}\lambda}{4} = \phi_c^{1/2}\left[\frac{1}{1-\phi_c} + \frac{1}{m\phi_c}\right] \approx 2/m^{1/2} \rightarrow \alpha_c \approx 0.52/m^{1/3}.$$

8 Voorn M. J., "Phase separation in polymer solutions", *Fortschr. Hochpolym.-Forsch., Bd.* 1. S. 192–233 (1959).

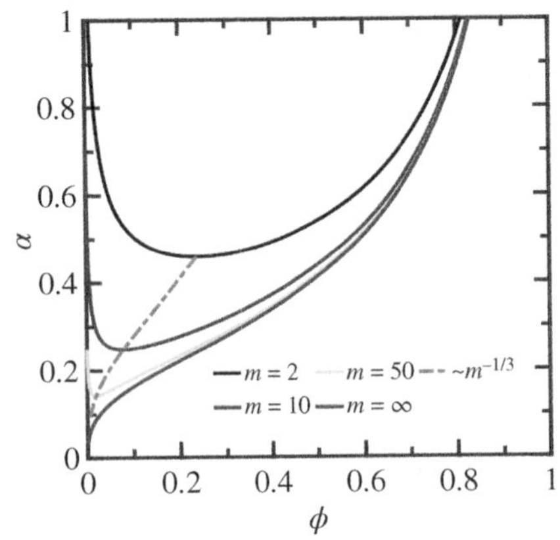

Figure 8.2 The spinodal line of complex coacervation in symmetric polyelectrolyte solutions. The dash-dotted line represents the locus of the critical condition corresponding to different polymer chain length (OVTheory.m).

(iv) Figure 8.2 shows the spinodal line of the system for several values of the polymer chain length m.
Qualitatively, the phase behavior resembles that predicted by the Flory–Huggins theory except that the Flory parameter is now replaced by the degree of ionization. The polymer density at the coacervate phase depends on the polymer chain length and charge density.
The Overbeek–Voorn theory predicts that coacervation takes place at lower polymer concentration as the chain length increases. For a typical polymer with $m = 1000$, $\phi_c \approx 0.52m^{-1/3}$ suggests that the concentration of the coacervate phase is extremely small (~ 0.001).
Although the modified Flory–Huggins theory captures the essential features of polyelectrolyte coacervation, its application to short chains is problematic because, at ambient conditions, they do not exhibit electrostatic-driven LLPS.

(v) At $m = \infty$, the polymer concentration in the dilute phase vanishes. Therefore, the equilibrium coacervation concentration is determined by $\mu_s = \mu_0$, which leads to, according to part (ii),

$$\ln(1-\phi) + \phi + \frac{\lambda\alpha^{3/2}\phi^{3/2}}{2} = 0.$$

In terms of the degree of ionization, rearranging the above equation gives

$$\alpha = \frac{1}{\phi}\left[\frac{-2\ln(1-\phi) - 2\phi}{\lambda}\right]^{3/2} = 0.$$

For small ϕ, we have $\alpha \approx \phi^{1/3}/\lambda^{2/3} \approx 0.27\phi^{1/3}$.

(vi) Figure 8.3 shows the LLPS phase diagrams of the polyelectrolyte systems with the polymer chain length $m = 2, 5, 10$ (A) and $m = 1000, 3000, \infty$ (B). The numerical results were obtained from the conventional thermodynamic equations of phase equilibrium.

Problem 8.11 Consider a polyelectrolyte solution with n_i polyionic chains of species i each with $m_i = v_i/v_0$ segments where v_i is the volume of the polyion and v_0 is the molecular volume of the solvent.

(i) Using the Flory–Huggins theory for the entropy of mixing and the Debye–Hückel theory for the electrostatic free energy, show that the Overbeek–Voorn theory (Problem 8.10) predicts the free energy of mixing per lattice site

$$\frac{\Delta f v_0}{k_B T} = \sum_i \frac{\phi_i}{m_i}\ln(\phi_i) + \left(1 - \sum_i \phi_i\right)\ln\left(1 - \sum_i \phi_i\right) - \lambda\left(\sum_i \alpha_i\phi_i\right)^{3/2},$$

where $\lambda \equiv 2\sqrt{\pi/v_0}l_B^{3/2}/3$, and α_i is the degree of ionization of polymer i.

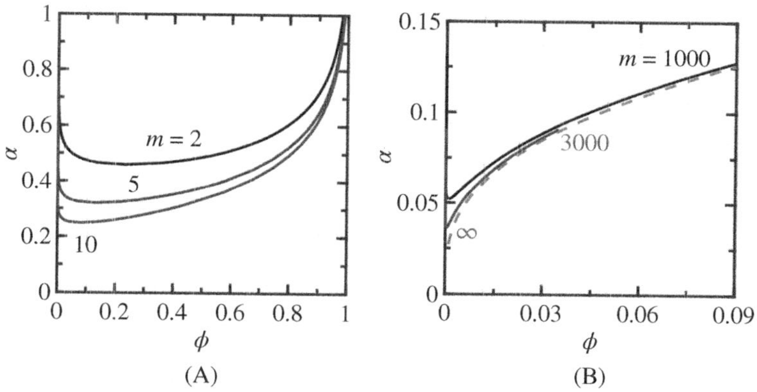

Figure 8.3 The LLPS phase diagrams for short- (A) and long-chain (B) symmetric polyelectrolyte solutions. The dashed line represents the equilibrium coacervation concentration at $m = \infty$ (OVTheory.m).

(ii) Derive analytical expressions for the chemical potential of each ionic species and the chemical potential of the solvent.

(iii) What are conditions of equilibrium for the LLPS in this system?

Solution:

(i) The free energy of mixing includes an entropic contribution that is the same as that predicted by the Flory–Huggins theory, and an electrostatic term given by the Debye–Hückel theory:

$$\frac{\Delta f v_0}{k_B T} = -\frac{\Delta S}{n_T k_B} - \frac{\kappa^3 v_0}{12\pi}.$$

According to the Flory–Huggins theory, the entropy of mixing per site is

$$\frac{\Delta S}{n_T k_B} = -\sum_i \frac{\phi_i}{m_i} \ln(\phi_i) - \left(1 - \sum_i \phi_i\right) \ln\left(1 - \sum_i \phi_i\right).$$

The Debye screening parameter is defined by

$$\kappa^2 = 4\pi l_B \sum_i \alpha_i \rho_i Z_i^2 = \frac{4\pi l_B}{v_0} \sum_i \alpha_i \phi_i,$$

where $l_B \equiv e^2/(4\pi\epsilon\epsilon_0 k_B T)$ is the Bjerrum length, $\rho_i = n_i m_i / v_0 n_T = \phi_i / v_0$ and α_i are the segment density and the degree of ionization for polyion i, respectively, and $|Z_i| = 1$ is the valence of each ionized segment.

Adding the two contributions together yields

$$\frac{\Delta f v_0}{k_B T} = \sum_i \frac{\phi_i}{m_i} \ln \phi_i + \left(1 - \sum_i \phi_i\right) \ln\left(1 - \sum_i \phi_i\right) - \lambda\left(\sum_i \alpha_i \phi_i\right)^{3/2},$$

where $\lambda \equiv 2\sqrt{\pi/v_0} l_B^{3/2}/3$.

(ii) The chemical potential of each ionic species is given by

$$\beta\mu_i = \beta\mu_i^0 + \left(\frac{\partial \Delta\beta F}{\partial n_i}\right)_{T,n_0,n_{j\neq i}}$$

$$= \beta\mu_i^0 + \ln\phi_i + 1 - m_i\sum_j \frac{\phi_j}{m_j} - m_i\left(1 - \sum_j \phi_j\right)$$

$$- \frac{\lambda m_i}{2}\left[3\alpha_i - \sum_j \alpha_j\phi_j\right]\left(\sum_j \alpha_j\phi_j\right)^{1/2},$$

where μ_p^0 denotes the reference chemical potential, $n_p = n_+ + n_-$ represents the number of polymer chains, and n_0 is the number of solvent molecules. Similarly, the solvent chemical potential is

$$\beta\mu_s = \beta\mu_s^0 + \left(\frac{\partial\Delta\beta F}{\partial n_s}\right)_{T,n_0}$$

$$= \beta\mu_s^0 + \ln\left(1 - \sum_i \phi_i\right) + \sum_j \phi_j(1 - 1/m_j) + \frac{\lambda}{2}\left(\sum_j \alpha_j\phi_j\right)^{3/2}.$$

(iii) The conditions of equilibrium are determined by

$$\mu_s^p = \mu_s^s \quad \text{and} \quad \mu_i^p + Z_i m_i e\psi^p = \mu_i^s + Z_i m_i e\psi^s,$$

where superscripts "p" and "s" denote polymer-rich phase and solvent-rich phase, respectively; e stands for the unit charge, and ψ represents the electric potential (viz., the Donnan potential). In comparison to conventional phase-equilibrium calculations for neutral systems, the extra variables ψ^p and ψ^s above can be solved from the electrical neutrality conditions

$$\sum_i Z_i\phi_i^s = 0 \quad \text{and} \quad \sum_i Z_i\phi_i^p = 0.$$

Problem 8.12 The Flory–Huggins theory is useful to predict the solubility of crystalline drugs in a polymer matrix. Such calculations are often carried out for the polymer materials design to improve bioavailability.

(i) Show that, for a given mixture, the Flory parameter can be estimated from experimental data for the melting point depression (MPD) of the drug dissolved in the polymer. The experimental data can be measured by using differential scanning calorimetry.
(ii) Sketch how the drug solubility varies with temperature.
(iii) Assume that the glass transition temperature of the binary mixture varies linearly with the polymer composition, sketch a temperature-solubility "phase diagram" that may be used as a guide for the formulation of drug delivery systems.

Solution:
This problem is adapted from Altamimi and Neau.[9]

(i) At the melting point, the chemical potential of the drug in the solid phase (S) is the same as that in the polymer melt (P)

$$\mu_1^S = \mu_1^P,$$

where subscript 1 stands for the drug compound. The Flory–Huggins theory predicts that the drug chemical potential in the polymer melt is

$$\beta\mu_1^P = \mu_1^0 + \ln\phi_1 + (1 - 1/m)\phi_2 + \phi_2^2\chi_F,$$

where ϕ_2 is the polymer volume fraction.

9 Altamimi M. A. and Neau S. H., "Use of the Flory–Huggins theory to predict the solubility of nifedipine and sulfamethoxazole in the triblock, graft copolymer Soluplus", *Drug Dev. Ind. Pharm.* 42(3), 446–455 (2016).

Meanwhile, the drug chemical potential in the liquid reference state is related to that in the solid phase through a thermodynamic cycle linking these two states to the melting point of the pure drug (viz., normal melting temperature T_m)

$$\beta\mu_1^0 = \beta\mu_1^S + \frac{\Delta H_{fus}}{k_B T_m}(T_m/T - 1) + \frac{\Delta C_p}{k_B}\left[\ln(T_m/T) - T_m/T + 1\right],$$

where ΔH_{fus} is the heat of fusion per drug molecule, and $\Delta C_p = C_p^0 - C_p^S$ is the difference between the per molecule heat capacities of the drug in liquid and solid phases.

Assuming $\ln(T_m/T) - T_m/T + 1 \approx 0$, we can predict the MPD from

$$-\frac{\Delta H_{fus}}{k_B T_m}(T_m/T - 1) \approx \beta\mu_1^S - \beta\mu_1^0 = \ln\phi_1 + (1 - 1/m)\phi_2 + \phi_2^2\chi_F. \quad \text{(C)}$$

(ii) Eq. (C) predicts the dependence of the drug freezing temperature as a function of the composition

$$\ln\phi_1 = -\frac{\Delta H_{fus}}{k_B T_m}(T_m/T - 1) - (1 - 1/m)\phi_2 - \phi_2^2\chi_F.$$

On the other hand, the drug solubility in the polymer melt can be predicted by the Flory–Huggins theory for the liquid–liquid equilibrium (LLE), i.e., the coexistence of two liquid states, one rich in polymer and the other rich in drug molecules (see Figure 8.4).

(iii) Schematically, Figure 8.4 shows a temperature-solubility "phase diagram." The plot may be used as a guide for the formulation of drug delivery system. Because the drug solubility in the polymer at the glass state is negligible, the drug formulation is confined in the region beyond the glass transition temperature and the LLE coexistence curve.

Problem 8.13 A modified Flory–Huggins theory was utilized to predict the association of telechelic polymers (viz., polymers with associating ends) in solution and help the development of polymer-based mist suppressors for jet fuels.[10] The modification follows an earlier theoretical work proposed by Goldstein to account for the association among polymer chains.[11] The mixing free energy per lattice site consists of contributions from the solvent molecules, polymer monomers, as well as all possible multi-chain aggregates

$$\frac{\Delta f v_0}{k_B T} = (1-\phi)\ln(1-\phi) + \sum_i \frac{\phi_i}{m_i}\ln\phi_i + \frac{\beta Z}{2}\left[(1-\phi)^2 f_{ss} + \phi^2 f_{pp} + 2\phi(1-\phi)f_{ps}\right],$$

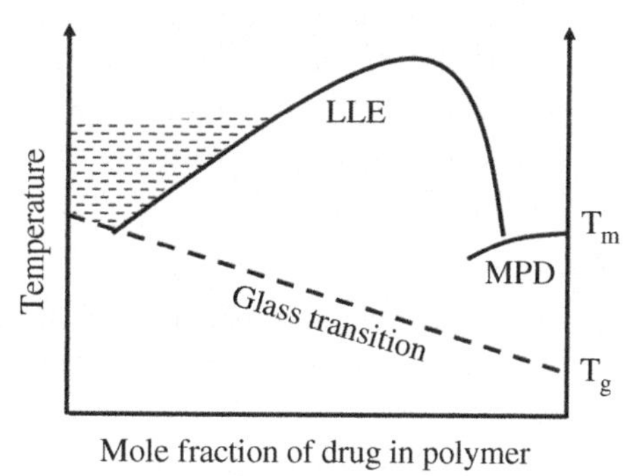

Figure 8.4 Phase diagram of drug dispersion in a polymer melt. The glass-transition temperature (dashed line) is assumed linearly dependent on polymer composition. The liquid–liquid equilibrium (LLE) and melting point depression (MPD) curves can be correlated with the Flory–Huggins theory, and the shaded area is suitable for drug formulation.

10 Wei M. H. et al., "Megasupramolecules for safer, cleaner fuel by end association of long telechelic polymers", *Science* 350, 72 (2015).

11 Goldstein R. E., "Model for phase equilibria in micellar solutions of nonionic surfactants", *J. Chem. Phys.* 84, 3367 (1986).

where ϕ_i represents the volume fraction of polymer aggregate with $i = 1, 2, \ldots$ chains, $m_i = i \times m$ is the number of lattice sites occupied by each polymer aggregate, $\phi = \sum_i \phi_i$ is the total polymer volume fraction, Z is the coordination number of the lattice, and f_{ij} are energies associated with the interactions of various species in the system.

(i) Show that the chemical potentials of the solvent and i-chain aggregates are, respectively,

$$\mu_s = \mu_s^0 + k_B T \left[\ln(1-\phi) + \phi - \sum_i \frac{\phi_i}{m_i} \right] - w\phi^2,$$

$$\mu_i = \mu_i^0 + k_B T \left[\ln \phi_i - (1-\phi)m_i + 1 - m_i \sum_j \frac{\phi_j}{m_j} \right] - m_i w (1-\phi)^2,$$

where superscript 0 denotes a reference system, $w \equiv w_{pp} + w_{ss} - 2w_{ps}$ with $w_{ij} = Z f_{ij}/2$.

(ii) The size distribution of multi-chain structures are determined by the condition of chemical equilibrium $\mu_i = i\mu_1$. Show that the mass-action relation leads to

$$\phi_i = \phi_1^i e^{-\beta \Delta G_i^0},$$

where $\Delta G_i^0 \equiv \mu_i^0 - i\mu_1^0 - (i-1)k_B T$ corresponds to the free energy of association.

(iii) The free energy of association proposed by Goldstein includes contributions due to the polymer-polymer association, the contact energy between the polymer aggregates and the solvent, and the change in polymer configuration due to the formation of aggregates:

$$\beta \Delta G_i^0 = -\ln(im) + i \ln m + i \left[m\epsilon + \sigma(m^2/i)^{1/3} + c(i^2/m)^{1/3} - 3/2 \right],$$

where terms $\ln(im)$ and $\ln m$ arise from the change of the concentration units from volume fraction to mole fraction, ϵ represents the reduced association energy, σ and c are model parameters related to the aggregate-solvent surface free energy and the geometry of polymer packing, respectively. Show that the concentration of aggregates with i polymer chains satisfies

$$\phi_i = im \left(\frac{\phi_1 e^{-\delta_0}}{m} \right)^i,$$

where $\delta_0 = m\epsilon + \sigma(m^2/i)^{1/3} + c(i^2/m)^{1/3} - 3/2$.

(iv) Using parameters $\sigma \approx 3.0$ for the reduced surface free energy and $c \approx 1.0$ for the geometric factor, plot the size distribution in terms of ϕ_i for $m = 20$ and $\phi_1 = 0.004$ at several representative values of the reduced association energy ($\epsilon \sim -1.0$). How does the most probably aggregate size change with the reduced association energy ϵ and with the polymer chain length m?

Solution:

(i) The mean-field approximation assumes that the energy of mixing is unaffected by polymer aggregation, i.e., it is same as that given by the original Flory–Huggins theory

$$\begin{aligned}
\Delta U &= \frac{Z}{2n_T} \left[n_s^2 f_{ss} + (mn_p)^2 f_{pp} + 2n_s m n_p f_{ps} \right] - \frac{Zn_s}{2} f_{ss} - \frac{Zmn_p}{2} f_{pp} \\
&= \frac{Z}{2} \left[n_s(\phi_s - 1) f_{ss} + mn_p(\phi_p - 1) f_{pp} + 2n_T \phi_p \phi_s f_{ps} \right] \\
&= -\frac{n_s m n_p}{n_T} \left[w_{ss} + w_{pp} - 2w_{ps} \right] = -\frac{n_s m n_p w}{n_T} = k_B T n_T \phi_p \phi_s \chi_F,
\end{aligned}$$

where n_p is the total number of polymer chains, $n_T = n_s + n_p m$, $w_{ij} = Zf_{ij}/2$ and $w \equiv w_{ss} + w_{pp} - 2w_{ps} = -k_B T \chi_F$. The entropy of mixing corresponds to that for a multicomponent polymer mixture. Thus, the solvent chemical potential is

$$\begin{aligned}
\mu_s &= \left(\frac{\partial F}{\partial n_s}\right)_{T,n_i} \\
&= \mu_s^0 + k_B T \frac{\partial}{\partial n_s}\left\{ n_s \ln(1-\phi) + \sum_i n_i \ln \phi_i - \frac{\beta n_s m n_p}{n_T} w \right\} \\
&= \mu_s^0 + k_B T \left\{ \ln(1-\phi) + \frac{n_s}{\phi_s}\frac{\partial \phi_s}{\partial n_s} + \sum_i \frac{n_i}{\phi_i}\frac{\partial \phi_i}{\partial ns} \right\} - \frac{m n_p}{n_T} w + \frac{n_s m n_p}{n_T^2} w \\
&= \mu_s^0 + k_B T \left[\ln(1-\phi) + \phi - \sum_i \frac{\phi_i}{m_i} \right] - w\phi^2,
\end{aligned}$$

where μ_s^0 is the solvent chemical potential at the reference state (viz., the pure solvent at system temperature and pressure), n_i is the number of aggregates that each consists of m_i polymer chains, and

$$\frac{n_s}{\phi_s} = n_T, \quad \frac{\partial \phi_s}{\partial n_s} = \frac{1}{n_T} - \frac{n_s}{n_T^2} = \frac{\phi}{n_T}, \quad \frac{n_i}{\phi_i} = \frac{n_T}{m_i}, \quad \text{and} \quad \frac{\partial \phi_i}{\partial n_s} = \frac{n_i m_i}{n_T^2} = \frac{\phi_i}{n_T}.$$

Similarly, the chemical potential of i-chain aggregates is

$$\begin{aligned}
\mu_i &= \left(\frac{\partial F}{\partial n_i}\right)_{T,n_s,n_{j\neq i}} = \mu_i^0 + k_B T \frac{\partial}{\partial n_i}\left\{ n_s \ln(1-\phi) + \sum_j n_j \ln \phi_j - \frac{\beta n_s m n_p}{n_T} w \right\} \\
&= \mu_i^0 + k_B T \left\{ \frac{n_s}{\phi_s}\frac{\partial \phi_s}{\partial n_i} + \ln \phi_i + \sum_j \frac{n_j}{\phi_j}\frac{\partial \phi_j}{\partial n_i} \right\} - \frac{n_s m}{n_T} w \left[\frac{\partial n_p}{\partial n_i} - \frac{n_p m_i}{n_T} \right] \\
&= \mu_i^0 + k_B T \left\{ \frac{n_s}{\phi_s}\frac{\partial \phi_s}{\partial n_i} + \ln \phi_i + \sum_j \frac{n_j}{\phi_j}\frac{\partial \phi_j}{\partial n_i} \right\} - m w \phi_s (i - \phi i) \\
&= \mu_i^0 + k_B T \left[\ln \phi_i - (1-\phi) m_i + 1 - m_i \sum_j \frac{\phi_j}{m_j} \right] - m_i w (1-\phi)^2,
\end{aligned}$$

where μ_i^0 is the chemical potential of the pure i-chain aggregates at system temperature and pressure. In deriving the above equation, we have used $n_p = \sum_i i n_i$, $n_T = n_s + \sum_i n_i m_i$, $m_i = im$, and the following relations:

$$\frac{n_s}{\phi_s} = n_T, \quad \frac{\partial \phi_s}{\partial n_i} = -\frac{n_s m_i}{n_T^2} = -\frac{(1-\phi) m_i}{n_T}, \quad \frac{n_j}{\phi_j} = \frac{n_T}{m_j},$$

$$n_i \frac{\partial \ln \phi_i}{\partial n_i} = \frac{n_i}{\phi_i}\left[\frac{m_i}{n_T} - \frac{m_i n_i}{n_T^2} m_i \right] = \frac{n_T m_i}{m_i}\left[\frac{1}{n_T} - \frac{m_i n_i}{n_T^2} \right] = 1 - \phi_i,$$

$$\frac{\partial \phi_{j\neq i}}{\partial n_i} = -\frac{n_j m_j}{n_T^2}\frac{\partial n_T}{\partial n_i} = -\frac{\phi_j m_i}{n_T} \quad \text{and} \quad n_j \frac{\partial \ln \phi_{j\neq i}}{\partial n_i} = \frac{n_j}{\phi_j}\frac{\partial \phi_{j\neq i}}{\partial n_i} = -m_i \phi_j / m_j.$$

(ii) From $\mu_i = i\mu_1$, we have

$$\begin{aligned}
&\mu_i^0 + k_B T \left[\ln \phi_i - (1-\phi) m_i + 1 - m_i \sum_{j=1}^{\infty} \frac{\phi_j}{m_j} \right] - m_i w (1-\phi)^2 \\
&= i\mu_1^0 + i k_B T \left[\ln \phi_1 - (1-\phi) m + 1 - m \sum_{j=1}^{\infty} \frac{\phi_j}{m_j} \right] - i m w (1-\phi)^2.
\end{aligned}$$

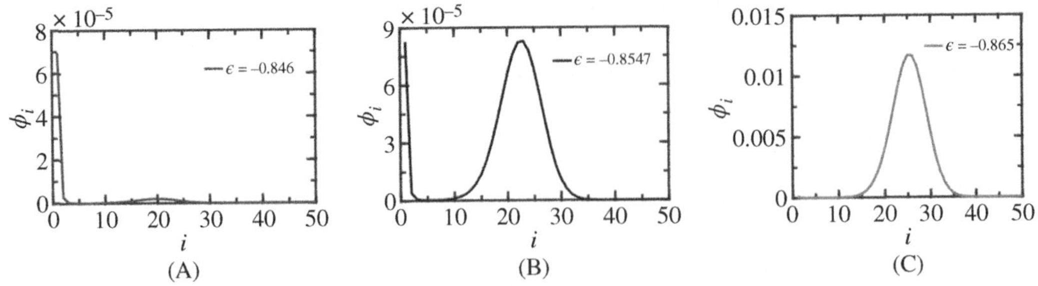

Figure 8.5 The size distribution of polymer aggregates for three values of the reduced association energy ϵ. (A) The polymer exists as individual chains when the association energy is sufficiently weak. (B) Aggregates and single chains have the same concentration at the critical association energy. (C) Polymer chains mainly exist as aggregates beyond the critical association energy (FHmicelles.m). Source: Adapted from Wu et al.[94]

Plugging $m_i = im$ into the above equation leads to

$$\ln \phi_i - i \ln \phi_1 = \beta(i\mu_1^0 - \mu_i^0) + i - 1.$$

Therefore, we obtain

$$\phi_i = \phi_1^i e^{-\beta \Delta G_i^0},$$

where $\Delta G_i^0 \equiv \mu_i^0 - i\mu_1^0 - (i-1)k_B T$ corresponds to the free energy of association.

(iii) Plugging

$$\beta \Delta G_i^0 = -\ln(im) + i \ln m + i\left[m\epsilon + \sigma(m^2/i)^{1/3} + c(i^2/m)^{1/3} - 3/2\right]$$

into the mass-action relation leads to

$$\phi_i = \phi_1^i \frac{im}{m^i} e^{-i\delta_0} = im\left(\frac{\phi_1 e^{-\delta_0}}{m}\right)^i,$$

where $\delta_0 = m\epsilon + \sigma(m^2/i)^{1/3} + c(i^2/m)^{1/3} - 3/2$.

(iv) Figure 8.5 illustrates the size distribution of polymer aggregates at three representative values of the reduced association energy. The most probable aggregates become slightly larger when the association energy increases. Approximately, the aggregate size linearly increases with the chain length (not shown).

Problem 8.14 The conventional thermodynamic description of polymer gels is mostly based on the Flory–Rehner model, which is an extension of the Flory–Huggins theory by incorporating chain elasticity and crosslink distribution (Figure 8.6). To get the essential ideas, consider a polymer gel made of N chains each with m segments. The polymer gel is surrounded by a bulk solvent with fixed temperature T and bulk chemical potential μ_s. The polymer and solvent inside the gel constitute a

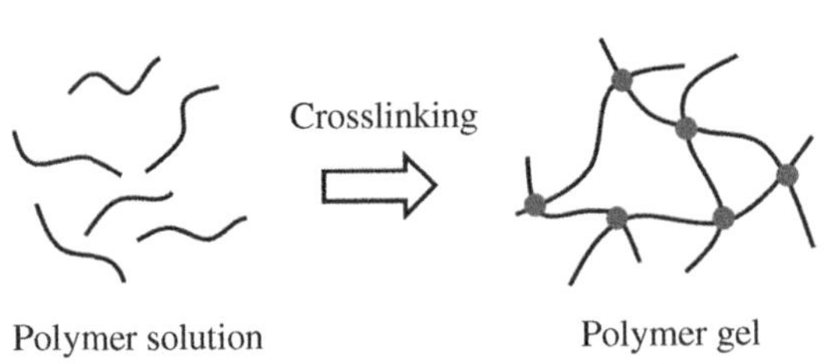

Figure 8.6 A polymer gel can be modeled as a network of N polymer chains permanently joined by $N/2$ crosslinks (black dots).

semi-grand canonical ensemble with the free energy depending on temperature and solvent chemical potential, i.e., $\mathcal{F} = \mathcal{F}(N, T, \mu_s)$.

(i) If the polymer gel could be treated as a single chain (viz., a super-chain with mN segments), show that the free energy of mixing would become

$$\Delta\beta F_0 = N_s \ln \phi_s + \chi_F(N_s + mN)\phi_s\phi,$$

where N_s is the number of solvent molecules inside the gel, $\phi_s = N_s/(N_s + mN)$ and $\phi = mN/(N_s + mN)$ are the solvent and polymer volume fractions, respectively. The Flory parameter χ_F accounts for the mean-field energy due to the polymer-solvent interactions.

(ii) Assuming that the polymer gel consists of a random network of N Gaussian chains, show that the change in the elastic energy in response to the gel swelling from a fully relaxed state with volume V_0 (which is achieved at the preparation condition) to volume V in the solvent can be written as

$$\Delta\beta F_{elas} = \frac{N}{2}(3\alpha^{2/3} - 3 - \ln \alpha),$$

where $\alpha = V/V_0$ denotes the swelling ratio.

(iii) As the boundary of the polymer network may serve as a semi-permeable membrane, validate the following expression for the osmotic pressure inside the gel based on the osmotic equilibrium

$$\Pi(T,\phi) = \frac{k_B T}{v_s}\left\{\frac{\phi_0}{m}\left[\frac{\phi}{2\phi_0} - \left(\frac{\phi}{\phi_0}\right)^{1/3}\right] - \phi - \ln(1-\phi) - \chi_F\phi^2\right\},$$

where v_s represents the molecular volume of the solvent, and $\phi_0 = \alpha\phi$ is the polymer volume fraction in the gel at the preparation condition.

Solution:

(i) For the binary mixture of one "super-chain" (viz., polymer network) in the solvent, the Flory–Huggins theory, predicts that the free energy of mixing is

$$\Delta\beta F_0 = N_s \ln \phi_s + \ln \phi + \chi_F(N_s + mN)\phi_s\phi.$$

Because N_s is macroscopic yet $0 < \phi < 1$, $\ln \phi$ is negligible in comparison with other terms in the above equation. Hence, we have

$$\Delta\beta F_0 = N_s \ln \phi_s + \chi_F(N_s + mN)\phi_s\phi.$$

(ii) The gel swelling from volume V_0 to V leads to changes in the entropy of the crosslink sites and the elastic free energy of individual chains. Because we assume that the crosslinks are randomly distributed in the space within the polymer network, the change in the crosslink entropy is the same as that predicted by the ideal-gas model

$$\Delta S_{crosslinks}/k_B = (N/2)\ln(V/V_0),$$

where $N/2$ is the number of the crosslink sites in the gel.

Meanwhile, the expansion of a Gaussian chain from length R_0 at the fully relaxed state to length R at the swollen state leads to a change in the elastic free energy of $\Delta\beta f_{chain} = 3[(R/R_0)^2 - 1]/2$ (see Eq. 3.166). Hence the change in the free energy of N Gaussian chains due to the gel swelling is

$$\Delta\beta F_{chains} = 3N[(R/R_0)^2 - 1]/2.$$

Because $V/V_0 = \alpha$ and $R/R_0 = \alpha^{1/3}$, a combination of these two contributions gives

$$\Delta\beta F_{elas} = \Delta\beta F_{chains} - \Delta S_{crosslinks}/k_B = \frac{N}{2}(3\alpha^{2/3} - 3 - \ln \alpha).$$

(iii) Osmotic equilibrium requires that the solvent chemical potential inside the gel is the same as that of the pure solvent in the bulk, i.e.,

$$\mu_s^{gel}(T, P + \Pi) = \mu_s^{pure}(T, P),$$

where P is the pressure of the pure solvent defined by T and μ_s. From parts (i) and (ii), we can obtain the free energy of the swollen gel

$$\mathcal{F} = F_{gel} + F_{pure} + \Delta F_0 + \Delta F_{elas},$$

where F_{gel} stands for the free energy without the solvent, and F_{pure} is the free energy of the pure solvent at system temperature and pressure.
Thus, the solvent chemical potential inside the gel can be derived from

$$\mu_s^{gel}(T, \Pi + P) = \left(\frac{\partial \mathcal{F}}{\partial N_s}\right)_{T,N} = \mu_s^{pure}(T, P + \Pi) + \left[\frac{\partial(\Delta F_0 + \Delta F_{elas})}{\partial N_s}\right]_{T,N},$$

where $\mu_s(T, P + \Pi) = \partial F_{sol}/\partial N_s$ is the chemical potential of the bulk solvent. Using the thermodynamic relation $\mu_s^{pure}(T, \Pi + P) - \mu_s^{pure}(T, P) = \Pi v_s$ and the condition of osmotic equilibrium mentioned above, we have

$$\Pi = -\frac{1}{v_s}\left[\frac{\partial(\Delta F_0 + \Delta F_{elas})}{\partial N_s}\right]_{T,N}. \quad \text{(D)}$$

To evaluate the partial derivative in Eq. (D), let

$$\begin{aligned}
\mathfrak{f} &\equiv \frac{\Delta\beta F_0 + \Delta\beta F_{elas}}{V} \\
&= \frac{1}{(N_s + mN)v_s}\left[N_s \ln \phi_s + \chi_F(N_s + mN)\phi_s\phi + \frac{N}{2}(3\alpha^{2/3} - 3 - \ln \alpha)\right] \\
&= \frac{1}{v_s}\left\{(1 - \phi)\ln(1 - \phi) + \chi_F(1 - \phi)\phi + \frac{\phi}{2m}\left[3\left(\frac{\phi_0}{\phi}\right)^{2/3} - 3 - \ln\left(\frac{\phi_0}{\phi}\right)\right]\right\},
\end{aligned}$$

and rewrite the osmotic pressure in terms of the partial derivative with respect to the polymer volume fraction

$$\begin{aligned}
\beta\Pi &= -\frac{1}{v_s}\left[\frac{\partial(N_s + mN)v_s\mathfrak{f}}{\partial N_s}\right]_{T,N} = -\left[\frac{\partial(N_s + mN)\mathfrak{f}}{\partial\phi^{-1}}\frac{\partial\phi^{-1}}{\partial N_s}\right]_{T,N} \\
&= -\left[\frac{\partial(N_s + mN)\mathfrak{f}}{\partial\phi^{-1}}\frac{1}{mN}\right]_{T,N} = -\left[\frac{\partial\phi^{-1}\mathfrak{f}}{\partial\phi^{-1}}\right]_{T,N}.
\end{aligned}$$

Carrying out the partial derivatives leads to

$$\beta\Pi(T, \phi) = \frac{1}{v_s}\left\{\frac{\phi_0}{m}\left[\frac{\phi}{2\phi_0} - \left(\frac{\phi}{\phi_0}\right)^{1/3}\right] - \phi - \ln(1 - \phi) - \chi_F\phi^2\right\}.$$

Problem 8.15 Poly(*N*-isopropylacrylamide) (PNIPAM) microgel particles may undergo a drastic volume change in liquid water when the temperature varies near 33 °C, the lower critical solution temperature (LCST) of the aqueous solutions of PNIPAM. As shown in Figure 8.7, the variation in the microgel radius can be quantitatively described by a modified Flory–Rehner model

$$\beta\Pi(T, \phi) = \frac{1}{v_s}\left\{\frac{\phi_0}{m}\left[\left(\frac{\phi}{\phi_0}\right)^{1/3} - \left(\frac{\phi}{\phi_0}\right)^{5/3} + \frac{\phi}{2\phi_0}\right] - \phi - \ln(1 - \phi) - \chi_F\phi^2\right\},$$

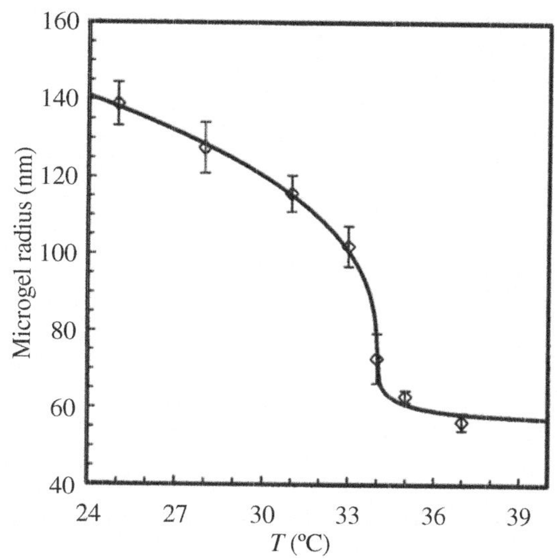

Figure 8.7 Radii of PNIPAM particles in liquid water versus temperature. The symbols are experimental data with the error bars reflecting the difference between dynamic and static light scattering measurements. The line is the prediction of a modified Flory–Rehner model.

where the Flory parameter is empirically given as a function of absolute temperature T (in Kelvin) and polymer volume fraction ϕ by fitting the phase diagram of PNIPAM aqueous solutions

$$\chi_F = \frac{3}{1 - 0.65\phi}\left\{2\ln\left[\frac{5001}{1 + 5000\exp(-2458.867/T)}\right] - \frac{4566.468}{T}\right\}.$$

Reproduce Figure 8.7 using the parameters from the published work,[12] i.e., the polymer fraction at the condition of preparation $\phi_0 = 0.0884$, the average number of segments between neighboring cross-linking points $m = 34$, and the particle radius at the preparation condition $R_0 = 125.8$ nm. Why is the original Flory–Rehner model problematic to fit the experimental results for PNIPAM microgel particles?

Solution:
At swelling equilibrium, the pressure inside the microgel is balanced by that of the surrounding solvent such that the osmotic pressure vanishes. Accordingly, the polymer volume fraction can be solved from $\Pi(T, \phi) = 0$. Figure 8.8 shows the radii of PNIPAM microgel particles calculated from $R = R_0(\phi_0/\phi)^{1/3}$, which reproduces the published results.
With the same set of parameters, the original Flory–Rehner model would predict the particle size increasing with temperature. It fails to fit the experimental data because the Gaussian chain model is flawed in describing the elasticity of PNIPAM chains in liquid water. In the modified Flory–Rehner model, the elastic free energy accounts for the size effects of polymer segments.

Problem 8.16 The canonical partition function for a one-component system of nonlinear rigid polymeric molecules can be written as

$$Q = \frac{Q^{IG}}{V^N \Theta^N}\int d\mathbf{r}^N \int d\varpi^N \exp[-\beta\Phi(\mathbf{r}^N, \varpi^N)],$$

where Q^{IG} is the ideal-gas partition function, $\Theta = 8\pi^2$, and $\Phi(\mathbf{r}^N, \varpi^N)$ is the total potential energy. Using the pairwise additivity assumption for the potential energy and thermodynamic relations

$$U = \left(\frac{\partial \beta F}{\partial \beta}\right)_{V,N} \quad \text{and} \quad P = -\left(\frac{\partial F}{\partial V}\right)_{T,N},$$

12 Wu J., Huang G. and Hu Z., "Interparticle potential and the phase behavior of temperature-sensitive microgel dispersions", *Macromolecules* 36, 440–448 (2003).

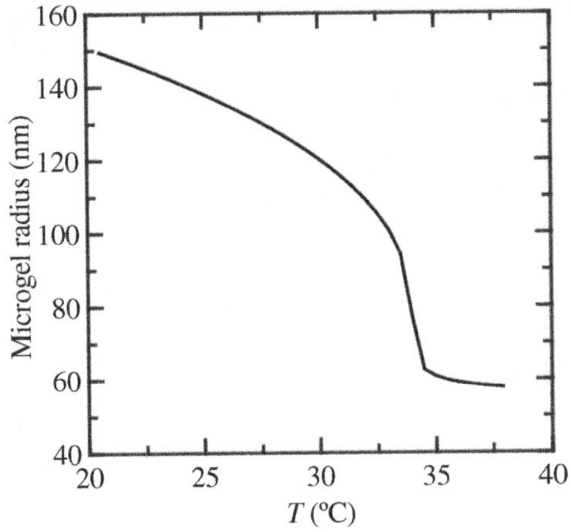

Figure 8.8 Radii of PNIPAM microgel particles versus temperature predicted by the modified Flory–Rehner model.

derive the following equations that relate the internal energy and pressure to the radial distribution function

$$U = U^{IG} + \frac{N^2}{2V}\int d\mathbf{r}\langle u(r,\varpi_1,\varpi_2)g(r,\varpi_1,\varpi_2)\rangle_{\varpi_1,\varpi_2},$$

$$P = \rho k_B T - \frac{\rho^2}{6}\int d\mathbf{r}r\langle u'(r,\varpi_1,\varpi_2)g(r,\varpi_1,\varpi_2)\rangle_{\varpi_1,\varpi_2},$$

where $\rho = N/V$ is the number density of the polymeric molecules, $u(r,\varpi_1,\varpi_2)$ denotes the pair potential, the prime sign means derivative with respect to distance, and $\langle\cdots\rangle$ means average over the Euler angles.

Solution:
The Helmholtz energy of the rigid chain system can be written as

$$F = -k_B T\ln Q = -k_B T\ln Q^{IG} - k_B T\ln Q_c = F^{IG} - k_B T\ln Q_c,$$

where the configurational partition function is defined as

$$Q_c = \frac{1}{V^N\Theta^N}\int d\mathbf{r}^N\int d\varpi^N\exp[-\beta\Phi(\mathbf{r}^N,\varpi^N)].$$

With the pairwise additivity assumption for the potential energy, the excess internal energy is related to the radial distribution function

$$\begin{aligned}
U^{ex} &= U - U^{IG} = -\left(\frac{\partial\ln Q_c}{\partial\beta}\right)_{V,N}\\
&= \frac{1}{Q_c}\frac{1}{V^N\Theta^N}\int d\mathbf{r}^N d\varpi^N\Phi(\mathbf{r}^N,\varpi^N)\exp[-\beta\Phi(\mathbf{r}^N,\varpi^N)]\\
&= \frac{1}{Q_c}\frac{1}{V^N\Theta^N}\int d\mathbf{r}^N\int d\varpi^N\frac{1}{2}\sum_{i\neq j}u(|\mathbf{r}_i-\mathbf{r}_j|,\varpi_i,\varpi_j)\exp[-\beta\Phi(\mathbf{r}^N,\varpi^N)]\\
&= \frac{1}{2}\int d\mathbf{r}\int d\mathbf{r}'\int d\varpi\int d\varpi'\sum_{i\neq j}\\
&\quad\times\langle\delta(\mathbf{r}-\mathbf{r}_i)\delta(\mathbf{r}'-\mathbf{r}_j)\delta(\varpi-\varpi_i)\delta(\varpi'-\varpi_j)u(|\mathbf{r}_i-\mathbf{r}_j|,\varpi_i,\varpi_j)\rangle\\
&= \frac{\rho^2}{2\Theta^2}\int d\mathbf{r}\int d\mathbf{r}'\int d\varpi\int d\varpi' u(|\mathbf{r}-\mathbf{r}'|,\varpi,\varpi')g(|\mathbf{r}-\mathbf{r}'|,\varpi,\varpi')\\
&= \frac{N^2}{2V}\int d\mathbf{r}\langle u(r,\varpi_1,\varpi_2)g(r,\varpi_1,\varpi_2)\rangle_{\varpi_1,\varpi_2}.
\end{aligned}$$

To derive the excess pressure $P^{ex} = P - \rho k_B T$, let $\mathbf{r}^* \equiv \mathbf{r}/V^{1/3}$. Using the identities $d\mathbf{r}/dV = \mathbf{r}/3V$, $d\mathbf{r}^{*N} = d\mathbf{r}^N/V^N$, and the algebraic relation

$$\frac{\partial \Phi(\mathbf{r}^N, \varpi^N)}{\partial V} = \frac{1}{2}\sum_{i\neq j} \frac{\partial u(|\mathbf{r}_{ij}|, \varpi_i, \varpi_j)}{\partial \mathbf{r}_{ij}} \frac{\partial \mathbf{r}_{ij}}{\partial V}$$
$$= \frac{1}{6V}\sum_{i\neq j} \frac{\partial u(|\mathbf{r}_{ij}|, \varpi_i, \varpi_j)}{\partial \mathbf{r}_{ij}} \cdot \mathbf{r}_{ij} = \frac{1}{6V}\sum_{i\neq j} \frac{\partial u(|\mathbf{r}_{ij}|, \varpi_i, \varpi_j)}{\partial r_{ij}} r_{ij}$$

where $\mathbf{r}_{ij} = \mathbf{r}_i - \mathbf{r}_j$ and $r_{ij} = |\mathbf{r}_{ij}|$, we obtain

$$\beta P^{ex} = \left(\frac{\partial \ln Q_c}{\partial V}\right)_{T,N}$$
$$= -\frac{1}{Q_c \Theta^N} \int d\mathbf{r}^{*N} \int d\varpi^N \frac{\partial \beta\Phi(\mathbf{r}^N, \varpi^N)}{\partial V} \exp[-\beta\Phi(\mathbf{r}^N, \varpi^N)]$$
$$= -\frac{1}{6VQ_c\Theta^N} \int d^{*N} \int d\varpi^N \sum_{i\neq j} \frac{\partial \beta u(|\mathbf{r}_{ij}|, \varpi_i, \varpi_j)}{\partial r_{ij}} r_{ij} \exp[-\beta\Phi(\mathbf{r}^N, \varpi^N)]$$
$$= -\frac{1}{6V} \int d\mathbf{r} \int d\mathbf{r}' \int d\varpi \int d\varpi' \sum_{i\neq j}$$
$$\times \langle \delta(\mathbf{r} - \mathbf{r}_i)\delta(\mathbf{r}' - \mathbf{r}_j)\delta(\varpi - \varpi_i)\delta(\varpi' - \varpi_j)\beta u'(|\mathbf{r}_i - \mathbf{r}_j|, \varpi_i, \varpi_j) r_{ij}\rangle$$
$$= -\frac{\rho^2}{6V\Theta^2} \int d\mathbf{r} \int d\mathbf{r}' \int d\varpi \int d\varpi' |\mathbf{r} - \mathbf{r}'| u'(|\mathbf{r} - \mathbf{r}'|, \varpi, \varpi') g(|\mathbf{r} - \mathbf{r}'|, \varpi, \varpi')$$
$$= -\frac{\rho^2}{6} \int d\mathbf{r} r \langle \beta u'(r, \varpi_1, \varpi_2) g(r, \varpi_1, \varpi_2)\rangle_{\varpi_1, \varpi_2}.$$

Problem 8.17 Starting from the grand-canonical ensemble for one-component polymeric molecules

$$\Xi = \sum_{N=0}^{\infty} \frac{1}{N!\Lambda^N} \int d\mathbf{R}^N \exp\left[-\beta\Phi(\mathbf{R}^N) + N\beta\mu\right],$$

where $\mathbf{R} \equiv (\mathbf{r}_1, \mathbf{r}_2, \ldots, \mathbf{r}_m)$ stands for the molecular configuration, $\Phi(\mathbf{R}^N)$ is the total potential energy (including both intra- and inter-molecular interactions), and other symbols have their usual meanings, show the following statistical-mechanical equations:

(i) the average number of molecules in the system

$$\langle N \rangle = \frac{\partial \ln \Xi}{\partial \beta\mu};$$

(ii) the mean-square deviation in the average number of molecules

$$\langle N^2 \rangle - \langle N \rangle^2 = \frac{\partial \langle N \rangle}{\partial \beta\mu};$$

(iii) the compressibility equation

$$\left(\frac{\partial \rho}{\partial \beta P}\right)_T = 1 + \frac{1}{\langle N \rangle} \int d\mathbf{R} \int d\mathbf{R}' \rho(\mathbf{R})\rho(\mathbf{R}') h(\mathbf{R}, \mathbf{R}');$$

(iv) for rigid molecules, the compressibility equation reduces to

$$\left(\frac{\partial \rho}{\partial \beta P}\right)_T = 1 + \rho \int d\mathbf{r} \langle h(|\mathbf{r} - \mathbf{r}'|, \varpi, \varpi')\rangle_{\varpi, \varpi'};$$

(v) the compressibility equation can be expressed in terms of any pair of the site-site total correlation function

$$\left(\frac{\partial\rho}{\partial\beta P}\right)_T = 1 + \rho\int d\mathbf{r}h_{ij}(r),$$

or the Fourier transform of the density-density correlation function $\chi_{ij}(r)$ at $k = 0$

$$\left(\frac{\partial\rho}{\partial\beta P}\right)_T = \hat{\chi}_{ij}(0)/\rho.$$

Solution:

(i) The average number of polymeric molecules in the system is given by

$$\langle N\rangle = \frac{1}{\Xi}\sum_{N=0}^{\infty}\frac{1}{N!\Lambda^N}\int d\mathbf{R}^N N\exp\left[-\beta\Phi(\mathbf{R}^N) + N\beta\mu\right]$$
$$= \frac{1}{\Xi}\sum_{N=0}^{\infty}\frac{1}{N!\Lambda^N}\int d\mathbf{R}^N\frac{\partial\exp\left[-\beta\Phi(\mathbf{R}^N) + N\beta\mu\right]}{\partial\beta\mu} = \frac{\partial\ln\Xi}{\partial\beta\mu}.$$

(ii) The second-order derivative of $\ln\Xi$ with respect to $\beta\mu$ yields

$$\frac{\partial\langle N\rangle}{\partial\beta\mu} = \frac{1}{\Xi}\sum_{N=0}^{\infty}\frac{1}{N!\Lambda^N}\int d\mathbf{R}^N\frac{\partial^2\exp\left[-\beta\Phi(\mathbf{R}^N) + N\beta\mu\right]}{\partial(\beta\mu)^2}$$
$$- \frac{1}{\Xi^2}\left\{\sum_{N=0}^{\infty}\frac{1}{N!\Lambda^N}\int d\mathbf{R}^N\frac{\partial\exp\left[-\beta\Phi(\mathbf{R}^N) + N\beta\mu\right]}{\partial\beta\mu}\right\}^2$$
$$= \langle N^2\rangle - \langle N\rangle^2.$$

(iii) From the Gibbs–Duhem equation, $(\partial\beta P/\partial\rho)_T = \rho(\partial\beta\mu/\partial\rho)_T$, we have

$$\left(\frac{\partial\rho}{\partial\beta P}\right)_T = \frac{1}{\langle N\rangle}\left(\frac{\partial\langle N\rangle}{\partial\beta\mu}\right)_{T,V} = \frac{1}{\langle N\rangle}(\langle N^2\rangle - \langle N\rangle^2).$$

Note that the average number of molecules in the system is related to the one-body density

$$\langle N\rangle = \left\langle\int d\mathbf{R}\sum_{i=1}^{N}\delta(\mathbf{R} - \mathbf{R}_i)\right\rangle = \int d\mathbf{R}\rho(\mathbf{R}),$$

and that

$$\langle N^2\rangle = \left\langle\int d\mathbf{R}\int d\mathbf{R}'\sum_{i=1}^{N}\delta(\mathbf{R} - \mathbf{R}_i)\sum_{j=1}^{N}\delta(\mathbf{R}' - \mathbf{R}_j)\right\rangle$$
$$= \int d\mathbf{R}\int d\mathbf{R}'[\rho(\mathbf{R},\mathbf{R}') + \delta(\mathbf{R} - \mathbf{R}')\rho(\mathbf{R})]$$
$$= \int d\mathbf{R}\int d\mathbf{R}'\rho(\mathbf{R},\mathbf{R}') + \langle N\rangle.$$

A combination of the three equations above leads to

$$\left(\frac{\partial\rho}{\partial\beta P}\right)_T = 1 + \frac{1}{\langle N\rangle}\int d\mathbf{R}\int d\mathbf{R}'[\rho(\mathbf{R},\mathbf{R}') - \rho(\mathbf{R})\rho(\mathbf{R}')]$$
$$= 1 + \frac{1}{\langle N\rangle}\int d\mathbf{R}\int d\mathbf{R}'\rho(\mathbf{R})\rho(\mathbf{R}')[g(\mathbf{R},\mathbf{R}') - 1]$$
$$= 1 + \frac{1}{\langle N\rangle}\int d\mathbf{R}\int d\mathbf{R}'\rho(\mathbf{R})\rho(\mathbf{R}')h(\mathbf{R},\mathbf{R}'),$$

where $g(\mathbf{R}, \mathbf{R}') \equiv \rho(\mathbf{R}, \mathbf{R}')/\rho(\mathbf{R})\rho(\mathbf{R}')$ and $h(\mathbf{R}, \mathbf{R}') \equiv g(\mathbf{R}, \mathbf{R}') - 1$ are the molecular radial distribution function and molecular total correlation function, respectively. Like their counterparts for simple fluids, both g and h are dimensionless.

(iv) For a uniform system of rigid molecules, the molecular configuration can be specified by $\mathbf{R} = (\mathbf{r}, \varpi)$, and the average molecular density can be expressed as $\rho(\mathbf{R}) = \rho/\Theta$ (see Problem 8.16). Thus, the compressibility equation becomes

$$\left(\frac{\partial \rho}{\partial \beta P}\right)_T = 1 + \frac{\rho^2}{\langle N\rangle \Theta^2} \int d\mathbf{r} \int d\mathbf{r}' \int d\varpi \int d\varpi' [g(|\mathbf{r} - \mathbf{r}'|, \varpi, \varpi') - 1]$$

$$= 1 + \rho \int d\mathbf{r} \langle h(|\mathbf{r} - \mathbf{r}'|, \varpi, \varpi')\rangle_{\varpi,\varpi'}.$$

(v) The site-site total correlations are defined by the intermolecular total correlation function

$$h_{ij}(|\mathbf{r} - \mathbf{r}'|) \equiv \frac{1}{\rho^2} \int d\mathbf{R} \int d\mathbf{R}' \rho(\mathbf{R})\rho(\mathbf{R}')h(\mathbf{R}, \mathbf{R}')\delta(\mathbf{r} - \mathbf{r}_i)\delta(\mathbf{r}' - \mathbf{r}_j),$$

where $\mathbf{r}_i$ represents the site i position for a molecule with configuration $\mathbf{R}$, and $\mathbf{r}_j$ is the site j position for a molecular with configuration $\mathbf{R}'$.
Because the above equation implies

$$V \int d\mathbf{r} h_{ij}(r) = \int d\mathbf{r} \int d\mathbf{r}' h_{ij}(|\mathbf{r} - \mathbf{r}'|) = \frac{1}{\rho^2} \int d\mathbf{R} \int d\mathbf{R}' \rho(\mathbf{R})\rho(\mathbf{R}')h(\mathbf{R}, \mathbf{R}'),$$

we can express the compressibility equation in terms of any pair of the site-site total correlation function

$$\left(\frac{\partial \rho}{\partial \beta P}\right)_T = 1 + \frac{1}{\langle N\rangle} \int d\mathbf{R} \int d\mathbf{R}' \rho(\mathbf{R})\rho(\mathbf{R}')h(\mathbf{R}, \mathbf{R}') = 1 + \rho \int d\mathbf{r} h_{ij}(r).$$

In terms of the density-density correlation function $\chi_{ij}(r)$, the compressibility equation can be written as

$$\left(\frac{\partial \rho}{\partial \beta P}\right)_T = \hat{\chi}_{ij}(0)/\rho,$$

where $\hat{\chi}_{ij}(0)$ stands for the Fourier transform of $\chi_{ij}(r)$ at $k = 0$.

Problem 8.18 The following exercise may help better understand the dielectric properties of materials from a molecular perspective. According to macroscopic electrostatics, the electric potential due to a charge Q in a continuous medium can be described by Coulomb's law

$$\psi(r) = \frac{Q}{4\pi\varepsilon_0 \varepsilon r},$$

where ε_0 is the permittivity of free space, ε is the dielectric constant, and r is the distance from the charge center. Conversely, from a molecular perspective, the electric potential may be decomposed into three terms

$$\psi(r) = \frac{Q}{4\pi\varepsilon_0 r} + \psi_c(r) + \psi_{ind}(r),$$

where the first term on the right corresponds to the electric potential in a vacuum, $\psi_c(r)$ arises from the partial charges of individual atoms in the dielectric medium, and $\psi_{ind}(r)$ represents an additional contribution due to atomic polarization. With $\psi_c(r)$ and $\psi_{ind}(r)$ expressed in terms of the atomic densities $\rho_i(r)$, a comparison of the continuous and atomistic descriptions of the electric potential allows us to predict the dielectric constant ε from atomic charge q_i and polarizability α_i. Show that, as $Q \to 0$,

(i) the atomic density near charge Q can be predicted from

$$\rho_i(\mathbf{r}) = \rho_i^b - \sum_j \frac{\beta Q q_j}{4\pi\varepsilon_0} \int d\mathbf{r}' \frac{\chi_{ij}(|\mathbf{r}-\mathbf{r}'|)}{|\mathbf{r}'|},$$

where ρ_i^b is the number density of atomic particle i in the uniform system (viz., the continuous medium in the absence of charge Q), and $\chi_{ij}(r)$ is the density-density correlation function for the atomic pair of i and j;

(ii) the local electric potential due to the redistribution of atomic particles in the dielectric medium is given by

$$\psi_c(r) = -\sum_{i,j} \frac{\beta Q q_i q_j}{(4\pi\varepsilon_0)^2} \int d\mathbf{r}' \int d\mathbf{r}'' \frac{\chi_{ij}(|\mathbf{r}'-\mathbf{r}''|)}{|\mathbf{r}''||\mathbf{r}-\mathbf{r}'|};$$

(iii) the atomic polarization contributes an additional electric potential at distance r

$$\psi_{ind}(r) = \sum_i \int d\mathbf{r}' \mathbf{P}_i(\mathbf{r}') \cdot \frac{(\mathbf{r}-\mathbf{r}')}{4\pi\varepsilon_0|\mathbf{r}-\mathbf{r}'|^3},$$

where $\mathbf{P}_i(\mathbf{r}') = \alpha_i\rho_i(\mathbf{r}')\mathbf{E}(\mathbf{r}')$ stands for the local density of point dipoles due to the polarization of atomic particle i, and $\mathbf{E}(\mathbf{r}')$ is the local electric field;

(iv) the local electric field is coupled with the atomic density and the local dipole density

$$\mathbf{E}(\mathbf{r}) = Q\mathbf{r}/(4\pi\varepsilon_0 r^3) - \nabla\psi_c(r) - \sum_i \int d\mathbf{r}' \mathbf{P}_i(\mathbf{r}') \cdot \mathbb{T}(\mathbf{r}-\mathbf{r}')/(4\pi\varepsilon_0),$$

where $\mathbb{T} = \mathbb{I}/r^3 - 3\mathbf{rr}/r^5$ stands for the dipole field tensor, and $\mathbb{I}$ denotes the unit tensor;

(v) without atomic polarization (viz., $\alpha_i = 0$), the dielectric constant can be predicted from

$$\varepsilon^{-1} = 1 + \frac{2\pi\beta}{3\varepsilon_0} \sum_{i,j} q_i q_j \int_0^\infty dr r^4 \chi_{ij}(r);$$

(vi) if the dielectric medium consists of nonpolarizable molecules with a rigid conformation, the dielectric constant satisfies the following equation

$$(1 - 1/\varepsilon)\varepsilon_0/\beta = \frac{\rho_b d^2}{3} + \rho_b^2 \Delta\hat{h}^{(2)},$$

where d denotes the molecular dipole moment, ρ_b represents the number density of molecules in the bulk, $\Delta\hat{h}^{(2)} \equiv \sum_{i,j} q_i q_j \hat{h}_{ij}^{(2)}$, and $\hat{h}_{ij}^{(2)}$ is defined by the Taylor expansion of the total correlation function in the Fourier space, i.e., $\hat{h}_{ij}(k) = \hat{h}_{ij}(0) + \hat{h}_{ij}^{(2)}k^2 + \cdots$;

(vii) for $\rho_b \to 0$, the dielectric constant approaches the ideal-gas expression $\varepsilon = 1 + \beta\rho_b d^2/(3\varepsilon_0)$.

Hint: The density-density correlation function may be understood in terms of the change in local atomic density $\rho_i(\mathbf{r})$ in response to an external potential $V_j^{ext}(\mathbf{r}')$

$$\chi_{ij}(|\mathbf{r}-\mathbf{r}'|) \equiv \langle \delta\rho_i(\mathbf{r})\delta\rho_j(\mathbf{r}')\rangle = -\delta\rho_i(\mathbf{r})/\delta\beta V_j^{ext}(\mathbf{r}').$$

Problem 8.23 shows a derivation of the above relation. In a molecular system, the density-density correlation function is related to the intramolecular and total correlation functions, i.e., $\chi_{ij}(r) = \rho_b\omega_{ij}(r) + \rho_b^2 h_{ij}(r)$.

Solution:

This problem is based on a publication by Chandler[13]

(i) For each atomic particle with partial charge q_j, the external potential due to the presence of charge Q is

$$V_j^{ext}(\mathbf{r}) = \frac{Qq_j}{4\pi\varepsilon_0 r}.$$

Accordingly, the linear response theory predicts a variation of the local atomic density

$$\delta\rho_i(\mathbf{r}) = -\sum_j \int d\mathbf{r}'\chi_{ij}(|\mathbf{r}-\mathbf{r}'|)\delta\beta V_j^{ext}(\mathbf{r}') = -\sum_j \int d\mathbf{r}'\chi_{ij}(|\mathbf{r}-\mathbf{r}'|)\frac{\beta Qq_j}{4\pi\varepsilon_0|\mathbf{r}'|}.$$

Rewriting the above equation leads to

$$\rho_i(\mathbf{r}) = \rho_i^b - \sum_j \frac{\beta Qq_j}{4\pi\varepsilon_0}\int d\mathbf{r}'\frac{\chi_{ij}(|\mathbf{r}-\mathbf{r}'|)}{|\mathbf{r}'|}.$$

(ii) Because each particle i carries a partial charge q_i, the density inhomogeneity results in a local electric potential

$$\begin{aligned}
\psi_c(r) &= \sum_i \int d\mathbf{r}'\frac{\rho_i(\mathbf{r}')q_i}{4\pi\varepsilon_0|\mathbf{r}-\mathbf{r}'|}\\
&= \sum_i \int d\mathbf{r}'\frac{\delta\rho_i(\mathbf{r}')q_i}{4\pi\varepsilon_0|\mathbf{r}-\mathbf{r}'|} + \sum_i \rho_i^b q_i \int d\mathbf{r}'\frac{1}{4\pi\varepsilon_0|\mathbf{r}-\mathbf{r}'|}\\
&= -\sum_{i,j}\int d\mathbf{r}'\int d\mathbf{r}''\chi_{ij}(|\mathbf{r}'-\mathbf{r}''|)\frac{\beta Qq_j}{4\pi\varepsilon_0|\mathbf{r}''|}\times\frac{q_i}{4\pi\varepsilon_0|\mathbf{r}-\mathbf{r}'|}\\
&= -\sum_{i,j}\frac{\beta Qq_iq_j}{(4\pi\varepsilon_0)^2}\int d\mathbf{r}'\int d\mathbf{r}''\frac{\chi_{ij}(|\mathbf{r}'-\mathbf{r}''|)}{|\mathbf{r}''||\mathbf{r}-\mathbf{r}'|},
\end{aligned}$$

where $\rho_i(\mathbf{r}) = \rho_i^b + \delta\rho_i(\mathbf{r})$, and $\sum_i \rho_i^b q_i = 0$ from the charge neutrality.

(iii) The induced dipole for each atom at position $\mathbf{r}$ depends only on the local electric field, i.e., $\mathbf{d}_i^{ind} = \alpha_i\mathbf{E}(\mathbf{r})$. Because each induced dipole at position $\mathbf{r}'$ generates an electric potential at $\mathbf{r}$,

$$\frac{\mathbf{d}_i^{ind}\cdot(\mathbf{r}-\mathbf{r}')}{4\pi\varepsilon_0|\mathbf{r}-\mathbf{r}'|^3},$$

the electric potential due to all induced dipoles is

$$\psi_{ind}(r) = \sum_i \int d\mathbf{r}'\rho_i(\mathbf{r}')\frac{\mathbf{d}_i^{ind}\cdot(\mathbf{r}-\mathbf{r}')}{4\pi\varepsilon_0|\mathbf{r}-\mathbf{r}'|^3} = \sum_i \int d\mathbf{r}'\mathbf{P}_i(\mathbf{r}')\cdot\frac{(\mathbf{r}-\mathbf{r}')}{4\pi\varepsilon_0|\mathbf{r}-\mathbf{r}'|^3},$$

where $\mathbf{P}_i(\mathbf{r}') = \alpha_i\rho_i(\mathbf{r}')\mathbf{E}(\mathbf{r}')$.

(iv) The local electric field includes contributions from the permanent charges and the induced dipoles and can be calculated from the gradient of the corresponding electric potentials

$$\mathbf{E}(\mathbf{r}) = Q\mathbf{r}/(4\pi\varepsilon_0 r^3) - \nabla\psi_c(r) - \sum_i \int d\mathbf{r}'\mathbf{P}_i(\mathbf{r}')\cdot\mathbb{T}(\mathbf{r}-\mathbf{r}')/(4\pi\varepsilon_0),$$

where the dipole field tensor is obtained by noting that $\mathbb{T}(\mathbf{r}) = -\nabla\mathbf{r}/r^3$.

13 Chandler D., "The dielectric constant and related equilibrium properties of molecular fluids: Interaction site cluster theory analysis", *J. Chem. Phys.* 67, 1113–1124 (1977).

(v) Without atomic polarization (viz., $\alpha_i = 0$), a comparison of the continuous and atomic descriptions of the local electric potential leads to

$$\psi_c(r) = \frac{Q}{4\pi\varepsilon_0 r}(1/\varepsilon - 1) = -\sum_{i,j} \frac{\beta Q q_i q_j}{(4\pi\varepsilon_0)^2} \int d\mathbf{r}' \int d\mathbf{r}'' \frac{\chi_{ij}(|\mathbf{r}' - \mathbf{r}''|)}{|\mathbf{r}''||\mathbf{r} - \mathbf{r}'|}.$$

Because the continuous description is macroscopic, the relation is strictly valid only at $r \to \infty$. Rearranging the above equation leads to

$$\frac{1 - 1/\varepsilon}{r} = \sum_{i,j} \frac{\beta q_i q_j}{4\pi\varepsilon_0} \int d\mathbf{r}' \int d\mathbf{r}'' \frac{\chi_{ij}(|\mathbf{r}' - \mathbf{r}''|)}{|\mathbf{r}''||\mathbf{r} - \mathbf{r}'|}. \tag{E}$$

Next, applying the 3D-Fourier transform on both sides of Eq. (E) gives

$$\frac{4\pi(1 - 1/\varepsilon)}{k^2} = \sum_{i,j} \frac{\beta q_i q_j}{4\pi\varepsilon_0} \hat{\chi}_{ij}(k) \frac{(4\pi)^2}{k^4}.$$

Hence, the dielectric constant can be calculated from

$$(\varepsilon_0 - \varepsilon_0/\varepsilon)/\beta = \lim_{k\to 0} \sum_{i,j} q_i q_j \frac{\hat{\chi}_{ij}(k)}{k^2}.$$

As the density-density correlation function of a bulk system $\chi_{ij}(r)$ depends on distance r, we may express $\hat{\chi}_{ij}(k)$ in the Taylor series of k^2

$$\hat{\chi}_{ij}(k) = \hat{\chi}_{ij}(0) + k^2 \hat{\chi}_{ij}^{(2)} + \cdots,$$

where

$$\hat{\chi}_{ij}^{(2)} = \int_0^{2\pi} d\phi \int_0^{\pi} d\theta \int_0^{\infty} dr r^2 \sin\theta \chi_{ij}(r) \left(-\frac{r^2 \cos^2\theta}{2}\right) = -\frac{2\pi}{3} \int_0^{\infty} dr r^4 \chi_{ij}(r).$$

Plugging the Taylor series into the preceding equation results in

$$(\varepsilon_0 - \varepsilon_0/\varepsilon)/\beta = \lim_{k\to 0} \sum_{i,j} \frac{q_i q_j}{k^2} [\hat{\chi}_{ij}(0) + k^2 \hat{\chi}_{ij}^{(2)} + \cdots].$$

Because $\hat{\chi}_{ij}(0)$ is related to the system compressibility independent of ij (see Problem 8.17), the leading term in the summation disappears due to charge neutrality, and thus

$$(1/\varepsilon - 1)\varepsilon_0/\beta = -\sum_{i,j} q_i q_j \hat{\chi}_{ij}^{(2)} = \frac{2\pi}{3} \sum_{i,j} q_i q_j \int_0^{\infty} dr r^4 \chi_{ij}(r)$$

or

$$\frac{1}{\varepsilon} = 1 + \frac{2\pi\beta}{3\varepsilon_0} \sum_{i,j} q_i q_j \int_0^{\infty} dr r^4 \chi_{ij}(r). \tag{F}$$

As $\chi_{ij}(r)$ can be easily obtained from liquid-state theories, molecular simulation or scattering experiment, Eq. (F) provides a theoretical basis for understanding dielectric materials based on molecular considerations.

(vi) For a molecular system with rigid conformation, we may express the density-density correlation function in terms of the intramolecular and total correlation functions, i.e.,

$$\chi_{ij}(r) = \rho_b \omega_{ij}(r) + \rho_b^2 h_{ij}(r).$$

Thus, the dielectric constant equation from part (v) can be rewritten as

$$(1 - 1/\varepsilon)\varepsilon_0/\beta = \sum_{i,j} q_i q_j \hat{\chi}_{ij}^{(2)} = \rho_b \sum_{i,j} q_i q_j \hat{\omega}_{ij}^{(2)} + \rho_b^2 \sum_{i,j} q_i q_j \hat{h}_{ij}^{(2)}.$$

Meanwhile, we can express the Fourier transform of the intramolecular correlation function (Eq. 8.81 in Section 8.3) in terms of a Taylor series

$$\hat{\omega}_{ij}(k) = \delta_{ij} + (1-\delta_{ij})\frac{\sin(kl_{ij})}{kl_{ij}}$$
$$= \delta_{ij} + (1-\delta_{ij})[1-(kl_{ij})^2/6+\cdots] = 1 + (\delta_{ij}-1)(kl_{ij})^2/6+\cdots,$$

where δ_{ij} is the Kronecker function, and l_{ij} represents the bond length or the separation between atomic particles i and j. Substituting $\hat{\omega}_{ij}^{(2)} = (\delta_{ij}-1)l_{ij}^2/6$ into the preceding equation yields

$$(1-1/\varepsilon)\varepsilon_0/\beta = \rho_b \sum_{i,j} q_i q_j (\delta_{ij}-1)l_{ij}^2/6 + \rho_b^2 \sum_{i,j} q_i q_j \hat{h}_{ij}^{(2)}. \quad \text{(G)}$$

On the other hand, the charge neutrality condition for the rigid molecule requires $\sum_i q_i = 0$. Therefore, the intramolecular term can be written as

$$\sum_{i,j} q_i q_j (\delta_{ij}-1)l_{ij}^2 = -\sum_{i,j} q_i q_j |\mathbf{r}_i - \mathbf{r}_j|^2 = 2\sum_{i,j} q_i q_j (\mathbf{r}_i - \mathbf{r}_c)\cdot(\mathbf{r}_j - \mathbf{r}_c)$$
$$= 2\sum_i q_i(\mathbf{r}_i - \mathbf{r}_c)\cdot \sum_j q_j(\mathbf{r}_j - \mathbf{r}_c) = 2\mathbf{d}\cdot\mathbf{d} = 2d^2,$$

where $\mathbf{d} = \sum_i q_i(\mathbf{r}_i - \mathbf{r}_c)$ is the molecular dipole moment, $\mathbf{r}_c$ represents the molecular center of mass, and $d = |\mathbf{d}|$. Plugging the above relation into Eq. (G) yields

$$(1-1/\varepsilon)\varepsilon_0/\beta = \tfrac{\rho_b d^2}{3} + \rho_b^2 \sum_{i,j} q_i q_j \hat{h}_{ij}^{(2)} = \frac{\rho_b d^2}{3} + \rho_b^2 \Delta\hat{h}^{(2)}, \quad \text{(H)}$$

where $\Delta\hat{h}^{(2)} \equiv \sum_{i,j} q_i\, q_j\, \hat{h}_{ij}^{(2)}$.

(vii) As $\rho_b \to 0$, $\hat{h}_{ij}^{(2)} = 0$, and Eq. (H) predicts $1-1/\varepsilon = \beta\rho_b d^2/(3\varepsilon_0) \equiv 3y$. Because $y \to 0$, $\varepsilon = 1/(1-3y) \approx 1+3y$. In terms of dimensionless parameter y, we can rewrite Eq. (H) as

$$1-1/\varepsilon = 3y + \tfrac{\beta\rho_b^2}{\varepsilon_0}\Delta\hat{h}^{(2)}. \quad \text{(I)}$$

Rearranging Eq. (I) leads to the formula originally derived by Høye and Stell for the dielectric constant of rigid nonpolarizable molecules[14]

$$\Delta\hat{h}^{(2)} = \frac{\varepsilon_0}{\beta\rho_b^2}(1-1/\varepsilon-3y) = \frac{d^2}{9y\rho_b}(1-1/\varepsilon-3y) = \frac{d^2}{9\rho_b}\left(\frac{\varepsilon-1}{y\varepsilon}-3\right).$$

Problem 8.19 Consider association between two spherical particles to form a dimer in a vacuum and that in a fluid. Figure 8.9 shows schematically the association processes. Prove that, at fixed temperature and pressure of the fluid phase, the difference between the excess chemical potential of the dimer and that of the monomers is related to the cavity correlation function between the two associating particles in the fluid

$$\mu_2^{ex} - 2\mu^{ex} = -k_B T \ln y(L),$$

where L is the center-to-center distance between the two particles in the dimer, and subscript "2" denotes the dimer. How would the above equation be extended to the formation of an m-mer in the fluid phase?

---

14 Høye J. S. and Stell G., "Dielectric constant in terms of atom–atom correlation functions", *J. Chem. Phys.* 65, 18–22 (1976).

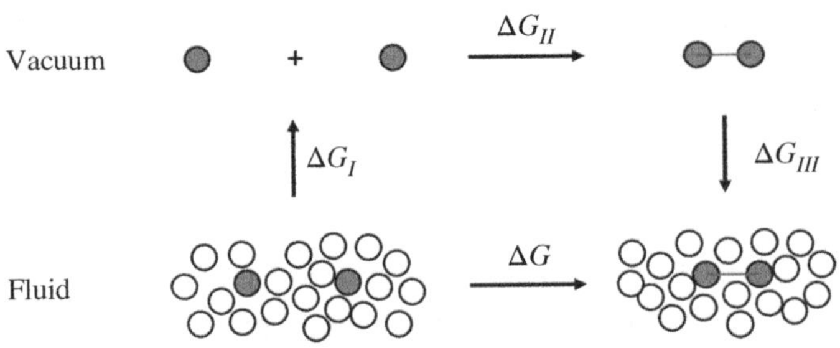

Figure 8.9 A thermodynamic cycle for association between two particles in a vacuum and that in a uniform fluid. The open cycles represent fluid molecules.

Solution:
The original derivation was reported by Zhou and Stell.[15] The same relation can be derived by using the canonical partition function (Problem 7.9). The thermodynamic cycle shown in Figure 8.9 suggests that the Gibbs energy of association can be written as

$$\Delta G = \Delta G_I + \Delta G_{II} + \Delta G_{III}, \tag{J}$$

where ΔG_{II} is the same as the pair potential between the associating particles $u(L)$, ΔG_I and ΔG_{III} are related to the solvation free energies of individual segments and the dimer in the fluid

$$\Delta G_I + \Delta G_{III} = \mu_2^{ex} - 2\mu^{ex}.$$

The solvation free energies and the excess chemicals are identical at infinite dilution of the particles because both represent the reversible work to transfer a particle (or a dimer) from the vacuum to the fluid. As discussed in Section 7.2, ΔG corresponds to the potential of mean force between the two associating particles

$$\Delta G = -k_B T \ln g(L).$$

Substituting the above relations to Eq. (J) yields

$$-k_B T \ln g(L) = u(L) + \mu_2^{ex} - 2\mu^{ex}.$$

The desired result can be obtained by rearrangement of the above equation

$$\begin{aligned}\mu_2^{ex} - 2\mu^{ex} &= -k_B T \ln g(L) - k_B T \ln \exp[\beta u(L)]\\ &= -k_B T \ln\{g(L)\exp[\beta u(L)]\} = -k_B T \ln y(L),\end{aligned}$$

where $y(L) = g(L)\exp[\beta u(L)]$.
The thermodynamic cycle is equally applicable to the formation of m-mer with rigid configuration $\mathbf{R} = (\mathbf{r}_1, \mathbf{r}_2, \dots, \mathbf{r}_m)$

$$\mu_m^{ex}(\mathbf{R}) - m\mu^{ex} = -k_B T \ln y(\mathbf{R}). \tag{K}$$

Eq. (K) is identical to that predicted by the thermodynamic perturbation theory (TPTm) for chain formation.

Problem 8.20 Consider association between monomeric species A and B to form dimer AB in a condensed phase at given temperature and pressure, $A + B \rightleftharpoons AB$. Show that the (apparent) equilibrium constant $K \equiv \rho_{AB}/(\rho_A \rho_B)$ can be expressed in terms of the cavity correlation function

$$K/K_0 = y_{AB}(L),$$

15 Zhou Y. and Stell G., "Chemical association in simple models of molecular and ionic fluids. III. The cavity function", *J. Chem. Phys.* 96, 1507–1515 (1992).

where L is the center-to-center distance between A and B in dimer AB, $y_{AB}(r)$ is the cavity correlation function of particles A and B in the condensed phase, and $K_0 \equiv \exp(-\beta\Delta G^0_{rxn})$ is the thermodynamic equilibrium constant defined by

$$\Delta G^0_{rxn} = \mu^0_{AB} - \mu^0_A - \mu^0_B,$$

where $\mu^0_{i=A,B,AB}$ is the chemical potential of pure i as an ideal gas at the system temperature and unit density ρ^0_i.

Solution:

The chemical equilibrium for association between monomeric species A and B is established when

$$\mu_{AB} = \mu_A + \mu_B.$$

For each chemical species $i = AB, A$ or B, the chemical potential can be written as

$$\mu_i = \mu^0_i + k_B T \ln \rho_i + \mu^{ex}_i,$$

where μ^0_i is the reference chemical potential, i.e., the chemical potential of pure i as an ideal gas at the system temperature and unit density ρ^0_i. Hence, the condition of equilibrium leads to the familiar mass action equation

$$K \equiv \frac{\rho_{AB}}{\rho_A \rho_B} = K_0 \exp[-\beta(\mu^{ex}_{AB} - \mu^{ex}_A - \mu^{ex}_B)],$$

where $K_0 \equiv \exp[-\beta(\mu^0_{AB} - \mu^0_A - \mu^0_B)] = \exp(-\beta\Delta G^0_{rxn})$ is the equilibrium constant.

Applying the thermodynamic cycle discussed in Problem 8.19 to the association between A and B gives

$$\mu^{ex}_{AB} - \mu^{ex}_A - \mu^{ex}_B = -k_B T \ln y_{AB}(L).$$

A combination of the above two equations leads to

$$K/K_0 = y_{AB}(L).$$

Note that $y_{AB}(L)$ is the cavity correlation function in the binary mixture of associating species A and B in the condensed phase, i.e., A and B interact with each other through both bonding and nonbonding potentials.

Problem 8.21 The second virial coefficient for hard-sphere chains may be understood in the context of polymers in a good solvent. According to the Flory–Huggins theory, the osmotic second virial coefficient is $A_2 = 1/2$ when the Flory parameter $\chi_F = 0$.

(i) In statistical mechanics, the second virial coefficient is conventionally defined through $Z = 1 + B_2\rho_m + \cdots$, where Z denotes the osmotic compressibility factor, and ρ_m is the number density of polymer chains. Show that the osmotic second virial coefficient A_2 defined in the Flory–Huggins theory is related to its conventional form

$$A_2 = B_2/(v_s m^2),$$

where v_s is the volume of a solvent molecule (viz., volume per lattice site), and m is the number of polymer segments.

(ii) Compare the virial coefficient predicted by the Flory–Huggins theory with that from the equations of state for hard-sphere chains discussed in Section 8.4 in the limit of large m

$$B_2/(m^2\sigma^3) \approx k_1,$$

where σ is the segment diameter, and k_1 is a constant between 0.38 and 1.33.

(iii) In a good solvent, polymer chains repel each other much like two hard spheres of radius R_g, where R_g stands for the polymer radius of gyration. Show that the conventional second virial coefficient may be estimated from

$$B_2 \approx \frac{16\pi R_g^3}{3}.$$

(iv) Scaling analysis indicates that for a polymer chain in a good solvent, $R_g \approx bm_F^{\nu}$, where b is the Kuhn length, and $\nu_F \approx 0.588$ is the Flory exponent. Show that the scaling analysis is consistent with experimental observation $A_2 \sim m^{-0.236}$.

Solution:

(i) According to the Flory–Huggins theory, the osmotic second virial coefficient A_2 is defined in terms of osmotic pressure Π

$$\frac{\Pi v_s}{k_B T} = \frac{\phi}{m} + A_2\phi^2 + \cdots. \tag{L}$$

For a binary system, the lattice model asserts that the polymer volume fraction ϕ is related to the number density of the polymer segments ρ and that of the solvent molecules ρ_s, i.e., $\phi = \rho/(\rho + \rho_s)$. Noting $(\rho + \rho_s)v_s = 1$, we multiply both sides of Eq. (L) with $\rho + \rho_s$ and find

$$\frac{\Pi}{k_B T} = \frac{\rho}{m} + A_2\frac{\rho^2}{\rho + \rho_s} + \cdots = \rho_m + (A_2 v_s m^2)\rho_m^2 + \cdots,$$

where $\rho_m = \rho/m$ is the number density of polymer chains. The osmotic compressibility factor is thus given by

$$Z = \frac{\Pi}{k_B T \rho_m} = 1 + (A_2 v_s m^2)\rho_m + \cdots = 1 + B_2\rho_m + \cdots$$

with $B_2 = A_2 v_s m^2$.

(ii) For $\chi_F = 0$, the Flory–Huggins theory predicts $A_2 = 0.5$, $v_s \approx \pi\sigma^3/6$ implies that it agrees reasonably well with $B_2/(m^2\sigma^3) \approx k_1$.

(iii) If the polymer chains interact with each other like hard spheres that cannot overlap, the second virial coefficient is then given by

$$B_2 = -\frac{1}{2}\int d\mathbf{r}[e^{-\beta u^{HS}(r)} - 1] = \frac{4\pi}{2}\int_0^{2R_g} drr^2 = \frac{16\pi R_g^3}{3},$$

where we have used $2R_g$ for the hard-sphere diameter.

(iv) With $R_g \approx bm_F^{\nu}$, we have

$$A_2 = B_2/(v_s m^2) = \frac{16\pi R_g^3}{3v_s m^2} = \frac{16\pi b^3 m^{3\nu_F}}{3v_s m^2} \sim m^{\alpha},$$

where $\alpha \equiv 3\nu_F - 2 \approx -0.236$ for $\nu_F \approx 0.588$.

Problem 8.22 Müller and Gubbins[16] derived the following equation of state for tangent hard-sphere chains based on the second-order thermodynamic perturbation theory (TPT2)

$$Z \equiv \frac{P}{\rho k_B T} = m\frac{1 + \eta + \eta^2 - \eta^3}{(1-\eta)^3} + (1-m)\frac{2 + 2\eta - \eta^2}{(2-\eta)(1-\eta)} + Z_2, \tag{M}$$

16 Müller E. A. and Gubbins K. E., "Simulation of hard triatomic and tetratomic molecules", *Mol. Phys.*, 80(4), 957–976 (1993).

where ρ stands for the number density of hard-sphere chains, $\eta = \pi m \rho \sigma^3/6$ is the segment packing fraction, σ is the hard-sphere diameter, and m is the number of segments per chain. The second-order term is given by

$$Z_2 = \frac{4\lambda - m(1+4\lambda) + [m^2(1+4\lambda) - 4\lambda]^{1/2}}{2\lambda(1+4\lambda)} \eta \left(\frac{\partial \lambda}{\partial \eta}\right)$$

with $\lambda = 0.233633\eta(1+0.472\eta)$.

(i) Show that Eq. (M) reduces to the ideal-gas law at the low density of hard-sphere chains;

(ii) Derive the second virial coefficient from Eq. (M) and compare it with the scaling relation $B_2/(m^2\sigma^3) \sim m^{-0.236}$ obtained from scaling analysis;

(iii) Show that the compressibility factor is linearly proportional to m as $m \to \infty$;

(iv) Compare the compressibility factor predicted by TPT1 and that by TPT2 for $m = 2000$ and discuss why the second-order term is relatively insignificant for tangent hard-sphere chains.

Solution:

(i) When $\eta \to 0$, we have

$$\frac{1+\eta+\eta^2-\eta^3}{(1-\eta)^3} \to 1 \quad \text{and} \quad \frac{2+2\eta-\eta^2}{(2-\eta)(1-\eta)} \to 1.$$

For the second-order term, $\lambda \to 0$ as $\eta \to 0$, and $\eta\left(\frac{\partial \lambda}{\partial \eta}\right) \to \lambda \to 0$. Hence, we have

$$\lim_{\eta \to 0} Z_2 = \frac{-m + [m^2]^{1/2}}{2\lambda}\lambda = 0.$$

Therefore, the Müller–Gubbins equation predicts

$$\lim_{\eta \to 0} Z = m + (1-m) + Z_2 = 1.$$

(ii) Expanding the Müller–Gubbins equation in terms of the Taylor series of η gives

$$Z_0 = m\frac{1+\eta+\eta^2-\eta^3}{(1-\eta)^3} = m[1+4\eta+\mathcal{O}(\eta^2)]$$

$$Z_1 = (1-m)\frac{2+2\eta-\eta^2}{(2-\eta)(1-\eta)} = (1-m)[1+2.5\eta+\mathcal{O}(\eta^2)]$$

$$Z_2 = \left[-\frac{(m-1)^2}{m} - \frac{1-6m^2+8m^3-3m^4}{m^3}\lambda + \mathcal{O}(\lambda^2)\right] a(\eta+2b\eta^2)$$

$$= -\frac{(m-1)^2}{m}a\eta + \mathcal{O}(\eta^2),$$

where $a = 0.233633$, and $b = 0.472$. Because $\eta = \pi m \rho \sigma^3/6$, a comparison of the Müller–Gubbins equation with $Z = 1 + B_2\rho + \mathcal{O}(\rho^2)$ gives the reduced second virial coefficient

$$\frac{B_2}{\sigma^3 m^2} = \frac{\pi}{6m}\left[4m + 2.5(1-m) - a(m-1)^2/m\right]$$

$$= \frac{\pi}{6}\left[1.5 + 2.5/m - a(1-1/m)^2\right] = 0.6631 + 1.5537/m - 0.1223/m^2.$$

As expected, TPT2 shows the mean-field behavior for B_2, which is qualitatively different from the scaling relation $B_2/(m^2\sigma^3) \sim m^{-0.236}$.

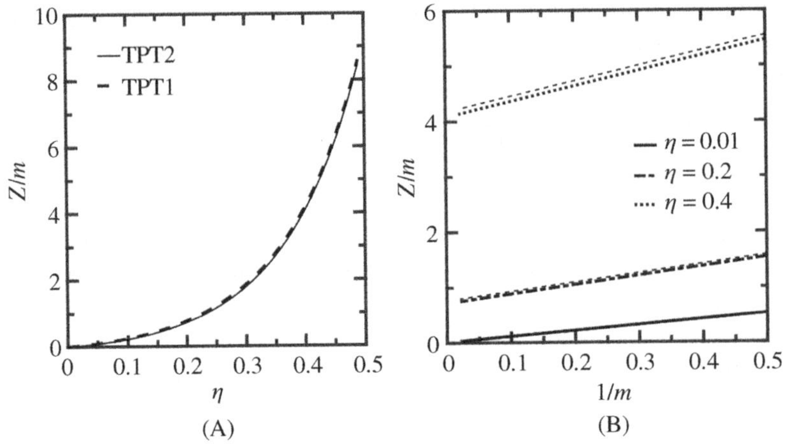

Figure 8.10 The compressibility factor for hard-sphere chains predicted by TPT1 (dashed lines) and that by TPT2 (other lines). (A) Z/m versus η at fixed chain length ($m = 2000$); (B) Z/m versus $1/m$ at three values of packing fractions (HSCTPT2.m).

(iii) For $m \to \infty$, the Müller–Gubbins equation becomes

$$\frac{Z}{m} = \frac{1+\eta+\eta^2-\eta^3}{(1-\eta)^3} - \frac{2+2\eta-\eta^2}{(2-\eta)(1-\eta)} + \frac{-(1+4\lambda)+\sqrt{1+4\lambda}}{2\lambda(1+4\lambda)}\eta\left(\frac{\partial \lambda}{\partial \eta}\right)$$

$$= \frac{\eta(3+4\eta-7\eta^2+2\eta^3)}{(2-\eta)(1-\eta)^3} + \frac{1-\sqrt{1+4\lambda}}{2\lambda\sqrt{1+4\lambda}}\eta\left(\frac{\partial \lambda}{\partial \eta}\right).$$

As $m \to \infty$, the compressibility factor is proportional to m. The prediction is intuitively understandable because the molecular volume scales linearly with the chain length.

(iv) Figure 8.10 compares the compressibility factor for hard-sphere chains predicted by TPT1 and TPT2. The contribution from the second-order term is relatively insignificant at all conditions because the correlation due to hard-sphere interactions is short-ranged, i.e., the correlation between the next nearest neighbor segments accounted for by the second-order term is negligible at most conditions.

Problem 8.23 The grand canonical partition function for a binary mixture of homopolymers A and B can be written as

$$\Xi = \sum_{\nu} \exp\{-\beta E_{\nu} - \int d\mathbf{r}[\beta V_A^{ext}(\mathbf{r})\tilde{\rho}_A(\mathbf{r}) + \beta V_B^{ext}(\mathbf{r})\tilde{\rho}_B(\mathbf{r})]\},$$

where E_{ν} stands for the potential energy of the polymer chains due to inter- and intra-chain interactions, $V_{i=A,B}^{ext}(\mathbf{r})$ is a one-body external potential applying to individual polymer segments, $\tilde{\rho}_{i=A,B}(\mathbf{r})$ is the instantaneous density of polymer segments, and the summation over microstates is a shorthand notation for

$$\sum_{\nu} = \sum_{N_A=0}^{\infty}\sum_{N_B=0}^{\infty} \frac{e^{\beta(N_A\mu_A+N_B\mu_B)}}{N_A!N_B!\Lambda^{3N_A}\Lambda^{3N_B}} \int \mathbf{R}_A^{N_A} \int \mathbf{R}_B^{N_B}$$

with all variables following their usual meanings.

(i) Show that the segment densities are given by

$$\rho_i(\mathbf{r}) = \langle \tilde{\rho}_i(\mathbf{r}) \rangle = \frac{\delta \beta \Omega}{\delta \beta V_i^{ext}(\mathbf{r})},$$

where $\beta\Omega = -\ln \Xi$.

(ii) Show that the density-density correlation function is related to the response of the segment density with respect to the one-body potential

$$\chi_{ij}(\mathbf{r},\mathbf{r}') \equiv \langle[\rho_i(\mathbf{r}) - \tilde{\rho}_i(\mathbf{r})][\rho_j(\mathbf{r}') - \tilde{\rho}_j(\mathbf{r}')]\rangle = -\frac{\delta\rho_i(\mathbf{r})}{\delta\beta V_j^{ext}(\mathbf{r})}.$$

Solution:

(i) The functional derivative of the grand potential with respect to the one-body external potential gives

$$\frac{\delta\Omega}{\delta V_i^{ext}(\mathbf{r})} = -\frac{1}{\Xi}\sum_\nu[-\tilde{\rho}_i(\mathbf{r})]\exp\left\{-\beta E_\nu - \int d\mathbf{r}[\beta V_A^{ext}(\mathbf{r})\tilde{\rho}_A(\mathbf{r}) + \beta V_B^{ext}(\mathbf{r})\tilde{\rho}_B(\mathbf{r})]\right\}$$
$$= \langle\tilde{\rho}_i(\mathbf{r})\rangle = \rho_i(\mathbf{r}).$$

(ii) The functional derivative of the segment density with respect to the one-body potential gives

$$\frac{\delta\rho_i(\mathbf{r})}{\delta\beta V_j^{ext}(\mathbf{r})} = \frac{1}{\Xi^2}\sum_\nu\tilde{\rho}_i(\mathbf{r})\tilde{\rho}_j(\mathbf{r})\exp\left\{-\beta E_\nu - \int d\mathbf{r}[\beta V_A^{ext}(\mathbf{r})\tilde{\rho}_A(\mathbf{r}) + \beta V_B^{ext}(\mathbf{r})\tilde{\rho}_B(\mathbf{r})]\right\}$$
$$-\frac{1}{\Xi}\sum_\nu\tilde{\rho}_i(\mathbf{r})\tilde{\rho}_j(\mathbf{r})\exp\left\{-\beta E_\nu - \int d\mathbf{r}[\beta V_A^{ext}(\mathbf{r})\tilde{\rho}_A(\mathbf{r}) + \beta V_B^{ext}(\mathbf{r})\tilde{\rho}_B(\mathbf{r})]\right\}$$
$$= \rho_i(\mathbf{r})\rho_j(\mathbf{r}) - \langle\tilde{\rho}_i(\mathbf{r})\tilde{\rho}_j(\mathbf{r}')\rangle = -\langle[\rho_i(\mathbf{r}) - \tilde{\rho}_i(\mathbf{r})][\rho_j(\mathbf{r}') - \tilde{\rho}_j(\mathbf{r}')]\rangle$$
$$= -\chi_{ij}(\mathbf{r},\mathbf{r}').$$

Problem 8.24 As discussed in Section 7.3, the static structure factor of a thermodynamic system can be determined from small-angle neutron scattering (SANS) or small-angle X-ray scattering (SAXS) measurements. In their application to multi-component systems, the intensity of the scattered wave at the detector is given by

$$I(k) = \sum_{ij}\sigma_i\sigma_j\hat{\chi}_{ij}(k), \tag{N}$$

where σ_i is atomic scattering factor (i.e., a parameter related to the beam interaction with the atomic scattering point i), and $\hat{\chi}_{ij}(k)$ is the Fourier transform of the density-density correlation function between atomic sites i and j. For a polymer melt with only two types of segments A and B, the assumption of incompressibility leads to a simplified expression for the scattering intensity

$$I(k) = k_\sigma\hat{\chi}(k). \tag{O}$$

(i) Derive Eq. (O) from Eq. (N) and relate $\hat{\chi}(k)$ and k_σ to $\hat{\chi}_{ij}(k)$ and σ_i;

(ii) Show that the random-phase approximation (RPA) predicts the scattering intensity for a binary mixture of A/B homopolymers

$$I(k) = k_\sigma\rho_t\left\{\frac{1}{\phi_A m_A f(kR_{g,A})} + \frac{1}{\phi_B m_B f(kR_{g,B})} - 2\chi_F\right\}^{-1},$$

where ρ_t is the total number density of polymer segments, ϕ_i and m_i are the volume fraction and the degree of polymerization for polymer i, χ_F is the Flory parameter, R_g stands for the radius of gyration, and $f(x) = 2(e^{-x^2} - 1 + x^2)/x^4$ is the Debye function.

(iii) Show that in the high-k limit, the scattering intensity can be approximated by

$$I(k) \approx \frac{12k_\sigma\rho_t\phi_A\phi_B}{k^2b^2},$$

where b is the Kuhn length for both polymers, and in the low-k limit,

$$I(k) \approx \frac{k_\sigma \rho_t}{2(\chi_{F,s} - \chi_F)},$$

where $\chi_{F,s}$ is the value of χ_F at the spinodal condition.

Solution:

(i) The assumption of incompressibility implies $\hat{\chi}_{AA}(k) = \hat{\chi}_{BB}(k) = -\hat{\chi}_{AB}(k) = -\hat{\chi}_{BA}(k)$. Substituting these relations into Eq. (N) indicates

$$I(k) = \sigma_A^2 \hat{\chi}_{AA}(k) + \sigma_B^2 \hat{\chi}_{BB}(k) + 2\sigma_A\sigma_B\hat{\chi}_{AB}(k) = (\sigma_A^2 - 2\sigma_A\sigma_B + \sigma_B^2)\hat{\chi}_{AA}(k) = k_\sigma \hat{\chi}(k),$$

where $k_\sigma \equiv \sigma_A^2 - 2\sigma_A\sigma_B + \sigma_B^2$ and $\hat{\chi}(k) \equiv \hat{\chi}_{AA}(k)$.

(ii) For a binary mixture of A/B homopolymers, the RPA predicts (Eq. 8.175)

$$\hat{\chi}(k) = \left\{ \frac{1}{\rho_A m_A f(kR_{g,A})} + \frac{1}{\rho_B m_B f(kR_{g,B})} - \frac{2\chi_F}{\rho_t} \right\}^{-1},$$

where ρ_i the number density of segment i, and m_i is the degree of polymerization. Therefore, the scattering intensity is

$$I(k) = k_\sigma \rho_t \left\{ \frac{1}{\phi_A m_A f(kR_{g,A})} + \frac{1}{\phi_B m_B f(kR_{g,A})} - 2\chi_F \right\}^{-1},$$

where $\phi_i = \rho_i/\rho_t$.

(iii) In the high-k limit, the Debye function can be approximated by $f(x) \approx 2/x^2$. Using $R_{g,i}^2 = m_i b^2/6$ for ideal chains, we can rewrite the scattering intensity as

$$I(k) \approx k_\sigma \rho_t \left\{ \frac{k^2 R_{g,A}^2}{2\phi_A m_A} + \frac{k^2 R_{g,B}^2}{2\phi_B m_B} \right\}^{-1}$$

$$= k_\sigma \rho_t \left\{ \frac{k^2 b^2}{12\phi_A} + \frac{k^2 b^2}{12\phi_B} \right\}^{-1} = \frac{12 k_b \rho_t}{k^2 b^2} \left\{ \frac{1}{\phi_A} + \frac{1}{\phi_B} \right\}^{-1} = \frac{12 k_\sigma \rho_t \phi_A \phi_B}{k^2 b^2}.$$

In the low-k limit, $f(x) \approx 1$ thus

$$I(k) \approx k_\sigma \rho_t \left\{ \frac{1}{\phi_A m_A} + \frac{1}{\phi_B m_B} - 2\chi_F \right\}^{-1} = \frac{k_b \rho_t}{2(\chi_{F,s} - \chi_F)},$$

where $\chi_{F,s} = \left(\frac{1}{\phi_A m_A} + \frac{1}{\phi_B m_B} \right)/2$.

Problem 8.25 The density-density correlation function derived from the random-phase approximation (RPA) can be utilized to account for correlation effects ignored in a typical mean-field theory. To elucidate, consider a binary mixture of A/B homopolymers as discussed in Section 8.6. It can be shown from functional integration that the free-energy density due to the density-density correlation can be calculated from

$$f_{corr} = \frac{1}{2} \sum_{ij} \int_0^1 d\lambda \int \frac{d\mathbf{k}}{(2\pi)^3} \hat{u}_{ij}(k) \hat{\chi}_{ij}(k, \lambda), \quad \text{(P)}$$

where $\hat{u}_{ij}(k)$ represents the Fourier transform of the pair potential between segments $u_{ij}(r)$. Using the response function predicted by RPA

$$\hat{\chi}_{ij}^{RPA}(k, \lambda) = [\hat{\chi}_0^{-1}(k) + \lambda \hat{\mathbf{u}}]_{ij}^{-1},$$

show that the correlation effect contributes to a free-energy density given by

$$f_{corr}^{RPA} = \frac{1}{2}\int \frac{d\mathbf{k}}{(2\pi)^3}\mathrm{Tr}\ln[\mathbf{I} + \hat{\chi}_0(k)\cdot\hat{\mathbf{u}}(k)],$$

where Tr stands for the matrix trace, and $\mathbf{I}$ is a 2×2 matrix.
Hint: The double summation with respect to segment indices i and j in Eq. (P) can be written as $\mathrm{Tr}(\hat{\mathbf{u}}^T\cdot\hat{\chi})$ and $\hat{\mathbf{u}}^T = \hat{\mathbf{u}}$.

Solution:
This problem is adapted from Qin and Morse.[17] Inserting the response function from RPA into Eq. (P) leads to

$$\begin{aligned} f_{corr} &= \frac{1}{2}\sum_{ij}\int_0^1 d\lambda\int\frac{d\mathbf{k}}{(2\pi)^3}\hat{u}_{ij}(k)\hat{\chi}_{ij}(k,\lambda) = \frac{1}{2}\int_0^1 d\lambda\int\frac{d\mathbf{k}}{(2\pi)^3}\mathrm{Tr}[\hat{\mathbf{u}}^T\cdot\hat{\chi}(k,\lambda)]\\ &= \frac{1}{2}\int_0^1 d\lambda\int\frac{d\mathbf{k}}{(2\pi)^3}\mathrm{Tr}\left[\frac{\hat{\mathbf{u}}}{\hat{\chi}_0^{-1}(k)+\lambda\hat{\mathbf{u}}}\right] = \frac{1}{2}\int\frac{d\mathbf{k}}{(2\pi)^3}\mathrm{Tr}\int_0^1 d\lambda\left[\frac{\hat{\mathbf{u}}}{\hat{\chi}_0^{-1}(k)+\lambda\hat{\mathbf{u}}}\right]\\ &= \frac{1}{2}\int\frac{d\mathbf{k}}{(2\pi)^3}\mathrm{Tr}\left\{\ln[\hat{\chi}_0^{-1}(k)+\hat{\mathbf{u}}] - \ln\hat{\chi}_0^{-1}(k)\right\} = \frac{1}{2}\int\frac{d\mathbf{k}}{(2\pi)^3}\mathrm{Tr}\ln[\mathbf{I}+\hat{\chi}_0(k)\hat{\mathbf{u}}(k)]. \end{aligned}$$

The answer is equivalent to Eq. (10) in the original publication because $\ln\det\mathbf{A} = \mathrm{Tr}\ln\mathbf{A}$ for any square nonsingular matrix $\mathbf{A}$.[18]

Problem 8.26 The random-phase approximation (RPA) is often used to interpret experimental data from X-ray/neutron scattering measurements of block copolymer systems.[19] As discussed in Section 8.6, the density-density correlation functions among polymer segments are defined as

$$\chi_{ij}(\mathbf{r},\mathbf{r}') = \langle\delta\tilde{\rho}_i(\mathbf{r})\delta\tilde{\rho}_j(\mathbf{r}')\rangle,$$

where i and j denote the types of polymer segments, $\delta\tilde{\rho}_i(\mathbf{r}) \equiv \tilde{\rho}_i(\mathbf{r}) - \rho_i$, $\tilde{\rho}_i(\mathbf{r})$ and ρ_i are the instantaneous and average number densities of segment i, respectively.

(i) Show that, for a diblock copolymer of Gaussian chains, the Fourier transforms of the response functions are linearly proportional to those of the intra-chain correlation functions (Problem 3.33)

$$\hat{\chi}_{ij}^0(k) = \frac{Nm_im_j}{V}\hat{\omega}_{ij}(k),$$

where N is the number of polymer chains in the system, V is the total volume, m_i and m_j represent the number of Kuhn segments in block i and block j, respectively.

(ii) In the presence of a weak external field $V_i^{ext}(\mathbf{r})$, RPA assumes a linear response of the local segment densities

$$\delta\hat{\rho}_i(\mathbf{k}) = -\sum_j\beta\hat{\chi}_{ij}(k)\hat{V}_j^{ext}(\mathbf{k}) \approx -\sum_j\beta\hat{\chi}_{ij}^0(k)[\hat{V}_j^{ext}(\mathbf{k}) + \hat{u}_j^{MF}(\mathbf{k})],$$

where $\hat{\chi}_{ij}^0(k)$ corresponds to that of the diblock Gaussian chains, and the one-body mean-field potentials are determined self-consistently from the density deviations

$$\hat{u}_i^{MF}(\mathbf{k}) = -\frac{Z}{\rho_t}\sum_{j=1}^{2}\epsilon_{ij}\delta\hat{\rho}_j(\mathbf{k}),$$

---

17 Qin J. and Morse D. C., "Renormalized one-loop theory of correlations in polymer blends", *J. Chem. Phys.*, 130, 224902 (2009).
18 Withers C. S. and Nadarajah S., "log det A = tr log A", *Int. J. Math. Educ. Sci. Technol.* 41, 1121–1124 (2010).
19 Ronald J. L. et al. "Chain conformation in ultrathin polymer films", *Nature* 400, 146–149 (1999).

where $\rho_t = \rho_1 + \rho_2$ represents the total segment density, and Z is the coordination number. Show that, with the assumption of incomprehensibility, the density-density correlation function of a diblock copolymer melt can be calculated from

$$\hat{\chi}(\mathbf{k}) = \left\{ \frac{\hat{\chi}^0_{11}(k) + 2\hat{\chi}^0_{12}(k) + \hat{\chi}^0_{22}(k)}{\hat{\chi}^0_{11}(k)\hat{\chi}^0_{22}(k) - [\hat{\chi}^0_{21}(k)]^2} - \frac{2\chi_F}{\rho_t} \right\}^{-1},$$

where $\hat{\chi}(\mathbf{k}) = \hat{\chi}_{11}(\mathbf{k}) = \hat{\chi}_{22}(\mathbf{k}) = -\hat{\chi}_{12}(\mathbf{k}) = -\hat{\chi}_{21}(\mathbf{k})$ and $\chi_F = Z\beta(\epsilon_{11} + \epsilon_{22} - 2\epsilon_{12})/2$ is the Flory parameter.

(iii) Plot the structure factor of a symmetric diblock polymer, $\hat{S}(k) = \hat{\chi}(k)/\rho_t$, versus k with $m =$ 1000, 2000, and 5000 Kuhn segments per chain. Assume the Kuhn length $b = 0.5$ nm and the Flory parameter $\chi_F = 0$ (e.g., segments differ by isotope replacement of certain elements).

(iv) Assume $\chi_F = 0$, derive the asymptotic limit of $\hat{S}(k)$ as $k \to 0$ and that as $k \to \infty$.

Solution:

(i) Given one type i polymer segment fixed at position $\mathbf{r}$, the instantaneous density of type j segments at position $\mathbf{r}'$ can be expressed in terms of the intra-chain and inter-chain contributions

$$\tilde{\rho}_j(\mathbf{r}') = \tilde{\rho}_j^{intra}(\mathbf{r}') + \tilde{\rho}_j^{inter}(\mathbf{r}').$$

When $|\mathbf{r} - \mathbf{r}'| \to \infty$, the intra-chain contribution vanishes. Hence, $\rho_j = \langle \tilde{\rho}_j(\mathbf{r}') \rangle = \langle \tilde{\rho}_j^{inter}(\mathbf{r}') \rangle$ at any position remote from $\mathbf{r}$, where $\langle \cdots \rangle$ denotes the ensemble average.

For a uniform system of Gaussian chains, only intra-chain segments are correlated. Therefore, the density-density correlation functions can be written as

$$\begin{aligned}
\chi^0_{ij}(\mathbf{r}, \mathbf{r}') &= \langle \tilde{\rho}_i(\mathbf{r})\tilde{\rho}_j(\mathbf{r}') \rangle - \rho_i\rho_j = \langle \tilde{\rho}_i(\mathbf{r})\tilde{\rho}_j^{intra}(\mathbf{r}') \rangle + \langle \tilde{\rho}_i(\mathbf{r})\tilde{\rho}_j^{inter}(\mathbf{r}') \rangle - \rho_i\rho_j \\
&= \langle \tilde{\rho}_i(\mathbf{r})\tilde{\rho}_j^{intra}(\mathbf{r}') \rangle + \langle \tilde{\rho}_i(\mathbf{r}) \rangle \langle \tilde{\rho}_j^{inter}(\mathbf{r}') \rangle - \rho_i\rho_j \\
&= \langle \tilde{\rho}_i(\mathbf{r})\tilde{\rho}_j^{intra}(\mathbf{r}') \rangle = N \sum_{n=1}^{m_i} \sum_{m=1}^{m_j} \langle \delta(\mathbf{r}_n - \mathbf{r})\delta(\mathbf{r}_m - \mathbf{r}') \rangle \\
&= \frac{N m_i m_j}{V} \omega_{ij}(|\mathbf{r} - \mathbf{r}'|),
\end{aligned}$$

where N is the number of polymer chains in the system, V is the total volume, and the Fourier transforms of the intra-chain correlation functions are given by (Problem 3.33)

$$\hat{\omega}_{ij}(k) = \begin{cases} 2(e^{-x_i} - 1 + x_i)/x_i^2, & i = j, \\ (e^{-x_i} - 1)(e^{-x_i} - 1)/(x_i x_j), & i \neq j, \end{cases} \tag{Q}$$

with $x_i = m_i k^2 b^2/6$.

(ii) We may express the density deviations for different types of segments as a vector and the density-density correlation function of the Gaussian chains in the matrix form

$$\delta\hat{\rho}(\mathbf{k}) = -\hat{\chi}_0(k) \cdot \beta[\hat{\mathbf{V}}^{ext}(\mathbf{k}) + \hat{\mathbf{u}}^{MF}(\mathbf{k})]. \tag{R}$$

Multiplying $\hat{\chi}_0^{-1}(k)$ on both sides of Eq. (R) gives

$$\hat{\chi}_0^{-1}(k) \cdot \delta\hat{\rho}(\mathbf{k}) = -\beta\hat{\mathbf{V}}^{ext}(\mathbf{k}) - \beta\hat{\mathbf{u}}^{MF}(\mathbf{k}).$$

Using the explicit expression for the inverse matrix

$$\hat{\chi}_0^{-1}(k) = \frac{1}{\det \hat{\chi}_0(k)} \begin{bmatrix} \hat{\chi}^0_{22}(k) & -\hat{\chi}^0_{21}(k) \\ -\hat{\chi}^0_{12}(k) & \hat{\chi}^0_{11}(k) \end{bmatrix},$$

and the incomprehensibility assumption, $\delta\hat{\rho}_1(\mathbf{k}) = -\delta\hat{\rho}_2(\mathbf{k})$, we have

$$[\hat{\chi}^0_{22}(k) + \hat{\chi}^0_{21}(k)]\delta\hat{\rho}_1(\mathbf{k})/\det\hat{\chi}_0(k) = -\beta\hat{V}^{ext}_1(\mathbf{k}) - \beta\hat{u}^{MF}_1(\mathbf{k}),$$

$$-[\hat{\chi}^0_{12}(k) + \hat{\chi}^0_{11}(k)]\delta\hat{\rho}_1(\mathbf{k})/\det\hat{\chi}_0(k) = -\beta\hat{V}^{ext}_2(\mathbf{k}) - \beta\hat{u}^{MF}_2(\mathbf{k}).$$

Combining these two equations with

$$\beta[\hat{u}^{MF}_1(\mathbf{k}) - \hat{u}^{MF}_2(\mathbf{k})] = -\frac{z\beta}{\rho_t}[\epsilon_{11} - 2\epsilon_{12} + \epsilon_{22}]\delta\hat{\rho}_1(\mathbf{k}) = -\frac{2\chi_F}{\rho_t}\delta\hat{\rho}_1(\mathbf{k})$$

leads to

$$\left[\frac{\hat{\chi}^0_{22}(k) + \hat{\chi}^0_{21}(k) + \hat{\chi}^0_{12}(k) + \hat{\chi}^0_{11}(k)}{\det\hat{\chi}_0(k)} - \frac{2\chi_F}{\rho_t}\right]\delta\hat{\rho}_1(\mathbf{k}) = -\beta\hat{V}^{ext}_1(\mathbf{k}) + \beta\hat{V}^{ext}_2(\mathbf{k}). \tag{S}$$

The incomprehensibility assumption implies $\hat{\chi}_{11}(\mathbf{k}) = \hat{\chi}_{22}(\mathbf{k}) = -\hat{\chi}_{12}(\mathbf{k}) = -\hat{\chi}_{21}(\mathbf{k})$. On the other hand, $\hat{\chi}_0(k)$ is a symmetric matrix, i.e., $\hat{\chi}^0_{21}(k) = \hat{\chi}^0_{12}(k)$. A comparison of Eq. (S) with the linear response relation

$$\delta\hat{\rho}_1(\mathbf{k}) = -\beta\hat{\chi}_{11}(k)\hat{V}^{ext}_1(\mathbf{k}) - \beta\hat{\chi}_{12}(k)\hat{V}^{ext}_2(\mathbf{k}) = -\beta\hat{\chi}_{11}(k)[\hat{V}^{ext}_1(\mathbf{k}) - \hat{V}^{ext}_2(\mathbf{k})]$$

yields

$$\hat{\chi}(k) \equiv \hat{\chi}_{11}(k) = \left\{\frac{\hat{\chi}^0_{11}(k) + 2\hat{\chi}^0_{12}(k) + \hat{\chi}^0_{22}(k)}{\hat{\chi}^0_{11}(k)\hat{\chi}^0_{22}(k) - [\hat{\chi}^0_{21}(k)]^2} - \frac{2\chi_F}{\rho_t}\right\}^{-1}. \tag{T}$$

(iii) The partial structure factors are defined as

$$\hat{S}^0_{ij}(k) \equiv \frac{\hat{\chi}^0_{ij}(k)}{\rho_t} = \frac{m_i m_j}{m_i + m_j}\omega_{ij}(k). \tag{U}$$

Accordingly, Eq. (T) predicts that the structure factor is given by

$$\hat{S}(k) = \frac{\hat{\chi}(k)}{\rho_t} = \left\{\frac{\hat{S}^0_{11}(k) + 2\hat{S}^0_{12}(k) + \hat{S}^0_{22}(k)}{\hat{S}^0_{11}(k)\hat{S}^0_{22}(k) - [\hat{S}^0_{21}(k)]^2} - 2\chi_F\right\}^{-1}. \tag{V}$$

For a symmetric diblock copolymer melt, $m_1 = m_2 = m/2$ where m is the number of Kuhn segments per chain. Therefore, Eq. (U) predicts

$$\hat{S}^0_{ij}(k) = \frac{m_i m_j}{m_i + m_j}\hat{\omega}_{ij}(k) = \frac{m}{4} \times \begin{cases} 2(e^{-x_i} - 1 + x_i)/x_i^2, & i = j, \\ (e^{-x_i} - 1)(e^{-x_i} - 1)/(x_i x_j), & i \neq j. \end{cases}$$

where $x_i = x_j = mk^2b^2/12$. With $\chi_F = 0$, the structure factor becomes

$$\hat{S}(k) = \frac{\hat{S}^0_{11}(k)\hat{S}^0_{22}(k) - [\hat{S}^0_{21}(k)]^2}{\hat{S}^0_{11}(k) + 2\hat{S}^0_{12}(k) + \hat{S}^0_{22}(k)} = \frac{m}{4}\frac{\hat{\omega}^0_{11}(k)\hat{\omega}^0_{22}(k) - [\hat{\omega}^0_{21}(k)]^2}{\hat{\omega}^0_{11}(k) + 2\hat{\omega}^0_{12}(k) + \hat{\omega}^0_{22}(k)}.$$

Figure 8.11 shows $\hat{S}(k)$ versus k for three symmetric diblock copolymer melts with $\chi_F = 0$ and $m = 1000, 2000$ and 5000. The peak positions reflect correlation between polymer segments on the length scale comparable with the radius of gyration $R_{g,i} = \sqrt{m/12}b \approx 4.6, 6.5$ and 7.9 nm per polymer block.

(iv) As $k \to 0$, we have from Eq. (Q) for $x_i \to 0$ and $x_j \to 0$

$$\hat{\omega}_{ij}(k) \approx \begin{cases} 1 - x_i/3, & i = j, \\ 1 - x_i/2 - x_j/2, & i \neq j. \end{cases}$$

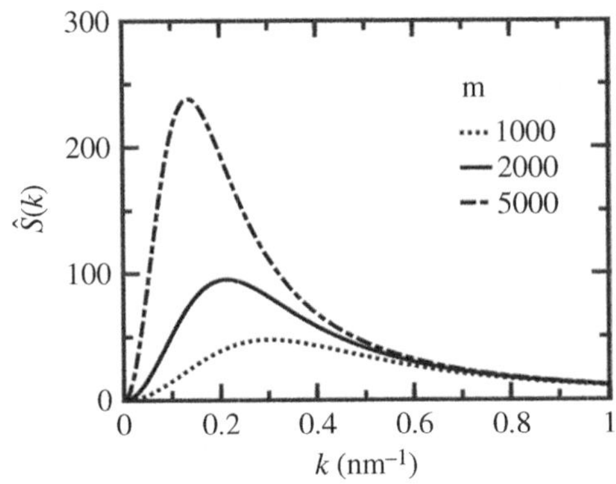

Figure 8.11 The structure factor of symmetric diblock copolymers with $m = 1000, 2000$, and 5000. Here, the Kuhn length is $b = 0.5$ nm, and the Flory parameter is $\chi_F = 0$ (BlockcopolymerRPA.m).

Hence, Eq. (U) predicts that the partial structure factors become

$$\hat{S}^0_{ij}(k) = \frac{m_i m_j}{m} \times \begin{cases} 1 - x_i/3, & i = j, \\ 1 - x_i/2 - x_j/2, & i \neq j, \end{cases}$$

where $m = m_i + m_j$. Using the simplified expressions for

$$\hat{S}^0_{11}(k)\hat{S}^0_{22}(k) - [\hat{S}^0_{21}(k)]^2 \approx \frac{m_1^2 m_2^2}{m^2}\left\{(1 - x_1/3)(1 - x_2/3) - (1 - x_1/2 - x_2/2)^2\right\}$$

$$= \frac{m_1^2 m_2^2}{m^2}\left\{\frac{2x_1}{3} + \frac{2x_2}{3}\right\} = \frac{m_1^2 m_2^2}{m^2}\frac{2mk^2b^2}{18} = \frac{m_1^2 m_2^2}{m}\frac{k^2b^2}{9}$$

and

$$\hat{S}^0_{11}(k) + \hat{S}^0_{22}(k) + 2\hat{S}^0_{21}(k) \approx \frac{m_1^2}{m} + \frac{m_2^2}{m} + \frac{2m_1 m_2}{m} = \frac{(m_1 + m_2)^2}{m} = m,$$

we obtain the structure factor from Eq. (V) in the limit of $k \to 0$

$$\hat{S}(k) \approx \frac{m_1^2 m_2^2}{m^2}\frac{k^2b^2}{9} = \frac{2m\alpha_1^2\alpha_2^2 k^2 R_g^2}{3},$$

where $\alpha_i = m_i/m$ and $R_g^2 = mb^2/6$.
On the other hand, as $k \to \infty$, we have

$$\hat{\omega}_{ij}(k) \approx \begin{cases} 2/x_i, & i = j, \\ 1/(x_1 x_2), & i \neq j, \end{cases}$$

Hence

$$\hat{S}^0_{11}(k)\hat{S}^0_{22}(k) - [\hat{S}^0_{21}(k)]^2 \approx \frac{m_1^2 m_2^2}{m^2}\frac{4}{x_1 x_2} = \frac{m_1^2 m_2^2}{m^2}\frac{4 \times 6^2}{m_1 m_2 b^4 k^4} = \frac{144\alpha_1\alpha_2}{b^2k^2}$$

and

$$\hat{S}^0_{11}(k) + \hat{S}^0_{22}(k) + 2\hat{S}^0_{21}(k)] \approx \frac{m_1^2}{m}\frac{2}{x_1} + \frac{m_2^2}{m}\frac{2}{x_2} + \frac{2m_1 m_2}{m}\frac{2}{x_1 x_2} = \frac{(m_1 + m_2)^2}{m}$$

$$\approx \frac{m_1^2}{m}\frac{12}{m_1 b^2 k^2} + \frac{m_2^2}{m}\frac{12}{m_2 b^2 k^2} = \frac{12}{b^2k^2}.$$

Plugging the above expressions into Eq. (V) yields

$$\hat{S}(k) \approx \frac{144\alpha_1\alpha_2}{b^2k^2} \times \frac{b^2k^2}{12} = \frac{12}{b^2k^2} = \frac{2m\alpha_1\alpha_2}{k^2R_g^2}.$$

Problem 8.27 This exercise applies the field theoretical method to an idealized system of electrolytes where the ionic species are described as point charges and the solvent is represented by a dielectric medium. Consider an ionic system with N_+ cations and N_- anions of equal valency $Z = Z_+ = -Z_-$. Assume that the interaction between unit charges can be represented by the reduced Coulomb potential

$$\upsilon_c(r) = l_B/r,$$

where r is the charge-to-charge distance, and $l_B = \beta e^2/(4\pi\varepsilon\varepsilon_0)$ is the Bjerrum length. As usual, $\beta = 1/(k_BT)$, e is the unit charge, ε is the dielectric constant of the medium, ε_0 is the permittivity of the free space. In the presence of a background charge with a fixed local density for the number of unit charges $\rho_c^0(\mathbf{r})$, the canonical partition function is given by

$$Q = \frac{1}{N_+!N_-!\Lambda^{3N}} \int d\mathbf{r}^N \exp\left\{-\frac{1}{2}\int d\mathbf{r} \int d\mathbf{r}' \tilde{\rho}_c(\mathbf{r})\upsilon_c(|\mathbf{r}-\mathbf{r}'|)\tilde{\rho}_c(\mathbf{r}') + NZ^2\upsilon_c(0)\right\}, \tag{W}$$

where $N = (N_+ + N_-)$ is the total number of ions in the system, Λ is the thermal wavelength of cations and anions, and $\tilde{\rho}_c(\mathbf{r})$ represents the local charge density

$$\tilde{\rho}_c(\mathbf{r}) \equiv \rho_c^0(\mathbf{r}) + Z_+\tilde{\rho}_+(\mathbf{r}) + Z_-\tilde{\rho}_-(\mathbf{r}),$$

where $\rho_c^0(\mathbf{r})$ is related to the background charge, and $\tilde{\rho}_\pm(\mathbf{r})$ are instantaneous ionic densities.

(i) Show that the partition function can be expressed in terms of the field variables

$$Q = \int D\rho_+ \int D\rho_- \int D\omega_+ \int D\omega_- \exp\left\{-\frac{1}{2}\int d\mathbf{r} \int d\mathbf{r}' \rho_c(\mathbf{r})\upsilon_c(|\mathbf{r}-\mathbf{r}'|)\rho_c(\mathbf{r}') + i\int d\mathbf{r}[\omega_+(\mathbf{r})\rho_+(\mathbf{r}) + \omega_-(\mathbf{r})\rho_-(\mathbf{r})] - \beta F_0(\omega_+, \omega_-)\right\}, \tag{X}$$

where $\rho_c(\mathbf{r}) \equiv \rho_c^0(\mathbf{r}) + Z_+\rho_+(\mathbf{r}) + Z_-\rho_-(\mathbf{r})$, F_0 stands for the Helmholtz energy of a binary system of noninteracting particles (viz., an ideal-gas mixture) in the presences of an effective one-body potential $\beta V_\pm^{ext}(\mathbf{r}) = i\omega_\pm(\mathbf{r}) - Z_\pm^2\upsilon_c(0)$.

(ii) Verify the following expression for the Helmholtz energy of the noninteracting system

$$\beta F_0 = \sum_{n=\pm} \int d\mathbf{r}\rho_n(\mathbf{r}) \left\{\ln[\rho_n(\mathbf{r})\Lambda^3] - 1 + \beta V_n^{ext}(\mathbf{r})\right\}.$$

(iii) Show that the saddle-point approximation leads to

$$\beta\mu_n = \ln[\rho_n(\mathbf{r})\Lambda^3] + Z_n \int d\mathbf{r}' \upsilon_c(|\mathbf{r}-\mathbf{r}'|)\rho_c(\mathbf{r}') - Z_n^2\upsilon_c(0).$$

(iv) Derive the Boltzmann distribution for ionic species

$$\rho_\pm(\mathbf{r}) = \rho_\pm^0 \exp[-\beta Z_\pm e\psi(\mathbf{r})],$$

where $\rho_\pm^0$ represent the number densities of ionic species in the bulk, and the local electrical potential $\psi(\mathbf{r})$ satisfies the Poisson equation

$$\nabla^2\psi(\mathbf{r}) = -\frac{e\rho_c(\mathbf{r})}{\varepsilon\varepsilon_0}.$$

(v) Show that the imaginary field is given by

$$i\omega_n(\mathbf{r}) = \beta Z_n e\psi(\mathbf{r}) + Z_n^2 v_c(0).$$

(vi) Derive the mean-field Helmholtz energy from the saddle-point approximation

$$\beta F_{MF} = \sum_{n=\pm}\int d\mathbf{r}\rho_n(\mathbf{r})[\ln\rho_n(\mathbf{r}) - 1] + \beta e\int d\mathbf{r}\rho_c(\mathbf{r})\psi(\mathbf{r}).$$

(vii) Discuss how the mean-field Helmholtz energy may be improved by using the random-phase approximation (RPA).

Solution:

(i) The field variables can be introduced by inserting the mathematical identity

$$1 = \int D\rho_\pm\delta(\rho_\pm - \tilde{\rho}_\pm) = \int D\rho_\pm\int D\omega_\pm \exp\left\{i\int d\mathbf{r}[\omega_\pm(\mathbf{r})[\rho_\pm(\mathbf{r}) - \tilde{\rho}_\pm(\mathbf{r})]\right\}$$

into the canonical partition function. Replacing the instantaneous density $\tilde{\rho}_\pm(\mathbf{r})$ in Eq. (W) with the corresponding density fields $\rho_\pm(\mathbf{r})$ leads to

$$\begin{aligned} Q = &\int D\rho_\pm\int D\omega_\pm \exp\\ &\times\left\{-\frac{1}{2}\int d\mathbf{r}\int d\mathbf{r}'\rho_c(\mathbf{r})v_c(|\mathbf{r}-\mathbf{r}'|)\rho_c(\mathbf{r}') + i\sum_{n=\pm}\int d\mathbf{r}\omega_n(\mathbf{r})\rho_n(\mathbf{r})\right\}\\ &\times\frac{1}{N_+!N_-!\Lambda^{3N}}\int d\mathbf{r}^N \exp\left\{-\sum_{n=\pm}\int d\mathbf{r}\left[i\omega_n(\mathbf{r}) - \frac{Z_n^2}{2}v_c(0)\right]\tilde{\rho}_n(\mathbf{r})\right\}. \end{aligned}$$

Thus, we derive Eq. (X), noting that the Helmholtz energy of noninteracting particles is given by

$$e^{-\beta F_0} = Q_0 = \frac{1}{N_+!N_-!\Lambda^{3N}}\int d\mathbf{r}^N \exp\left\{-\sum_{n=\pm}\int d\mathbf{r}\beta V_n^{ext}(\mathbf{r})\tilde{\rho}_n(\mathbf{r})\right\},$$

where $\beta V_n^{ext}(\mathbf{r}) = i\omega_n(\mathbf{r}) - Z_n^2 v_c(0)$.

(ii) See Appendix 8D for evaluating the Helmholtz energy of noninteracting particles.

(iii) We can express the grand partition function of the ionic system in terms of the field variables

$$\Xi = \sum_{N_\pm=0}^{\infty} Q\exp\left\{\sum_{n=\pm}\int d\mathbf{r}\tilde{\rho}_n(\mathbf{r})\beta\mu_n\right\} = \int D\rho_\pm\int D\omega_\pm e^{-\beta\mathcal{H}},$$

where μ_n denotes the ionic chemical potential. The "field" Hamiltonian $\mathcal{H}$ is defined as

$$\beta\mathcal{H} = \frac{1}{2}\int d\mathbf{r}\int d\mathbf{r}'\rho_c(\mathbf{r})v_c(|\mathbf{r}-\mathbf{r}'|)\rho_c(\mathbf{r}') - i\sum_{n=\pm}\int d\mathbf{r}\omega_n(\mathbf{r})\rho_n(\mathbf{r}) + \beta\Omega_0,$$

and the grand potential of noninteracting particles is

$$\begin{aligned} \beta\Omega_0 &\equiv \beta F_0 - \sum_{n=\pm}\int d\mathbf{r}\rho_n(\mathbf{r})\beta\mu_n\\ &= \sum_{n=\pm}\int d\mathbf{r}\rho_n(\mathbf{r})\left\{\ln[\rho_n(\mathbf{r})\Lambda^3] - 1 + \beta V_n^{ext}(\mathbf{r}) - \beta\mu_n\right\}. \end{aligned}$$

In the saddle-point approximation, the grand partition function is evaluated by maximizing the integrand of the path integral or, equivalently, by minimizing the Hamiltonian $\mathcal{H}$. The functional derivative of $\beta\mathcal{H}$ with respect to $\rho_n(\mathbf{r})$ leads to

$$\frac{\delta\beta\mathcal{H}}{\delta\rho_n} = Z_n \int d\mathbf{r}' v_c(|\mathbf{r}-\mathbf{r}'|)\rho_c(\mathbf{r}') - i\omega_n(\mathbf{r}) + \frac{\delta\beta\Omega_0}{\delta\rho_n} = 0. \tag{Y}$$

Because

$$\frac{\delta\beta\Omega_0}{\delta\rho_n} = \ln[\rho_n(\mathbf{r})\Lambda^3] + \beta V_n^{ext}(\mathbf{r}) - \beta\mu_n, \tag{Z}$$

and $\beta V_n^{ext}(\mathbf{r}) = i\omega_n(\mathbf{r}) - Z_n^2 v_c(0)$, substituting Eq. (Z) into (Y) gives

$$\begin{aligned}\beta\mu_n &= \ln[\rho_n(\mathbf{r})\Lambda^3] + \beta V_n^{ext}(\mathbf{r}) + Z_n \int d\mathbf{r}' v_c(|\mathbf{r}-\mathbf{r}'|)\rho_c(\mathbf{r}') - i\omega_n(\mathbf{r}) \\ &= \ln[\rho_n(\mathbf{r})\Lambda^3] + Z_n \int d\mathbf{r}' v_c(|\mathbf{r}-\mathbf{r}'|)\rho_c(\mathbf{r}') - Z_n^2 v_c(0).\end{aligned} \tag{AA}$$

(iv) Note that for a point charge of valence Z_n at position $\mathbf{r}$, the electrical energy can be expressed in terms of the mean electrical potential $\psi(\mathbf{r})$

$$Z_n \int d\mathbf{r}' v_c(|\mathbf{r}-\mathbf{r}'|)\rho_c(\mathbf{r}') - Z_n^2 v_c(0) \equiv \beta Z_n e\psi(\mathbf{r}). \tag{AB}$$

Because the self-energy cancels with the divergence point in the integration over $\rho_c(\mathbf{r}')$, $\psi(\mathbf{r})$ should exclude the contribution due to the point charge placed at $\mathbf{r}$, i.e., it is exclusively related to *other* charges in the system. Applying the Fourier transformation to both sides of Eq. (AB) leads to

$$Z_n \frac{4\pi l_B \rho_c(\mathbf{k})}{k^2} = \beta Z_n e\psi(\mathbf{k}). \tag{AC}$$

The Poisson equation can be derived by using the explicit expression for $l_B = \beta e^2/(4\pi\varepsilon\varepsilon_0)$ and applying the inverse Fourier transformation to $4\pi l_B \rho_c(\mathbf{k}) = \beta e k^2 \psi(\mathbf{k})$

$$\nabla^2\psi(\mathbf{r}) = -\frac{e\rho_c(\mathbf{r})}{\varepsilon\varepsilon_0}.$$

Meanwhile, the Boltzmann equation is obtained from Eq. (AA) and the ideal-gas relation $\beta\mu_n = \ln[\rho_n^0\Lambda^3]$ for the bulk system:

$$\rho_n(\mathbf{r}) = \rho_n^0 \exp[-\beta Z_n e\psi(\mathbf{r})].$$

(v) The functional derivative of $\beta\mathcal{H}$ with respect to $\omega_n(\mathbf{r})$ leads to

$$\frac{\delta\beta H}{\delta\omega_n} = -i\rho_n(\mathbf{r}) + \frac{\delta\beta\Omega_0}{\delta\beta V_n^{ext}(\mathbf{r})}\frac{\delta\beta V_n^{ext}(\mathbf{r})}{\delta\omega_n} = -i\rho_n(\mathbf{r}) + i\rho_n(\mathbf{r}) = 0. \tag{AD}$$

Eq. (AD) indicates that the saddle-point condition is automatically satisfied. For the reference system with noninteracting particles in an external field, the density profiles are exactly known

$$\rho_n(\mathbf{r}) = \rho_n^0 \exp[-\beta V_n^{ext}(\mathbf{r})] = \rho_n^0 \exp[-i\omega_n(\mathbf{r}) + Z_n^2 v_c(0)].$$

A comparison of the above expression with the Boltzmann distribution gives

$$i\omega_n(\mathbf{r}) = \beta Z_n e\psi(\mathbf{r}) + Z_n^2 v_c(0).$$

(vi) Substituting the saddle-point density and potential into the field Hamiltonian leads to

$$\beta F_{MF} = \beta F_0 + \frac{1}{2}\int d\mathbf{r}\int d\mathbf{r}'\rho_c(\mathbf{r})\upsilon_c(|\mathbf{r}-\mathbf{r}'|)\rho_c(\mathbf{r}') - i\sum_{n=\pm}\int d\mathbf{r}\omega_n(\mathbf{r})\rho_n(\mathbf{r})$$
$$= \sum_{n=\pm}\int d\mathbf{r}\rho_n(\mathbf{r})\left\{\ln[\rho_n(\mathbf{r})\Lambda^3] - 1\right\} + \beta e\int d\mathbf{r}\rho_c(\mathbf{r})\psi(\mathbf{r}),$$

where the self-energy term has been omitted with the understanding that it has been subtracted from the double integration.

(vii) A number of tutorial reviews are available for applying RPA to ionic systems (e.g., see Wang[20]).

20 Wang Z. G., "Variational methods in statistical thermodynamics—A pedagogical introduction", In *Variational Methods in Molecular Modeling*, Wu J. (Ed.), 2016, Springer, pp. 1–29.

9

Solvation, Electrolytes, and Electric Double Layer

Problem 9.1 The Kirkwood–Buff (KB) fluctuation theory of solutions[1] can be understood as an extension of the compressibility equation for one-component systems (Section 7.3)

$$\left(\frac{\partial \rho}{\partial \beta P}\right)_T = \frac{1}{\rho}\left(\frac{\partial \rho}{\partial \beta \mu}\right)_T = 1 + \rho \int d\mathbf{r} h(r), \tag{A}$$

to mixtures and solutions. The key results can be derived by considering, as discussed in Problem 2.22, the correlated fluctuations of the molecular numbers for individual species in an open system that are described by the grand canonical ensemble

$$\left\langle \delta N_i \delta N_j \right\rangle = \left(\frac{\partial \left\langle N_i \right\rangle}{\partial \beta \mu_j}\right)_{V,T,\mu_{k\neq j}}. \tag{B}$$

Because $\left\langle \delta N_i \delta N_j \right\rangle$ is related to the total correlation function (TCF) between molecular species i and j (Problem 8.17)

$$\left\langle \delta N_i \delta N_j \right\rangle = \langle N_i \rangle \delta_{ij} + \frac{\langle N_i \rangle \langle N_j \rangle}{V} \int d\mathbf{r} h_{ij}(r), \tag{C}$$

where δ_{ij} is the Kronecker delta-function, the thermodynamic properties of a multicomponent system can be predicted from the KB integrals $G_{ij} \equiv \int d\mathbf{r} h_{ij}(r)]$. In Eqs. (B) and (C), subscripts i and j denote the chemical species in the system, and the TCF is defined as $h_{ij}(r) \equiv \langle g_{ij}(\mathbf{R}_i, \mathbf{R}_j)\rangle - 1$ where $g_{ij}(\mathbf{R}_i, \mathbf{R}_j)$ represents the corresponding radial distribution function (RDF), $\langle \cdots \rangle$ denotes average over all possible molecular configurations, and r stands for the intermolecular distance. The latter can be determined from molecular configurations $\mathbf{R}_i$ and $\mathbf{R}_j$.

(i) Let

$$A_{ij} \equiv \left(\frac{\partial \beta \mu_i}{\partial \rho_j}\right)_{T,\rho_{k\neq j}} \quad \text{and} \quad B_{ij} \equiv \left(\frac{\partial \rho_i}{\partial \beta \mu_j}\right)_{T,\mu_{k\neq j}}, \tag{D}$$

where $\rho_i = \langle N_i \rangle / V$. Show that, in the matrix form, A_{ij} can be determined from

$$\hat{\mathbf{A}} = \hat{\mathbf{B}}^{-1}.$$

(ii) Show $B_{ij} = \rho_i(\delta_{ij} + \rho_j G_{ij})$.

1 Kirkwood J. G. and Buff F. P., "The statistical mechanical theory of solutions", *J. Chem. Phys.* 19, 774 (1951).

Solutions Manual to Fundamentals and Practice in Statistical Thermodynamics,
First Edition. Jianzhong Wu and John M. Prausnitz.

(iii) Practical applications of the KB theory are often concerned with quantities related to the variation of the chemical potential with the molecular number of a chemical species at a fixed pressure. Show that

$$\mu_{ij} \equiv \left(\frac{\partial \beta\mu_i}{\partial N_j}\right)_{T,P,N_{k\neq j}} = \frac{1}{V}\left(A_{ij} - \frac{\beta\overline{V}_i\overline{V}_j}{\kappa_T}\right). \tag{E}$$

Here, $N_i = \langle N_i \rangle$ for briefness, $\overline{V}_i$ stands for the partial volume of component i, and κ_T is the isothermal compressibility of the system.

(iv) Verify that the partial volume can be expressed as

$$\overline{V}_i = k_B T\kappa_T \sum_k \rho_k A_{ki}, \tag{F}$$

Hint: At constant temperature, the Gibbs–Duhem equation is given by $\sum_i N_i d\mu_i = VdP$.

(v) Based on $V = \sum_i N_i\overline{V}_i$, verify

$$\kappa_T^{-1} = k_B T\sum_{i,j}\rho_i\rho_j A_{ij}. \tag{G}$$

(vi) Based on Eqs. (E), (F), and (G), show

$$\mu_{ij} = \frac{1}{V}\frac{\sum_{nk}\rho_n\rho_k[A_{nk}A_{ij} - A_{ni}A_{kj}]}{\sum_{n,k}\rho_n\rho_k A_{nk}}. \tag{H}$$

Solution:

The derivations below follow Ben Naim[2]

(i) At fixed T, we can express the total derivative of multivariable function $\mu_i = \mu_i(\{\rho_k\})$ as

$$d\mu_i = \sum_k \left(\frac{\partial \mu_i}{\partial \rho_k}\right)_{T,\rho_{n\neq k}} d\rho_k.$$

Dividing both sides of the above equation by $d\mu_j$ with $\mu_{m\neq j}$ fixed leads to

$$\left(\frac{\partial \mu_i}{\partial \mu_j}\right)_{T,\mu_{k\neq j}} = \delta_{ij} = \sum_k \left(\frac{\partial \beta\mu_i}{\partial \rho_k}\right)_{T,\rho_{n\neq k}} \left(\frac{\partial \rho_k}{\partial \beta\mu_j}\right)_{T,\mu_{m\neq j}} = \sum_k A_{ik}B_{kj}.$$

Hence, $\hat{\mathbf{A}} = \hat{\mathbf{B}}^{-1}$.

(ii) From Eqs. (B) and (C), we have

$$B_{ij} = \left(\frac{\partial \rho_i}{\partial \beta\mu_j}\right)_{T,\mu_{k\neq j}} = \langle \delta N_i \delta N_j \rangle / V = \rho_i(\delta_{ij} + \rho_j G_{ij}).$$

(iii) At fixed T and $N_{k\neq j}$, the total derivative of multivariable function $\mu_i = \mu_i(P, N_j)$ gives

$$d\mu_i = \left(\frac{\partial \mu_i}{\partial P}\right)_{T,\{N_k\}} dP + \left(\frac{\partial \mu_i}{\partial N_j}\right)_{T,P,N_{k\neq j}} dN_j.$$

With fixed total volume V, dividing both sides of the above equation by dN_j leads to

$$\left(\frac{\partial \mu_i}{\partial N_j}\right)_{T,V,N_{k\neq j}} = \left(\frac{\partial \mu_i}{\partial P}\right)_{T,\{N_k\}} \left(\frac{\partial P}{\partial N_j}\right)_{T,V,N_{k\neq j}} + \left(\frac{\partial \mu_i}{\partial N_j}\right)_{T,P,N_{k\neq j}}. \tag{I}$$

---

2 Ben Naim A., *Molecular Theory of Solutions*, Oxford University Press, Oxford, 2009. Chapter 4.

Using the thermodynamic identity

$$\left(\frac{\partial \mu_i}{\partial P}\right)_{T,\{N_k\}} = \left(\frac{\partial V}{\partial N_i}\right)_{T,P,N_{k\neq i}} = \overline{V}_i,$$

and the chain rule of differentiation

$$\left(\frac{\partial P}{\partial N_j}\right)_{T,V,N_{k\neq j}} = -\left(\frac{\partial P}{\partial V}\right)_{T,\{N_k\}}\left(\frac{\partial V}{\partial N_j}\right)_{T,P,N_{k\neq j}} = \frac{\overline{V}_j}{\kappa_T V}, \tag{J}$$

where $\kappa_T = -(1/V)(\partial V/\partial P)_{T,\{N_k\}}$, we obtain from Eq. (I)

$$\mu_{ij} \equiv \left(\frac{\partial \beta\mu_i}{\partial N_j}\right)_{T,P,N_{k\neq j}} = \frac{1}{V}\left(\frac{\partial \beta\mu_i}{\partial \rho_j}\right)_{T,\rho_{k\neq j}} - \frac{\beta\overline{V}_i\overline{V}_j}{V\kappa_T} = \frac{1}{V}\left(A_{ij} - \frac{\beta\overline{V}_i\overline{V}_j}{\kappa_T}\right).$$

(iv) From $dP = \sum_k \rho_k d\mu_k$, we obtain

$$\left(\frac{\partial P}{\partial N_i}\right)_{T,V,N_{j\neq i}} = \sum_k \rho_k\left(\frac{\partial \mu_k}{\partial N_i}\right)_{T,V,N_{j\neq i}}.$$

Substituting Eq. (J) and the definition of A_{ki} into the above equation leads to

$$\frac{\overline{V}_i}{\kappa_T V} = \frac{1}{\beta V}\sum_k \rho_k\left(\frac{\partial \beta\mu_k}{\partial \rho_i}\right)_{T,\rho_{j\neq i}} = \frac{1}{\beta V}\sum_k \rho_i A_{ki}.$$

Thus, we have

$$\overline{V}_i = k_B T\kappa_T \sum_k \rho_k A_{ki}.$$

(v) Substituting Eq. (F) into $1 = \sum_i \rho_i \overline{V}_i$ results in $1 = k_B T\kappa_T \sum_i\sum_k \rho_i\rho_k A_{ki}$. Hence

$$\kappa_T^{-1} = k_B T\sum_{i,j}\rho_i\rho_j A_{ij}.$$

(vi) Plugging Eqs. (F) and (G) into Eq. (E) leads to

$$\mu_{ij} = \frac{\kappa_T}{\beta V}\left(\frac{\beta A_{ij}}{\kappa_T} - \frac{\beta^2\overline{V}_i\overline{V}_j}{\kappa_T^2}\right) = \frac{\kappa_T}{\beta V}\left(\frac{\beta A_{ij}}{\kappa_T} - \sum_n \rho_n A_{ni}\sum_k \rho_k A_{kj}\right)$$
$$= \frac{1}{V}\frac{\sum_{nk}\rho_n\rho_k[A_{nk}A_{ij} - A_{ni}A_{kj}]}{\sum_{n,k}\rho_n\rho_k A_{nk}}.$$

Problem 9.2 Regarding the applications of the Kirkwood–Buff (KB) theory to formulation chemistry, Shimizu stated *"Not only can statistical thermodynamics guide us through experimental design, analysis and interpretation but also helps us keep confusion at bay."*[3] To elucidate how the KB theory helps the rational design of experiments and better understanding cosolvent effects, consider chemical processes involving a solute molecule (denoted as u) in solvent (1) and cosolvent (2) shown schematically in Figure 9.1.

3 Shimizu S., "Formulating rationally via statistical thermodynamics", *Curr. Opin. Colloid Interface Sci.* 48, 53–64 (2020).

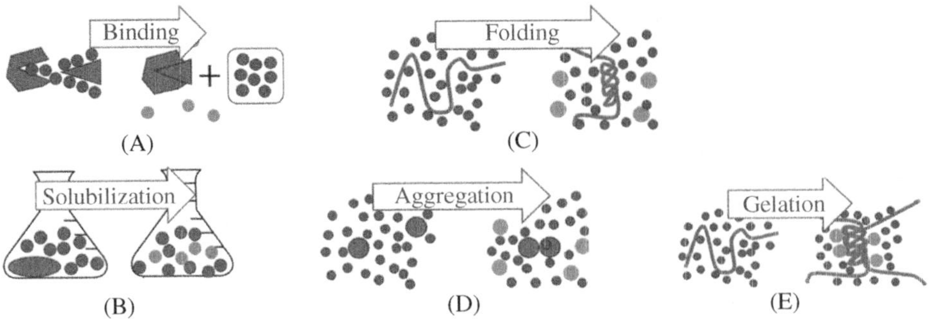

Figure 9.1 Cosolvent plays an important role in diverse chemical processes such as binding between biomacromolecules (A), solubilization of pharmaceuticals (B), protein folding (C), colloidal aggregation (D), and gelation (E). Source: Adapted from Shimizu.[3]

(i) Show that the equilibrium constant K for solute transition from an initial state (I) to its final state (F) $u_I \Longrightarrow u_F$ varies with the solvent activity a_1 according to

$$\left(\frac{\partial \ln K}{\partial \ln a_1}\right)_{T,P,\rho_u\to 0} = \rho_1(\Delta G_{u1} - \Delta G_{u2}), \tag{K}$$

where ρ_1 is the number density of the solvent molecules, $\rho_u \to 0$ means the solute at infinite dilution, ΔG_{u1} and ΔG_{u2} stand for the changes of the KB integrals for the solvent and cosolvent, respectively.

(ii) Show that the volume change accompanying the solute transition $u_I \Longrightarrow u_F$ is given by

$$\Delta V = -\rho_1 \overline{V}_1 \Delta G_{u1} - \rho_2 \overline{V}_2 \Delta G_{u2}. \tag{L}$$

(iii) Discuss how Eqs. (K) and (L) offer insights into experimental design to study cosolvent effects.

Solution:

(i) For the ternary system at constant T and P, the Gibbs–Duhem equation can be written as

$$-N_u d\mu_u = N_1 d\mu_1 + N_2 d\mu_2. \tag{M}$$

As the equilibrium constant for transition $u_I \Longrightarrow u_F$ is defined by the change in the solute chemical potential at infinite dilution, $\Delta G^0 = \mu^0_{u_F} - \mu^0_{u_I}$, we may consider one solute molecule $N_u = 1$ in a macroscopic volume V and compare Eq. (M) for the solute before and after the transition:

$$-d\Delta G^0 = \Delta N_1 d\mu_1 + \Delta N_2 d\mu_2. \tag{N}$$

Dividing both sides of Eq. (N) by $d\mu_1 = k_B T d\ln a_1$ gives

$$-\frac{1}{k_B T}\left(\frac{\partial \Delta G^0}{\partial \ln a_1}\right)_{T,P,\rho_u\to 0} = \Delta N_1 + \Delta N_2 \left(\frac{\partial \mu_2}{\partial \mu_1}\right)_{T,P,\rho_u\to 0}. \tag{O}$$

At infinite dilution, Eq. (M) reduces to $N_1 d\mu_1 + N_2 d\mu_2 = 0$, which implies

$$\left(\frac{\partial \mu_2}{\partial \mu_1}\right)_{T,P,\rho_u\to 0} = \frac{\rho_1}{\rho_2}, \tag{P}$$

where $\rho_i = N_i/V$. Plugging Eq. (P), along with $\Delta G^0 = -k_B T \ln K$ and $\Delta N_i = \rho_i \Delta G_{ui}$, into Eq. (O) gives the desired result

$$\left(\frac{\partial \ln K}{\partial \ln a_1}\right)_{T,P,\rho_u\to 0} = \rho_1 \Delta G_{u1} - \rho_2 \Delta G_{u2}\left(\frac{\rho_1}{\rho_2}\right) = \rho_1(\Delta G_{u1} - \Delta G_{u2}).$$

(ii) The volume change due to the solute transition can be found from the integration of the Maxwell relation $\frac{\partial V}{\partial N_u} = \frac{\partial \mu_u}{\partial P}$:

$$\Delta V = \left(\frac{\partial \Delta G^0}{\partial P}\right)_{T,x_2,\rho_u\to 0} = -\Delta N_1\left(\frac{\partial \mu_1}{\partial P}\right)_{T,x_2,\rho_u\to 0} - \Delta N_2\left(\frac{\partial \mu_2}{\partial P}\right)_{T,x_2,\rho_u\to 0}$$
$$= -\rho_1\overline{V}_1\Delta G_{u1} - \rho_2\overline{V}_2\Delta G_{u2}.$$

where x_2 stands for the cosolvent mole fraction.

(iii) Eqs. (K) and (L) allow for the prediction of KB integrals ΔG_{u1} and ΔG_{u2} thereby the solvation behavior from experimental data for variations of the equilibrium constant in response to the changes in the solvent activity or pressure. The equations can be used to clarify how complementary experimental techniques can be used to quantify solute-solvent and solute-cosolvent interactions. Importantly, these equations predict how many experiments are necessary in order to account for all pertinent parameters, which is essential for the experimental design. Because the microscopic details underpinning solute-solvent interactions are not easily accessible through conventional experiments, the KB theory offers a powerful tool to investigate such effects indirectly without approximations.

Problem 9.3 To formulate the McMillan–Mayer theory using the grand canonical ensemble, consider a liquid solution as a binary mixture with the grand partition function

$$\Xi = \sum_{\nu} \exp\{-\beta[E_\nu - N_0\mu_0 - N\mu]\}, \tag{Q}$$

where microstate ν is defined by the number of solvent molecules N_0 and that of solute molecules N as well as their atomic degrees of freedom, E_ν represents the total energy, μ_0 and μ are the solvent and solute chemical potentials, respectively.

(i) Show that Ξ can be written as the McMillan–Mayer partition function

$$\tilde{\Xi} \equiv \Xi/\Xi_0 = \sum_{\nu_N} \exp[-\beta u_{\nu_N} + \beta N\mu], \tag{R}$$

where Ξ_0 is the grand partition function of the pure solvent, ν_N denotes the microstates of N solute molecules, and u_{ν_N} represents the solvated energy of N solute molecules.

(ii) At infinite dilution, the osmotic pressure is described by the van't Hoff law

$$\Pi = \frac{k_BT}{V}\ln\tilde{\Xi} = \frac{\langle N\rangle k_BT}{V}. \tag{S}$$

Solution:

(i) As discussed in Section 9.1, the total energy of the system can be written as

$$E_\nu = E_{\nu_0} + E_{\nu_N},$$

where E_{ν_0} corresponds to the energy of the pure solvent at microstate ν_0, and E_{ν_N} is the solute energy. The latter encompasses not only the energy of the solute molecules but also the potential energy due to the solute-solvent interactions, and thus it depends on the microstates of both solvent and solute molecules. Accordingly, the grand partition function is given by

$$\Xi = \sum_{\nu_N}\sum_{\nu_0} \exp\{-\beta(E_{\nu_0} + E_{\nu_N} - N_0\mu_0 - N\mu)\}$$
$$= \Xi_0 \sum_{\nu_N} e^{\beta N\mu} \frac{\sum_{\nu_0} \exp\{-\beta(E_{\nu_0} + E_{\nu_N} - N_0\mu_0)\}}{\sum_{\nu_0} \exp\{-\beta(E_{\nu_0} - N_0\mu_0)\}}$$
$$= \Xi_0 \sum_{\nu_N} e^{\beta N\mu}\langle \exp(-\beta E_{\nu_N})\rangle_0 = \Xi_0 \sum_{\nu_N} \exp[-\beta u_{\nu_N} + \beta N\mu],$$

where $u_{\nu_N} = -k_B T \ln \langle \exp(-\beta E_{\nu_N}) \rangle_0$ represents the solvated energy of the solute molecules. Note that $\tilde{\Xi} \equiv \Xi/\Xi_0$ is formally identical to the grand partition function for the solute molecules in the gas phase except that the molecular energy in the vacuum is replaced by the solvated energy.

(ii) At infinite dilution, we have

$$u_{\nu_N} = \sum_{i=1}^{N} u^0_{\nu_i},$$

where $u^0_{\nu_i}$ is the solvated energy of a solute molecule at microstate ν_i^0. Thus, the McMillan–Mayer partition function becomes

$$\tilde{\Xi} = \sum_{\nu_N} \exp[-\beta \sum_{i=1}^{N} u^0_{\nu_i} + \beta N \mu] = \sum_{N=0}^{\infty} \frac{1}{N!} \left\{ \sum_{\nu_1} \exp[-\beta u^0_{\nu_1} + \beta \mu] \right\}^N = e^{\tilde{\xi}},$$

where $\tilde{\xi} \equiv \sum_{\nu_1} \exp[-\beta u^0_{\nu_1} + \beta \mu] = e^{\beta \mu} \tilde{q}_0$, and $\tilde{q}_0$ represents a single-molecule partition function. From $PV = k_B T \ln \Xi$, we find that the osmotic pressure satisfies

$$\Pi = P(T, V, \mu_0, \mu) - P(T, V, \mu_0) = \frac{k_B T}{V} \ln \tilde{\Xi} = \frac{k_B T}{V} \tilde{\xi}.$$

The above equation is the same as the van't Hoff law because the average number of solute molecules in the system is

$$\langle N \rangle = \frac{\partial \ln \Xi}{\partial \beta \mu} = \frac{\partial \ln \tilde{\Xi}}{\partial \beta \mu} = \frac{\partial \tilde{\xi}}{\partial \beta \mu} = \tilde{\xi}.$$

Problem 9.4 Consider the McMillan–Mayer partition function for a colloidal dispersion of spherical particles

$$\tilde{\Xi} = \sum_{\nu_N} \exp[-\beta u_{\nu_N} + \beta N \mu], \tag{T}$$

where ν_N stands for the microstates of N colloidal particles, μ is the colloidal chemical potential, and $u_{\nu_N} \equiv -k_B T \ln \langle \exp(-\beta E_{\nu_N}) \rangle_0$ is the solvated energy of N particles at microstate u.

(i) Show that, in terms of the conventional phase-space variables, Eq. (T) can be written as

$$\tilde{\Xi} = \sum_{N=0}^{\infty} \frac{e^{\beta N(\mu - F_1)}}{N! \Lambda^{3N}} \int d\mathbf{r}^N \exp[-\beta W(\mathbf{r}^N)], \tag{U}$$

where Λ is the thermal wavelength, F_1 is the single-particle solvation free energy, and $W(\mathbf{r}^N)$ is the potential of mean force for N particles at positions $\mathbf{r}^N \equiv (\mathbf{r}_1, \mathbf{r}_2, \ldots, \mathbf{r}_N)$.

(ii) Assume pairwise additivity for the potential of mean force $W(\mathbf{r}^N) \approx \sum_{i<j} w(|\mathbf{r}_i - \mathbf{r}_j|)$, show the virial expansion for the osmotic pressure

$$Z \equiv \frac{\Pi V}{\langle N \rangle k_B T} = 1 + B_2 \rho + B_3 \rho^3 + \cdots, \tag{V}$$

where $\rho = \langle N \rangle / V$, and the osmotic virial coefficients B_2, B_3, and $\cdots$ are defined by the Mayer function $f_{ij} \equiv \exp[-\beta w(|\mathbf{r}_i - \mathbf{r}_j|)] - 1$:

$$B_2 = -\frac{1}{2} \int d\mathbf{r}_{12} f_{12}, \quad B_3 = -\frac{1}{3} \int d\mathbf{r}_{12} \int d\mathbf{r}_{13} f_{12} f_{23} f_{13}, \quad \ldots,$$

(iii) Verify the osmotic virial equation

$$\Pi = \rho k_B T - \frac{\rho^2}{6} \int d\mathbf{r} \frac{\partial w(r)}{\partial r} r g(r), \tag{W}$$

where $g(r)$ is the radial distribution function (RDF) for colloidal particles.

(iv) Verify the osmotic compressibility equation

$$(\partial \rho / \partial \beta \Pi)_T = 1 + \rho \int d\mathbf{r} h(r), \tag{X}$$

where $h(r)$ is the total correlation function (TCF).

Solution:

(i) For spherical particles in a solvent, the solved energy can be written as

$$u_{v_N} = -k_B T \ln \langle \exp\{-\beta[K_N + \Phi(\mathbf{r}^N)]\}\rangle_0 = K_N - k_B T \ln \langle \exp[-\beta \Phi(\mathbf{r}^N)]\rangle_0,$$

where $K_N = \sum_i^N p_i^2/2m$ stands for the kinetic energy of colloidal particles, p_i is the particle momentum and m is the particle mass. $\Phi(\mathbf{r}^N)$ represents the potential energy due to the particle-particle and particle-solvent interactions.

At infinite dilution, the solved energy for each particle includes the kinetic energy due to the particle motion and a potential energy due to its interaction with the solvent

$$u_{v_i} = -k_B T \ln \langle \exp\{-\beta[K_i + \Phi_i(\mathbf{r})]\}\rangle_0 = K_i - k_B T \ln \langle \exp[-\beta \Phi_1(\mathbf{r})]\rangle_0 = K_i + F_1,$$

where $K_i = p_i^2/2m$, $\Phi_i(\mathbf{r})$ denotes the potential energy due to the particle interaction with the solvent molecules, and F_1 is the solvation free energy.

Because $\langle K_N \rangle_K = \sum_i \langle K_i \rangle_K$ and the potential of mean force is defined as

$$W(\mathbf{r}^N) = \langle u_{v_N} \rangle_K - \sum_i \langle u_{v_i} \rangle_K = -k_B T \ln \langle \exp[-\beta \Phi(\mathbf{r}^N)]\rangle_0 - N F_1,$$

where $\langle \cdots \rangle_K$ represents average over the kinetic degrees of freedom, the McMillan–Mayer partition function becomes

$$\begin{aligned}\tilde{\Xi} &= \sum_{N=0}^{\infty} \frac{1}{N! h^{3N}} \int d\mathbf{p}^N \int d\mathbf{r}^N \exp[-\beta K_N + \ln \langle \exp[-\beta \Phi(\mathbf{r}^N)]\rangle_0 + \beta N \mu] \\ &= \sum_{N=0}^{\infty} \frac{e^{\beta N(\mu - F_1)}}{N! \Lambda^{3N}} \int d\mathbf{r}^N \exp[-\beta W(\mathbf{r}^N)],\end{aligned}$$

where $\mathbf{p}^N \equiv (\mathbf{p}_1, \mathbf{p}_2, \ldots, \mathbf{p}_N)$, and $\Lambda = \sqrt{h^2/2\pi m k_B T}$ arises from the integration over the kinetic energy.

(ii) The derivation is the same as that for Problem 7.3 except $z \equiv e^{\beta(\mu - F_1)}/\Lambda^3$.

(iii) See Section 7.3.

(iv) See Section 7.3.

Problem 9.5 In the McMillan–Mayer (MM) theory, the thermodynamic state of a solution is specified by, in addition to temperature T and total volume V, the solvent chemical potential μ_0 and the number of molecules for each solute species $\{N_i\}$. However, most experiments are carried out in the Lewis–Randall (LR) framework where temperature, pressure, and chemical compositions are used as the controllable variables. Because of the difference in macroscopic constraints, thermodynamic quantities presented in the MM framework may not be the same as those in the LR framework or those in the Kirkwood–Buff (KB) theory where thermodynamic variables are derived from the grand canonical ensemble.

(i) Show that, in the MM framework, the Gibbs–Duhem equation is given by

$$\sum_i N_i d\mu_i = Vd\Pi,$$

where subscripts $i = 1,2, \ldots$ stand for solutes, and Π stands for the osmotic pressure.

(ii) Based on the equations derived in Problem 9.1, show

$$\left(\frac{\partial \Pi}{\partial \rho_i}\right)_{T,\mu_0,\rho_{j\neq i}} = k_B T \sum_i \rho_i A'_{ij},$$

where subscript "0" denotes the solvent, A'_{ij} corresponds to the matrix inverse of $B'_{ij} \equiv \rho_i(\delta_{ij} + \rho_j G_{ij})$, and G_{ij} are the KB integrals. Different from A_{ij} and B_{ij} defined in Problem 9.1, here subscripts i and j in A'_{ij} and B'_{ij} are indices only for the solute species.

(iii) In the KB framework, the activity coefficient of each solute is defined by $\mu_i = \mu_i^0 + k_B T \ln(x_i \gamma_i^{KB})$, where x_i is the solute mole fraction, and μ_i^0 refers to the chemical potential at system temperature T, solvent chemical potential μ_0, and the infinite dilution of the solute $\rho_i \to 0$. Show

$$\rho_i \left(\frac{\partial \ln \gamma_i^{KB}}{\partial \rho_j}\right)_{T,\rho_{k\neq j}} = x_i - \delta_{ij} + \rho_i A_{ij},$$

where δ_{ij} is the Kronecker delta function, and A_{ij} is defined in Problem 9.1 with both solute and solvent components considered on an equal footing.

(iv) In defining the solute activity coefficient in the MM framework, the reference state refers to an infinite dilution of the solute at system temperature and the chemical potential of the solvent. Show

$$\rho_i \left(\frac{\partial \ln \gamma_i^{MM}}{\partial \rho_j}\right)_{T,\mu_0,\rho_{k\neq j}} = x_i - \delta_{ij} + \rho_i A'_{ij}.$$

(v) In the LR framework, the reference state for the solute refers to the solute at infinite dilution at system temperature and pressure. Show

$$N_i \left(\frac{\partial \ln \gamma_i^{LR}}{\partial N_j}\right)_{T,P,N_{k\neq j}} = x_i - \delta_{ij} + \rho_i \left(A_{ij} - \frac{\beta \overline{V}_i \overline{V}_j}{\kappa_T}\right),$$

where $\overline{V}_i$ stands for the partial molar volume, and κ_T is the isothermal compressibility.

(vi) Relate the different definitions of the reference chemical potentials.

Solution:

(i) This problem is adapted from Lee.[4] In the MM framework, temperature T and solvent chemical potential μ_0 are fixed. Therefore, the pressure of the pure solvent in osmotic equilibrium with the solution, P_0, is a constant. In that case, the Gibbs–Duhem equation for the solution becomes

$$N_0 d\mu_0 + \sum_i N_i d\mu_i = Vd(\Pi + P_0) \quad \Rightarrow \quad \sum_i N_i d\mu_i = Vd\Pi.$$

(ii) From the Gibbs–Duhem equation derived from part (i), we have

$$\left(\frac{\partial \Pi}{\partial \rho_i}\right)_{T,\mu_0,\rho_{j\neq i}} = k_B T \sum_j \rho_j \left(\frac{\partial \beta\mu_j}{\partial \rho_i}\right)_{T,\mu_0,\rho_{j\neq i}},$$

4 Lee L. L., "Thermodynamic consistency and reference scale conversion in multisolvent electrolyte solutions", *J. Mol. Liq.* 87, 129–147 (2000).

where $\rho_i = N_i/V$. Similar to part (i) and (ii) in Problem 9.1, with T and μ_0 fixed, the total derivative of multivariable function $\mu_i = \mu_i(\{\rho_k\})$ gives

$$d\mu_i = \sum_k \left(\frac{\partial \mu_i}{\partial \rho_k}\right)_{T,\mu_0,\rho_{n\neq k}} d\rho_k.$$

With $\mu_{m\neq j}$ fixed, dividing both sides of the above equation by $d\mu_j$ leads to

$$\left(\frac{\partial \mu_i}{\partial \mu_j}\right)_{T,\mu_0,\mu_{m\neq j}} = \delta_{ij} = \sum_k \left(\frac{\partial \beta\mu_i}{\partial \rho_k}\right)_{T,\mu_0,\rho_{n\neq k}} \left(\frac{\partial \rho_k}{\partial \beta\mu_j}\right)_{T,\mu_0,\mu_{m\neq j}} = \sum_k A'_{ik}B'_{kj},$$

where

$$A'_{ij} = \left(\frac{\partial \beta\mu_i}{\partial \rho_j}\right)_{T,\mu_0,\rho_{k\neq j}} \quad \text{and} \quad B'_{ij} = \left(\frac{\partial \rho_i}{\partial \beta\mu_j}\right)_{T,\mu_0,\mu_{k\neq j}} = \rho_i(\delta_{ij} + \rho_j G_{ij}).$$

Hence, A'_{ij} corresponds to the matrix inverse of B'_{ij} and

$$\left(\frac{\partial \Pi}{\partial \rho_i}\right)_{T,\mu_0,\rho_{j\neq i}} = k_B T \sum_j \rho_j A'_{ij}.$$

(iii) In the KB framework, the system temperature and volume are fixed. Hence

$$\left(\frac{\partial \ln \gamma_i^{KB}}{\partial \rho_j}\right)_{T,\rho_{k\neq j}} = \left(\frac{\partial \beta\mu_i}{\partial \rho_j}\right)_{T,\rho_{k\neq j}} - \left(\frac{\partial \ln x_i}{\partial \rho_j}\right)_{T,\rho_{k\neq j}} = A_{ij} - \delta_{ij}/\rho_i + 1/\rho_T, \tag{Y}$$

where δ_{ij} is the Kronecker delta function, $\rho_T = \sum_k \rho_k$ and, as discussed in Problem 9.1,

$$A_{ij} = \left(\frac{\partial \beta\mu_i}{\partial \rho_j}\right)_{T,\rho_{k\neq j}}.$$

Multiplying both sides of Eq. (Y) with ρ_i leads to

$$\rho_i \left(\frac{\partial \ln \gamma_i^{KB}}{\partial \rho_j}\right)_{T,\rho_{k\neq j}} = x_i - \delta_{ij} + \rho_i A_{ij}.$$

(iv) In the MM framework, temperature, total volume, and solvent chemical potential are fixed. Thus,

$$\left(\frac{\partial \ln \gamma_i^{MM}}{\partial \rho_j}\right)_{T,\mu_0,\rho_{k\neq j}} = \left(\frac{\partial \beta\mu_i}{\partial \rho_j}\right)_{T,\mu_0,\rho_{k\neq j}} - \left(\frac{\partial \ln x_i}{\partial \rho_j}\right)_{T,\mu_0,\rho_{k\neq j}}$$
$$= A'_{ij} - \delta_{ij}/\rho_i + 1/\rho_T, \tag{Z}$$

where, as discussed above,

$$A'_{ij} = \left(\frac{\partial \beta\mu_i}{\partial \rho_j}\right)_{T,\mu_0,\rho_{k\neq j}}.$$

Multiplying both sides of Eq. (Z) with ρ_i leads to

$$\rho_i \left(\frac{\partial \ln \gamma_i^{MM}}{\partial \rho_j}\right)_{T,\mu_0,\rho_{k\neq j}} = x_i - \delta_{ij} + \rho_i A'_{ij}.$$

(v) In the LR framework, the system temperature and pressure are fixed. Thus

$$\left(\frac{\partial \ln \gamma_i^{LR}}{\partial N_j}\right)_{T,P,N_{k\neq j}} = \left(\frac{\partial \beta\mu_i}{\partial N_j}\right)_{T,P,\rho_{k\neq j}} - \delta_{ij}/N_i + 1/N_T, \tag{AA}$$

where $N_T = \sum_k N_k$. Following the equations discussed in Problem 9.1, we have

$$\left(\frac{\partial \beta\mu_i}{\partial N_j}\right)_{T,P,N_{k\neq j}} = \frac{1}{V}\left(A_{ij} - \frac{\beta\overline{V}_i\overline{V}_j}{\kappa_T}\right).$$

Multiplying both sides of Eq. (AA) with N_i leads to

$$N_i\left(\frac{\partial \ln \gamma_i^{LR}}{\partial N_j}\right)_{T,P,N_{k\neq j}} = x_i - \delta_{ij} + \rho_i\left(A_{ij} - \frac{\beta\overline{V}_i\overline{V}_j}{\kappa_T}\right).$$

(vi) Different definitions of the reference chemical potentials are related through conventional thermodynamic relation

$$\mu_i^0(T,P,\rho_i \to 0) = \mu_i^0(T,P_V,\rho_i \to 0) + \int_{P_V}^{P}\left(\frac{\partial \mu_i^0(T,P,\rho_i \to 0)}{\partial P}\right)_{T,\rho_i\to 0} dP$$
$$= \mu_i^0(T,V,\mu_0,\rho_i \to 0) + \int_{P_V}^{P}\overline{V}_i^{\infty}\, dP = \mu_i^0(T,V,\mu_0,\rho_i \to 0) + \Pi\overline{V}_i^{\infty},$$

where P_V is the pressure of the pure solvent defined by the system temperature T and solvent chemical potential μ_0, $P = P(T,V,\mu_i)$ is the system pressure, $\overline{V}_i^{\infty}$ is the partial molar volume of i at infinite dilution, and $\Pi = P - P_V$ is the osmotic pressure. Because $\mu_i^0(T,V,\mu_0,\rho_i \to 0) = \mu_i^0(T,V,\mu_0,\mu_i \to -\infty)$, the MM and KB frameworks have the same reference state.

Problem 9.6 Verify the Kirkwood charging formula for calculating the solvation free energy[5]

$$F_1 = \int_0^1 d\lambda \int d\mathbf{r}\rho(\mathbf{r},\lambda)\bar{u}(\mathbf{r}), \tag{AB}$$

where $\rho(\mathbf{r},\lambda)$ stands for the local density of solvent molecules near a solute with the solvent-solute potential scaled by a factor $0 \le \lambda \le 1$, and $\bar{u}(\mathbf{r})$ is the configurationally averaged solvent-solute potential. Can Eq. (AB) be used for nonspherical solutes?

Solution:

Consider, in general, a semi-grand canonical ensemble with a single solute molecule dissolved in a pure solvent. As discussed in Section 9.1, the semi-grand partition function can be written as

$$\mathbb{Z} = \sum_{\nu=\nu_1+\nu_0} e^{-\beta(\Delta E_{\nu_1} + E_{\nu_0} - \mu_0 N_{0,\nu})}. \tag{AC}$$

Because the kinetic energy of the solute molecule is not influenced by solvation, we may replace ΔE_{ν_1} with the change in the potential energy of the solute molecule $\Delta\Phi_{\nu_1}$ due to the solvation of a solute molecule. With the solvent-solute potential scaled by a factor $0 \le \lambda \le 1$, Eq. (AC) becomes

$$\mathbb{Z}(\lambda) = \sum_{\nu} e^{-\beta(\lambda\Delta\Phi_{\nu_1} + E_{\nu_0} - \mu_0 N_{0,\nu})}. \tag{AD}$$

5 Kirkwood J. G., "Statistical mechanics of fluid mixtures", *J. Chem. Phys.* 3, 300 (1935).

The solvation free energy corresponds to the change in the free energy of the system due to the insertion of a solute molecule

$$F_1 = \mathbb{F}(T, V, 1, \mu_0) - \mathbb{F}(T, V, \mu_0) = -k_B T \ln \frac{\mathbb{Z}(\lambda = 1)}{\mathbb{Z}(\lambda = 0)}.$$

The right side of the above equation can be evaluated by integration from $\lambda = 0$ to 1 (viz., the Kirkwood "charging" process)

$$-k_B T \ln \frac{\mathbb{Z}(\lambda = 1)}{\mathbb{Z}(\lambda = 0)} = -k_B T \int_0^1 d\lambda \frac{\partial \ln \mathbb{Z}(\lambda)}{\partial \lambda} = \int_0^1 d\lambda \frac{\sum_\nu \Delta\Phi_{\nu_1} e^{-\beta(\lambda \Delta\Phi_{\nu_1} + E_{\nu_0} - \mu_0 N_{0,\nu})}}{\sum_\nu e^{-\beta(\lambda \Delta\Phi_{\nu_1} + E_{\nu_0} - \mu_0 N_{0,\nu})}}$$
$$= \int_0^1 d\lambda \langle \Delta\Phi_{\nu_1} \rangle_\lambda,$$

where $\langle \cdots \rangle_\lambda$ represents the ensemble average with the solute-solvent potential scaled by λ. Assuming that the solute-solvent potential is pairwise additive

$$\Delta\Phi_{\nu_1} = \sum_i u(\mathbf{R}, \mathbf{R}_{0,i}),$$

where $\mathbf{R}$ and $\mathbf{R}_{0,i}$ stand for the configurations of the solute and solvent molecule i, we obtain the Kirkwood charging formula by noting that

$$\langle \Delta\Phi_{\nu_1} \rangle_\lambda = \left\langle \sum_i \hat{u}(\nu_1, \nu_{0,i}) \right\rangle_\lambda = \int d\mathbf{R} \int d\mathbf{R}_0 \left\langle \delta(\mathbf{R} - \mathbf{R}_{\nu_1}) \sum_i \delta(\mathbf{R}_0 - \mathbf{R}_{0,i}) u(\nu_1, \nu_{0,i}) \right\rangle_\lambda$$
$$= \int d\mathbf{R} \int d\mathbf{R}_0 u(\mathbf{R}, \mathbf{R}_0) \left\langle \delta(\mathbf{R} - \mathbf{R}_{\nu_1}) \sum_i \delta(\mathbf{R}_0 - \mathbf{R}_{0,i}) \right\rangle_\lambda$$
$$= \int d\mathbf{R} \int d\mathbf{R}_0 u(\mathbf{R}, \mathbf{R}_0) \rho(\mathbf{R}, \mathbf{R}_0, \lambda) = \int d\mathbf{r} \rho(\mathbf{r}, \lambda) \bar{u}(\mathbf{r}), \tag{AE}$$

where $\rho(\mathbf{R}, \mathbf{R}_0, \lambda)$ is the two-body density-density correlation function. In writing the last line of the above equation, we define the average density of solvent molecules and the average solute-solvent potential at position $\mathbf{r} \equiv (\mathbf{r}_{\mathbf{R}_0} - \mathbf{r}_\mathbf{R})$

$$\rho(\mathbf{r}, \lambda) \equiv \int d\mathbf{R} \int d\mathbf{R}_0 \delta(\mathbf{r} - (\mathbf{r}_{\mathbf{R}_0} - \mathbf{r}_\mathbf{R})) \rho(\mathbf{R}, \mathbf{R}_0, \lambda),$$

and

$$\bar{u}(\mathbf{r}) \equiv \frac{\int d\mathbf{R} \int d\mathbf{R}_0 \delta(\mathbf{r} - (\mathbf{r}_{\mathbf{R}_0} - \mathbf{r}_\mathbf{R})) u(\mathbf{R}, \mathbf{R}_0) \rho(\mathbf{R}, \mathbf{R}_0, \lambda)}{\int d\mathbf{R} \int d\mathbf{R}_0 \delta(\mathbf{r} - (\mathbf{r}_{\mathbf{R}_0} - \mathbf{r}_\mathbf{R})) \rho(\mathbf{R}, \mathbf{R}_0, \lambda)},$$

where $\mathbf{r}_\mathbf{R}$ and $\mathbf{r}_{\mathbf{R}_0}$ represent the solute and solvent positions, respectively. If both solute and solvent molecules are spherical, we would have

$$\rho(\mathbf{r}, \lambda) = \rho(r, \lambda) \quad \text{and} \quad \bar{u}(\mathbf{r}) = u(r),$$

where $u(r)$ is the pair potential.
While the Kirkwood charging formula is applicable to arbitrary molecular systems, the calculations of $\rho(\mathbf{r}, \lambda)$ and $\bar{u}(\mathbf{r})$ become significantly more difficult as the complexity of molecular configurations increases.

Problem 9.7 The Setschenow equation is commonly used to describe salt effects on the solvation free energy of a neutral solute in an aqueous solution

$$\Delta F_1 \equiv F_1(C_s) - F_1(C_s = 0) = k_s C_s \left(\frac{N_A k_B T}{\log e} \right),$$

where C_s stands for the molar concentration of the added salt, N_A is the Avogadro constant, $\log e \approx 0.434$, and k_s is called the Setschenow coefficient, a constant having the units of liter/mol (i.e., the units of C_s^{-1}) and depending on the solute–salt interactions. In most cases, k_s is positive, suggesting the reduction of the solubility known as the salting-out effect. Conversely, the salting-in effect refers to situations when $k_s < 0$.

(i) Typically, a neutral solute does not strongly interact with salt ions. Assuming that ions in water behavior like charged hard spheres, discuss how the presence of salt ions affects the solvation free energy.

(ii) Discuss possible reasons leading to the salting-in effects.

Solution:

Salt effects on the hydration free energy of chemical species have been subject to numerous theoretical studies but the microscopic details are not yet fully understood even for relatively simple solutes. The difficulties are mostly related to the unique properties of liquid water at both atomic and macroscopic scales. The following analysis intends to offer only a qualitative understanding of the salting-out and salting-in effects in terms of the solution packing density and solute–solvent interactions. The interested readers are encouraged to come up with their own understandings of this problem.

(i) The scaled-particle theory (SPT) predicts that if ions in water behavior like charged hard spheres, their influence on the solvation free energy will be mostly manifested in the cavity formation term. As discussed in Section 9.2, the cavity formation free energy for a solute of radius a is given by

$$W(R) = 4\pi\kappa_0 + 4\pi R\kappa + 4\pi R^2\gamma + \frac{4\pi R^3}{3}P,$$

where $R = a_0 + a$ is the cavity radius, and a_0 denotes the radius of a solvent molecule. See Eq. (9.53) for the physical meanings of other parameters. At fixed temperature T and pressure P, the addition of salt in liquid water increases the overall packing fraction η due to the increased density, leading to a larger cavity formation free energy

$$\beta\Delta W(R) = f(\eta, R/a_0) - f(\eta_0, R/a_0),$$

with

$$f(\eta, R/a_0) = -\ln(1-\eta) + \frac{9}{2}\left(\frac{\eta}{1-\eta}\right)^2 - \left(\frac{R}{a_0}\right)\left[\frac{6\eta}{1-\eta} + 18\left(\frac{\eta}{1-\eta}\right)^2\right] + \left(\frac{R}{a_0}\right)^2\left[\frac{12\eta}{1-\eta} + 18\left(\frac{\eta}{1-\eta}\right)^2\right].$$

At the ambient condition ($T = 25\,°\text{C}$ and $P = 1$ atm), the packing fraction of liquid water is $\eta_0 = 4\pi\rho_0 a_0^3/3 \approx 0.371$.

Figure 9.2A shows the dependence of $\beta\Delta W(R)$ on $\Delta\eta = \eta - \eta_0$ for several values of R/a_0, the ratio of the cavity radius to the solvent radius. The positive correlation explains the salting-out effects commonly observed in experiments. Because the solution density is (approximately) linearly dependent on the salt concentration, the linear relation between $\beta\Delta W(R)$ and $\Delta\eta$ provides a justification of the Setschenow equation. The salt effects become noticeably more significant as the solute size increases, which is consistent with experimental observations.

(ii) We may estimate the van der Waals energy between the solute and solvent molecules using the mean-field approximation (Eq. 9.58)

$$F_1^{vdw} \approx -\frac{32\pi}{9}\rho R^3\epsilon_{LJ} = -\frac{8\eta\epsilon_{LJ}}{3}\left(\frac{R}{a_0}\right)^3.$$

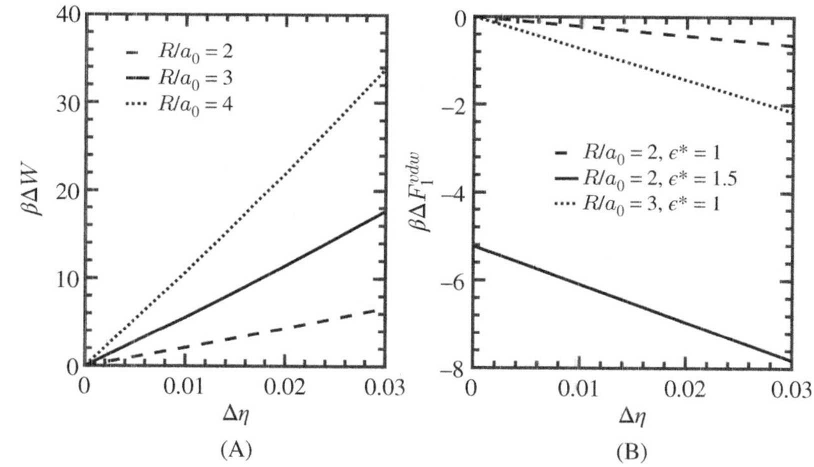

Figure 9.2 Theoretical underpinnings of the Setschenow equation. A. The salting-out effects may be attributed to cavity formation and the increase of the solution density. B. Solute-solvent attraction is responsible for salting-in effects. Here, $\epsilon^* \equiv \epsilon_{LJ}/\epsilon_{LJ}^0$ (SPT.m).

The salt effects on the attractive part of the solvation free energy are shown in the changes of ϵ_{LJ} and η

$$\Delta F_1^{vdw} = -\frac{8}{3}(\eta\epsilon_{LJ} - \eta_0\epsilon_{LJ}^0)\left(\frac{R}{a_0}\right)^3.$$

Figure 9.2B shows that the attractive contribution to the solvation free energy is also linearly dependent on the solution density. For typical nonpolar solutes, the dominant contribution comes from cavity formation, leading to the salting-out effect. Nevertheless, a strong attraction between the solute-solvent molecules will lead to the salting-in effects. Figure 9.2B indicates that the salting-in effects are enhanced as the solute size increases.

With reasonable parameters for the solute and solvent molecules, SPT is able to provide a near quantitative description of the hydration free energies of small nonpolar solutes in KOH solutions.[6]

Problem 9.8 Estimate Henry's constants of the following inert gases in liquid water and in benzene at $T = 25\,°\text{C}$ and $P = 1$ atm using the parameters shown in Table 9.1.[7]

Compare the theoretical results with experimental data and explain why, at similar conditions, nonpolar solutes are much less soluble in water than in a nonpolar solvent.

Solution:

As discussed in Section 9.1, Henry's constant K_H is related to the solvation free energy F_1 through

$$\ln(\beta K_H/\rho_0) = \beta F_1,$$

where ρ_0 is the number density of the pure solvent, and $\beta = 1/(k_B T)$.

---

6 Shoor S. K. and Gubbins K. E., "Solubility of nonpolar gases in concentrated electrolyte solutions", *J. Phys. Chem.*, 73, 498 (1969).

7 Pierotti R. A., "A scaled particle theory of aqueous and nonaqueous solutions", *Chem. Rev.* 76, 717–726 (1976).

Table 9.1 Parameters for the scaled-particle theory of solvation.

	ϵ/k_B, K	σ, Å	η	v (cm^3/mol)
H_2O	79.3	2.77	0.371	18.07
C_6H_6	496	5.26	0.513	89.40
He	6.03	2.63		
Ne	35.7	2.79		
Ar	125	3.41		
Kr	169	3.67		
Xe	217	3.96		
Rn	290	4.23		

According to the scaled-particle theory (SPT), the solvation free energy includes contributions due to cavity formation and van der Waals attraction

$$\beta F_1 = -\ln(1-\eta) + \left(\frac{3\eta}{1-\eta}\right)\Re + \left[\frac{3\eta}{1-\eta} + \frac{9}{2}\left(\frac{\eta}{1-\eta}\right)^2\right]\Re^2 + \frac{\beta P\eta}{\rho_0}\Re^3 - \frac{8\eta\epsilon_{LJ}}{3}(1+\Re)^3$$

where $\Re = \sigma/\sigma_0$ is the diameter ratio of the solute and solvent molecules, and $\epsilon_{LJ} \approx \sqrt{\epsilon\epsilon_0}$ is the Lennard–Jones (LJ) energy parameter for the solute-solvent interaction.

Figure 9.3A presents Henry's constants for 6 inert gases in liquid water and in benzene at $T = 25\,°\mathrm{C}$ and $P = 1$ atm predicted by SPT. For comparison, the experimental value for argon in water is $\ln(K_H/P) = 10.56$, which compares well with the theoretical result, $\ln(K_H/P) = 10.09$. As shown in Figure 9.3B, nonpolar solutes are much less soluble in water than in a nonpolar solvent because of higher cavity formation energy and smaller solute-solvent van der Waals attraction.

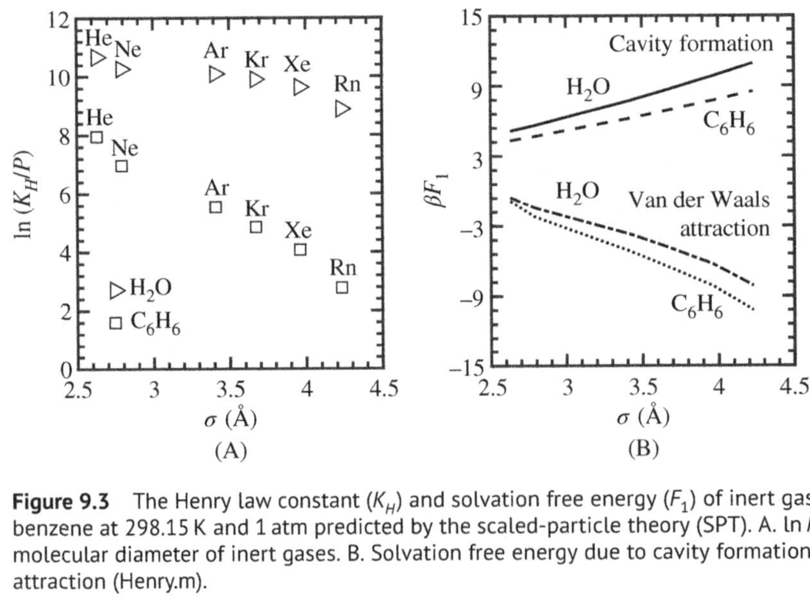

Figure 9.3 The Henry law constant (K_H) and solvation free energy (F_1) of inert gases in liquid water and in benzene at 298.15 K and 1 atm predicted by the scaled-particle theory (SPT). A. ln K_H versus σ, the molecular diameter of inert gases. B. Solvation free energy due to cavity formation and van der Waals attraction (Henry.m).

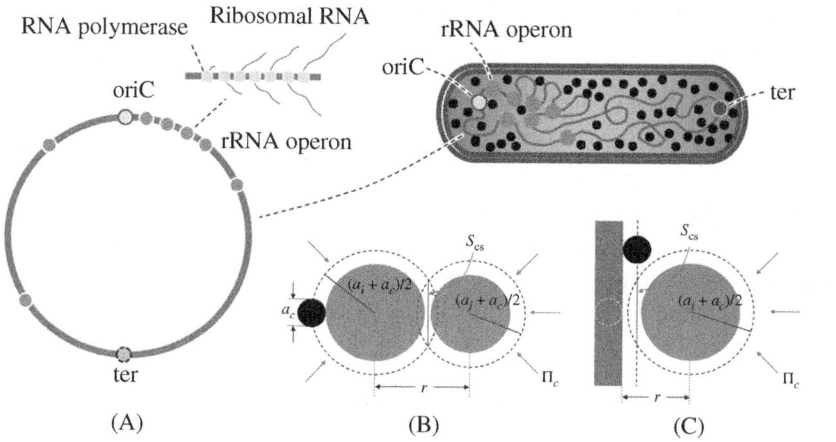

Figure 9.4 Entropic forces for chromosome organization in bacterial cells. A. *E. coli* chromosome consists of seven ribosomal RNA (rRNA) operons, one replication origin (oriC), and the terminus (ter). B. and C. Depletion of "crowders" (black spheres) from two rRNA operons (gray spheres) (B) and from an rRNA operon and the cellular wall (C). Source: Adapted from Jung and Ha.[8]

Problem 9.9 Biomolecular crowding refers to steric effects on biological processes due to the presence of macromolecules such as proteins and ribosomes that do not directly contribute to the activity.[8] The so-called "crowders" occupy about 30% of the cellular space and their effects can be understood in terms of the depletion potentials as described by the Asakura–Oosawa (AO) theory. In the toy model shown schematically in Figure 9.4 for chromosome organization in bacterial cells, we consider rRNA operons as big spheres and other biomolecules inside the cell as hard spheres of uniform size (viz., crowders).

(i) Show that the overlap volume for two spheres of radius R with center-to-center distance $0 < r < 2R$ is given by

$$\Delta V = \frac{4\pi R^3}{3}\left[1 - \frac{3r}{4R} + \frac{r^3}{16R^3}\right].$$

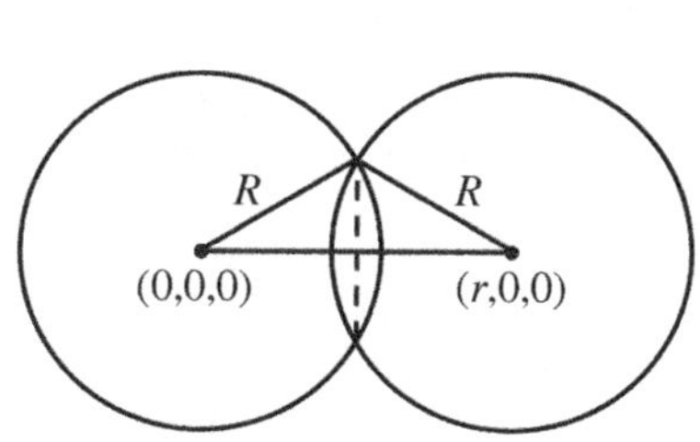

(ii) Show that the depletion force between two rRNA operons of diameter a_i is given by

$$\mathbf{f}_{dep}(r) = -\frac{\pi\Pi_c}{4}\left[(a_i + a_c)^2 - r^2\right]$$

where Π_c is the osmotic pressure due to the presence of crowders with diameter a_c.

---

8 Jung Y. and Ha B. Y., "Collapse transition of a heterogeneous polymer in a crowded medium", *J. Chem. Phys.* 155, 054902 (2021).

(iii) Show that the depletion potential between an rRNA operon of diameter a_i and the cellular wall is given by

$$u^w_{dep}(r) = -\frac{\pi\Pi_c}{24}(a_i + 2a_c - 2r)^2(2a_i + a_c + 2r),$$

where $a_i/2 < r < a_i/2 + a_c$ is the distance between the center of the rRNA operon and the wall surface.

(iv) Predict the contact potential between two rRNA operons and that between an rRNA operon and the cellular wall using the Carnahan–Starling equation of state for Π_c. Assume $a_c = 5$ nm, $a_i = 50$ nm, and the volume fraction of crowders is $\eta = 0.3$.

(v) Comment the implications of the theoretical results.

Solution:

(i) The overlap volume corresponds to the volume of two identical spherical caps of height $h = R - r/2$:

$$\Delta V = 2 \times \frac{\pi}{3}h^2(3R - h) = \frac{2\pi}{3}(R - r/2)^2 \times (3R - R + r/2)$$
$$= \frac{\pi}{12}(2R - r)^2(4R + r) = \frac{4\pi R^3}{3}\left[1 - \frac{3r}{4R} + \frac{r^3}{16R^3}\right].$$

(ii) The potential between rRNA operons is proportional to the overlap volume and the osmotic pressure of the crowder solution Π_c

$$u_{dep}(r) = -\Delta V\Pi_c = -\frac{4\pi R^3}{3}\left[1 - \frac{3r}{4R} + \frac{r^3}{16R^3}\right]\Pi_c.$$

Thus, the depletion force is

$$\mathbf{f}_{dep}(r) = -\frac{\partial u_{dep}(r)}{\partial r} = \frac{4\pi R^3}{3}\left[-\frac{3}{4R} + \frac{3r^2}{16R^3}\right]\Pi_c = -\frac{\pi\Pi_c}{4}\left[(a_i + a_c)^2 - r^2\right],$$

where $R = (a_i + a_c)/2$.

(iii) Similar to that for interaction between two rRNA operons, the depletion potential between an rRNA operon and the cellular wall is proportional to the overlap volume and the osmotic pressure Π_c

$$u^w_{dep}(r) = -\Delta V_w\Pi_c.$$

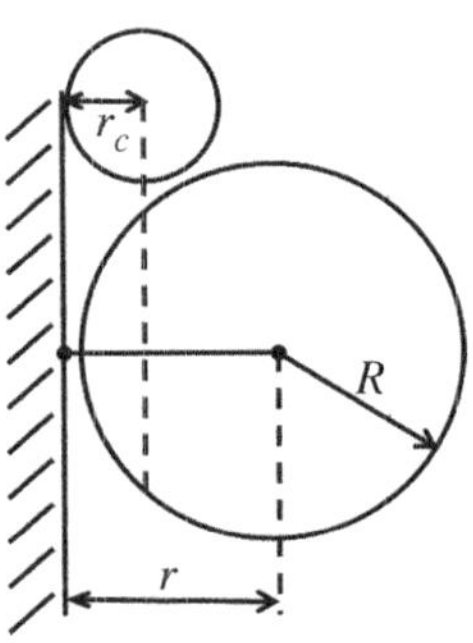

As shown above, the overlap volume corresponds to the volume of a spherical cap of height $h = r_c - (r - R) = r_c + R - r$:

$$\Delta V_w = \frac{\pi}{3}h^2(3R - h) = \frac{\pi}{3}(r_c + R - r)^2(3R - r_c + r - R)$$
$$= \frac{\pi}{24}(2a_c + a_i - 2r)^2(2a_i + a_c + 2r),$$

where $r_c = a_c/2$ and $R = (a_i + a_c)/2$. Therefore, the wall depletion potential is

$$u^w_{dep}(r) = -\Delta V_w \Pi_c = -\frac{\pi \Pi_c}{24}(a_i + 2a_c - 2r)^2(2a_i + a_c + 2r).$$

(iv) According to the Carnahan–Starling equation of state, the osmotic pressure due to the crowders is

$$\beta\Pi_c/\rho_c = \frac{1+\eta+\eta^2-\eta^3}{(1-\eta)^3}.$$

The reduced potential between two rRNA operons is thus given by

$$\beta u_{dep}(r) = -\frac{4\pi\rho_c R^3}{3}\left[1 - \frac{3r}{4R} + \frac{r^3}{16R^3}\right] \times \frac{1+\eta+\eta^2-\eta^3}{(1-\eta)^3}$$

$$= -(1 + a_i/a_c)^3\left[1 - \frac{3r}{2(a_i+a_c)} + \frac{r^3}{2(a_i+a_c)^3}\right]\frac{\eta+\eta^2+\eta^3-\eta^4}{(1-\eta)^3}.$$

For $a_i/a_c = 50/5 = 10$ and $r = a_i$, we have

$$\beta u_{dep}(a_i) = -(1 + a_i/a_c)^3\left[1 - \frac{3a_i}{2(a_i+a_c)} + \frac{a_i^3}{2(a_i+a_c)^3}\right]\frac{\eta+\eta^2+\eta^3-\eta^4}{(1-\eta)^3}$$

$$= -(1+10)^3\left[1 - \frac{3}{2(1+0.1)} + \frac{1}{2(1+0.1)^3}\right] \times \frac{0.3+0.3^2+0.3^3-0.3^4}{(1-0.3)^3} \approx -19.1.$$

The potential between one rRNA operon and the cellular wall is

$$\beta u^w_{dep}(r) = -\frac{\pi\beta\Pi_c}{24}(a_i + 2a_c - 2r)^2(2a_i + a_c + 2r)$$

$$= -\frac{1}{4}(a_i/a_c + 2 - 2r/a_c)^2(2a_i/a_c + 1 + 2r/a_c)\frac{\eta+\eta^2+\eta^3-\eta^4}{(1-\eta)^3}.$$

At $r = a_i/2$ and $a_i/a_c = 10$, we have

$$\beta u^w_{dep}(r = a_i/2) = -\frac{1}{4}(a_i/a_c + 2 - a_i/a_c)^2(2a_i/a_c + 1 + a_i/a_c)\frac{\eta+\eta^2+\eta^3-\eta^4}{(1-\eta)^3}$$

$$= -\frac{1}{4}(10 + 2 - 10)^2(2 \times 10 + 1 + 10) \times \frac{0.3+0.3^2+0.3^3-0.3^4}{(1-0.3)^3} \approx -37.0.$$

(v) Although the interaction between rRNA operons in *E. coli* depends on many factors, the strong depletion force is at least partially responsible for the chromosome collapse into a highly condensed state. Besides, the strong rRNA operon-wall attraction explains the important confinement effects. More insightful discussions can be found from the original article and pertinent literature.

Problem 9.10 Estimate NaBr solubility in water–methanol mixtures at 25 °C using

$$\ln x_s \approx -l_B/\sigma, \tag{AF}$$

and compare the theoretical results with the experimental data shown in Table 9.2. Comment the discrepancy between the theoretical and experimental results and suggest possible ways of improvement. The radii of Na and Br atoms are 1.54 and 1.82 Å, respectively. See Section 9.3 for the notations.

Table 9.2 Dielectric constants (ε) of water–methanol mixtures and NaBr solubility at 25 °C.

w_1	0	10	20	30	40	50	60	70	80	90	100
ε	78.48	74.18	69.99	65.55	60.94	56.28	51.67	47.11	42.6	37.91	32.66
w_s	0.129	0.188	0.235	0.283	0.326	0.359	0.383	0.411	0.437	0.462	0.486

*w_1 represents the mass fraction of methanol in the water-methanol mixture free of salt, and w_s is the mass fraction of NaBr in the solution. Solubility data are from Pinho S. P. and Macedo E. A., "Solubility of NaCl, NaBr, and KCl in water, methanol, ethanol, and their mixed solvents", *J. Chem. Eng. Data* 50, 29–32 (2005); and the dielectric constants from Albright P. S. and Gosting L. J., "Dielectric Constants of the Methanol-Water System from 5 to 55 °", *J. Am. Chem. Soc.* 68, 1061–1063 (1946).

Solution:

The molar fraction solubility x_s and weight fraction solubility w_s are related by

$$w_s = \frac{n_s M_s}{n_s M_s + n_0 M_0} = \frac{x_s}{x_s + (1 - x_s) M_0 / M_s}, \tag{AG}$$

where n_s and n_0 stand for the moles of the salt and the solvent, respectively; M_s and M_0 are the corresponding molar masses. For a mixture of solvent 1 and 2, the molar mass is given by

$$M_0 = \left(\frac{w_1}{W_1} + \frac{w_2}{W_2} \right)^{-1},$$

where w_1 and w_2 are salt-free mass fractions.

With $\sigma \approx 3.36$Å estimated from the radii of Na and Br atoms and the Bjerrum length l_B calculated from the dielectric constants given by Table 9.2, it is straightforward to predict x_s from Eq. (AF), which can be subsequently converted to w_s using Eq. (AG).

Figure 9.5 compares the theoretical results and experimental data. The agreement is far from perfect! The discrepancy increases with the methanol content, suggesting that the theoretical results would be improved by including the van der Waals solvation energy. The dashed line in Figure 9.5 is obtained by assuming that the van der Waals energy is linearly dependent on the solvent composition.

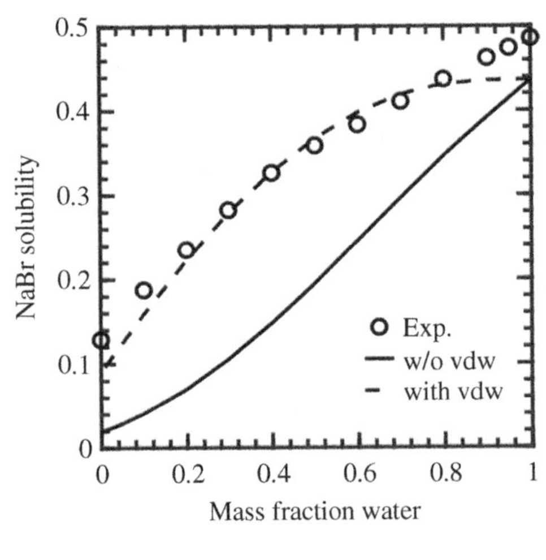

Figure 9.5 The solubility of NaBr (mass fraction) in water-methanol mixtures at 25 °C. The solid line is predicted by Eq. (AF) and the dashed line is after the correction with van der Waals attraction (NaBr.m).

Problem 9.11 Verify that the Debye–Hückel theory satisfies the Stillinger–Lovett second moment condition

$$\sum_i \sum_j Z_i Z_j \rho_i^0 \rho_j^0 \int d\mathbf{r} r^2 g_{ij}(r) = -\frac{3}{2\pi l_B}.$$

Solution:

The Debye–Hückel theory predicts that the radial distribution functions (RDFs) of a dilute electrolyte solution are given by

$$g_{ij}(r) \approx 1 - \frac{Z_i Z_j l_B e^{-\kappa r}}{r}.$$

Substituting the RDFs into the Stillinger–Lovett second moment condition yields

$$\sum_i \sum_j Z_i Z_j \rho_i^0 \rho_j^0 \int d\mathbf{r} r^2 g_{ij}(r)$$

$$= \sum_i \sum_j Z_i Z_j \rho_i^0 \rho_j^0 \int d\mathbf{r} r^2 \left[1 - \frac{Z_i Z_j l_B e^{-\kappa r}}{r}\right]$$

$$= -l_B \sum_i \sum_j Z_i^2 Z_j^2 \rho_i^0 \rho_j^0 \int_0^\infty dr 4\pi r^4 \left(\frac{e^{-\kappa r}}{r}\right) = -4\pi l_B \times \sum_i \sum_j Z_i^2 Z_j^2 \rho_i^0 \rho_j^0 \times \int_0^\infty dr r^3 e^{-\kappa r}$$

$$= -4\pi l_B \times \left(\sum_i Z_i^2 \rho_i^0\right)^2 \times \left(\frac{6}{\kappa^4}\right) = -4\pi l_B \times \left(\frac{\kappa^2}{4\pi l_B}\right)^2 \times \left(\frac{6}{\kappa^4}\right) = -\frac{3}{2\pi l_B}.$$

In the above derivations, we have used the mathematical relations

$$\sum_i Z_i \rho_i^0 = 0, \qquad \int_0^\infty dr r^3 e^{-\kappa r} = \frac{6}{\kappa^4} \quad \text{and} \quad \kappa^2 = 4\pi l_B \sum_i Z_i^2 \rho_i^0.$$

Problem 9.12 Find a relation between the mean activity coefficient of a monovalent salt (e.g., NaCl) defined in terms of molality m (mol/kg of solvent) and that in terms of the number density of ions, $\rho_\pm$.

Solution:

The ionic chemical potential can be expressed in terms of either number density ρ_i or molality m_i

$$\mu_i = \mu_i^0 + k_B T \ln(\rho_i \gamma_i / \rho_i^0) = \mu_i^{0,m} + k_B T \ln(m_i \gamma_i^m / m_i^0),$$

which implies

$$\beta\mu_i^0 - \beta\mu_i^{0,m} = \ln \frac{m_i \gamma_i^m / m_i^0}{\rho_i \gamma_i / \rho_i^0} = \ln \frac{\gamma_i^m}{\gamma_i} + \ln \frac{m_i}{\rho_i} - \ln \frac{m_i^0}{\rho_i^0}. \tag{AH}$$

In terms of molality, the number density is

$$\rho_i = \frac{m_i N_A d}{1 + 0.001 m_i M_i},$$

where d in kg/m^3 is the mass density of the solution, N_A is the Avogadro number, M_i in g/mol stands for the molecular weight of ionic species i. Because the activity coefficients are unity at infinite dilution, i.e., $\gamma_i^m = \gamma_i = 1$ when $m_i \to 0$ and $\rho_i \to 0$, we have

$$\beta\mu_i^0 - \beta\mu_i^{0,m} = \lim_{m_i, \rho_i \to 0} \ln \frac{m_i}{\rho_i} - \ln(m_i^0/\rho_i^0) = -\ln(N_A d_0) - \ln(m_i^0/\rho_i^0), \tag{AI}$$

where d_0 in kg/m^3 is the mass density of the pure solvent.

Combining Eqs. (AH) and (AI) leads to the relation between the molality and number density activity coefficients

$$\gamma_i^m = \gamma_i \frac{d}{d_0(1 + 0.001 m_i M_i)}.$$

For a monovalent salt, the mean activity coefficient is

$$\gamma_\pm^m = \sqrt{\gamma_+^m \gamma_-^m} = \gamma_\pm \left(\frac{d}{d_0(1 + 0.001 m_+ M_+)} \times \frac{d}{d_0(1 + 0.001 m\_M\_)} \right)^{1/2}.$$

At low density, $0.001 m_i M_i \ll 1$ and $d \approx d_0$, the above equation is reduced to

$$\gamma_\pm^m \approx \gamma_\pm.$$

In practice, the reference states are defined in terms of the infinite dilute of salt instead of individual ions. Accordingly, the mean activity coefficients are related by

$$\gamma_\pm^m \approx \gamma_\pm \frac{d}{d_0(1 + 0.001 m_s M_s)},$$

where m_s is the salt molality, and M_s is the molecular weight.

Problem 9.13 Verify that the Debye–Hückel theory predicts the following equation for the activity coefficient of water in a dilute electrolyte solution

$$\ln \gamma_w = \frac{v_w \kappa^3}{24\pi},$$

where v_w is the molecular volume of liquid water, and κ is the Debye screening parameter.

Solution:
The Debye–Hückel theory predicts that the osmotic pressure depends on the total number density of ions ρ_0 and the Debye screening parameter κ (Eq. 9.114)

$$\beta \Pi = \rho_0 - \frac{\kappa^3}{24\pi}.$$

Meanwhile, the water activity is related to the osmotic pressure through

$$\ln a_w = -\beta \Pi v_w,$$

where $v_w = 1/\rho_w$, and ρ_w is the number density of water molecules. For a dilute electrolyte solution, $\ln x_w = \ln[\rho_w/(\rho_w + \rho_0)] \approx -\rho_0/\rho_w$. Thus, the activity coefficient of water is given by

$$\ln \gamma_w = \ln(a_w/x_w) = -v_w \rho_0 + \frac{v_w \kappa^3}{24\pi} + \rho_0/\rho_w = \frac{v_w \kappa^3}{24\pi}.$$

Problem 9.14 Consider a single electrolyte solution at moderate ion concentrations.

(i) Show that the solvent activity a_s can be related to the osmotic pressure through

$$\ln a_s \approx -\beta \Pi v_s^0 \approx -\beta \Pi / \rho_s,$$

where v_s^0 is the molecular volume of the pure solvent at system temperature T and pressure P, and ρ_s is the number density of solvent molecules in the electrolyte solution. Explain approximations used in the above equation explicitly.

(ii) Verify the following relation between the osmotic coefficient (Φ) and the mean ionic activity coefficient ($\gamma_\pm$)

$$\ln \gamma_\pm = \Phi - 1 + \int_0^{\rho_0} d\rho_0 (\Phi - 1)/\rho_0,$$

where ρ_0 is the total ion density.

(iii) Prove the mean ionic activity coefficient derived from the Pitzer equation, Eq. (9.144).

Solution:

(i) The osmotic equilibrium requires the equality of the solvent chemical potential

$$\mu_s(T, P) = \mu_s^0(T, P) + k_B T \ln a_s = \mu_s^0(T, P - \Pi),$$

where superscript "0" denotes the pure solvent, and Π is the osmotic pressure of the solution. The pressure effect on the solvent chemical potential can be calculated from

$$\mu_s^0(T, P) = \mu_s^0(T, P - \Pi) + \int_{P-\Pi}^{P} v_s^0 dP.$$

Assuming that v_s^0 can be approximated by $v_s^0 \approx v_s^0(T, P)$, we can evaluate the above integration analytically

$$\ln a_s \approx -\beta \Pi v_s^0.$$

Typically, the change in the pure solvent volume is negligible if the osmotic pressure is not exceedingly large. While the assumption of constant solvent molar volume is well justified, the assumption $v_s^0 \approx 1/\rho_s$ is problematic at high electrolyte concentrations.

(ii) At constant temperature and pressure, the Gibbs–Duham equation for a binary mixture of an electrolyte (E) and a solvent (s) can be written as

$$\rho_s d \ln a_s + \rho_E d \ln a_E = 0$$

where ρ_s and ρ_E are the number densities of solvent molecules and the electrolyte (viz., ion pairs). As discussed above, the solvent activity is related to the osmotic pressure as well as the osmotic coefficient (Φ)

$$\ln a_s = -\beta \Pi v_s^0 = -(\nu_+ + \nu_-)\rho_E \Phi v_s^0 = -\nu \rho_E \Phi v_s^0 = -\rho_0 \Phi v_s^0$$

where $\nu_\pm$ are the stoichiometric coefficients of the ionic species, $\nu = \nu_+ + \nu_-$, and $\rho_0 = \nu \rho_E$. Meanwhile, the activity of the electrolyte is given by

$$a_E = (a_\pm)^\nu \quad \text{with} \quad a_\pm = \rho_\pm^0 \gamma_\pm = \rho_E (\nu_+^{\nu_+} \nu_-^{\nu_-})^{1/\nu} \gamma_\pm.$$

In terms of Φ and $\gamma_\pm$, the Gibbs–Duham equation thus becomes

$$-\rho_s v_s^0 \rho_0 d\Phi - \rho_s v_s^0 \Phi d\rho_0 + \nu \rho_E d \ln \rho_E + \nu \rho_E d \ln \gamma_\pm = 0.$$

Because $\rho_0 = \nu \rho_E$ and $d \ln \rho_0 = d \ln \rho_E$, rearranging the above equation leads to

$$\ln \gamma_\pm = \rho_s v_s^0 d\Phi + (\rho_s v_s^0 \Phi - 1) d \ln \rho_0 = \rho_s v_s^0 d\Phi + [(\rho_s v_s^0 \Phi - 1)/\rho_0] d\rho_0.$$

At moderate electrolyte concentrations, $\rho_s v_s^0 \approx 1$ and, as $\rho_0 \to 0$, $\Phi = 1$ and $\gamma_\pm = 1$. Integrating the Gibbs–Duhem equation results in

$$\ln \gamma_\pm = \Phi - 1 + \int_0^{\rho_0} d\rho_0 (\Phi - 1)/\rho_0 = \Phi - 1 + \int_0^{\rho_0} (\Phi - 1) d \ln \rho_0.$$

(iii) The Pitzer equation for the osmotic coefficient is

$$\Phi = 1 + \frac{2}{3}\pi\rho_0 a^3 - \frac{l_B|Z_+Z_-|}{12a}\left[\frac{2\kappa a}{1+\kappa a} - \left(\frac{\kappa a}{1+\kappa a}\right)^2\right].$$

Plugging Φ into the equation derived in part (ii) yields

$$\begin{aligned}\ln\gamma_\pm &= \Phi - 1 + \int_0^{\rho_0}(\Phi-1)d\ln\rho_0\\ &= \Phi - 1 + \frac{2}{3}\pi\rho_0 a^3 - \frac{l_B|Z_+Z_-|}{6a}\int_0^x\left[\frac{2x}{1+x} - \left(\frac{x}{1+x}\right)^2\right]d\ln x\\ &= \Phi - 1 + \frac{2}{3}\pi\rho_0 a^3 - \frac{l_B|Z_+Z_-|}{6a}\left[\frac{x}{1+x} + \ln(1+x)\right],\end{aligned}$$

where $x = \kappa a$ and $d\ln\rho_0 = 2d\ln x$.

Problem 9.15 In applying the extended Debye–Hückel (DH) theory to identify the immiscibility boundaries of metal oxides in silicate melts and glasses (e.g., Na_2O–SiO_2, Li_2O–SiO_2, SrO–SiO_2 etc.), we start with the molar Gibbs energy of mixing

$$\Delta g_{mix} \equiv \Delta G_{mix}/(RT) = x_1\ln a_1 + x_2\ln a_2, \tag{AJ}$$

where x_1 and x_2 are the mole fractions of the solvent (e.g., SiO_2) and the solute (e.g., Na_2O or other metal oxides), a_1 and a_2 are the activities of the solvent and solute, respectively.

(i) Show that the spinodal line is determined by

$$\frac{\partial^2\Delta g_{mix}}{\partial x_2^2} = \frac{1}{x_1}\frac{\partial\ln a_2}{\partial x_2} = 0.$$

(ii) Verify Eq. (9.138), i.e., the mean ionic activity coefficient predicted by the modified DH theory is given by

$$\ln\gamma_\pm = -\frac{\kappa l_B|Z_+Z_-|}{2(1+\kappa a)}.$$

(iii) Show that, in combination with the quasi-chemical ion-pairing model, the DH theory predicts the solute activity coefficient

$$\ln\gamma_\pm = -\frac{\kappa_B l_B|Z_+Z_-|}{2(1+\kappa_B a)} + \ln\alpha,$$

where α is the fraction of the solute dissociating into ionic species (i.e., $C_{\nu_+}A_{\nu_-} \to \nu_+C^{Z_+} + \nu_-A^{Z_-}$), $\kappa_B = [4\pi(\nu_+Z_+^2 + \nu_-Z_-^2)\alpha l_B\rho_2]^{1/2}$ is the modified Debye screening parameter, ρ_2 is the number density of the solute molecules, $\nu_\pm$ and $Z_\pm$ are the stoichiometric coefficients and valence defined by the solute dissociation.

(iv) What is the activity of the solute predicted by the extended DH theory?

(v) Assuming that the degree of dissociation α is independent of the solution composition, show that the solute composition at the spinodal line satisfies

$$x_2 = \left[\frac{A - 4B \pm \sqrt{A(A-8B)}}{4B^2}\right]^2,$$

where $A \equiv |Z_+Z_-|l_B\kappa_B/(2x_2^{1/2})$ and $B \equiv \kappa_B a/x_2^{1/2}$.

(vi) Show that the extended DH theory predicts the critical temperature and the critical composition

$$T_c = \frac{|Z_+Z_-|e^2}{64\pi\varepsilon\varepsilon_0 k_B a} \quad \text{and} \quad x_{2,c} = \frac{M}{64\pi\alpha_c(\nu_+ + \nu_-)dN_A a^3},$$

where M and d are the average molecular weight and the mass density of the mixture, respectively, and other symbols have their usual meanings.

(vii) Find the degree of dissociation at the critical condition, α_c, using the Fuoss quasi-chemical ion-pairing model for the association constant

$$K_A = \frac{2\pi\nu a^3}{3}\exp\left[\frac{|Z_+Z_-|l_B}{a(1+\kappa_B a)}\right].$$

Hint: To a good approximation, the condensed phase is incompressible. Accordingly, the mole fraction of the solute may be assumed proportional to the number density, i.e., $x_2 \propto \rho_2$.

Solution:

This problem and the solution are adapted from McGahay and Tomozawa.[9]

(i) The spinodal line is defined by the second derivative of the Gibbs energy of mixing

$$\frac{\partial^2 \Delta g_{mix}}{\partial x_2^2} = 0.$$

Taking the first derivative of Eq. (AJ) leads to

$$\frac{\partial \Delta g_{mix}}{\partial x_2} = \left[x_1\frac{\partial \ln a_1}{\partial x_2} + x_2\frac{\partial \ln a_2}{\partial x_2}\right] - \ln a_1 + \ln a_2 = -\ln a_1 + \ln a_2,$$

where the terms inside the square brackets vanish because of the Gibbs–Duhem relation. The second derivative is

$$\frac{\partial^2 \Delta g_{mix}}{\partial x_2^2} = -\frac{\partial \ln a_1}{\partial x_2} + \frac{\partial \ln a_2}{\partial x_2} = \frac{x_2}{x_1}\frac{\partial \ln a_2}{\partial x_2} + \frac{\partial \ln a_2}{\partial x_2} = \frac{1}{x_1}\frac{\partial \ln a_2}{\partial x_2}.$$

(ii) The mean ionic activity coefficient predicted by the modified DH theory is given by

$$\ln\gamma_\pm = \frac{\nu_+\ln\gamma_+ + \nu_-\ln\gamma_-}{\nu_+ + \nu_-} = -\frac{\kappa l_B}{2(1+\kappa a)}\left[\frac{\nu_+Z_+^2 + \nu_-Z_-^2}{\nu_+ + \nu_-}\right]$$

$$= -\frac{\kappa l_B \nu_+^2 Z_+^2}{2(1+\kappa a)}\left[\frac{1/\nu_+ + 1/\nu_-}{\nu_+ + \nu_-}\right] = -\frac{\kappa l_B \nu_+^2 Z_+^2}{2(1+\kappa a)\nu_+\nu_-} = -\frac{\kappa l_B|Z_+Z_-|}{2(1+\kappa a)}.$$

(iii) The quasi-chemical ion-pairing model asserts that ion binding reduces the number of free ions described by the DH theory. Therefore, the solute activity coefficient is given by

$$\ln\gamma_\pm = \ln\frac{\rho_2^D\gamma_\pm^D}{\rho_2} = -\frac{\kappa_B l_B|Z_+Z_-|}{2(1+\kappa_B a)} + \ln\alpha,$$

where $\alpha = \rho_2^D/\rho_2$ is the fraction of the solute dissociation into ionic species, and $\gamma_\pm^D$ is the mean ionic activity coefficient for the free ions. The latter can be predicted from the modified DH theory with the Debye parameter calculated from $\kappa_B = [4\pi(\nu_+Z_+^2 + \nu_-Z_-^2)\alpha l_B\rho_2]^{1/2}$.

9 McGahay V. and Tomozawa M., "Phase separation in silicate melts and glasses", *J. Non-Cryst. Solids* 109, 27–34 (1989).

(iv) Like typical electrolyte solutions, the solute activity is defined through dissociation reaction $C_{v_+}A_{v_-} \rightarrow v_+C^{Z_+} + v_-A^{Z_-}$

$$\mu_2 = \mu_2^0 + k_BT\ln a_2 = v_+\mu_+ + v_-\mu_- = v_+\mu_+^0 + v_-\mu_-^0 + v_+\ln a_+ + v_-\ln a_-,$$

where superscript 0 denotes the reference state. Let $\mu_2^0 = v_+\mu_+^0 + v_-\mu_-^0$ and $a_2 = a_\pm^v$ where $v = v_+ + v_-$, we have

$$a_\pm = \left[(\rho_+\gamma_+)^{v_+}(\rho_-\gamma_-)^{v_-}\right]^{1/v} = \rho_2\left[(v_+\gamma_+)^{v_+}(v_-\gamma_-)^{v_-}\right]^{1/v} = \rho_2\left[v_+^{v_+}v_-^{v_-}\right]^{1/v}\gamma_\pm.$$

Hence, the solute activity is given by

$$\ln a_2 = v\ln a_\pm = v\ln\rho_2 + \ln\left[v_+^{v_+}v_-^{v_-}\right] + v\ln\gamma_\pm. \quad \text{(AK)}$$

(v) At the spinodal line, the solute composition satisfies

$$\frac{\partial^2\Delta g_{mix}}{\partial x_2^2} = \frac{1}{x_1}\frac{\partial\ln a_2}{\partial x_2} = v\frac{\partial\ln\rho_2}{\partial x_2} + v\frac{\partial\ln\gamma_\pm}{\partial x_2} = 0. \quad \text{(AL)}$$

The number density of the solute is related to the mass density of the solution d and average molecular weight M

$$\rho_2 = x_2dN_A/M \equiv x_2\rho_2^0,$$

where N_A is Avogadro's number, and $\rho_2^0 \equiv dN_A/M$ is a constant according to the assumption that ρ_2 is proportional to x_2, which implies

$$\partial\ln\rho_2/\partial x_2 = 1/x_2.$$

Meanwhile, the mean ionic activity coefficient can be written as

$$\ln\gamma_\pm = -\frac{\kappa_Bl_B|Z_+Z_-|}{2(1+\kappa_Ba)} + \ln\alpha = -\frac{\kappa_B^0\sqrt{x_2}l_B|Z_+Z_-|}{2(1+\kappa_B^0\sqrt{x_2}a)} + \ln\alpha$$

where $\kappa_B^0 \equiv [4\pi(v_+Z_+^2 + v_-Z_-^2)\alpha l_B\rho_2^0]^{1/2}$. With $\kappa_B = \kappa_B^0\sqrt{x_2}$ and the degree of dissociation α treated as a constant, we can derive the derivative of the mean ionic activity coefficient with respect to the solute mole fraction

$$\frac{\partial\ln\gamma_\pm}{\partial x_2} = -\frac{\partial}{\partial\kappa_B}\left[\frac{\kappa_Bl_B|Z_+Z_-|}{2(1+\kappa_Ba)}\right]\times\frac{\partial\kappa_B}{\partial x_2} = -\left[\frac{l_B|Z_+Z_-|}{2(1+\kappa_Ba)} - \frac{\kappa_Bal_B|Z_+Z_-|}{2(1+\kappa_Ba)^2}\right]\times\frac{\kappa_B^0}{2\sqrt{x_2}}$$

$$= -\frac{\kappa_B^0}{4\sqrt{x_2}}\frac{l_B|Z_+Z_-|}{(1+\kappa_B^0\sqrt{x_2}a)^2} = -\frac{1}{2\sqrt{x_2}}\frac{A}{(1+B\sqrt{x_2})^2},$$

where $A \equiv |Z_+Z_-|l_B\kappa_B^0/2$ and $B \equiv \kappa_B^0a$. From Eq. (AL), we obtain

$$\frac{\partial\ln\rho_2}{\partial x_2} + \frac{\partial\ln\gamma_\pm}{\partial x_2} = \frac{1}{x_2} - \frac{1}{2\sqrt{x_2}}\frac{A}{(1+B\sqrt{x_2})^2} = 0,$$

which results in

$$2B^2x_2 + (4B - A)\sqrt{x_2} + 2 = 0.$$

The above quadratic equation has the solutions of

$$x_2 = \left[\frac{A - 4B \pm \sqrt{A(A-8B)}}{4B^2}\right]^2.$$

(vi) When the temperature is below the critical point, the spinodal line yields two compositions. At the critical temperature, however, there is only one solution, which is satisfied by

$$A = 8B \quad \Rightarrow \quad |Z_+Z_-|l_B\kappa_B^0/2 = 8\kappa_B^0 a \quad \Rightarrow \quad |Z_+Z_-|l_B/(16a) = 1.$$

Hence, the critical temperature is given by

$$T_c = \frac{|Z_+Z_-|e^2}{64\pi\varepsilon\varepsilon_0 k_B a}.$$

The critical composition can be found from $|Z_+Z_-|l_B = 16a$ and the spinodal condition

$$x_{2,c} = \left[\frac{A-4B}{4B^2}\right]^2 = \frac{1}{B^2} = \frac{1}{(\kappa_B^0 a)^2} = \frac{1}{4\pi(\nu_+Z_+^2 + \nu_-Z_-^2)\alpha_c\rho_2^0 l_B a^2}$$

$$= \frac{M}{4\pi\nu|Z_+Z_-|\alpha_c dN_A l_B a^2} = \frac{M}{4\pi\nu\alpha_c dN_A a^2} \times \frac{1}{16a} = \frac{M}{64\pi\alpha_c\nu dN_A a^3},$$

where $\nu_+Z_+^2 + \nu_-Z_-^2 = \nu|Z_+Z_-|$ is obtained from the charge neutrality. At the critical point, the solute number density is $\rho_{2,c}^0 = 1/(64\pi\alpha_c\nu a^3)$.

(vii) According to the Fuoss quasi-chemical ion-pairing model for ion association constant, the degree of dissociation can be calculated from

$$\frac{1-\alpha}{\alpha^2} = K_A\rho_2^0 = \frac{2\pi\nu a^3\rho_2^0}{3}\exp\left[\frac{|Z_+Z_-|l_B}{a(1+\kappa_B a)}\right].$$

At the critical point, we have $\nu a^3\rho_{2,c}^0 = 1/(64\pi\alpha_c)$, $|Z_+Z_-|l_B/a = 16$ and

$$x_{2,c} = \frac{1}{(\kappa_B^0 a)^2} = \frac{x_{2,c}}{(\kappa_B a)^2} \quad \Rightarrow \quad \kappa_B a = 1.$$

Hence, the degree of dissociation is

$$\frac{1-\alpha_c}{\alpha_c} = \frac{2\pi}{3} \times \frac{1}{64\pi} \times \exp\left(\frac{16}{1+1}\right) = \frac{1}{96}e^8 \approx 31.05,$$

which yields $\alpha_c \approx 0.0312$.

Problem 9.16 Consider the restricted primitive model (RPM) for a symmetric electrolyte solution containing cations and anions of the same diameter σ.

(i) Show that the excess internal energy and the osmotic coefficient derived from the mean spherical approximation (MSA) can be expressed as

$$\beta U^{MSA}/N = -\frac{x^2 + x - x\sqrt{1+2x}}{24\eta}$$

$$\Phi = \frac{1+\eta+\eta^2-\eta^3}{(1-\eta)^3} - \frac{1}{72\eta}\left[2(1+2x)^{3/2} - 3x(1+2x)^{1/2} - 3x - 2\right],$$

where $N = N_+ + N_-$ is the total number of cations and anions, $\eta = \pi\rho_0\sigma^3/6$ is the ionic volume fraction, $\rho_0 = N/V$ is the total ion density, and $x \equiv \kappa\sigma$.

(ii) Compare the MSA results with those predicted by the modified Debye–Hückel (DH) theory

$$\beta U^{DH}/N = -\frac{x^3}{48\eta(1+xa/\sigma)}$$

$$\Phi = 1 - \frac{(\sigma/a)^3}{48\eta}\left[1 + xa/\sigma - (1+xa/\sigma)^{-1} - 2\ln(1+xa/\sigma)\right],$$

with $a = \sigma$. Repeat the comparison with $a = \sigma/2$ and comment which relation between a and σ is more reasonable.

Solution:

(i) Within the RPM, MSA predicts (Eq. 9.174)

$$\beta U^{MSA}/V = -l_B \sum_i \frac{\rho_i^0 Z_i^2 \Gamma}{1+\Gamma\sigma} = -\frac{2l_B I\Gamma}{1+\Gamma\sigma} = -\frac{\Gamma\kappa^2}{4\pi(1+\Gamma\sigma)},$$

where $\kappa^2 = 8\pi l_B I$. Using the following relations for the MSA screening parameter

$$\Gamma = \frac{\kappa}{2(1+\Gamma\sigma)} \quad \text{and} \quad \Gamma\sigma = \frac{\sqrt{1+2\kappa\sigma}-1}{2} \Rightarrow (\Gamma\sigma)^2 = \frac{1+x-\sqrt{1+2x}}{2},$$

we obtain the excess energy per ion

$$\beta U^{MSA}/N = -\frac{\Gamma\kappa^2}{4\pi\rho_0(1+\Gamma\sigma)} = -\frac{\Gamma^2\kappa}{2\pi\rho_0} = -\frac{(\Gamma\sigma)^2 x}{12\eta} = -\frac{x^2+x-x\sqrt{1+2x}}{24\eta}.$$

Similarly, MSA predicts that the electrostatic part of the osmotic coefficient is given by (Eq. 9.180)

$$\begin{aligned}\Phi^{MSA} &= -\frac{\Gamma^3}{3\pi\rho_0} = -\frac{(\Gamma\sigma)^3}{18\eta} = -\frac{1}{144\eta}\left[(1+2x)^{3/2} + 3(1+2x)^{1/2} - 3(1+2x) - 1\right] \\ &= -\frac{1}{72\eta}\left[2(1+2x)^{3/2} - 3x(1+2x)^{1/2} - 3x - 2\right].\end{aligned}$$

(ii) We may first compare the MSA and DH results through the Taylor expansions. For the terms related to the internal energy, we have

$$x^2 + x - x\sqrt{1+2x} = \frac{x^3}{2} - \frac{x^4}{2} + \frac{5x^5}{8} + \cdots \text{(MSA)}$$

$$x^3/(2(1+x)) = \frac{x^3}{2} - \frac{x^4}{2} + \frac{x^5}{2} + \cdots (a = \sigma, \text{ designated as DH1})$$

$$x^3/(2(1+x/2)) = \frac{x^3}{2} - \frac{x^4}{4} + \frac{x^5}{8} + \cdots (a = \sigma/2, \text{ designated as DH2}),$$

The above equations indicate that the results from MSA are closer to those from DH1 (i.e., when $\sigma = a$). The same conclusion can be reached from the analysis of the osmotic coefficient. In RPM, the thermodynamic properties depend on two variables, η and x. The latter is dependent on the ion concentration and the Bjerrum length

$$\begin{aligned}x^2 &= (\kappa\sigma)^2 = 8\pi l_B \sigma^2 I = 8\pi l_B \sigma^2 \times \frac{\rho_E}{2}(\nu_+ Z_+^2 + \nu_- Z_-^2) = 8\pi l_B \sigma^2 \times \frac{\rho_E}{2}\nu|Z_+ Z_-| \\ &= 8\pi l_B \sigma^2 \times \frac{\rho_0}{2}|Z_+ Z_-| = 24(l_B/\sigma)|Z_+ Z_-|\eta = l_B^* \eta,\end{aligned}$$

where ρ_E is the number density of electrolyte $C_{\nu_+}A_{\nu_-}$, and $l_B^* \equiv 24(l_B/\sigma)|Z_+ Z_-|$.

For an aqueous electrolyte containing monovalent ions at room temperature, $l_B \approx 7.14$ Å, and $\sigma \approx 4.25$ Å, which predicts $l_B^* \sim 40$. For an organic electrolyte with dielectric constant $\varepsilon \sim 20$, $l_B^* \sim 160$.

Figure 9.6 presents the numerical results over a concentration range approximately from 0 to 6 M. Although MSA and the modified DH theories approach the same limiting results at low concentrations, their differences are substantial at high ion concentrations. A comprehensive comparison of MSA and the DH theory was reported by Maribo-Mogensen et al.[10]

Problem 9.17 According to the cell model, a double-strained (ds)-DNA chain can be represented by a rigid cylinder of radius $a \approx 1$ nm with the closest separation between negative charges

10 Maribo-Mogensen B., Kontogeorgis G. M., and Thomsen K., "Comparison of the Debye–Hückel and the mean spherical approximation theories for electrolyte solutions", *Ind. Eng. Chem. Res.* 51, 5353–5363 (2012).

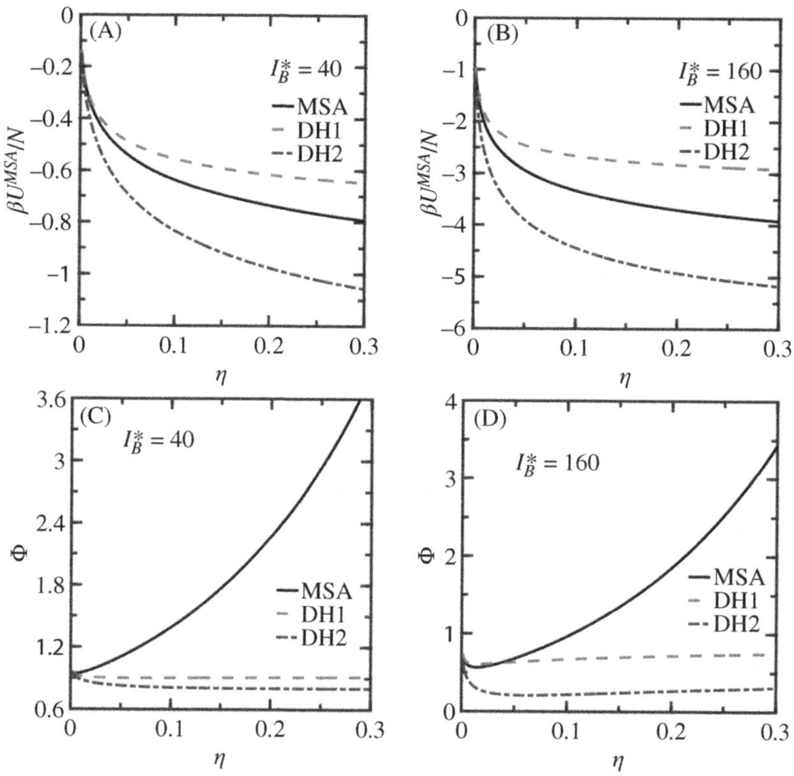

Figure 9.6 Comparison of the reduced internal energies (A&B) and osmotic coefficients (C&D) predicted by MSA and by the modified DH theory. DH1 denotes the modified DH theory with $a = \sigma$, and DH2 denotes that for $a = \sigma/2$ (MSA-RPM.m).

$b \approx 0.17$ nm. In an aqueous solution of monovalent ions, the Poisson–Boltzmann (PB) equation can be written as

$$\frac{1}{r}\frac{d}{dr}\left(\frac{d\varphi}{dr}\right) = \kappa^2 \sinh \varphi,$$

where $\varphi \equiv \beta e\psi$ is the reduced electric potential, and κ is the Debye screening parameter.

(i) Find an analytical solution for the linearized PB equation (viz., the Debye–Hückel approximation):

$$\frac{1}{r}\frac{d}{dr}\left(r\frac{d\varphi}{dr}\right) = \kappa^2\varphi;$$

(ii) Compare the analytic solution with numerical results from the nonlinear PB equation for $\kappa a = 0.1$ and 0.5; What are the corresponding salt concentrations?

Solution:

(i) Let $x \equiv \kappa r$ and write the linearized PB equation as

$$\varphi'' + \varphi'/x - \varphi = 0.$$

The ordinary differential equation has the solution of

$$\varphi(x) = CK_0(x),$$

where $K_0(x)$ is the zeroth-order modified Bessel function of the second kind, and parameter C can be fixed by the boundary condition at $x_0 = \kappa a$

$$\varphi'(x_0) = -CK_1(x_0) = 2\xi/x_0,$$

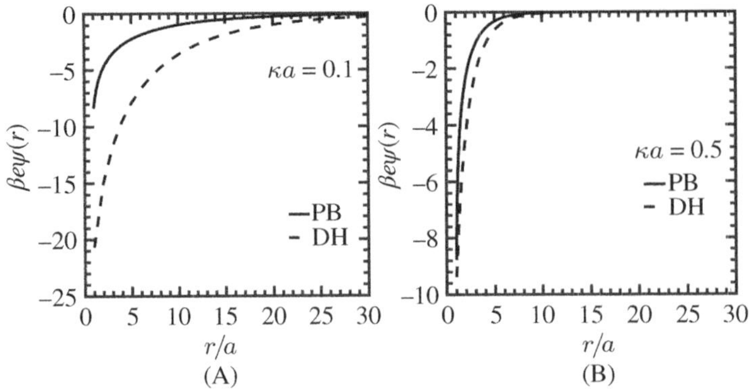

Figure 9.7 Reduced electric potential near a ds-DNA chain predicted by the PB (solid lines) and the DH (dashed lines) equations for $\kappa a = 0.1$ (A) and 0.5 (B) (cell_model.m).

where $K_1(x)$ stands for the first-order modified Bessel function of the second kind, and $\xi \equiv l_B/b$. Substituting with $C = -2\xi/x_0K_1(x_0)$ yields the analytic solution

$$\varphi(x) = -\frac{2\xi K_0(x)}{x_0K_1(x_0)}.$$

(ii) Figure 9.7 compares the analytic solution with numerical results from the nonlinear PB equation. The salt concentration is related to the ionic strength I, which in turn can be determined from the Debye screening parameter

$$\kappa^2 = 8\pi l_B I.$$

For a monovalent electrolyte, the ionic strength is the same as the ion concentration so that the molar concentration of the salt is given by

$$C_s(\text{M}) = \frac{(\kappa a)^2}{8\pi l_B a^2 N_A} = \frac{(\kappa a)^2}{8\pi \times 7.14 \times 10^2 \times 6.02214 \times 10^{23-30+3}} \approx \frac{(\kappa a)^2}{10.8066}.$$

The above equation predicts $C_s \approx 9.25 \times 10^{-4}$ M for $\kappa a = 0.1$ and $C_s \approx 0.0231$ M for $\kappa a = 0.5$. Because the ds-DNA chain has a large surface charge density, the difference between the numerical and analytical results is substantial even at low salt concentrations. The linearized PB equation greatly over-predicts the electric potential near the DNA surface.

Problem 9.18 Consider Katchalsky's cell model for an aqueous solution of ds-DNA chains with monovalent ions. At high salt concentrations, the Poisson–Boltzmann (PB) equation can be simplified by neglecting the counterions

$$\frac{1}{r}\frac{d}{dr}\left(\frac{d\varphi}{dr}\right) = \kappa^2 \sinh\varphi,$$

where $\kappa^2 = 8\pi C_s l_B N_A$ depends only on the salt concentration C_s.

(i) Calculate the electric potential near the DNA surface from the approximated PB equation at the volume fraction of ds-DNA $\phi = 0.01$ and $C_s = 1$ M.

(ii) Compare the results with that predicted by the modified Bessel functions discussed in Problem 9.17

$$\varphi(x) = -\frac{2\xi K_0(x)}{x_0K_1(x_0)}. \qquad \text{(AM)}$$

(iii) What is the osmotic pressure of the system relative to that of a buffer solution of the same salt concentration?

Solution:

(i) According to Katchalsky's cell model, the cell radius R can be deduced from the volume fraction of ds-DNA chains

$$R = a/\sqrt{\phi} \approx 1/\sqrt{0.01} \approx 10 \text{ nm},$$

where $a \approx 1$ nm is the cross-section radius of a ds-DNA chain. As discussed in Problem 9.17, at $C_s = 1$ M, the reduced Debye screening parameter is

$$x_0 \equiv \kappa a \approx \sqrt{10.8066} \approx 3.2873.$$

The electric potential near the ds-DNA surface can be numerically solved from the approximated PB equation with the boundary conditions

$$\varphi'(x_0) = 2\xi/x_0 \quad \text{and} \quad \varphi'(\kappa R) = 0,$$

where $\xi = l_B/b = 4.2$ and $\kappa R = x_0 R/a = 32.873$.

(ii) Figure 9.8 presents the numerical results in comparison with that from the single-chain model, Eq. (AM). Because of the strong screening effects at high salt concentrations, the ion distributions in different cells are nearly independent to each other. As a result, the local electric potential can be approximated by that corresponding to an isolated DNA chain (the dashed line).

(iii) The zero electric potential at the cell boundary ($R/a = 10$) implies that the system has no osmotic pressure relative to that of a buffer solution of the same salt concentration.

Problem 9.19 Consider an aqueous solution of ds-DNA chains at room temperature containing only monovalent counterions (i.e., salt free). Solve the Poisson–Boltzmann (PB) equation used in Katchalsky's cell model

$$\frac{1}{r}\frac{d}{dr}\left(\frac{d\varphi}{dr}\right) = -\frac{\kappa_c^2}{2}e^{-\varphi},$$

where $\kappa_c = \sqrt{8\pi\rho_c^0 l_B}$ is the Debye screening parameter. Assume that the volume fraction of ds-DNA chains is $\phi = 0.1$. What is the osmotic pressure of the DNA solution relative to pure liquid water?

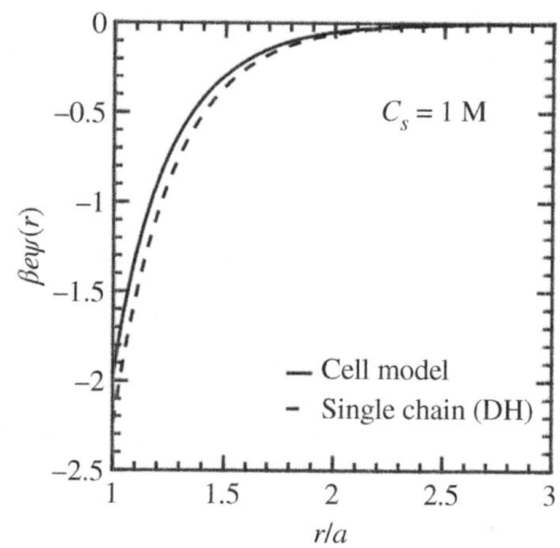

Figure 9.8 Reduced electric potential near a ds-DNA chain predicted by the cell model. The dashed line corresponds to the result from the DH approximation (cell_highSalt.m).

Solution:
The cell radius R can be calculated from the volume fraction of ds-DNA chains

$$R = a/\sqrt{\phi} \approx 1/\sqrt{0.1} \approx 3.1623 \text{ nm},$$

where $a \approx 1$ nm is the cross-section radius of a ds-DNA chain. As discussed in Section 9.8, the PB equation has an analytical solution (Eq. 9.212)

$$\varphi(r) = 2\ln\left\{\frac{\kappa_c r}{2|\lambda|}\cos[\lambda \ln(r/R_M)]\right\},$$

where parameters κ_c, λ, and R_M are solved from

$$1 - \lambda\tan[\lambda\ln(R/R_M)] = 0, \quad 1 - \lambda\tan[\lambda\ln(a/R_M)] = \xi \quad \text{and} \quad (k_c R)^2 = 4(1+\lambda^2).$$

To solve the above equations, we can first calculate λ from Eq. (9.219)

$$\lambda\ln(a/R) = \arctan[(1-\xi)/\lambda] - \arctan(1/\lambda).$$

Plugging $a = 1$ nm, $\xi = l_B/b = 4.2$, and $R = 3.1626$ nm into the above equation gives $\lambda \approx 1.4961$. Hence, we have

$$k_c = 2\sqrt{1+\lambda}/R = 2 \times \sqrt{1+1.4961}/3.1626 \approx 0.9991 \text{ nm}^{-1},$$
$$R_M = R\exp[-\arctan(1/\lambda)/\lambda] \approx 2.1331 \text{ nm}.$$

From $\kappa_c = \sqrt{8\pi\rho_c^0 l_B}$, we can find the counterion concentration at the cell boundary ρ_c^0. Subsequently, the osmotic pressure of the DNA solution is given by

$$\Pi \approx \rho_c^0 k_B T = \frac{\kappa_c^2 k_B T}{8\pi l_B} = \frac{0.9991^2 \times 10^{18} \times 1.381 \times 10^{-23} \times 298.15}{8\pi \times 7.14 \times 10^{-10} \times 101325} \approx 2.2604 \text{ atm}.$$

Problem 9.20 Show that, for a salt-free polyelectrolyte solution at a small polymer volume fraction, the osmotic coefficient predicted by Katchalsky's cell model can be approximated by

$$\Phi \approx \frac{1}{2\xi}\left[1 + \frac{\pi^2}{\ln^2(R/a)}\right].$$

When $R \to \infty$, the osmotic coefficient is reduced to that from the Oosawa–Manning theory $\Phi_\infty = 1/(2\xi)$.

Solution:
According to Katchalsky's cell model, the osmotic coefficient for a salt-free polyelectrolyte solution is given by

$$\Phi = \frac{\rho_c^0}{\bar{\rho}_c} = \frac{\kappa_c^2}{8\pi l_B} \times \pi R^2 b = \frac{1+\lambda^2}{2\xi},$$

where $(k_c R)^2 = 4(1+\lambda^2)$, $\xi = l_B/b$ and λ can be obtained from

$$1 - \lambda\tan[\lambda\ln(R/R_M)] = 0 \quad \text{and} \quad 1 - \lambda\tan[\lambda\ln(a/R_M)] = \xi.$$

After some manipulation of the above equations, we can obtain

$$\lambda\ln(a/R) = \arctan[(1-\xi)/\lambda] - \arctan(1/\lambda).$$

At low polyelectrolyte (PE) volume fractions, a/R is a small number while the arctangent functions are bounded between $-\pi/2$ and $\pi/2$. The divergence of the logarithm function requires $\lambda \to 0$, suggesting $\arctan(1/\lambda) \approx \pi/2$ and $\arctan[(1-\xi)/\lambda] \approx -\pi/2$. In other words, we may approximate λ as

$$\lambda \approx \frac{\pi}{\ln(R/a)}.$$

Hence, the osmotic coefficient becomes

$$\Phi = \frac{1+\lambda^2}{2\xi} \approx \frac{1}{2\xi}\left[1+\frac{\pi^2}{\ln^2(R/a)}\right].$$

When $R \to \infty$, the above equation is the same as that predicted by the Oosawa–Manning theory $\Phi_\infty = 1/(2\xi)$.

Problem 9.21 One of the simplest models of electric double layer (EDL) concerns a charged surface in contact with a dielectric medium containing only the counterions. Figure 9.9 shows schematically the average local ion density $\rho(z)$ and the mean electric potential $\psi(z)$. Assume that the dielectric constant is uniform throughout the system and that the counterions can be represented by point charges of valence Z.

(i) Verify that the counterion density at position $z = 0$, $\rho_0 \equiv \rho(0)$ is related to the surface charge density σ:

$$\rho_0 = 2\pi l_B(\sigma/e)^2,$$

where $l_B \equiv e^2/(4\pi\varepsilon\varepsilon_0 k_B T)$ is the Bjerrum length, and e is the unit charge.

(ii) Show that the Poisson–Boltzmann (PB) equation can be written as

$$\frac{\partial^2\varphi(z)}{\partial z^2} = \frac{2\exp[-\varphi(z)]}{\lambda_{GC}^2},$$

where $\varphi(z) \equiv \beta Z e\psi(z)$ is in the dimensionless electric potential, and $\lambda_{GC} \equiv e/(2\pi l_B|\sigma Z|)$ is called the Gouy–Chapman length.

(iii) Solve the PB equation with the boundary conditions

$$\psi(0) = \psi_s \qquad \text{and} \qquad \left.\frac{d\psi(z)}{dz}\right|_{z=0^+} = -\frac{\sigma}{\varepsilon\varepsilon_0}.$$

(iv) Verify that the ion distribution satisfies the charge neutrality, i.e., the surface charge is balanced by that of the counterions.

(v) What is the differential capacitance of the system predicted by the point charge model?

(vi) How would the results be different if the counterion distribution is restricted to $z > a > 0$?

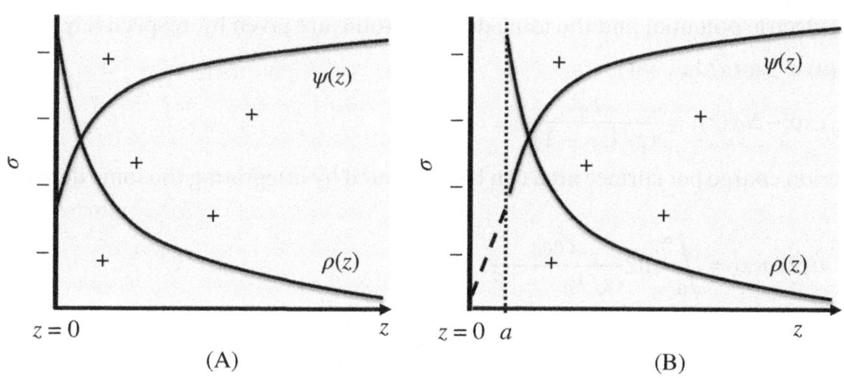

Figure 9.9 Schematic of a charged surface in contact with the diffusive counterions in a dielectric medium. Here, σ represents the surface charge density, $\rho(z)$ and $\psi(z)$ stand for the local ion density and electric potential, respectively. A. Ion distribution starts from $z = 0$. B. Ion distribution starts at $z = a$.

Solution:

(i) The contact density of ionic species can be obtained from the contact-value theorem (for the salt-free case, the osmotic pressure in the bulk is $\Pi = 0$)[11]

$$k_B T \rho_0 = \varepsilon\varepsilon_0 E^2/2,$$

where E is the magnitude of the electric field at the liquid side of the surface. Meanwhile, Gauss' law predicts that the electric field is related to the surface charge density

$$E = -\left.\frac{d\psi}{dz}\right|_{z=0^+} = \frac{\sigma}{\varepsilon\varepsilon_0}.$$

Hence, a combination of these two equations gives the counterion density at contact

$$\rho_0 = \frac{\varepsilon\varepsilon_0}{2k_BT}\left(\frac{\sigma}{\varepsilon\varepsilon_0}\right)^2 = 2\pi\left(\frac{e^2}{4\pi\varepsilon\varepsilon_0 k_B T}\right)(\sigma/e)^2 = 2\pi l_B(\sigma/e)^2.$$

(ii) The local density of counterions in the diffuse layer follows the Boltzmann distribution

$$\rho(z) = \rho_0 \exp[-\Delta\varphi(z)],$$

where $\Delta\varphi(z) \equiv \varphi(z) - \varphi(0)$. Plugging the local charge density $q(z) = \rho(z)Ze$ into the Poisson equation $\nabla^2\psi = -q/\varepsilon\varepsilon_0$ leads to

$$\frac{\partial^2\psi(z)}{\partial z^2} = -\frac{\rho_0 Ze\exp[-\Delta\varphi(z)]}{\varepsilon\varepsilon_0} = -\frac{2\pi l_B(\sigma/e)^2 Ze\exp[-\Delta\varphi(z)]}{\varepsilon\varepsilon_0}.$$

Multiplying both sides of the above equation with βZe leads to

$$\frac{\partial^2\varphi(z)}{\partial z^2} = -\frac{2\pi\beta l_B(\sigma Z)^2\exp[-\Delta\varphi(z)]}{\varepsilon\varepsilon_0} = -\frac{8\pi^2\beta e^2 l_B(\sigma Z)^2\exp[-\Delta\varphi(z)]}{4\pi\varepsilon\varepsilon_0 e^2}$$
$$= -\frac{2(2\pi l_B\sigma Z)^2\exp[-\Delta\varphi(z)]}{e^2} = -\frac{2\exp[-\Delta\varphi(z)]}{\lambda_{GC}^2},$$

where $\lambda_{GC} \equiv e/(2\pi l_B|\sigma Z|)$ is called the Gouy–Chapman length.

(iii) Let $x \equiv z/\lambda_{GC}$ and $y \equiv \Delta\varphi(z)$. The PB equation becomes

$$y''(x) = -2e^y.$$

With the boundary conditions $y(0) = 0$ and $y'(0) = -\lambda_{GC}\beta Ze\sigma/(\varepsilon\varepsilon_0) = 2$ (note $\sigma Z < 0$), we can solve the second-order ordinary differential equation

$$y(x) = -2\ln(x+1).$$

Hence, the local electric potential and the ionic density profile are given by, respectively,

$$\varphi(z) = \varphi(0) + 2\ln(z/\lambda_{GC} + 1),$$

$$\rho(z) = \rho_0\exp[-\Delta\varphi(z)] = \frac{\rho_0}{(z/\lambda_{GC}+1)^2}.$$

(iv) The total counterion charge per surface area can be calculated by integrating the ionic density profile

$$\sigma_c = \int_0^\infty dz Ze\rho(z) = \int_0^\infty dz\frac{Ze\rho_0}{(z/\lambda_{GC}+1)^2}$$
$$= Ze\rho_0\lambda_{GC}\int_0^\infty dx\frac{1}{(x+1)^2} = Ze\rho_0\lambda_{GC} = \frac{Ze^2\times 2\pi l_B(\sigma/e)^2}{2\pi l_B|\sigma Z|} = -\sigma.$$

Therefore, the charge neutrality of the system is satisfied.

---

11 Henderson D. and Blum L., "Some exact results and the application of the mean spherical approximation to charged hard spheres near a charged hard wall", *J. Chem. Phys.* 69(12), 5441–5449 (1978).

(v) The differential capacitance per surface area can be calculated from the partial derivative of the surface charge with respect to the electric potential

$$C_d = \left.\frac{\partial \sigma}{\partial \psi}\right|_{z=0} = -\left.\frac{\partial \varepsilon\varepsilon_0 \psi'}{\partial \psi}\right|_{z=0} = -\varepsilon\varepsilon_0 \frac{\psi''(0)}{\psi'(0)} = -\varepsilon\varepsilon_0 \frac{-2/\lambda_{GC}^2}{\psi'(2/\lambda_{GC})} = \frac{\varepsilon\varepsilon_0}{\lambda_{GC}}.$$

Note that the differential capacitance of a salt-free EDL is inversely proportional to the Gouy–Chapman length, i.e., the capacitance per unit area increases with the surface charge density.

(vi) If the counterions are distributed slightly away from the surface with a smallest distance $a > 0$ (Figure 9.9B), the boundary conditions for the PB equation become

$$\psi(a) = \psi(0) + Ea = \psi(0) - \frac{\sigma a}{\varepsilon\varepsilon_0} \quad \text{and} \quad \left.\frac{d\psi(z)}{dz}\right|_{z=a} = -\frac{\sigma}{\varepsilon\varepsilon_0}.$$

Essentially, both the ion density profile and the local electric potential are shifted by a distance a. Because the electric potential is additive, the system is equivalent to two sequentially connected capacitors

$$\frac{1}{C_d} = \frac{\partial \psi(0)}{\partial \sigma} = \frac{\partial[\psi(a) - Ea]}{\partial \sigma} = \frac{1}{\partial \sigma}\left(\frac{\sigma a}{\varepsilon\varepsilon_0}\right) + \frac{\partial \psi(a)}{\partial \sigma} = \frac{a}{\varepsilon\varepsilon_0} + \frac{\lambda_{GC}}{\varepsilon\varepsilon_0} = \frac{a + \lambda_{GC}}{\varepsilon\varepsilon_0}.$$

Therefore, the modified model results in a differential capacitance per surface area of

$$C_d = \frac{\varepsilon\varepsilon_0}{a + \lambda_{GC}}.$$

Problem 9.22 Consider a charged surface similar to that discussed in Problem 9.22 but in contact with an electrolyte solution. Assume that both cations and anions can be represented by point charges of valence $Z = |Z_\pm|$ and that the solvent acts like a dielectric medium.

(i) Show that the Poisson–Boltzmann (PB) equation can be written as

$$\frac{\partial^2 \varphi(z)}{\partial z^2} = \kappa^2 \sinh(z), \tag{AN}$$

where $\varphi(z) \equiv \beta Z e \psi(z)$ is the dimensionless electric potential, and $\kappa = \sqrt{2\beta Z^2 e^2 \rho_0/\varepsilon\varepsilon_0}$ is the Debye screening parameter.

(ii) Verify that, with the boundary conditions $\varphi(0) = \varphi_s$ and $\varphi(\infty) = 0$, Eq. (AN) has the following solution

$$\varphi(\lambda) = 2\ln\left[\frac{1 + \xi_s e^{-\lambda}}{1 - \xi_s e^{-\lambda}}\right], \tag{AO}$$

where $\xi_s \equiv \tanh(\varphi_s/4)$ and $\lambda \equiv \kappa z$.

(iii) Show that the ionic density profiles are given by

$$\rho_+(\lambda)/\rho_0 = \left[\frac{1 - \xi_s e^{-\lambda}}{1 + \xi_s e^{-\lambda}}\right]^2 \quad \text{and} \quad \rho_-(\lambda)/\rho_0 = \left[\frac{1 + \xi_s e^{-\lambda}}{1 - \xi_s e^{-\lambda}}\right]^2. \tag{AP}$$

(iv) Plot $\varphi(\lambda)$ and $\rho_\pm^* \equiv \rho_\pm(\lambda)/\rho_0$ versus λ for $\xi_s = 0.1$ and 1. Comment the numerical results.

(v) Show that the PB equation predicts the surface charge density

$$\sigma = 4Ze\rho_0\kappa^{-1}\sinh(\varphi_s/2). \tag{AQ}$$

Eq. (AQ) is known as the Grahame equation.

(vi) Show that Eq. (AQ) results in the differential capacitance per unit area

$$C_d = \frac{\partial \sigma}{\partial \psi_s} = \varepsilon\varepsilon_0\kappa\cosh(\varphi_s/2). \tag{AR}$$

Solution:

(i) Based on the Boltzmann equation for the ionic density profiles, we can write the local charge density as

$$\rho_c(z) = \sum_{i=\pm} Z_i e \rho_i(z) = Ze\rho_0 \exp[-\varphi(z)] - Ze\rho_0 \exp[\varphi(z)] = -2Ze\rho_0 \sinh \varphi(z),$$

where ρ_0 is the ion density in the bulk (i.e., $z = \infty$). Plugging the local charge density into the Poisson equation becomes

$$\frac{\partial^2 \psi(z)}{\partial z^2} = -\frac{\rho_c(z)}{\varepsilon\varepsilon_0} = \frac{2Ze\rho_0 \sinh \varphi(z)}{\varepsilon\varepsilon_0}.$$

Multiplying both sides of the above equation with βZe leads to

$$\frac{\partial^2 \varphi}{\partial z^2} = \frac{2\beta(Ze)^2 \rho_0 \sinh \varphi(z)}{\varepsilon\varepsilon_0} = \kappa^2 \sinh \varphi,$$

where $\kappa^2 = 2\beta Z^2 e^2 \rho_0 / \varepsilon\varepsilon_0$.

(ii) We can solve Eq. (AN) analytically following a procedure similar to that used in Problem 5.19. Let $\lambda \equiv \kappa z$ and rewrite the differential equation as

$$\frac{\partial^2 \varphi}{\partial \lambda^2} = \sinh \varphi, \tag{AS}$$

with boundary conditions $\varphi(0) = \varphi_s$ and $\varphi(\infty) = 0$. Note that

$$\frac{\partial^2 \varphi}{\partial \lambda^2} = \frac{1}{2}\frac{1}{\partial \varphi}\left(\frac{\partial \varphi}{\partial \lambda}\right)^2.$$

Integrating Eq. (AS) with respect to φ gives

$$\frac{1}{2}\left(\frac{\partial \varphi}{\partial \lambda}\right)^2 = \int_0^{\varphi} d\varphi \sinh \varphi = \cosh \varphi - 1 = 2\sinh^2(\varphi/2),$$

where the integration constant can be fixed from $\partial\varphi/\partial\lambda = 0$ at $\lambda = \infty$. Because $\partial\varphi/\partial\lambda$ has a sign opposite to that of the surface potential φ_s (thus that of $\sinh(\varphi/2)$), taking the square root of the above equation leads to

$$\frac{\partial \varphi}{\partial \lambda} = -2\sinh(\varphi/2).$$

Integrating the differential equation

$$\int_0^{\lambda} d\lambda = -\int_{\varphi_s}^{\varphi} \frac{d\varphi/2}{\sinh(\varphi/2)},$$

gives

$$\lambda = -\ln\left[\frac{\tanh(\varphi/4)}{\tanh(\varphi_s/4)}\right]. \tag{AT}$$

Using the identity $\tanh(y) = (e^{2y} - 1)/(e^{2y} + 1)$, we can rearrange Eq. (AT) and find

$$\varphi(\lambda) = 2\ln\left[\frac{1 + \xi_s e^{-\lambda}}{1 - \xi_s e^{-\lambda}}\right],$$

where $\xi_s = \tanh(\varphi_s/4)$. Note that $|\xi_s| \leq 1$ and its sign is the same as that of the surface electric potential φ_s.

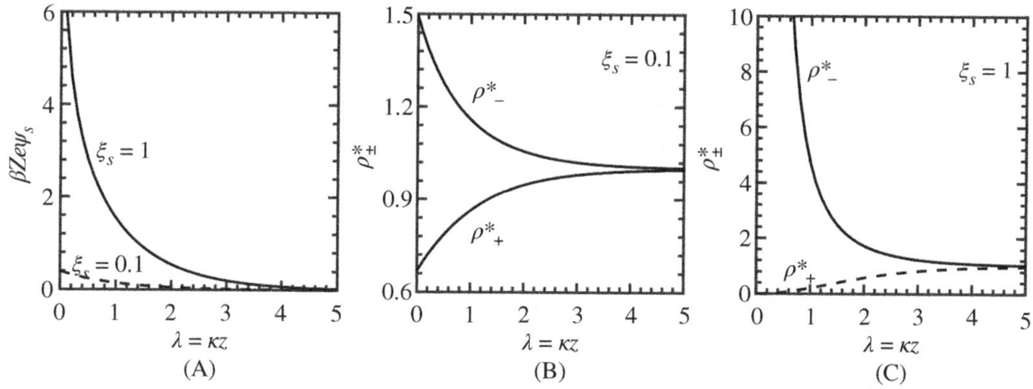

Figure 9.10 The local electric potential (A) and the ionic density profiles (B and C) near a positively charged surface. For $\xi_s = 0.1$, the electrical energy for each counterion at the surface is on the order of k_BT, and the surface energy is much larger than k_BT for $\xi_s = 1$ (GCEDL.m).

(iii) According to the Poisson–Boltzmann equation, the ionic density profiles are given by

$$\rho_+(z)/\rho_0 = \exp(-\varphi) = \left[\frac{1-\xi_s e^{-\lambda}}{1+\xi_s e^{-\lambda}}\right]^2,$$

$$\rho_-(z)/\rho_0 = \exp(\varphi) = \left[\frac{1+\xi_s e^{-\lambda}}{1-\xi_s e^{-\lambda}}\right]^2.$$

(iv) Figure 9.10 presents the local ion densities and electric potentials. At room temperature, the electric potential at the surface is $\psi_s \approx 0.01$ V for $\xi_s = 0.1$, which results in an electrical energy on the order of k_BT for each counterion. When $\xi_s = 1$, the results correspond to conditions that the surface potential is much larger than k_BT. Note that the thickness of the diffuse layer is not determined by the surface potential but by the Debye screening length $1/\kappa$.

(v) The surface charge density can be obtained from the boundary condition at $z = 0$

$$\sigma = -\varepsilon\varepsilon_0 \left.\frac{d\psi(z)}{dz}\right|_{z=0} = 4Ze\rho_0\kappa^{-1}\sinh(\varphi_s/2) = (8\varepsilon\varepsilon_0\rho_0 k_BT)^{1/2}\sinh(\varphi_s/2).$$

(vi) The differential capacitance per unit area is

$$C_d = \frac{\partial\sigma}{\partial\psi_s} = 2Ze\rho_0\kappa^{-1}\cosh(\varphi_s/2)\frac{\partial\varphi_s}{\partial\psi_s} = \varepsilon\varepsilon_0\kappa\cosh(\varphi_s/2),$$

where we have used $\varphi_s = \beta Ze\psi_s$ and $\kappa^2 = 2\beta Z^2e^2\rho_0/\varepsilon\varepsilon_0$.

Problem 9.23 The Gouy–Chapman (GC) model for the diffuse layer consists of a charged surface in contact with cations and anions represented by point charges. It assumes that the solvent has a dielectric constant ε which is uniform throughout the system.

(i) Show that the linearized Poisson–Boltzmann (PB) equation leads to

$$\frac{\partial^2\varphi(z)}{\partial z^2} = \kappa^2\varphi(z), \tag{AU}$$

where $\varphi(z) \equiv \beta e\psi(z)$ is the dimensionless electric potential, and κ is the Debye parameter.

(ii) Solve Eq. (AU) with the boundary conditions $\varphi(0) = \varphi_s$ and $\varphi(\infty) = 0$.

(iii) Derive the differential capacitance per unit area from the linearized PB equation.

Solution:

(i) According to the linearized Boltzmann equation, the ionic density profiles are given by

$$\rho_i(z) = \rho_i^0 \exp[-Z_i\varphi(z)] \approx \rho_i^0[1 - Z_i\varphi(z)],$$

where ρ_i^0 is the ion density in the bulk (i.e., $z = \infty$), and Z_i is the ion valence. Hence, the Poisson equation becomes

$$\frac{\partial^2\psi(z)}{\partial z^2} = -\frac{1}{\varepsilon\varepsilon_0}\sum_i Z_i e\rho_i(z) \approx -\frac{1}{\varepsilon\varepsilon_0}\sum_i Z_i e\rho_i^0[1 - Z_i\varphi(z)] = \frac{\varphi(z)}{\varepsilon\varepsilon_0}\sum_i Z_i^2 e\rho_i^0,$$

where $\sum_i Z_i e\rho_i^0 = 0$ due to the charge neutrality. Multiplying both sides of the above equation with βe leads to

$$\frac{\partial^2\varphi(z)}{\partial z^2} = \kappa^2\varphi(z),$$

where $\kappa^2 = \sum_i \beta Z_i^2 e^2\rho_i^0/\varepsilon\varepsilon_0$.

(ii) With the boundary conditions $\varphi(0) = \varphi_s$ and $\varphi(\infty) = 0$, Eq. (AU) can be solved analytically

$$\varphi(z) = \varphi_s \exp(-\kappa z).$$

(iii) Gauss's law predicts that the surface charge density is given by

$$\sigma = -\varepsilon\varepsilon_0 \left.\frac{d\psi(z)}{dz}\right|_{z=0} = \varepsilon\varepsilon_0\kappa\varphi_s.$$

Hence, the differential capacitance per unit area is

$$C_d = \frac{\partial\sigma}{\partial\psi_s} = \varepsilon\varepsilon_0\kappa.$$

Problem 9.24 In Stern's modification of the Gouy–Chapman (GC) model, ions in the EDL are distributed according to the Boltzmann equation with a nearest distance a. Schematically, Figure 9.11 shows the ionic distributions and the local electric potential. Assume that cations and anions are point charges of equal valence $Z = |Z_\pm|$ and that the dielectric constant ε is uniform throughout the system.

(i) Show that the Poisson–Boltzmann (PB) equation predicts the electrical potential in the diffuse layer ($z \geq a$)

$$\tanh\left[\frac{\beta Ze\psi(z)}{4}\right] = \frac{b}{1 + \sqrt{1 + b^2}} \exp[-\kappa(z - a)], \tag{AV}$$

where $b \equiv 1/(\lambda_{GC}\kappa)$, and other parameters have their usual meanings.

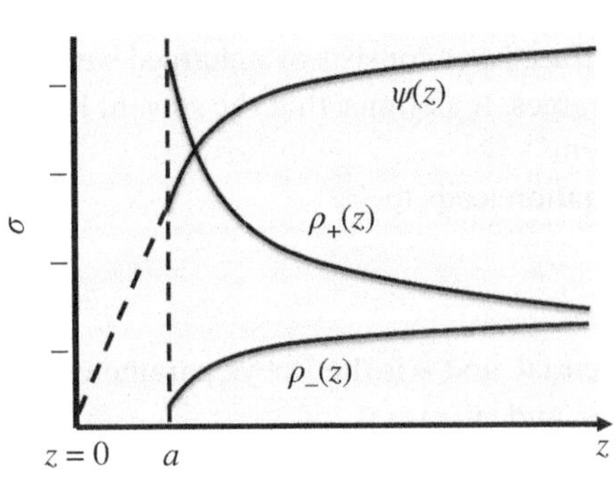

Figure 9.11 Stern's modification of the Gouy–Chapman (GC) model.

(ii) Show that the surface potential is given by

$$\psi_s = \frac{a\sigma}{\epsilon\epsilon_0} + \frac{2}{\beta Ze}\text{arcsinh}(b). \tag{AW}$$

The first term on the right side of Eq. (AW) corresponds to the potential difference across the Stern layer, which is often called the inner layer potential or the Helmholtz potential. The second term is referred to as the diffuse layer potential.

(iii) Show that the Stern model predicts the EDL differential capacitance

$$\frac{1}{C_d} = \frac{1}{\epsilon\epsilon_0}\left\{a + \frac{1}{\kappa\sqrt{1+b^2}}\right\}. \tag{AX}$$

(iv) Show that, at high electrolyte concentration and/or small surface charge density such that b is a small value, the differential capacitance can be approximated by those of two capacitors in series

$$\frac{1}{C_d} = \frac{a}{\epsilon\epsilon_0} + \frac{1}{\kappa\epsilon\epsilon_0}. \tag{AY}$$

(v) Plot the reduced capacitance $C_d^* \equiv C_d/(\kappa\epsilon\epsilon_0)$ versus the reduced surface potential φ_s and discuss the significance of the Stern layer thickness.

Solution:

(i) In the Stern model, the Poisson–Boltzmann (PB) equation has a solution identical to that given by the GC theory, i.e., Eq. (AT) in Problem 9.22. The only difference is that one of the boundary conditions is set at $\psi(a) = \psi_a$ instead of $\psi(0) = \psi_s$:

$$\lambda - \kappa a = -\ln\left[\frac{\tanh(\varphi/4)}{\tanh(\varphi_a/4)}\right], \tag{AZ}$$

where $\lambda = \kappa z$ and $\varphi(z) \equiv \beta Ze\psi(z)$. Because ions are excluded from the Stern layer, the electric field at $z = 0$ is the same as that at $z = a$ such that

$$\sigma = -\varepsilon\varepsilon_0 \left.\frac{d\psi(z)}{dz}\right|_{z=0} = -\varepsilon\varepsilon_0 \left.\frac{d\psi(z)}{dz}\right|_{z=a} = 4Ze\rho_0\kappa^{-1}\sinh(\varphi_a/2).$$

Now let

$$b \equiv \frac{\sigma\kappa}{4Ze\rho_0} = \kappa^2\frac{\sigma}{4Ze\rho_0\kappa} = \frac{2\beta Z^2e^2\rho_0}{\epsilon\epsilon_0}\frac{\sigma}{4Ze\rho_0\kappa} = \frac{\beta Ze\sigma}{2\epsilon\epsilon_0\kappa} = \frac{1}{\lambda_{GC}\kappa},$$

where λ_{GC} is the GC length. With the boundary condition $\sinh(\varphi_a/2) = b$ and mathematical identities

$$\tanh(x/2) = \frac{\sinh x}{1+\cosh x} \quad \text{and} \quad \cosh^2 x - \sinh^2 x = 1,$$

we can rewrite Eq. (AZ) as

$$\tanh(\varphi/4) = \frac{\sinh(\varphi_a/2)}{1+\cosh(\varphi_a/2)}\exp[-\kappa(z-a)] = \frac{b}{1+\sqrt{1+b^2}}\exp[-\kappa(z-a)].$$

(ii) The potential drop across the Stern layer can be calculated from the electric field $-\left.\frac{d\psi(z)}{dz}\right|_{z=0}$ and the Stern layer thickness a

$$\Delta\psi = -a\left.\frac{d\psi(z)}{dz}\right|_{z=0} = \frac{a\sigma}{\epsilon\epsilon_0}.$$

Hence, the surface potential is

$$\psi_s \equiv \psi(0) = \Delta\psi + \psi(a) = \frac{a\sigma}{\epsilon\epsilon_0} + \frac{2}{\beta Ze}\text{arcsinh}(b).$$

(iii) From the surface electric potential, we can derive the differential capacitance

$$\frac{1}{C_d} = \frac{\partial \phi_s}{\partial \sigma} = \frac{a}{\epsilon\epsilon_0} + \frac{2}{\beta Ze} \times \frac{\partial \text{arcsinh}(b)}{\partial b} \times \frac{\partial b}{\partial \sigma} = \frac{a}{\epsilon\epsilon_0} + \frac{2}{\beta Ze} \times \frac{1}{\sqrt{1+b^2}} \times \frac{\beta Ze}{2\epsilon\epsilon_0\kappa}$$

$$= \frac{1}{\epsilon\epsilon_0}\left\{a + \frac{1}{\kappa\sqrt{1+b^2}}\right\}.$$

(iv) At small values of $b = 1/(\lambda_{GC}\kappa)$, which correspond to high salt concentration and/or small surface charge density, we may expand the surface electric potential in a power series of b

$$\psi_s = \frac{a\sigma}{\epsilon\epsilon_0} + \frac{2}{\beta Ze}(b - \frac{b^3}{6} + \cdots) = \frac{a\sigma}{\epsilon\epsilon_0} + \frac{2}{\beta Ze} \times \frac{\beta Ze\sigma}{2\epsilon\epsilon_0\kappa} + \cdots = \frac{\sigma}{\epsilon\epsilon_0}(a + 1/\kappa + \cdots)$$

Accordingly, the differential capacitance is

$$\frac{1}{C_d} = \frac{\partial \psi_s}{\partial \sigma} \approx \frac{1}{\epsilon\epsilon_0}\left\{a + \frac{1}{\kappa} + \cdots\right\}. \tag{BA}$$

The Gouy–Chapman–Stern (GCS) model indicates that the EDL capacitance corresponds to those of two capacitors in series

$$C_{\text{GCS}}^{-1} = C_{\text{Stern}}^{-1} + C_{\text{diffuse}}^{-1},$$

where C_{Stern} and $C_{diffuse}$ are the Stern layer and diffuse layer capacitances, respectively. At high electrolyte concentration and/or small surface charge, these two contributions are given by

$$C_{\text{Stern}} = \frac{a}{\epsilon\epsilon_0} \quad \text{and} \quad C_{\text{diffuse}} = \frac{1}{\kappa\epsilon\epsilon_0}.$$

(v) We may find an indirect relation between C_d^* and φ_s through the dimensionless parameter b. According to Eq. (AX), the reduced capacitance, $C_d^* \equiv C_d/(\kappa\epsilon\epsilon_0)$, can be written as

$$\frac{1}{C_d^*} = \kappa a + \frac{1}{\sqrt{1+b^2}}.$$

Meanwhile, Eq. (AW) indicates that the reduced surface potential, $\varphi_s \equiv \beta Ze\psi_s$, is given by

$$\varphi_s = \frac{\beta Zea\sigma}{\epsilon\epsilon_0} + 2\,\text{arcsinh}(b) = 2a\kappa b + 2\,\text{arcsinh}(b).$$

Figure 9.12 presents the reduced capacitance versus the reduced surface potential for three values of κa. Because $C_d^* < 1/(\kappa a)$, the finite value of the Stern layer thickness prevents the capacitance divergence at high surface electric potentials.

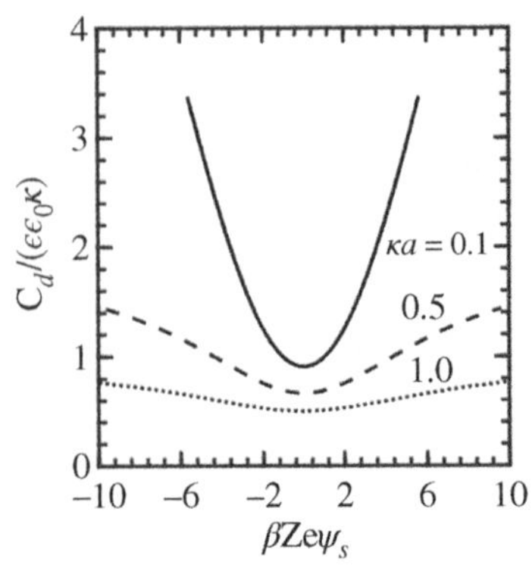

Figure 9.12 The differential capacitance versus the surface electric potential predicted by the Gouy–Chapman–Stern model (GCSEDL.m).

Problem 9.25 Based on the reduced electric potential near a charged surface in contact with a symmetric electrolyte solution discussed in Problem 9.22

$$\varphi(z) = 2\ln\left[\frac{1+\xi_s e^{-\kappa z}}{1-\xi_s e^{-\kappa z}}\right],$$

where $\varphi(z) = \beta Ze\psi(z)$ and $\xi_s = \tanh(\varphi_s/4)$, investigate the accumulated charge per surface area as a function of the distance from the surface

$$\sigma(z) \equiv \sigma + \int_0^z dz' \rho_c(z'),$$

where $\rho_c(z)$ stands for the local charge density.

(i) Verify $\sigma(\infty) = 0$, i.e., the system satisfies the charge neutrality.

(ii) Show

$$\frac{\sigma(z)}{\sigma} = \frac{1-\xi_s^2}{e^{\lambda} - \xi_s^2 e^{-\lambda}}$$

where $\lambda \equiv \kappa z$.

(iii) Show that, at infinite dilution ($\kappa \to 0$)

$$\frac{\sigma(z)}{\sigma} = \frac{1}{1+z/\lambda_{GC}},$$

where $\lambda_{GC} \equiv e/(2\pi l_B|\sigma Z|)$ is the Gouy–Chapman length. The surface charge is reduced by half when the distance is equal to λ_{GC}.

Solution:

(i) From the local charge density, local potential, and surface charge density derived from Problem 9.22

$$\rho_c(z) = -2Ze\rho_0 \sinh\varphi(z), \quad \frac{\partial\varphi}{\partial\lambda} = -2\sinh(\varphi/2), \quad \text{and} \quad \sigma = 4Ze\rho_0\kappa^{-1}\sinh(\varphi_s/2),$$

we find the total charge of ionic species in the system

$$\begin{aligned}\int_0^\infty dz' \rho_c(z') &= -2Ze\rho_0 \int_0^\infty dz \sinh\varphi(z) = 2Ze\rho_0 \int_0^{\varphi_s} d\varphi \sinh\varphi \frac{\partial z}{\partial\varphi}\\ &= \frac{2Ze\rho_0}{\kappa}\int_0^{\varphi_s} d\varphi \sinh\varphi \frac{1}{-2\sinh(\varphi/2)} = -\frac{2Ze\rho_0}{\kappa}\int_0^{\varphi_s} d\varphi \cosh(\varphi/2)\\ &= -\frac{4Ze\rho_0}{\kappa}\sinh(\varphi_s/2) = -\sigma.\end{aligned}$$

Therefore, the system satisfies the charge neutrality condition.

(ii) As shown in part (v) of Problem 9.22, Gauss's law predicts that, at position z, the accumulated charge density per surface area satisfies

$$\sigma(z) = 4Ze\rho_0 k^{-1}\sinh[\varphi(z)/2].$$

Hence

$$\frac{\sigma(z)}{\sigma} = \frac{\sinh[\varphi(z)/2]}{\sinh[\varphi_s/2]}. \tag{BB}$$

From

$$\tanh(\varphi/4) = \xi_s e^{-\lambda} = \frac{\sinh(\varphi/2)}{1+\cosh(\varphi/2)} = \frac{y}{1+\sqrt{1+y^2}},$$

where $y \equiv \sinh(\varphi/2)$, we find

$$y = \frac{2}{\xi_s^{-1}e^{\lambda} - \xi_s e^{-\lambda}}.$$

Substituting the above equation into Eq. (BB) leads to

$$\frac{\sigma(z)}{\sigma} = \frac{\xi_s^{-1} - \xi_s}{\xi_s^{-1}e^{\lambda} - \xi_s e^{-\lambda}} = \frac{1 - \xi_s^2}{e^{\lambda} - \xi_s^2 e^{-\lambda}}. \tag{BC}$$

(iii) From Eq. (AQ) in Problem 9.22, we have

$$\sinh(\varphi_s/2) = \frac{\sigma\kappa}{4Ze\rho_0} = \frac{1}{\lambda_{GC}\kappa},$$

where $\lambda_{GC} = 4Ze\rho_0/\sigma\kappa^2$ is the Gouy–Chapman length. Hence

$$\xi_s = \tanh(\varphi_s/4) = \frac{\sinh(\varphi/2)}{1 + \cosh(\varphi/2)} = \frac{1}{x + \sqrt{1 + x^2}},$$

where $x \equiv \lambda_{GC}\kappa$. As $\kappa \to 0$, we may expand $1 - \xi_s^2$ as a power series of x

$$1 - \xi_s^2 = 2x - 2x^2 + \cdots.$$

Noting that $\lambda = \kappa z = xz/\lambda_{GC}$, we can carry out a similar expansion for the denominator in Eq. (BC)

$$e^{\lambda} - \xi_s^2 e^{-\lambda} = (2 + 2z/\lambda_{GC})x - (2 + 2z/\lambda_{GC})x^2 + \cdots.$$

Keeping only the quadratic terms in both expansions leads to

$$\lim_{\kappa \to 0} \frac{\sigma(z)}{\sigma} = \frac{2x - 2x^2}{(2 + 2z/\lambda_{GC})x - (2 + 2z/\lambda_{GC})x^2} = \frac{1}{1 + z/\lambda_{GC}}. \tag{BD}$$

Eq. (BD) indicates that, as $\kappa \to 0$, the surface charge density is reduced by half when the distance is equal to the Gouy–Chapman length λ_{GC}.

Problem 9.26 Consider ion adsorption at a planar surface with charge density σ. Assume that cations and anions can be represented by point charges of equal valence $Z = |Z_\pm|$ and that the dielectric constant ε is uniform throughout the system.

(i) Show that the Poisson–Boltzmann (PB) equation predicts the surface excess of ionic species

$$\Gamma \equiv \int_0^\infty [\rho_+(z) + \rho_-(z) - 2\rho_0] = \frac{\sqrt{\sigma^2 + \varsigma^2} - \varsigma}{Ze},$$

where $\varsigma \equiv \sqrt{8\varepsilon\varepsilon_0\rho_0 k_B T}$.

(ii) Plot the reduced surface excess $\Gamma^* \equiv Ze\Gamma/\varsigma$ versus the reduced surface charge density $\sigma^* \equiv \sigma/\varsigma$ and find the asymptotic trends for the small and large values of σ.

Solution:

(i) Using the Boltzmann equation for ionic density profiles $\rho_\pm(z) = \rho_0 \exp[\mp\varphi(z)]$, we can write the surface excess as:

$$\Gamma \equiv \int_0^\infty [\rho_+(z) + \rho_-(z) - 2\rho_0] = 2\rho_0 \int_0^\infty dz[\cosh\varphi(z) - 1]$$

$$= -2\rho_0 \int_0^{\varphi_s} d\varphi \left(\frac{\partial z}{\partial \varphi}\right) [2\sinh^2(\varphi/2)].$$

According to the Gouy–Chapman theory,

$$\frac{\partial \varphi}{\partial z} = -2\kappa \sinh(\varphi/2).$$

Plugging it into the preceding equation gives

$$\Gamma = -2\rho_0 \int_0^{\varphi_s} d\varphi \frac{2\sinh^2(\varphi/2)}{[-2\kappa \sinh(\varphi/2)]} = 2\rho_0\kappa^{-1} \int_0^{\varphi_s} d\varphi \sinh(\varphi/2)$$
$$= 4\rho_0\kappa^{-1}[\cosh(\varphi_s/2) - 1].$$

Note that the surface electric potential φ_s and the surface charge density σ are related through the Grahame equation:

$$\sigma = 4Ze\rho_0\kappa^{-1} \sinh(\varphi_s/2).$$

With $\kappa = \sqrt{2\beta Z^2 e^2 \rho_0/\varepsilon\varepsilon_0}$ and $\varsigma \equiv 4Ze\rho_0/\kappa = \sqrt{8\epsilon\epsilon_0\rho_0 k_B T}$, we have

$$\sinh(\varphi_s/2) = \sigma/\varsigma \quad \text{and} \quad \cosh(\varphi_s/2) = \sqrt{1 + (\sigma/\varsigma)^2}.$$

Hence, the surface excess becomes

$$\Gamma = \frac{\varsigma}{Ze}[\sqrt{1 + (\sigma/\varsigma)^2} - 1] = \frac{1}{Ze}(\sqrt{\sigma^2 + \varsigma^2} - \varsigma). \tag{BE}$$

(ii) In dimensionless units, $\Gamma^* \equiv Ze\Gamma/\varsigma$ and $\sigma^* \equiv \sigma/\varsigma$, Eq. (BE) predicts that the surface excess of the ionic species follows a universal relation

$$\Gamma^* = (\sqrt{\sigma^{*2} + 1} - 1) = \begin{cases} \sigma^{*2}/2 & |\sigma^*| \ll 1, \\ |\sigma^*| & |\sigma^*| \gg 1. \end{cases}$$

Figure 9.13 shows the adsorption isotherm predicted by the Gouy–Chapman theory.

Problem 9.27 Huang[12] proposed that the potential drop due to the chemisorption of water molecules at the surface of a metal electrode can be estimated from the dipole moment and orientations of water molecules

$$\Delta\psi_{chem\text{-}water} = -\frac{N_{ad}\theta_w d_w}{\epsilon_0\epsilon_w}\langle\cos\alpha\rangle, \tag{BF}$$

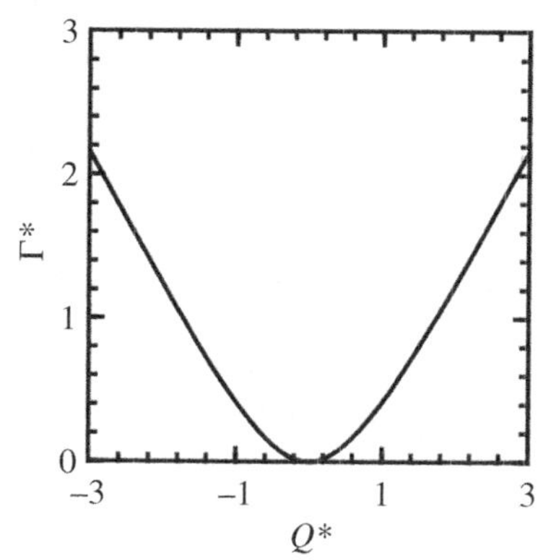

Figure 9.13 Ion adsorption predicted by the Gouy–Chapman theory.

12 Huang J., "Zooming into the inner Helmholtz plane of Pt(111)–aqueous solution interfaces: Chemisorbed water and partially charged ions", *JACS Au.* 3 (2), 550–564 (2023).

where N_{ad} is the number density of binding sites at the metal surface, θ_w is the surface coverage, ϵ_w is the dielectric constant of the water layer, ϵ_0 is the permittivity of the free space, d_w is the magnitude of the dipole moment of water molecules, and α is the angle between the directions of the electric field and the dipole moment. Suppose that the energy for each water molecule can be expressed in terms of the local electric field E_{local} and the dipole orientation

$$u_W = (d_w E_{local} + \delta\mu)\cos\alpha$$

where $\delta\mu$ represents the difference in the surface binding energy between the H-down and H-up adsorption of water molecules.

(i) Evaluate the canonical partition function for water molecules at the metal surface

$$Q = \left\{ \frac{1}{4\pi} \int_0^{2\pi} d\phi \int_0^{\pi} d\alpha \sin\alpha \exp(-\beta u_W) \right\}^{N_w},$$

where $N_w = N_{ad}\theta_w d_w$.

(ii) Show that the ensemble average of the dipole orientation is given by

$$\langle \cos\alpha \rangle = -\coth[\beta(d_w E_{local} + \delta\mu)] + \frac{1}{\beta(d_w E_{local} + \delta\mu)}.$$

Solution:

(i) The canonical partition function is

$$\begin{aligned} Q &= \left\{ \frac{1}{4\pi} \int_0^{2\pi} d\phi \int_0^{\pi} d\alpha \sin\alpha \exp(-\beta u_W) \right\}^{N_w} \\ &= \left\{ \frac{1}{2} \int_{-1}^{1} d\xi \exp[-\beta(d_w E_{local} + \delta\mu)\xi] \right\}^{N_w} \\ &= \left\{ \frac{\sinh[\beta(d_w E_{local} + \delta\mu)]}{\beta(d_w E_{local} + \delta\mu)} \right\}^{N_w}. \end{aligned}$$

(ii) The ensemble average of the dipole orientation is given by

$$\begin{aligned} \langle \cos\alpha \rangle &= \frac{\int_{-1}^{1} d\xi \xi \exp[-\beta(d_w E_{local} + \delta\mu)\xi]}{\int_{-1}^{1} d\xi \exp[-\beta(d_w E_{local} + \delta\mu)\xi]} \\ &= -\frac{\partial}{\partial[\beta(d_w E_{local} + \delta\mu)]} \ln\left\{ \frac{\sinh[\beta(d_w E_{local} + \delta\mu)]}{[\beta(d_w E_{local} + \delta\mu)]} \right\} \\ &= -\coth[\beta(d_w E_{local} + \delta\mu)] + \frac{1}{\beta(d_w E_{local} + \delta\mu)}. \end{aligned}$$